CALVIN C. ELGOT
SELECTED PAPERS

CALVIN C. ELGOT

Calvin C. Elgot
Selected Papers

Edited by Stephen L. Bloom

With a Foreword
by Dana S. Scott
and
'A Glimpse Back'
by Samuel Eilenberg

Springer-Verlag
New York Heidelberg Berlin

Calvin C. Elgot
Mathematical Sciences Department
IBM Watson Research Center
(1960-1980)
Yorktown Heights, New York 10598
USA

Editor

Stephen L. Bloom
Department of Pure and Applied Mathematics
Stevens Institute of Technology
Hoboken, New Jersey 07030
USA

Library of Congress Cataloging in Publication Data
Elgot, Calvin C.
Calvin C. Elgot selected papers.
Bibliography: p.
1. Mathematics--Collected works. 2. Recursion theory --Collected works. 3. Machine theory--Collected works.
I. Bloom, Stephen L. II. Title.
QA3.E5825 1982 510 82-19161

With One Halftone and 135 Line Illustrations.

Softcover reprint of the hardcover 1st edition 1982

9 8 7 6 5 4 3 2 1

ISBN-13: 978-1-4613-8179-2 e-ISBN-13: 978-1-4613-8177-8
DOI: 10.1007/978-1-4613-8177-8

Foreword

Cal Elgot was a very serious and thoughtful researcher, who with great determination attempted to find basic explanations for certain mathematical phenomena—as the selection of papers in this volume well illustrate. His approach was, for the most part, rather finitist and constructivist, and he was inevitably drawn to studies of the *process* of computation. It seems to me that his early work on decision problems relating automata and logic, starting with his thesis under Roger Lyndon and continuing with joint work with Büchi, Wright, Copi, Rutledge, Mezei, and then later with Rabin, set the stage for his attack on the theory of computation through the abstract treatment of the notion of a machine. This is also apparent in his joint work with A. Robinson reproduced here and in his joint papers with John Shepherdson. Of course in the light of subsequent work on decision problems by Büchi, Rabin, Shelah, and many, many others, the subject has been placed on a completely different plane from what it was when Elgot left the area. But I feel that his papers, results—and style—were very definitely influential at the time and may well have altered the course of the investigation of these problems.

As Sammy Eilenberg explains, the next big influence on Elgot's thinking was category theory, which gave him a way of expressing his ideas in a sharply algebraic manner. The joint book with Eilenberg is one illustration of this influence. Note, however, that Eilenberg himself went on in his own (and, one hopes, soon to be completed) treatise on automata theory to make that subject a substantial branch of algebra *without* very much use of category theory. What Elgot seems to have been seeking was rather a suitable part of *categorical logic*, which he found partly in Lawvere's algebraic theories plus his own addition of *iteration*. Eilenberg is perhaps more interested in the analysis of certain *structures* (as is a common concern in a large part of abstract algebra), while Elgot wanted to find a way of organizing certain *properties* that appear in the study of operations on abstract algorithms.

I first met Cal and his wife Jane in Berkeley in 1954–55 when we took part in the very stimulating set theory lectures of Alfred Tarski. The year with Tarski seems to have left a lasting impression on Elgot's work—especially as concerns the use of rigorous set-theoretical constructions and of model theory. I later visited Elgot many times at Yorktown Heights over the years. Our last two meetings were at the Mac Lane Symposium at Aspen, Colorado, and, finally, the next year at Oxford, a very few days before his most untimely death.

At both of these last meetings Cal spoke to me at length and with particular intensity about the rôle he saw for category-theoretic methods. Unfortunately, I do not think his publications give us clearly enough the picture he was in the process of forming in the last period of his research. He had come to a synthesis and unification through category theory, which perhaps the somewhat lengthy detail of his later papers does not quite let shine through. I wish I had written notes on our conversations so I could recapture the sweep of his plans. Alas, we planned to meet and correspond again soon, so I put off thinking about the details; and in Oxford I was distracted by many duties and personal concerns, so my memory now fails me. I wrote to both Bloom and Shepherdson, the latter of whom Cal was visiting for joint work in the weeks before leaving England, but neither felt he could recapture and state precisely Elgot's final philosophy.

A hint—but I feel only a preliminary hint—is given in the brief paper for the Mac Lane Symposium. One problem with that paper, as Elgot admits there, is that the examples given are rather slight. As this paper is not reprinted in this volume, it is perhaps worth repeating a few of Elgot's paragraphs. First, there is his motivation (p. 85):

> A couple of years ago, despairing that progress in the foundations of the theory of computation was taking place much too slowly, I came to the conclusion that what was needed was a "suitable" framework for such a theory. On the one hand sufficient detail and decisions would have to be worked out in such a way as to promise an attractive, perspicuous and incisive theory, and on the other hand, the framework would have to be sufficiently broad to allow many different kinds of questions to be asked and answered. Independent studies carried out within the framework would be insured a measure of cohesiveness. The perspective gained would permit judgements concerning the importance of and the relationships among the various questions. One conclusion concerning the nature of the framework which has been slowly evolving in my mind: *the language of categorical algebra would be essential.*

There is nothing to quarrel with in Elgot's statement of desires for the framework of the theory; the conclusion about categorical algebra, however, does not seem to follow from what he has just said about theory building. In fact, on the next page in a paragraph called "the anomaly," Elgot remarks:

> The importance that I attached (and still do) to the framework enterprise, together with the conclusion reached, immediately provided a pressing puzzle: only the most elementary and basic concepts of category theory were central for the framework. Indeed the oft-quoted dictum—which it seems to me is quite consistent with present category-theoretic activity—to the effect "the reason for defining 'category' is to define 'functor,' and the reason for defining 'functor' is to define 'natural transformation' . . . " seemed strikingly at odds with both those aspects of category theory which I found most useful in the past (cf. *Monadic computation and iterative algebraic theories*, reprinted in this volume) as well as those aspects which I anticipated would be most useful in the future. But these latter have been, for the most part, not sufficiently clear to attempt to elucidate. The conference at Aspen stimulated this first, very small, attempt to give these matters voice even though the ideas involved have not yet fully crystallized.

Elgot goes on to discuss the "new roles" for categorical algebra, namely: "(1) category theory as a tool for applied mathematics; (2) category theory as a common foundation for theoretic computer science and *finitely describable* mathematics." We will return in a moment to the discussion of the examples of the use of category theory that Elgot has in mind. Elgot ends the paper with thoughts on the "common foundation," first recalling some remarks of F. W. Lawvere, about which he comments:

> Lawvere's approach to the Foundation serves as a first approximation to the goal (2). It takes the point of view, however, that all categories pre-exist, while I would like to construct categories as required. Lawvere wants his foundation to be capable of defining and proving the usual things while I seek an incisive tool for re-examining much of mathematics that exists and which would serve as a guide in creating new mathematics (including theoretical computer science). On a more detailed level, I would like the underlying logic to be quantifier-free and regard the treatment of composition (in categories) as a ternary relation to be a distortion. The underlying language I have (only vaguely) in mind would gain power by having available the ability to describe general algorithmic procedures for introducing new functions and having available, for proofs, a principle (or principles) of mathematical induction.

I am fully in agreement with Elgot's desire for a workable *mathematical* language for algorithms which would allow proofs as well as definitions (since programming languages are too focused on the latter), and I can see reasons why the developments in category theory may be pointing the way, but—unless I have completely misunderstood his point—his strictures on *logic* seem far too harsh. Elgot ends the last section of the Mac Lane paper with a quote from Hermann Weyl on Brouwerian intuitionism and from Aleksandrov on abstract concepts in mathematics, and there is great justice in his suggestion that a balance is needed between being constructive and being abstract (and in his criticism of Lawvere's early hopes for category theory as a foundation for mathematics). Where I am unable to follow him is in the suggestion that the logic has to be *quantifier free*. It is true that if you have a powerful definition language, then a great portion of a (constructive) proof can be reduced to algebraic calculation plus free-variable inductive arguments. This has been demonstrated by a number of contributions by logicians to proof theory. But ordinary mathematical argument does not seem to proceed this way, and recent renewed interest in intuitionistic mathematics shows that such a restricted logic is not necessary for remaining constructive. But what is needed is a more extensive development of "programming logic." There are a number of starts but not as yet anything very comprehensive, I feel. It is a subject that interests me a great deal, and Elgot's work will certainly be significant for the future developments. I only wish I could discuss it with him.

Returning to Elgot's remarks about the desire for *constructing* categories, the example he gives in the paper is that of the *stack*. Without going into any mathematical details, it seems to me that the point he is making is that any many-sorted algebra can be viewed as a category with "objects" or "types," just those needed to explain what structures of the sorts are—and there are usually only finitely many needed. If he had recalled at this point something he knew about functors, the existence of many of these algebras could have been conveniently proved by invoking general theorems about initial objects in categories of algebras associated with these functors. And there are ways of expressing the constructive character of this existence proof. This remark may not at once justify the use of natural transformations, but it does go a long way to pointing out—in answer to Elgot's "anomaly"—why it is good to think in terms of functors. But surely Elgot knew this.

One of the biggest questions that confronts me in the programme of "re-doing" parts of theoretical computer science is finding the right *choice of category*. As the reader will see in the articles in this volume, Elgot made the choice of a type of category called "an iterative theory," which has quite nice algebraic properties; however, these properties have not yet come to a fully satisfactory proof theory, in my view. For instance, in the joint paper with Bloom and Wright reprinted here, it takes quite a bit of trouble to make the iteration operation everywhere defined in such a way that certain desirable laws are preserved. Such a theorem is just a start at seeing what sort of a logic we can have for these notions. My feeling is that the last word has not been said on the subject of "fixed-point algebra."

Another indicator of the problems concerning the question about the choice of category can be appreciated by reference to the recent paper by Jerzy Tiuryn [4]. Here it is shown that iterative theories (or, as the author prefers, iterative *algebras*) can be extended to partially ordered "regular" algebras in such a way that the iteration becomes a suitable *least* fixed point in the partial ordering. Tiuryn shows, however, that the extension (which can be made "minimal" and unique) is not always *faithful* in that a homomorphism of the given iterative algebra may be necessary. The consequences of this result for the proof theory must be significant. Are there equations true in the ordered algebras that are unreasonable for Elgot's algebra? Perhaps the examples are at hand, but I do not see them. In any case, more investigation of laws is surely needed.

As a further example of the scope for choice of category, we may refer to the recent papers of Arbib and Manes [1, 2, 3], which take good account of Elgot's writings. What I think that these authors have found is much more smooth-going categorical axiomatization of the diagramatic operations used by Elgot. They introduce a calculus of formal power series in their categories that seems very easy to use in proofs. Moreover, they show with the aid of these power-series expansions that there is a unique "canonical" fixed-point (or iteration) operator—where "canonical" has a simple categorical definition. These definitions and investigations seem to me, therefore, to fit exactly into Elgot's programme. Note, however, that Arbib and Manes use infinite expansions expressed in the usual mathematical way. Therefore, they are not giving induction axioms of the kind Elgot was asking for, since they use ordinary mathematical induction for their proofs in the standard style. Hence, there is still something to be done in making the proof theory more abstract. (These papers mentioned, by the way, are only a very small sample of the large literature that has grown up in the last 15 years. I did not mean to suggest that these are representative, but both sets of authors address specific concerns of Elgot as explained.)

One of the main objectives in my own research has been to have a category that is *cartesian closed*, that is, with function spaces. There are many reasons, especially for programming language semantics, why I also wish to solve "domain equations" involving the function-space functor. In papers too numerous to mention, it is shown that there are many categories for which this is possible. So again we have a problem of choice of category and of knowing the relation of the fixed-point theory in these categories to those Elgot was concerned with. I hope, therefore, that it will not be regarded as casting any shadow on Elgot's memory if I say that I view Elgot's work as but a first chapter of a theory of iteration and fixed points of which we will see many more future chapters from many hands, alas, without Cal's criticism and guidance.

DANA S. SCOTT

REFERENCES

1. M. A. Arbib and E. G. Manes, Partially additive categories and flow diagram semantics. *Journal of Algebra* **62** (1980), 203–227.

2. M. A. Arbib and E. G. Manes, Partially additive monoids, graph growing and the algebraic semantics of recursive calls. Springer-Verlag Lecture Notes in Computer Science, 73 (1979), 127–138.

3. M. A. Arbib and E. G. Manes, The pattern-of-calls expansion is the canonical fixpoint for recursive definitions. *Journal of the Association for Computing Machinery* **29** (1982), 577–602.

4. J. Tiuryn, Unique fixed points vs least fixed points. *Theoretical Computer Science* **12** (1980), 229–254.

Preface

Calvin C. Elgot (1922–1980) became a mathematician relatively late in life. He spent 1943–1945 in the U.S. Army, lectured at the Pratt Institute from 1945 to 1948 and received a B.S. in mathematics from the City College of New York in 1948. He worked for the U.S. Ordnance Lab for several years before entering graduate school at the University of Michigan in 1955. He wrote his dissertation, "Decision problems of finite automata design and related arithmetics", under the direction of Roger Lyndon, and was awarded the Ph.D. in 1960. From that time until his death he was a member of the Mathematical Sciences Department of the IBM Watson Research Center at Yorktown Heights.

In conversation Elgot was quiet, warm and generous. He was a pleasure to be with, especially at scientific meetings, where his delight in people and good food would cheer everyone. His easygoing manner was in striking contrast to the forcefulness with which he would defend his strong opinions on technical matters. The combination of his warm personality and his sharp opinions had a profound influence on those fortunate enough to know him. Elgot came to Stevens Institute in 1969 as an Adjunct Professor. He had recently finished "Recursiveness" with Samuel Eilenberg, and I attended his lectures, which were based on this book. It was not just his enthusiasm for this topic but the spare beauty of his mathematical formulations that impressed me. Looking at the mathematical world through category theoretic glasses was a new and strange experience for me, and Cal was extremely patient in showing me the benefits of this viewpoint. After a year or so I was "hooked" and became his collaborator on several papers. I learned only after reading Eilenberg's piece in this volume that Cal was responsible also for Eilenberg's becoming interested in the theory of computation. Of course, the influence was mutual, since Eilenberg seems to be the one who introduced Cal to categories.

Elgot's interests in the theory of computation clearly differed from those of most U.S. scientists. Sometimes he would describe his research as mathematics that arose from an analysis of certain aspects of computation. This work was not "doing" computer science but "redoing" it. He was aiming at finding the right framework, at good formulations, elegant proofs, concise explanations. On occasion he would criticize a piece of work because it did not "fit together gracefully". He admired the European computer science community where there are many mathematicians whose work seems motivated by the same concerns. (In turn, he was much appreciated by the Europeans, as evidenced by their dedicating the 1981 meeting on the "Fundamentals of Computation Theory" to his memory.) I am sure that reading through this collection of papers, one will be able to see the evidence of a fascinating and original mind.

The purpose of this selection is both to honor the memory of Calvin Elgot and to provide a service to the community of theoretical computer science. Collected in one place, together with three early works, are Elgot's recent papers concerning the theory of computation. Thus his articles on finite automata (except for the one with I. Copi and J. B. Wright), decision problems, recursive functions, lists, matricial theories and others, while dealing with relevant material, have not been included.

These works fall rather naturally into four groups. Part I consists of three early papers. The article written with Copi and Wright as well as the "RASP" paper with Abraham Robinson are extremely well known and have had a wide influence. "Abstract algorithms and diagram closure" deserves to be better known. It is one of the first studies of non-numerical algorithms. In the "RASP" paper a detailed model of the computer is given. Elgot later decided that a deeper mathematical analysis required first a study of a simpler model. The first two articles in Part II are a sampling of several papers in which he was led to the idea of the algebraic model he called "iterative algebraic theories". These theories are introduced in "Monadic computation and iterative algebraic theories". Papers 4 and 5 in Part II deal with problems connected with iterative theories. In paper 4 a detailed study of "trees" yields a concrete and natural description of the "free" iterative theories. In the last article, the role of "morphisms $1 \rightarrow 0$" in computation is examined (although this is not stated explicitly in the paper).

The articles in Part II deal with the semantics of flowchart computation. Those in Part III deal with syntax—the flowchart schemes themselves. "Structured programming with and without GO TO statements" contains many ideas: the usefulness of considering multi-entrance, multi-exit schemes; a few simple operations on schemes are shown to be sufficient to build any scheme from the atomic schemes; several classes of "structured" schemes defined by combinations of these operations are studied and the possibility of dealing with classes of structured schemes with "full power" is demonstrated. The two other papers in this section, coauthored with John Shepherdson, deal with the "reducible schemes". The reducible schemes are characterized by both a graph-theoretic property and by the property that they are precisely the schemes that can be constructed using certain simple operations on schemes. In the axiomatization paper, the reducible schemes are characterized in a more abstract way—as the free algebra in a certain equational class.

In the last years of his life, Elgot became interested in parallel computation and the joint paper with Ray Miller in Part IV deals with this subject. In this article, after a very precise definition of the "mutual exclusion problem", a solution to the problem is given. (The last two papers—"An equational axiomatization of the algebra of reducible flowchart schemes" and "On coordinated sequential processes" —have not been published previously.)

Elgot died much too early. He had planned to write a book on the foundations of the theory of computation, which would attempt to demonstrate the usefulness of the language of categorical algebra "as a common foundation for theoretical computer science and *finitely describable* mathematics". (The quotation is from his short note on this topic: "On new roles for categorical algebra".) He had also contemplated writing a history of mathematics, emphasizing the constructive aspects of various fields. Needless to say, his premature death was a great loss to the scientific community.

Before ending this preface I would like to thank some of the many people who assisted me in the preparation of this selection. Joe Rutledge and Ray Miller were very helpful in getting this project started; John Shepherdson, Eric Wagner, Michael Arbib and Ralph Tindell also offered useful suggestions. The contributions of Samuel Eilenberg and Dana Scott are gratefully acknowledged. Finally, I want to thank Jim Thatcher, for both his sage advice and constant encouragement.

STEPHEN L. BLOOM

Contents

Permissions

Springer-Verlag would like to thank the original publishers of C. C. Elgot's papers for granting permissions to reprint specific papers in this volume. The following list contains the credit lines for those articles that are under copyright.

[4] Copi, Irving M., Calvin C. Elgot, and Jesse B. Wright, *J. ACM.* **5**(2) April, 181–196. © 1958, Association for Computing Machinery, Inc., reprinted by permission.

[17] Elgot, Calvin C. and Abraham Robinson, *J. ACM.* **11**(4), 365–399, © 1964, Association for Computing Machinery, Inc., reprinted by permission.

[25] Elgot, Calvin C., in *Programming Languages* (edited by F. Genuys), 1–42, © 1968, Academic Press Inc. (London) Ltd., reprinted by permission.

[34] Elgot, Calvin C., in *Logic Colloquium* '73. Studies in Logic and the Foundations of Mathematics (Vol. 80) (edited by H. E. Rose and J. C. Shepherdson), 175–230, © 1975, North Holland/American Elsevier Publishing Company, reprinted by permission.

[38] Elgot, Calvin C., Stephen L. Bloom, and Ralph Tindell, *JCSS* **16**(3), 362–399, © 1978, Academic Press, New York and London, reprinted by permission.

[43] Bloom, Stephen L., Calvin C. Elgot, and Jesse B. Wright, *SIAM Journal of Computing* **9**(1), 25–45, © 1980, Society for Industrial and Applied Mathematics, reprinted by permission.

[44] Bloom, Stephen L., Calvin C. Elgot, and Jesse B. Wright, *SIAM Journal of Computing* **9**(3), 525–540, © 1980, Society for Industrial and Applied Mathematics, reprinted by permission.

[37] Elgot, C. C., *IEEE Transactions on Software Engineering* **SE-2**(1), 41–54, © 1976, The Institute of Electrical and Electronics Engineers, Inc., reprinted by permission.

[42] Elgot, Calvin C. and John C. Shepherdson, *Theoretical Computer Science* **8**(3), 325–357, © 1979, North-Holland Publishing Company, reprinted by permission.

A Glimpse Back

I first met Cal Elgot when he was a graduate student at Columbia, before he transferred to the University of Michigan to pursue his interests in mathematical logic.

We then met again in the mid sixties, when I was invited to give a colloquium talk to the mathematical community at IBM in Yorktown Heights. I talked about the extraordinary natural transformations that Max Kelly and I recently discovered. The main argument involved a rather pretty graph-theoretical construction which the audience seemed to like.

A few weeks later, Cal called up and made an appointment to see me. He explained that in his opinion the subject in which he and some of his colleagues (at IBM and elsewhere) were interested needed a substantial infusion of algebra of the type that I practiced. So, would I set up a more or less systematic schedule of visits to Yorktown Heights and get acquainted with the people and the problems. It sounded interesting and, out of sheer curiosity, I agreed.

When I went to Yorktown Heights for my first visit, I knew nothing. I did not know what an automaton or a formal language were. My ideas of recursion were hazy. My first contact was Jesse Wright and with his help (and with reading Robin-Scott and a few other papers) I learned rather fast. Out of these conversations grew our joint paper "Automata in General Algebras". This was my first intrusion into the field of theoretical computer sciences.

Cal was away at the time, and by the time he came back and we started talking I had already acquired some knowledge and intuition about the subject.

One theme kept recurring in our conversations rather persistently. Cal was convinced that iteration was a key operation in the theory of formal languages and that this was not sufficiently explicit. In recursion theory, the arithmetization of the subject allowed iteration to be disguised as induction.

We proceeded to remedy the situation and, by somewhat enlarging the setting, we managed to obtain a description of the phenomenon of recursiveness placing iteration in the forefront. Out of this grew our joint booklet on "Recursiveness".

Lawvere's theory of Algebraic Theories was then a relatively new subject; when I told Cal about it he recognized immediately that the notion of iteration belongs there, and gradually the notion of an Iterative Theory evolved. I believe that this invention was the fulfillment of one of Cal's dearest secret dreams.

Samuel Eilenberg

Bibliography

1. C. C. Elgot, "On single vs. triple-address computing machines," *J. ACM* (July, 1954).
2. C. C. Elgot, "Complex least squares," Abstract, *Bulletin of the AMS* (January, 1956).
3. J. R. Buchi, C. C. Elgot, and J. B. Wright, "Nonexistence of certain algorithms in finite automata theory," Abstract, *Notices* of the AMS, *5*, No. 2, (April, 1958) p. 30.
4. I. M. Copi, C. C. Elgot, and J. B. Wright, "Realization of events by logical nets," *J. ACM 5*, No. 2 (April, 1958) 181–196. *Sequential Machines* (E. F. Moore, Ed.) Addison-Wesley (1964).
5. J. R. Buchi and C. C. Elgot, "Decision problems of weak second-order arithmetics and finite automata, Part I," Abstract, *Notices* of the AMS, *5*, No. 7 (December, 1958) p. 834.
6. C. C. Elgot, "Lectures on switching theory and automata theory," Univ. of Michigan Technical Report 2755-4-P (1959).
7. C. C. Elgot and J. B. Wright, "Quantifier elimination in a problem of logic design," *Michigan Mathematical Journal 6* (1959), 65–69.
8. C. C. Elgot, "Decision problems of weak second-order arithmetics and finite automata, Part II," Abstract, *Notices* of the AMS, *6*, No. 1 (February, 1959) p. 48.
9. C. C. Elgot, "On equationally definable classes of algebras," *Notices* of the AMS, *6*, No. 1 (February, 1959) p. 48.
10. C. C. Elgot and J. B. Wright, "Series-parallel graphs and lattices," *Duke Mathematical Journal 26*, No. 2 (June, 1959) 325–338.
11. C. C. Elgot, "Decision problems of finite automata design and related arithmetics," *Transactions of the AMS 98*, No. 1 (January, 1961) 21–51.
12. C. C. Elgot and J. D. Rutledge, "RS-machines with almost blank tape," Abstract, *Notices* of the AMS (August, 1961). Full report, RC-870, IBM Research Center, Yorktown Heights, New York (December, 1962). *J. ACM. 11* No. 3 (July, 1964) pp. 313–337.
13. C. C. Elgot and J. D. Rutledge, "Operation on finite automata—Extended summary," *Proceedings* of the Second Annual Symposium on Switching Circuit Theory and Logical Design, AIEE (September, 1961) 129–132. Also, IBM Research Report, NC-168 (September, 1961).
14. C. C. Elgot, "Truth functions realizable by single threshold organs," RC-373, IBM Research Center, Yorktown Heights, New York (December, 1960). Also *Proceedings* of the Second Annual Symposium on Switching Circuit Theory and Logical Design, AIEE (September, 1961) 225–245.

15. C. C. Elgot and J. D. Rutledge, "Machine properties preserved under state minimization," RC-717, IBM Research Center, Yorktown Heights, New York (June, 1962). Also, *Proceedings* of the Third Annual Symposium on Switching Circuit Theory and Logical Design, AIEE (September, 1962) 61–70.

16. C. C. Elgot and J. E. Mezei, "Two-sided finite-state transductions," IBM Research Report, RC-1017 (June, 1963). Also, *Proceedings* Fourth Annual Symposium on Switching Circuit Theory and Logical Design, AIEE (October, 1963) in summary form. A modified version of this paper appeared as "On relations defined by generalized finite Automata" in *IBM Journal* (January, 1965).

17. C. C. Elgot and Abraham Robinson, "Random access-stored program machines, an approach to programming languages," IBM Research Report, RC-1101 (January, 1964). *J. ACM 11*, No. 4 (October, 1964) 365–399.

18. C. C. Elgot and M. O. Rabin, "On the first-order theory of generalized successor (Abstract)," presented at Meeting of AMS (January, 1963).

19. C. C. Elgot and M. O. Rabin, "Decision problems of extensions of second-order theory of successor," presented at Meeting of AMS (January, 1963).

20. C. C. Elgot, "Direction and instruction—controlled machines," IBM Research Note, NC-500. Also, *Proceedings* of Brooklyn Polytech. Symposium on System Theory (1965) pp. 121–126. Polytechnic Press of the Polytechnic Institute of Brooklyn.

21. C. C. Elgot and M. O. Rabin, "Decidability and undecidability of extensions of second (first) order theory of (generalized) successor," IBM Research Report, RC-1388 (March, 1965). Also, *Journal of Symbolic Logic*, *31* No. 2 (June, 1966) 169–181.

22. C. C. Elgot, "A perspective view of discrete automata and their design," IBM Research Report, RC-1261 (August, 1964). Also, *American Mathematical Monthly 72*, No. 2, Part II (February, 1965) 125–134.

23. C. C. Elgot, "Machine species and their computation languages," IBM Research Report, RC-1260 (August, 1964). Also, *Formal Language Description Languages for Computer Programming* (edited by T. B. Steel) North Holland (1966) 160–178.

24. C. C. Elgot, A. Robinson, and J. D. Rutledge, "Multiple control computer models," IBM Research Report, RC-1622 (March, 1966). Also, Systems and Comp. Sci., University of Toronto Press (1967) 60–76.

25. C. C. Elgot, "Algorithms abstrait et fermeture de diagrammes," Report of the University of Paris (October, 1966). Also as "Abstract algorithms and diagram closure," IBM Research Report, RC-1750 (January, 1967) and *Programming Languages* (Ed. F. Genuys) Academic Press (1968) 1–42.

26. C. C. Elgot, "A notion of interpretability of algorithms in algorithms," Report of IBM Laboratory, Vienna (October, 1966).

27. S. Eilenberg, C. C. Elgot, and J. C. Shepherdson, "Sets recognized by *n*-type automata," IBM Research Report, RC-1944 (November, 1967). *Journal of Algebra 13*, No. 4, (1969) 447–464.

28. S. Eilenberg and C. C. Elgot, "Iteration and recursion," IBM Research Report, RC-2148 (July, 1968). Also, *Proceedings of the National Academy of Sciences*, *61*, No. 2 (October, 1968) 378–379.

29. S. Eilenberg and C. C. Elgot, "Recursiveness," Monograph published by Academic Press (1970).

30. C. C. Elgot, "The external behavior of machines," *Proceedings* of the Third Hawaii International Conference on System Sciences (1970). IBM Research Report, RC-2740 (December, 1969).

31. C. C. Elgot, "The common algebraic structure of exit-automata and machines," IBM Research Report, RC-2744 (January, 1970). Also, *Computing 6*, (January, 1971) 349–370.

32. C. C. Elgot, "Algebraic theories and program schemes," IBM Research Report, RC-2925 (June, 1970). Also in *Proceedings* of Symposium on the Semantics of Algorithmic Languages (Ed. Erwin Engeler). Springer Verlag (1971).

33. C. C. Elgot, "Remarks on one argument program schemes," IBM Research Report, RC-3482 (August, 1971). Also in Courant Computer Science Symposium 2 *Formal Semantics of Programming Languages* (edited by Randall Rustin) Prentice Hall (1972).

34. C. C. Elgot, "Monadic computation and iterative algebraic theories," IBM Research Report, RC-4564 (October, 1973). Also, Logic Colloquium '73, Vol. 80 in *Studies in Logic and the Foundations of Mathematics* (edited by H. E. Rose and J. C. Shepherdson) North Holland/American Elsevier Publ. Co. (1975) 175–230. Abstract in *Journal of Symbolic Logic* (June, 1974).

35. C. C. Elgot, "Matricial theories," IBM Research Report, RC-4833 (May, 1974). *Journal of Algebra*, *42*, No. 2 (October, 1976) 391–421.

36. S. L. Bloom and C. C. Elgot, "The existence and construction of free iterative theories," IBM Research Report, RC-4937 (July, 1974). *JCSS 12*, No. 3 (June, 1976) 305–318.

37. C. C. Elgot, "Structured programming with and without GO TO statements," IBM Research Report, RC-5626 (September, 1975). *IEEE Transactions on Software Engineering, SE-2*, No. 1 (March, 1976) 41–54. Erratum and Corrigendum (September, 1976).

38. C. C. Elgot, S. L. Bloom, and R. Tindell, "On the algebraic structure of rooted trees," IBM Research Report, RC-6230 (October, 1976). *JCSS 16*, (1978) 362–399.

39. C. C. Elgot and L. Snyder, "On the many facets of Lists," IBM Research Report, RC-6449 (March, 1977, revised June, 1977). *Theoretical Computer Science*, *5* No. 3 (December, 1977) 275–305.

40. C. C. Elgot, "Finite automaton from a flowchart scheme point of view," IBM Research Report, RC-6517 (May, 1977). Also, *Proceedings* Math. Foundations of Computer Science, Tatranska Lomnica, High Tatras, Czechoslovakia (September 5–9, 1977).

41. C. C. Elgot, "Some geometrical categories associated with flowchart schemes," IBM Research Report, RC-6534 (May, 1977). Also, *Proceedings* of Conference on Fundamentals of Computation Theory, Poznan-Kornik, Poland (September 19–23, 1977).

42. C. C. Elgot and J. C. Shepherdson, "A semantically meaningful characterization of reducible flowchart schemes," IBM Research Report, RC-6656 (July, 1977). *Theoretical Computer Science 8*, No. 3 (June, 1979) 325–357.

43. S. L. Bloom, C. C. Elgot, and J. B. Wright, "Solutions of the iteration equation and extensions of the scalar iteration operations," IBM Research Report, RC-7029 (March, 1978), *SIAM Journal of Computing 9*, (February, 1980) 25–45.

44. S. L. Bloom, C. C. Elgot, and J. B. Wright, "Vector iteration in pointed iterative theories," IBM Research Report, RC-7322 (March, 1978), *SIAM Journal of Computing 9*, (February, 1980) 525–540.

45. C. C. Elgot, "A representative strong equivalence class for accessible flowchart schemes," IBM Research Report RC-7181 (June, 1978). Also, *Proceedings* of the International Conference on Mathematical Studies of Information Processing (August, 1978, Kyoto, Japan).

46. C. C. Elgot, "Assignment statements in the context of algebraic theories," *Proceedings* of the IBM Japan Symposium (August, 1978, Kobe, Japan). IBM Research Report, RC-7369, (November 13, 1979).

47. C. C. Elgot, "The multiplicity semiring of a boolean ring," IBM Research Report, RC-7540, (March, 1979).

48. C. C. Elgot and R. E. Miller, "On coordinated sequential processes," IBM Research Report, RC-7778, (July, 1979).

49. C. C. Elgot, "On new roles for categorical algebra," IBM Research Report, RC-7931, (October, 24, 1979). To appear in volume commemorating the Symposium on Algebra, held in Aspen, Colorado, May 23–27, 1979.

50. C. C. Elgot and J. C. Shepherdson, "An equational axiomatization of the algebra of reducible flowchart schemes," IBM Research Report, RC-8221 (April, 1980).

Reprinted from JOURNAL OF THE ASSOCIATION FOR COMPUTING MACHINERY
Volume 5, Number 2, April 1958

Realization of Events by Logical Nets*

IRVING M. COPI, CALVIN C. ELGOT, AND JESSE B. WRIGHT†

University of Michigan, Ann Arbor, Mich.

1. *Introduction*

In *Representation of Events in Nerve Nets and Finite Automata* [3], S. C. Kleene obtains a number of interesting results. The most important of these, his analysis and synthesis theorems (theorems 5 and 3), are obscured both by the complexity of his basic concepts and by the nature of the elements used in his nets. In this paper we give a new formulation and new proofs of Kleene's analysis and synthesis theorems, in which we presuppose no acquaintance with Kleene's work. We use simpler basic concepts and construct our nets out of more familiar and convenient elements (see section 4). The simplified formulation and proofs should make these important results more widely accessible. Some detailed comments on Kleene's ideas are made in section 7.

The problems of analysis and synthesis concern the relationship between structure and behavior of nets. It is characteristic of nets that for any time t the state of any output of the net is completely determined by the states of the inputs of that net between time 1 and time t. Thus it is customary to regard the behavior of an output of a net as a function or "transformation" which assigns to every finite sequence of input states the state which it produces in that output. Any specification of that transformation expresses the behavior of the net output. In this paper we adopt Kleene's method of expressing behavior by means of events, an event being a set of finite sequences of input states. (The interrelations of several different methods of expressing behavior are discussed in the appendix.) To understand how an event can express the behavior of a net output, let the output state *on* be assigned to each sequence of input states in the event and let the output state *off* be assigned to each sequence of input states not in the event. This assignment defines the transformation corresponding to the given event. In this way an event specifies a transformation and thus expresses the behavior of a net output. A net output is said to realize an event if and only if its behavior is given by the corresponding transformation.

These notions can be made more precise in the following way. If we express an input state of a k-input net by means of a k-tuple of zeros and ones, then we can define a *k-event* to be a set of finite sequences of k-tuples of zeros and ones,

* Received September, 1957. This research was supported by funds provided by the Office of Naval Research, Contract Nonr-1224(21), Project NR049–114, and by Project MICHIGAN, a Tri-Service contract (DA-36-039sc-57654) administered by the U. S. Army Signal Corps. Funds for the writing of this report were provided by the Office of Naval Research under Contract Nonr-1224(21).

† The authors wish to thank Don W. Warren for his significant contributions in the early stages of the writing of this paper. They also wish to thank J. Richard Buchi, Arthur W. Burks, and John H. Holland for their helpful criticism and suggestions.

$k \geqq 0$. And we define an *event* to be a k-event for some $k \geqq 0$. A k-event E is *realized* by an output of a k-input net just in case for every time l the output is on at l if and only if a sequence in E of length l is applied to the inputs over the interval from time 1 to time l. Certain events (see sections 2 and 3) are defined to be "regular" and it is proved in section 5 that all regular and only regular events can be realized by the outputs of nets. This provides a characterization of exactly those kinds of behavior which can be realized by nets. Regular events are described by formulas called "regular expressions". The proof of our synthesis theorem (2) gives an effective procedure for finding a net and an output of that net which realizes a given regular event, and the proof of our analysis theorem (1) gives an effective procedure for finding the regular event which is realized by a given output of a given net.

In section 6 we apply these analysis and synthesis algorithms (theorems 1 and 2) to illustrative examples.

2. *Regular Sets and Regular Expressions*

The notion of "regularity" is central to our entire discussion. In the present section we define "regularity" for expressions and (derivatively) for sets, and prove our first lemma.

Given any non-empty finite ordered set A of n objects, we may wish to consider sets of finite sequences of those objects. To do so it is convenient to introduce a language L_A containing three operator symbols "$\vee$", "$\cdot$", and "*"; an alphabet of $n + 1$ letters $\Lambda, A_1, A_2, \cdots, A_n$; and *formulas* defined by the following inductive rule:

1. Each single letter of L_A is a formula.
2. If Ω_1 and Ω_2 are formulas of L_A, then so are $(\Omega_1 \vee \Omega_2)$, $(\Omega_1 \cdot \Omega_2)$, and Ω_1^*.
3. Nothing is a formula of L_A unless its being so follows from these rules.

The language L_A has the following interpretation in terms of sets of sequences of objects of A:

1. The letter Λ denotes the empty set, and each letter A_i denotes the set whose only member is the ith element of A, regarded as a sequence of length 1.
2. If Ω_1 and Ω_2 are formulas, then
   a) the formula $(\Omega_1 \vee \Omega_2)$ denotes the union of the sets denoted by Ω_1 and Ω_2;
   b) the formula $(\Omega_1 \cdot \Omega_2)$ denotes that set which contains every sequence of length $l_1 + l_2$ ($l_1 > 0$ and $l_2 > 0$) whose first section, consisting of its first l_1 terms, is a sequence in the set denoted by Ω_1, and whose last section, consisting of its last l_2 terms, is a sequence in the set denoted by Ω_2 (obviously $(\Omega \cdot \Lambda) = (\Lambda \cdot \Omega) = \Lambda$ for any Ω);
   c) the formula Ω_1^* denotes the union of the sets denoted by Ω_1, $(\Omega_1 \cdot \Omega_1)$, $((\Omega_1 \cdot \Omega_1) \cdot \Omega_1)$, $\cdots$.

The formulas of L_A are defined to be *regular expressions*, and a set of sequences of objects from A will be called a *regular set* if and only if it is denoted by a regular expression.

Where R is any binary relation defined on a given set S, a sequence of objects from S, $a_1a_2 \cdots a_l$, is an *R-sequence* if and only if for every i ($i = 1, 2, \cdots, l - 1$) a_iRa_{i+1}. (Note that any sequence of length 1 is an R-sequence.) Now we are able to state and prove

LEMMA 1. If R is a binary relation defined on a finite set S containing a and b, then the set of all R-sequences beginning with a and ending with b is regular.

PROOF: By induction on n, the number of objects in the set S.

α) $n = 1$. Here $a = b$.

Case 1. If $\overline{aRa}$, then the only R-sequence beginning with a and ending with b is a itself regarded as a sequence of length 1. Here the set of all R-sequences beginning with a and ending with b is the unit set $A = \{a\}$, which is regular.

Case 2. If aRa, then the set of all R-sequences beginning with a and ending with b is $\{a, aa, aaa, \cdots\}$, which is the regular set A^*.

β) $n > 1$. Here we assume the lemma true for any non-empty set containing fewer than n objects, and consider any set containing n objects with relation R defined over that set.

Case 1. $a = b$. Any R-sequence beginning with a and ending with a can be written $a\alpha_1a\alpha_2a\alpha_3 \cdots a\alpha_ra$. Here $a\alpha_ka$ can be simply aa, in case aRa, otherwise α_k is an R-sequence containing no occurrence of a. Let $e_1, \cdots, e_g$ ($g \geqq 0$) be the objects e other than a such that aRe, and let $f_1, \cdots, f_h$ ($h \geqq 0$) be the objects f other than a such that fRa. Now any R-sequence α_k begins with an e, ends with an f, and contains no occurrence of a. By the induction hypothesis, the set of all R-sequences beginning with a particular e, ending with a particular f, and containing no occurrence of a is regular. Let $B_1, B_2, \cdots, B_{gh}$ be these regular sets of R-sequences. Now if aRa, the set of all possible R-sequences $a\alpha_k$ is $A \vee A\cdot(B_1 \vee B_2 \vee \cdots \vee B_{gh})$, which reduces to A if there are no B's, that is, if $gh = 0$. But if $\overline{aRa}$, the set of all possible R-sequences $a\alpha_k$ is $A\cdot(B_1 \vee B_2 \vee \cdots \vee B_{gh})$, which reduces to the empty set if there are no B's, that is, if $gh = 0$. In either case the set is regular: call it C. From the construction of C it is clear that every sequence in C, in C^*, and in $C^*\cdot A$ is an R-sequence. Hence every sequence in $A \vee C^*\cdot A$ is an R-sequence beginning with a and ending with b, and every R-sequence beginning with a and ending with b is in $A \vee C^*\cdot A$. The set of all R-sequences beginning with a and ending with b is a regular set, since it is denoted by a regular expression.

Case 2. $a \neq b$. Any R-sequence beginning with a and ending with b can be written $a\alpha_1a\alpha_2a\alpha_3 \cdots a\alpha_ra\beta b$. Here each α_k is as in case 1, and $a\beta b$ can be simply ab in case aRb, otherwise β is an R-sequence containing no occurrence of a. By the argument of case 1, the set of all possible R-sequences $a\alpha_k$ is the regular set C. By the inductive assumption the set, B, of R-sequences not containing a from an e to b is regular. If $g = 0$, then B is empty. Let $D = A\cdot B$. Then every sequence in D is an R-sequence; every sequence in $D \vee C^*\cdot D$ is an R-sequence beginning with

a and ending with b, and every R-sequence beginning with a and ending with b is in $D \vee C^* \cdot D$. Hence the set of all such R-sequences is regular since it is denoted by a regular expression.

3. *Sequences of k-tuples*

In the present section we consider sets of sequences of objects of a particular kind, namely sets of sequences of k-tuples of ones and zeros. For a fixed $k \geqq 0$, there are 2^k distinct k-tuples of ones and zeros. (By convention there is exactly one 0-tuple.) Where A^k is the set of all k-tuples of ones and zeros, the language L_{A^k} (also written L^k) will have its alphabet consist of the $1 + 2^k$ letters Λ, A_1, $A_2, \cdots, A_{2^k}$. Every regular expression or formula of L^k will denote a regular set of sequences of k-tuples of ones and zeros.

There are obvious interrelations between any two languages L^k and $L^{k'}$. Let us suppose for the sake of definiteness that $k < k'$. Where A^k is the set of all k-tuples and $A^{k'}$ is the set of all k'-tuples, since $k < k'$, each k-tuple of A^k will consist of the first k terms of a k'-tuple of $A^{k'}$. Given any k'-tuple, the *corresponding k-tuple* is that k-tuple which consists of the first k terms of the given k'-tuple. In general there will be $2^{k'-k}$ distinct k'-tuples that have the same corresponding k-tuple. The set $A^{k'}$ can be mapped onto the set A^k by mapping each k'-tuple of $A^{k'}$ onto the corresponding k-tuple of A^k.

This mapping can be extended to sequences of k'-tuples and sets of sequences of k'-tuples. Given any sequence of k'-tuples, the *corresponding sequence of k-tuples* is the sequence of corresponding k-tuples, that is, the sequence whose jth member is the corresponding k-tuple of the jth member of the original sequence. And given any set of sequences of k'-tuples, the *corresponding set of sequences of k-tuples* is the set of corresponding sequences of k-tuples.

We can now state and prove

LEMMA 2. If a given set of sequences of k'-tuples is regular then so is the corresponding set of sequences of k-tuples ($k < k'$).

PROOF: If the given set of sequences of k'-tuples is regular then it is denoted by a formula of $L^{k'}$. Now we map the formulas of $L^{k'}$ onto formulas of L^k according to the following rules:

1. Let the letter in $L^{k'}$ that denotes the empty set of k'-tuples be mapped onto the letter in L^k that denotes the empty set of k-tuples.
2. Let each letter in $L^{k'}$ that denotes a unit set containing a single k'-tuple regarded as a sequence of length 1 be mapped onto the letter in L^k that denotes the corresponding unit set containing the single corresponding k-tuple regarded as a sequence of length 1.
3. Where formulas Ω_1' and Ω_2' of $L^{k'}$ are mapped onto formulas Ω_1 and Ω_2 of L^k, then let the formulas $(\Omega_1' \vee \Omega_2')$, $(\Omega_1' \cdot \Omega_2')$, and $\Omega_1'^*$ of $L^{k'}$ be mapped onto formulas $(\Omega_1 \vee \Omega_2)$, $(\Omega_1 \cdot \Omega_2)$, and Ω_1^* of L^k.

Now if a given set of sequences of k'-tuples is denoted by a formula Ω' of $L^{k'}$, then by the mapping defined above the corresponding set of sequences of k-tuples is denoted by the formula Ω of L^k. Hence if the given set of

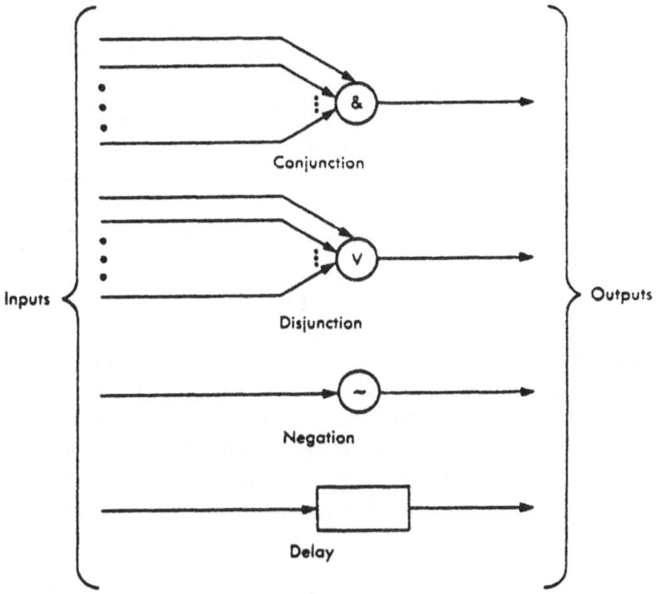

FIG. 4a

sequences of k'-tuples is regular so is the corresponding set of sequences of k-tuples.

4. *Nets*

We shall consider nets constructed out of four kinds of elements: conjunctions, disjunctions, negations, and delays. Each element has exactly one output, each negation and delay has exactly one input, and each conjunction and disjunction can have any number $n \geqq 1$ of inputs (see figure[1] 4a). The output of an element will sometimes be referred to as an "element output", and an input of an element will sometimes be referred to as an "element input."

At any discrete moment of time t ($t = 1, 2, 3, \cdots$) each input and each output is in one of two states, 1 or 0, *on* or *off*. Where i is an input or an output, that i is on (off) at time t is written $i(t) = 1$ (0). The output of a conjunction is on at time t if and only if all of its inputs are on at time t, the output of a disjunction is on at time t if and only if at least one of its inputs is on at time t. The output of a negation is on at time t if and only if its input is off at time t. These three kinds of elements, called switching elements, realize the indicated truth functions with no temporal delay. The output of a delay is on at time $t + 1$ if and only if its input was on at time t; at time 1 every delay output is off.

A combination of elements is a *net* if and only if it can be constructed from elements according to the following inductive rule:

1. Any element is a net.

[1] Figure numbers are related to section numbers.

2. If N_1 and N_2 are distinct nets, the first containing an output o and the second containing an input i which is not an output, then the result of identifying (connecting) o and i is a net.
3. If N is a net containing a delay output d and an input i which is not an output, then the result of identifying d and i is a net.
4. If N is a net containing inputs i_1 and i_2 which are not outputs, then the results of identifying i_1 and i_2 is a net.
5. If N_1 and N_2 are distinct nets then N_1 and N_2 taken together constitute a single net.

A *net output* of a net is any element output contained in that net. A *net delay output* of a net is an output of any delay element contained in that net. Each element output in a net is a distinct net output of that net, since an element output can be identified only with an input that is not an output. Since no two element outputs can ever be identified, a "net output" could equally well be defined as the result of identifying an element output with zero or more inputs.

A *net input* of a net is any input which is contained in that net which is not an output. Any net with one or more distinct net inputs can have its distinct net inputs ordered and labeled i_1, i_2, i_3, $\cdots$, and we shall assume that every such net has had its distinct net inputs so ordered and labeled. A *k input net* is a net with exactly $k \geqq 0$ distinct net inputs.

An *input state* of a k input net is an arbitrary assignment of ones and zeros to its inputs. Since the net inputs are ordered, each input state can be represented by a k-tuple of ones and zeros, and each of the 2^k distinct k-tuples represents a distinct input state of the net. We shall sometimes speak of the k-tuples as being the input states they represent.

If a k input net contains q delay outputs we order and label its net inputs and delay outputs i_1, i_2, $\cdots$, i_k, d_1, d_2, $\cdots$, d_q. A *complete state* of the net is an arbitrary assignment of ones and zeros to the net inputs and delay outputs. Since the net inputs and delay outputs are ordered, each complete state can be represented by a k'-tuple of ones and zeros ($k' = k + q$), and each of the $2^{k'}$ distinct k'-tuples represents a distinct complete state of the net. We shall sometimes speak of the k'-tuples as being the complete states they represent. Note that not all complete states are necessarily assumed by the net.

It should be remarked that the state of any output of a net at time t is (uniquely) determined by the complete state of the net at time t (cf. theorems VIII and IX of Burks and Wright [2]).

The interrelations between complete states and input states are fairly clear. Each input state of a net consists of the first k terms of a complete state of that net. Given the complete state of a net it will be convenient to refer to the input state contained in it as the input state *derived* from that complete state, or more briefly as its *derived input state.* The complete state of a k input net is a k'-tuple, and the derived input state of that net is the corresponding k-tuple—in the sense of "corresponding" explained in section 3.

We say that an input state or k-tuple is *applied* to a net at time t just in case the jth net input is on at time t if and only if the jth term of the k-tuple is a one.

A sequence of input states is called an *input sequence*. We say that an input sequence is *applied* to a net over the l' consecutive moments of time $l - l' + 1$, $l - l' + 2, \cdots, l$ if and only if the jth input state of the input sequence is applied to the net at time $l - l' + j$ ($j = 1, 2, \cdots, l'$).

An input sequence is said to be *derived* from a sequence of complete states if and only if both sequences contain the same number of terms (say l) and the jth term of the input sequence is derived from the jth term of the sequence of complete states ($j = 1, 2, \cdots, k$). And a set σ of input sequences is said to be derived from a set σ' if and only if every sequence in σ is derived from a sequence in σ' and every sequence in σ' has a derived sequence in σ.

Since a complete state usually includes states of delay outputs as well as states of net inputs, the number of complete states that a net can assume at any time $t > 1$ is in general greater than the number of input states it can have at that time. But that is not true for time $t = 1$. Let us use the term *initial complete state* for any complete state that a net can assume at time $t = 1$. Since every delay output is in state 0 at time $t = 1$, any initial complete state of a net is (uniquely) determined by its derived input state. And since the state of each delay output of a net at time $t + 1$ is determined by the complete state of the net at time t, the complete state of a net at time $t + 1$ is wholly determined by its complete state at time t and its input state at time $t + 1$. These remarks suffice to establish

LEMMA 3. Any given input sequence applied to a net over the first l moments of time determines a (unique) sequence of complete states which the net assumes during the first l moments of time.

Of the $2^{k'}$ distinct complete states of a net $a_1, a_2, \cdots, a_{2^{k'}}$, we say that a_i stands in the *direct-transition relation* to a_j if and only if the last q terms of a_j (the states of the q delay element outputs) are those that would be produced at time $t + 1$ if the net were in complete state a_i at time t (cf. Burks and Wang [1], p. 31). Next we define a sequence of complete states $a_1 a_2 \cdots a_l$ to be a *transition sequence* of complete states if and only if for every i ($i = 1, 2, \cdots, l - 1$) a_i stands in the direct-transition relation to a_{i+1}. (Any complete state regarded as a sequence of length 1 is a transition sequence.) Since the direct-transition relation is a binary relation defined on the finite set of complete states of a given net, transition-sequences of complete states are R-sequences of the kind discussed in section 2. It is obvious that any sequence of complete states assumed by a net over any l' consecutive moments of time must be a transition sequence of complete states.

5. *Analysis and Synthesis Theorems*

THEOREM 1. Any set of input sequences realized by a net output is regular.

PROOF: For any k input net there are exactly 2^k distinct input states and 2^k distinct initial complete states: $a_1, a_2, \cdots, a_{2^k}$. Let $b_1, b_2, \cdots, b_m$ ($0 \leq m \leq 2^{k'}$) be all the distinct complete states in which output o is on. Let σ' be the set of all transition sequences of complete states beginning with an a_i

and ending with a b_j. It is clear that $o(l) = 1$ if and only if the sequence of complete states assumed by the net over the first l moments of time belongs to σ'.

Where σ is the set of input sequences derived from σ', it follows by lemma 3 that the input sequence applied to the net over the first l moments of time belongs to σ if and only if the sequences of complete states assumed by the net over the first l moments of time belongs to σ'. Hence $o(l) = 1$ if and only if the input sequence applied to the net over the first l moments of time belongs to σ.

By lemma 1, taking as R the direct-transition relation, and as S the set of all complete states, the set of all transition sequences beginning with a particular a_i and ending with a particular b_j is regular. There are $m \cdot 2^k$ such regular sets of transition sequences, and their logical sum or disjunction, which is σ', is a regular set also. Since σ is derived from σ', lemma 2 assures us that σ is regular also.

Hence $o(l) = 1$ if and only if the input sequence applied to the net over the first l moments of time belongs to a regular set of input sequences. Which means that any set of input sequences realized by a net output is regular.

THEOREM 2. Any regular set of input sequences is realized by a net output.

PROOF: To prove theorem 2 it will be convenient to establish a somewhat stronger result in whose formulation we use the term "S-unit". An *S-unit* (starter) is defined to be a zero input net constructed from a delay element, a negation element, and a 2 input disjunction element, by identifying the negation output with one of the disjunction inputs, identifying the disjunction output with the delay input, and identifying the delay output with the negation input and with the other disjunction input. Output s of the S-unit is on only at time 1.

In what follows, every k input net N that has an output which realizes a regular expression of L^k will contain exactly one S-unit. And the output s of that S-unit will be identified with some input of an element of N not in the S-unit. Hence if the S-unit is detached from N the input which *was* identified with s is no longer identified with any output, and becomes the $k + 1$st net input i_{k+1} of the resulting $k + 1$ input net N'.

Now we can state the stronger result to be proved: Any formula Ω of any L^k is realized by an output o of a k input net N containing exactly one S-unit, whose detachment produces a $k + 1$ input net N' such that $o(l) = 1$ if and only if there is an l' such that

1) Ω denotes a regular set of input sequences which contains a sequence of l' k-tuples that is applied to the first k net inputs of N' over the l' consecutive moments of time $l - l' + 1, l - l' + 2, \cdots, l$; and
2) $i_{k+1}(l - l' + 1) = 1$. Our proof is by induction on the number of symbols in the formula, counting each occurrence of a letter or of $\vee$ or of $*$ as a single symbol.

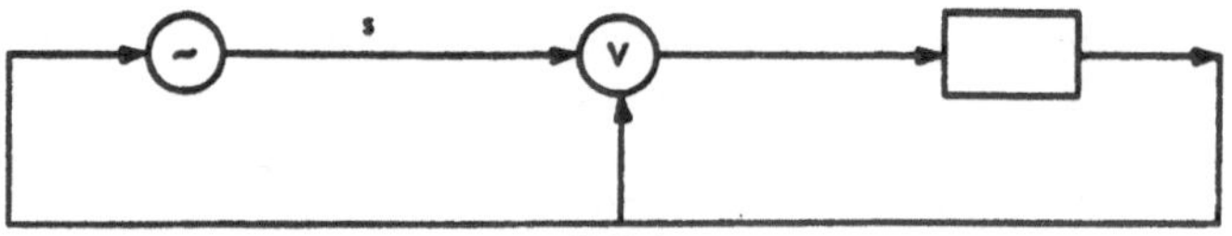

FIG. 5a

α) $n = 1$. Here the formula is one of the $1 + 2^k$ letters of L^k, and the regular set of sequences of k-tuples it denotes is either the empty set or a unit set containing a single k-tuple.

Case 1. If the formula is Λ it denotes the empty set, and is realized by the output o of a conjunction element whose $k + 2$ inputs are the k inputs of the net, the output of a negation element whose input is the first net input, and the output s of an S-unit. Since $o(t) = 0$ for every t, o realizes Λ.

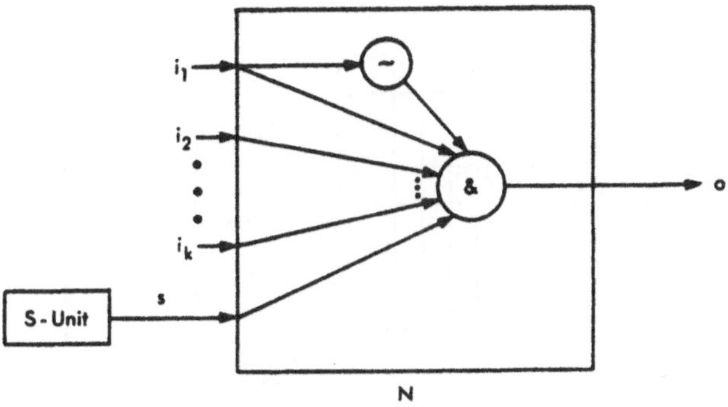

Fig. 5b

Case 2. If the formula is an A_i it denotes a unit set containing a single k-tuple, and is realized by the output o of a $k + 1$ conjunction element whose jth input is the jth net input i_j if the jth term of the k-tuple is 1, otherwise the output of a negation element whose input is the jth net input i_j, and whose $k + 1$st input is the output s of an S-unit. The following figure illustrates N where the k-tuple is $10 \cdots 10$:

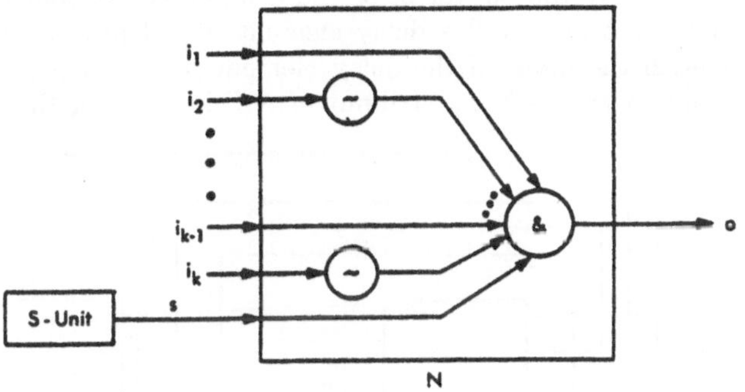

Fig. 5c

β) Here we assume the theorem true for any formula containing fewer than n symbols, and consider any formula Ω containing n (>1) symbols. Here $\Omega = (\Omega_1 \vee \Omega_2)$ or $\Omega = (\Omega_1 \cdot \Omega_2)$ or $\Omega = \Omega_1^*$.

Case 1. $\Omega = (\Omega_1 \vee \Omega_2)$. By the induction hypothesis, since Ω_1 and Ω_2 contain fewer than n symbols, they are realized by outputs o_1 and o_2 of k input nets N_1 and N_2 each containing exactly one S-unit, etc. Now Ω is realized by the output o of the k input net N constructed out of N_1', N_2', an S-unit, and a 2 input disjunction element whose output is o: by identifying o_1 with one input of the disjunction element and o_2 with the other, by identifying the output s of the S-unit with the k + 1st input of N_1' and the k + 1st input of N_2', and identifying input i_j of N_1' with input i_j of N_2' for $j = 1, 2, \cdots, k$.

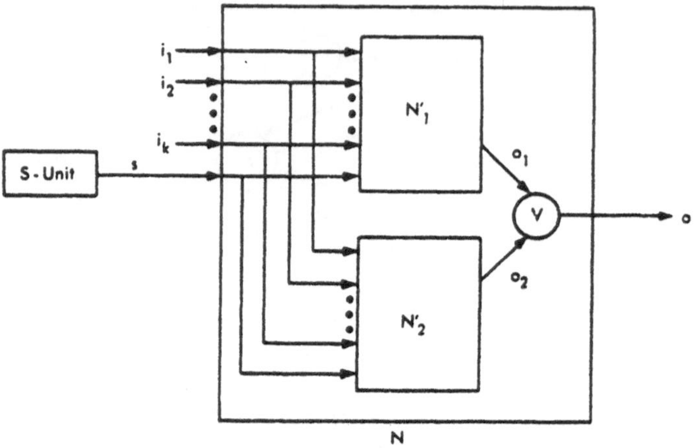

FIG. 5d

Case 2. $\Omega = (\Omega_1 \cdot \Omega_2)$. By the induction hypothesis, since Ω_1 and Ω_2 each contain fewer than n symbols, they are realized by outputs o_1 and o_2 of k input nets N_1 and N_2 each containing exactly one S-unit, etc. Now Ω is realized by the output o_2 of the k input net N constructed out of N_1', N_2', an S-unit, and a delay element: by identifying the output o_1 of N_1' with the input of the delay element, identifying the output s of the S-unit with the k + 1st input of N_1', identifying the delay output

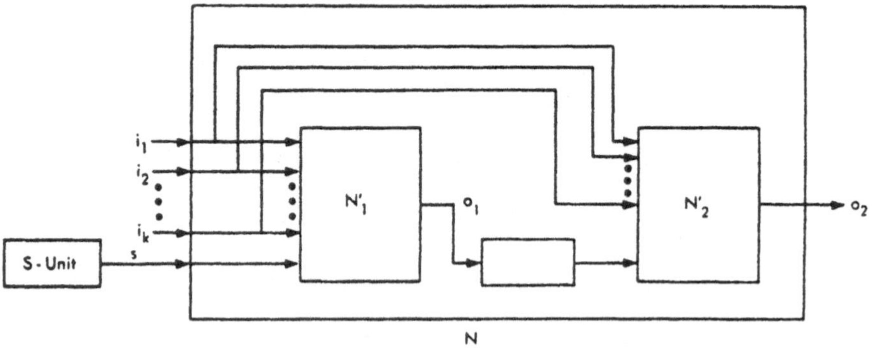

FIG. 5e

with the $k + 1$st input of N_2', and identifying input i_j of N_1' with input i_j of N_2' for $j = 1, 2, \cdots, k$.

Case 3. $\Omega = \Omega_1^*$. By the induction hypothesis, since Ω_1 contains fewer than n symbols, it is realized by output o_1 of k input net N_1 containing exactly one S-unit, etc. Now Ω is realized by the output o_1 of the k input net N constructed out of N_1', an S-unit, a delay element, and a 2 input disjunction element: by identifying the disjunction output with the $k + 1$st input of N_1', by identifying the output s of the S-unit with one input of the disjunction element and the delay output with the other, and identifying the output o_1 with the input of the delay element.

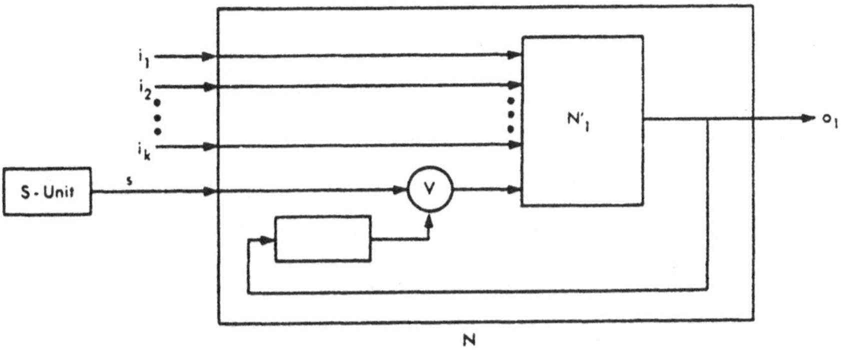

FIG. 5f

6. *Examples*

A. *Synthesis*. We use the algorithm embodied in the synthesis theorem to construct a net whose designated (in these cases its rightmost) output realizes $[(A \vee (B \cdot B)) \vee ((B \cdot A^*) \cdot B)]^*$, where $A = \{0\}$ and $B = \{1\}$.

If an output of a net N realizes F, we shall say that F' is realized by the corresponding output of the net N' formed by detaching the S-unit from N. Figures 6j and 6k are unabbreviated representations of the nets represented by figures 6g and 6i, respectively.

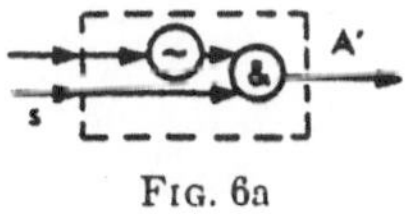

FIG. 6a

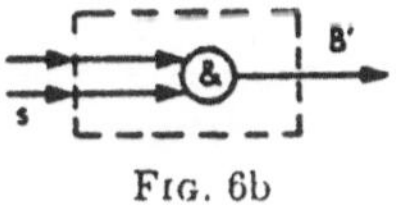

FIG. 6b

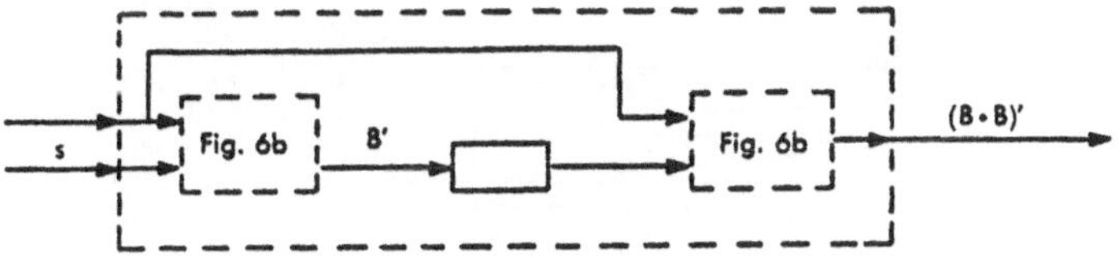

FIG. 6c

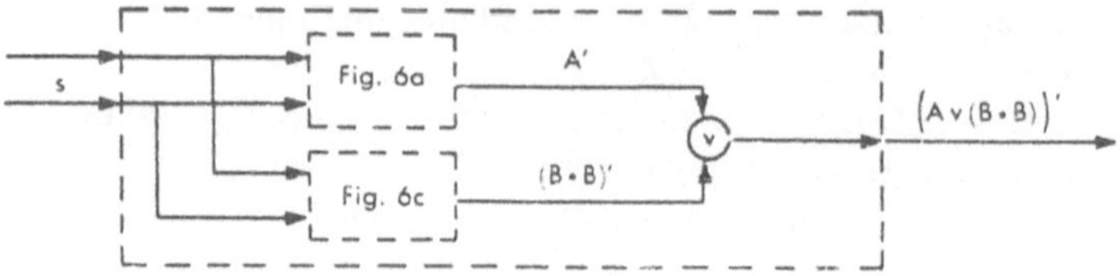

FIG. 6d

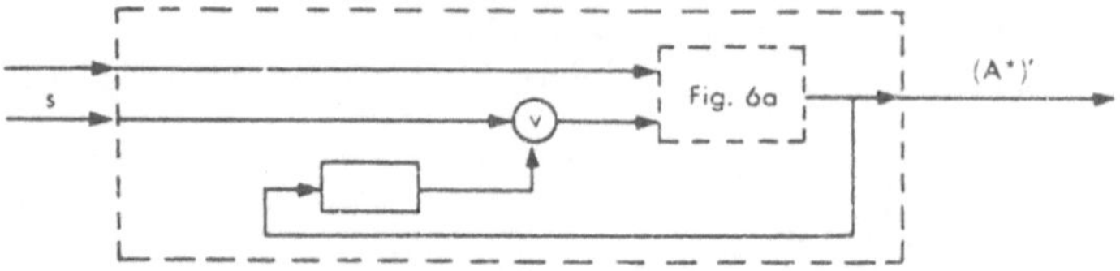

FIG. 6e

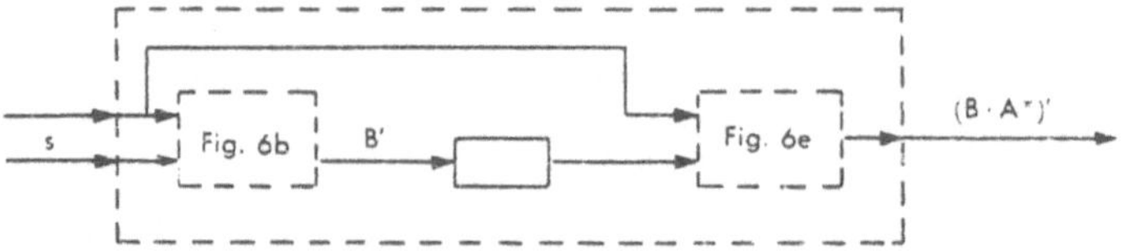

FIG. 6f

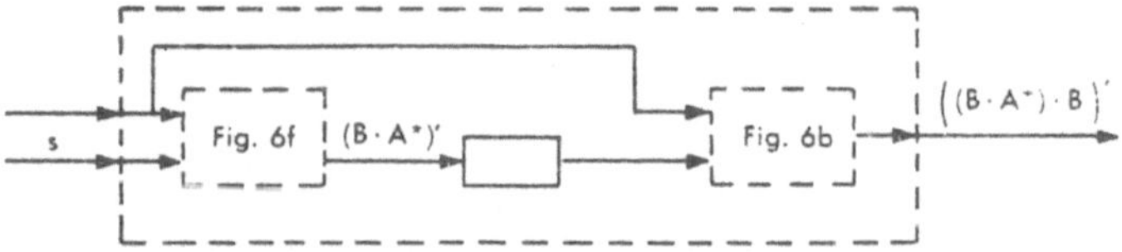

FIG. 6g

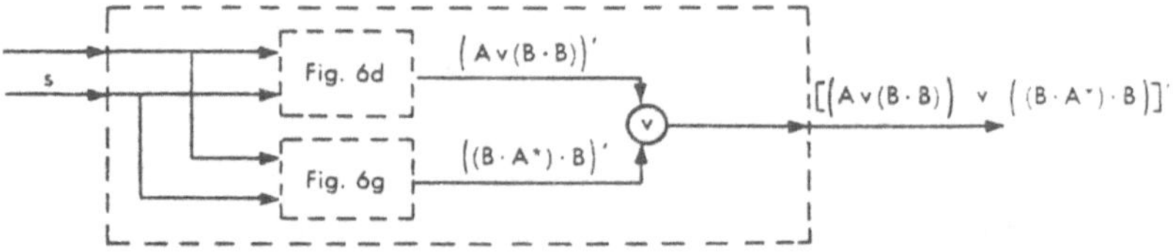

FIG. 6h

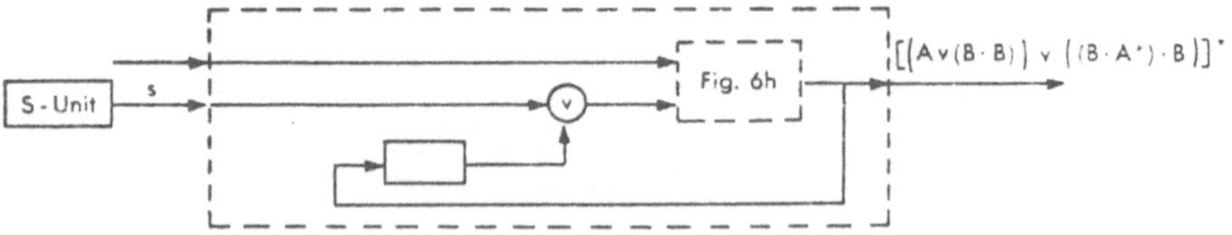

FIG. 6i

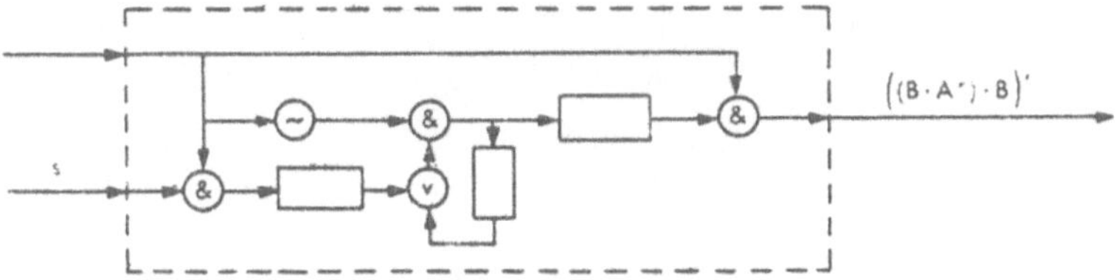

FIG. 6j

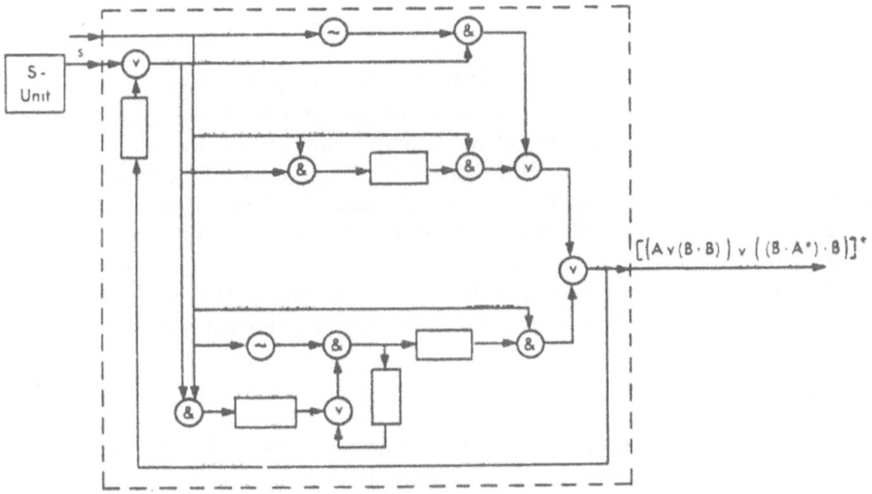

FIG. 6k

B. *Analysis.* We wish to find a regular expression which is realized by the designated output of the net below.

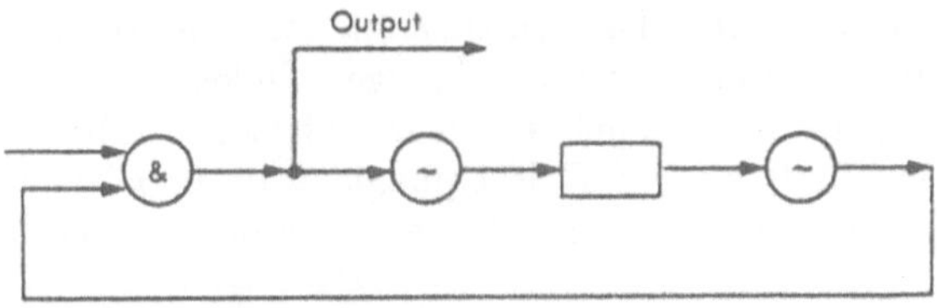

FIG. 6l

If we let the first bit represent the state of the input and the second bit represent the state of the delay output, then the direct transition relation (on complete states) is given by the table on page 194. The initial complete states are 00 and 10, while there is only one complete state in which the designated output is active, namely, 10. We, therefore, wish to find all R-sequences from 00 to 10 and also from 10 to 10.

In this particular example, it is easy to find the regular set by inspection.

R	
00	01
00	11
01	01
01	11
10	00
10	10
11	01
11	11

However, if we wish to use the algorithm embodied in lemma 1, we observe that the set of R-sequences from 11 to 10 in (i.e., constructed from elements of) $\{10, 11\}$ is empty since the "D" of the lemma is empty. For the same reason the set of R-sequences from 01 to 10 in $\{01, 10\}$ is empty. It follows that the set of R-sequences from 01 (11) to 10 in $\{01, 10, 11\}$ and from 00 to 10 in $\{00, 01, 10, 11\}$ is empty while from 10 to 10 in $\{00, 01, 10, 11\}$ is $\{10\} \vee \{10\}^* \cdot \{10\}$. (The "C" of the lemma is $\{10\}$.)

Applying now the projection of lemma 2, we obtain $\{1\} \vee \{1\}^* \cdot \{1\}$. The designated output of the net above realizes this regular expression.

7. *Comments*

Basic to Kleene's concept of event is his notion of tagged table. A table may be tagged in one of two ways, initial or non-initial. A set E of tagged k-columned tables represents a k-event. A k-columned table, m, of length l, is an *occurrence* of E if (1) m with the tag initial is an element of E, or (2) there exists an $r \geqq 1$ such that the table, tagged non-initial, consisting of rows $r, r + 1, \cdots, l - r + 1$ is an element of E. Two sets, E, F, of tagged k-columned tables are *equivalent*, i.e., represent the same event, if the set of all occurrences of E is the same as the set of all occurrences of F. An *event* (in the sense of Kleene) may be defined as an equivalence class of sets of tagged tables.

Three binary operations are defined on sets of tagged tables. It is not true in general, however, that, e.g., $E \cdot F$ is equivalent to $E' \cdot F'$ whenever the pairs E, E' and F, F' are equivalent. Thus these operations on sets do not carry over to events.

Kleene's regular expressions denote sets of tagged tables. Two regular expressions are *equal* if they denote the same set. Two regular expressions are *equivalent* if they denote equivalent sets.

Since the main complication in Kleene's theory is due to the concept of tagged tables, we sought to determine whether it was possible to obtain the theory without such a concept. This concept seemed to play an essential role in the synthesis theorem (Kleene's theorem 3, our theorem 2). We found that the concept can be eliminated as a primitive one and defined in terms of the concepts we adopt. The result is a great simplification in the notion of event, the fusion of the concepts equality and equivalence of regular expressions, and simpler operations $\vee$, $\cdot$, *.

Another complication in Kleene's development is contributed by the unit delay built into the elements of his nets. By separating logical units from delay units one obtains more flexible, in a sense stronger, nets. As a result our synthesis theorem is, in a sense, weaker than Kleene's; but it brings the essential nature of the result into sharper focus.

We use a singulary "*" operation rather than a binary one because the operation Kleene uses seems "essentially" singulary and because the singulary operation simplifies the algebra of regular events. It should be noted that the singulary and binary star operations are interdefinable.

APPENDIX

In general, there are $k \geqq 0$ inputs and $n > 0$ outputs in a net or finite automaton. Each input and each output may assume a finite number of states. If the states of every input for every moment of time (given by the natural numbers) are specified for a net, then the state of every output at every moment of time is determined. Assuming that the inputs (outputs) are ordered and that the ith input (output) is capable of assuming $a_i(b_i)$ states the input (output) over time t, $1 \leqq t < l + 1$, may be described by a k (n)-rowed matrix whose ith row entries denote any of $a_i(b_i)$ states, and with l columns, where l is a positive integer or infinity. The behavior of the net or automaton may then be described in the following alternative ways:

(a) A transformation (mapping, function) from k-rowed infinite matrices to n-rowed infinite matrices with the property P: whenever the first l columns of two k-rowed matrices are the same, the associated n-rowed matrices have the same first l columns.

(b) A transformation which associates with k-rowed, l-columned (k is fixed; l varies over all the positive integers) matrices n-rowed, l-columned matrices and which has property P.

(c) A transformation which associates with k-rowed finite matrices, n-rowed one-columned matrices.

(d) An n-tuple of (ordered) partitions of all the k-rowed finite matrices. The ith, $1 \leqq i \leq n$, (ordered) partition consists of a b_i-tuple of sets.

If T_a is a transformation of type (a) then a transformation of type (b), T_b, may be uniquely associated with it by defining $T_b(x_m) = (T_a(x))_m$ where x is an infinite matrix whose first m columns is x. With each T_b may be associated a transformation T_c, whose domain is the same as the domain of T_b, and whose value when applied to a m-columned matrix is the mth column of $T_b(x)$. With each T_c an n-tuple of (ordered) partitions of all the k-rowed finite matrices may be associated as follows:

The jth set, $1 \leqq j \leqq b_i$, of the ith entry, $1 \leqq i \leqq n$, of the n-tuple consists of all finite k-rowed matrices x such that the ith row of $T_c(x)$ is the jth output state. It is clear with a little reflection that these four modes of expressing behavior are in biunique correspondence.

Burks and Wright [2] use method (a) to express behavior with $n = 1$, all $a_i = 2$, $b_1 = 2$ and the infinite k-rowed matrices regarded as k-tuples of infinite sequences (of zeros and ones). Burks and Wang [1] use method (a) (among others) with all a_i, $b_i = 2$ but with the infinite matrices regarded as infinite sequences of k (n)-tuples. E. F. Moore [4] uses method (b) with $k = 1$ and $n = 1$ but with a_1 and b_1 arbitrary. S. C. Kleene [3], suitably interpreted, uses method (d) with $n = 1$, $b_1 = 2$ and the $a_i = 2$. G. H. Mealy [5] uses method (b) with the b_i, $b_j = 2$ and with the finite k (n)-rowed matrices interpreted as finite sequences of k (n)-tuples. We use, in this paper, (d) with $n = 1$, $b_1 = 2$ and the $a_i = 2$ and interpret the finite k-rowed matrices as finite sequences of k-tuples.

REFERENCES

[1] A. W. Burks and H. Wang, The logic of automata, *Jour. Assoc. Comp. Mach.*, Part I: *4* (1957), 193–218; Part II: *4* (1957), 279–297.

[2] A. W. Burks and J. B. Wright, Theory of logical nets, *Proc. IRE 41* (1953), 1357–1365.

[3] S. C. Kleene, Representation of events in nerve nets and finite automata, *Automata Studies*, edited by C. E. Shannon and J. McCarthy, Princeton University Press, 1956, pp. 3–41.

[4] E. F. Moore, Gedanken-experiments on sequential machines, *Automata Studies*, edited by C. E. Shannon and J. McCarthy, Princeton University Press, 1956, pp. 129–153.

[5] G. H. Mealy, A method for synthesizing sequential circuits, *The Bell System Technical Journal, 34* (1955), 1045–1079.

Random-Access Stored-Program Machines, an Approach to Programming Languages

CALVIN C. ELGOT AND ABRAHAM ROBINSON*

IBM Watson Research Center, Yorktown Heights, New York

Abstract. A new class of machine models as a framework for the rational discussion of programming languages is introduced. In particular, a basis is provided for endowing programming languages with semantics. The notion of Random-Access Stored-Program Machine (RASP) is intended to capture some of the most salient features of the central processing unit of a modern digital computer. An instruction of such a machine is understood as a mapping from states (of the machine) into states. Some classification of instructions is introduced. It is pointed out in several theorems that programs of finitely determined instructions are properly more powerful if address modification is permitted than when it is forbidden, thereby shedding some light on the role of address modification in digital computers. The relation between problem-oriented languages (POL) and machine languages (ML) is briefly considered.

1. *Introduction*

Programming for digital computers has led to the development of a variety of programming languages. These systems are classified, broadly, according to whether they are oriented toward the human user who wishes to prepare a given computational procedure in the simplest possible way for further automatic processing (problem-oriented languages, briefly, POL) or toward the detailed instructions carried out by the machine ultimately (machine languages, ML). In this paper we set up a framework which permits a rational discussion of such languages, of their functions, of the relations between them and of the ways in which they can be extended or improved.

Existing analyses of programming languages, such as can be found, for example, in the ALGOL 60 report [1], place the emphasis on the syntactical form of concepts or definitions. By contrast, the present approach concentrates on the role of the constituents of a language as commands or instructions when stored in a machine. This implies that the programming language is considered in conjunction with a machine, actual or ideal. Previous theoretical machine models (such as Turing machines and finite automata) which might be relevant here do not seem to be sufficiently concrete or, more precisely, lack some of the salient features of the concrete situation which are relevant for our purposes. For example, Turing machines lack the random-access feature and finite automata may be regarded as special cases of Turing machines. Other models try to imitate concrete machines more closely but are unsuitable for a general analysis of the type carried out here. In some cases the stored program feature is lacking.

A brief plan of this paper follows. In Section 2 we introduce a class of ideal

* Robinson at IBM and University of California, Los Angeles, California.

Journal of the Association for Computing Machinery, Vol. 11, No. 4 (October, 1964), pp. 365–399 machines, called *I*ntermediate *M*odels for *P*rogramming, briefly IMP's, which brings out some of the features of an actual machine in which we are interested but is still too primitive for our purposes. This leads up to the introduction, in Section 3, of our central notion, that of a Random-Access Stored-Program Machine or, briefly, a RASP which is intended to capture some of the most salient features of the central processing unit of a modern digital computer. As the name indicates the features of a RASP involve the assumption that the program as well as the data given for a particular computation are stored in the memory of the machine and that "immediate" access to all memory locations is available. In the same section we define the notion of an *instruction schema* which is basic for the sequel.

In Section 4, we construct a RASP which, in a sense that is made precise, can compute all recursive functions (where the initial content of the machine, including the stored program, depends on the particular function which is to be computed and, to a lesser extent, on the values of the argument or arguments). In Section 7 we show how a RASP can be used to compute finite sequences (of numbers, say) for an input of finite sequences without the intervention of arithmetization (Gödelization). In this context it is established (Theorem 7.4 and 7.5) that self-modifying programs are more powerful (can compute more sequential functions) than nonself-modifying ones, provided that individual instructions fall within a prescribed, rather broad, class of instructions called *finitely determined* (Section 3).

As we shall see the notion of a RASP involves a pair of functions which may or may not be computable (recursive). This raises the question: In what sense may a RASP itself be said to be computable (as we expect it to be, in some sense, if it is to correspond to a concrete computer)? The question is discussed in some detail in Section 5.

Section 6 is concerned with the relation between different RASP's. We define a notion of *definitional extension* which turns out to correspond to the passage from a "poorer" (e.g., machine) language to a "richer" (e.g., problem-oriented) language. However, the words "poorer" and "richer" have to be taken advisedly since it turns out that the enrichment of the language is achieved, not by an increase in the vocabulary, but by a change in the interpretation of the vocabulary. Within this setting we expect to define in the sequel precisely what is meant by the reduction of a program in the richer language to an equivalent program in the poorer language. The problem of achieving such a reduction is called the compiler problem.

2. *Intermediate Models For Programming (IMP)*

To introduce our detailed approach, let us consider the evolution which takes place in the memory of a digital computer from the moment when the program has been stored in it together with the initial data. At this stage, we rule out the possibility that new data or instructions are read into the machine *during* the subsequent computation. In order to describe the configuration in the memory at any time, we assign symbols $a_0, a_1, a_2, a_3, \cdots$ to the individual storage locations. We suppose that the locations are distinct and that, together, they exhaust the entire memory. We denote by $b_0, b_1, b_2, b_3, \cdots$ the units of information that can be stored in the a_j (which are realized in practice by the several physical states of which the locations are capable), and assume that any b_i can be stored in any a_j. Let $A = \{a_0, a_1, a_2, \cdots\}$, $B = \{b_0, b_1, b_2, \cdots\}$; then the configuration in the memory at a specified time can be represented by a function $k(x)$ from A into B, such that for any a_i, $k(a_i) = b_j$ yields the contents of the storage location denoted by a_i. Also, as mentioned already, we shall suppose throughout that we are dealing with a stored program. This implies that the transition of the memory from one configuration to another is controlled by an instruction which is taken by the machine from the contents of a certain location. At the same time, the contents of the memory at a given stage also tell the machine where to take its next instruction from (by means of a "go to" statement). We schematize the situation by the introduction of the following abstract concept, which is called an *intermediate model for programming* (IMP).

An IMP is an ordered sextuple $J = \langle A, B, b_o, K_o, g^1, g^2\rangle$ where A and B are (usually, countably infinite[1] and, possibly, overlapping or coinciding) sets of objects called *addresses* and *words*, respectively; and b_o is a specific element of B called the *empty word*. Some preparation is required in order to describe the remaining elements of J.

Let K be the set of all functions $k(x)$ which are defined on A and take values in B. The elements of K are called *content-functions*. The set of all $a \in A$ such that $k(a) \neq b_o$ is called the *support* of an element k of K. The set of all elements k of K whose support is finite is denoted by K_f.

Continuing our definition of an IMP, we require next that K_o be a subset of K. Of particular interest will be the case $K_o = K_f$ when the IMP is called *finitely supported*.

Let Σ be the Cartesian product of K and A, $\Sigma = K \times A$, so that Σ is the set of all ordered pairs $\langle k, a\rangle$ for $k \in K$, $a \in A$. Similarly, let $\Sigma_o = K_o \times A$, so that Σ_o is a subset of Σ. The elements of Σ_o are called the *states* of the IMP. (The state of an IMP at a given "instant" determines its future course independent of its earlier history.) Then g^1 is supposed to be a particular mapping from Σ_o into K_o, while g^2 is a mapping from Σ_o into A. Thus, g^1 and g^2 may also be regarded as functions of two variables $g^1 = g^1(x, y)$, $g^2 = g^2(x, y)$, whose arguments vary over K_o and A and which take values in K_o and A, respectively. We may combine g^1 and g^2 into a single function $g = \langle g^1, g^2\rangle$ which maps Σ_o into Σ_o. More precisely, g maps every state $\sigma = \langle k, a\rangle$ in Σ_o on the state $\langle g^1(k, a), g^2(k, a)\rangle = g(\sigma)$. Referring to the concrete case considered above, $\sigma = \langle k, a\rangle$ is supposed to describe the situation in the machine at a given time where $k = k(x)$ represents

[1] The cases which arise when at least one of A or B is finite are also of importance. It is not ruled out, however, that the cardinality of the sets in question exceeds that of the natural numbers.

the contents of the memory and a is (the address, or label, of) the control location.

An infinite sequence of states $\text{comp}(\sigma_0) = \langle \sigma_0, \sigma_1, \sigma_2, \cdots \rangle$ of a given IMP, J, will be called a *computation* if $\sigma_{i+1} = g(\sigma_i)$ for $i = 0, 1, 2, \cdots$. If $\sigma_{i+1} = g(\sigma_i) = \sigma_i$ for some i, then, clearly, $\sigma_{i+2} = \sigma_{i+1}$, $\sigma_{i+3} = \sigma_{i+2}$, etc. In this case σ_i is said to be a *stable* state of the given IMP.

Let E be a finite subset of A and let $\text{comp}(\sigma_0) = \langle \sigma_0, \sigma_1, \sigma_2, \cdots \rangle$, $\sigma_i = \langle k_i, a_i \rangle$. We define $\text{comp}_E(\sigma_0) = \text{comp}(\sigma_0)$ if for all i, $a_i \notin E$ and define $\text{comp}_E(\sigma_0) = \langle \sigma_0, \sigma_1, \cdots, \sigma_n \rangle$ if $a_n \in E$ and for all $i < n$, $a_i \notin E$. In this latter case $\text{comp}_E(\sigma_0)$ is finite and said to be *successful*, and $\text{comp}_E(\sigma_0)$ is said to *terminate in* σ_n. We also speak of n as being *the first instant that* $\text{comp}_E(\sigma_0)$ *reaches an element of* E. In case $E = \{e\}$, we write $\text{comp}_e(\sigma_0)$ for $\text{comp}_E(\sigma_0)$. In this case n *is the first instant that* $\text{comp}_e(\sigma_0)$ *reaches* e.

Although the notion of an IMP undoubtedly captures some of the features of an actual machine computation, it is deficient in some respects which are essential for our purposes. In particular, although it was mentioned that, for a given state $\sigma = \langle k, a \rangle$, the content of the storage location with address a determines, possibly together with a, the transition from σ to the next state, $g(\sigma)$, we have not actually mirrored this fact in the construction of the notion of an IMP. In the next section this deficiency is remedied by the introduction of the related concept of a random-access stored-program machine (RASP).

3. *Random-Access, Stored-Program Machines (RASP)*

Before introducing the notion of a RASP, let us think again in terms of an example. The following schema is typical of a rather general class of instructions which may appear in a program.

3.1: "Insert $F(a_0, \cdots, a_j, k(a_0), \cdots, k(a_j))$ in $G(a_0, \cdots, a_j, k(a_0), \cdots, k(a_j))$; if $R(a_0, \cdots, a_j, k(a_0), \cdots, k(a_j))$ is *true*, go to $H(a, a_0, \cdots, a_j)$, and if $R(a_0, \cdots, a_j, k(a_0), \cdots, k(a_j))$ is *false*, go to $\bar{H}(a, a_0, \cdots, a_j)$".

In this instruction k and a are supposed to constitute the given state σ, $\sigma = \langle k, a \rangle$, while R is a fixed relation whose meaning is known (interpreted); F is a fixed function with values in B, while G, H, $\bar{H}$ are fixed functions with values in A. The arguments of these functions vary over A and B as indicated. In spite of the notation used in 3.1 for the sake of simplicity, the variables in these functions and relation need not be the same; moreover, the functions (and relation) may be independent of some of their argument positions. The entire instruction 3.1, in the form of a code word, is supposed to appear as the content corresponding to the address a for the state σ, i.e., as $k(a)$; and the next state, $\sigma' = \langle k', a' \rangle$, is to be obtained from σ in the following way.

The next content function k' satisfies: $k'(x) = k(x)$ for all x except $x = G(a_0, \cdots, a_j, k(a_0), \cdots, k(a_j))$, for which $k'(x) = F(a_0, \cdots, a_j, k(a_0), \cdots, k(a_j))$. If $R(a_0, \cdots, a_j, k(a_0), \cdots, k(a_j))$ is true, then $a' = H(a, a_0, \cdots, a_j)$, while if $R(a_0, \cdots, a_j, k(a_0), \cdots, k(a_j))$ is false, then $a' = \bar{H}(a, a_0, \cdots, a_j)$. Symbolically:

3.2:

$$k'(G(a_0, \cdots, a_j, k(a_0), \cdots, k(a_j))) = F(a_0, \cdots, a_j, k(a_0), \cdots, k(a_j))$$
$$\wedge\ (\forall x)[x \neq G(a_0, \cdots, a_j, k(a_0), \cdots, k(a_j)) \rightarrow k'(x) = k(x)]$$
$$\bullet \wedge \bullet\ [R(a_0, \cdots, a_j, k(a_0), \cdots, k(a_j)) \rightarrow a' = H(a, a_0, \cdots, a_j)]$$
$$\wedge\ [\sim R(a_0, \cdots, a_j, k(a_0), \cdots, k(a_j)) \rightarrow a' = \bar{H}(a, a_0, \cdots, a_j)].$$

Thus, at any stage, the content $b = k(a)$ of the control location a determines the transformation from the given state, σ, to the next state, σ'. We incorporate this essential feature in our next model by means of the following definition.

A *random-access, stored-program machine* (briefly, a RASP) is defined as an ordered sextuple $P = \langle A, B, b_o, K_o, h^1, h^2\rangle$, where A and B are (possibly overlapping or coinciding) sets of abstract objects called addresses and words, respectively; and b_o is an element of B, called the empty word, as in the definition of an IMP. Defining K and K_f exactly as in Section 2, K_o is again supposed to be a subset of K, and the RASP is called *finitely supported* if $K_o = K_f$. Finally, with Σ and Σ_o defined as before, so that $\Sigma = K \times A$, $\Sigma_o = K_o \times A$, h^1 is a mapping from $\Sigma_o \times B$ into K_o, while h^2 is a mapping from $\Sigma_o \times B$ into A. We may write h^1 and h^2 also as functions of *three* variables $h^1(x, y, z)$, $h^2(x, y, z)$, taking values in K_o and A, respectively, as x varies over K_o, y varies over A and z varies over B. We impose the following condition or *axiom*.[2]

3.3: For any $k \in K_o$, $a \in A$, $h^1(k, a, b_o) = k$, $h^2(k, a, b_o) = a$.
This completes the definition of a RASP.

Combining h^1 and h^2 into a single mapping $h = \langle h^1, h^2\rangle$, we see that h maps the elements of $\Sigma_o \times B$ into Σ_o. In particular, if h is applied to the element $\langle\sigma, b\rangle$ of $\Sigma_o \times B$, where $\sigma = \langle k, a\rangle$, then the result is the state $\langle h^1(k, a, b), h^2(k, a, b)\rangle$. Writing h as a function of two variables, $h(x, y)$, where x varies over Σ_o and y varies over B; h^1, h^2, h are then connected by the identity

$$h(<x, y>, z) = \langle h^1(x, y, z), h^2(x, y, z)\rangle.$$

For any fixed $b \in B$, $h(x, b)$ defines a mapping from Σ_o into Σ_o or, in other words, it defines an element of $\Sigma_o^{\Sigma_o}$, the class of functions defined on Σ_o and taking values in Σ_o. Thus, h defines a mapping $B \rightarrow \Sigma_o^{\Sigma_o}$, the image of any $b \in B$ under this map being h_b which is defined by the requirement that $h_b(x) = h(x, b)$ for all x. For any $b \in B$ then, h_b defines a transition from one state (element of Σ_o) into another state, and this is precisely the task carried out by 3.1 in the concrete case considered above; h_b will be called an (atomic) *instruction* of the RASP $\langle A, B, b_o, K_o, h^1, h^2\rangle$. We also speak of an arbitrary mapping of Σ_o into Σ_o as an *instruction*. The condition 3.3 becomes, in terms of h,

3.4: For all $\sigma \in \Sigma_o$, $h(\sigma, b_o) = \sigma$.

It will also be convenient to speak of a *labeled instruction* as the mapping obtained from an instruction by fixing some element a. Thus, $h_{a,b}$ is a mapping from K_o into $K_o \times A$, $h^1_{a,b} : K_o \rightarrow K_o$, $h^2_{a,b} : K_o \rightarrow A$ and $h_{a,b}(k) = h_b(k, a)$, $h^1_{a,b}(k) = h_b^{\,1}(k, a)$, $h^2_{a,b}(k) = h_b^{\,2}(k, a)$.

[2] The axiom is a conceptual convenience. Its role is inessential.

By the *range of influence*, $RI(h_{a,b})$, *of a labeled instruction* $h_{a,b}$, we mean: $\{d \mid d \in A$, for some $k \in K_o$, $k'(d) \neq k(d)$, where $k' = h^1_{a,b}(k)\}$.

Every RASP P generates in a natural way an IMP, J, which is linked to it by the following rules. The first four elements of J coincide with the corresponding elements of P. Thus, the set of addresses, the set of words, the empty word and the set of content functions, K_o, is the same in J as in P. In order to complete the definition of J, we now define the functions g^1 and g^2 in terms of h^1 and h^2 by putting, for any $k \in K_o$ and $a \in A$,

3.5: $g^1(k, a) = h^1(k, a, k(a)), \qquad g^2(k, a) = h^2(k, a, k(a))$.

It is clear that these definitions yield an IMP as introduced in the preceding section. If P is finitely supported, then so is J. A computation of P will be understood to mean a computation of the corresponding IMP, J. It will be seen that the definition of J in terms of P does not, by any means, make use of all the information provided by P since the right-hand sides of 3.5 involve h^1 and h^2 only for very special sets of argument values. However, this may be said to correspond to the real situation.

A word b will be called *active* (in a given RASP) if for some state $\sigma \in \Sigma_o$, $h_b(\sigma) \neq \sigma$. All other words will be called *inactive* or *passive*. This terminology refers to the effect of a word when regarded as an instruction code. The empty word is passive as stated in 3.4. In a concrete case many other words may be expected to be passive. The further classification of instructions, or of words regarded as instructions (instructions which produce data in terms of given data or which produce instructions in terms of given data in certain ways or instructions from given instructions, etc.), is quite relevant to our problem. However, this is a very complex task, and we shall not attempt to carry it out systematically at the present stage. Instead, we proceed to specify some conditions on instructions which will turn out to be important in our subsequent argument.

Putting $h_b^1(x, y) = h^1(x, y, b)$ and $h_b^2(x, y) = h^2(x, y, b)$, we have $h_b(<x, y>) = \langle h_b^1(x, y), h_b^2(x, y)\rangle$ identically, for all $b \in B$. For a given b, we shall say that $h_b = \langle h_b^1, h_b^2\rangle$ (and also h_b^1) is *location independent* if $h_b^1(x, y)$ is independent of its second argument. In this case the content function produced by the action of h_b^1 on a given state depends only on the content function of that state and not on its control location. The condition is satisfied by h_{b_o} and by 3.1.

Now h_b^2 may always be regarded as a mapping from K_o into A^A. If h_b^1 is location independent, then it may be regarded as a mapping from K_o into K_o. If, in addition, $K_o = K = B^A$, then $h_b^1: B^A \to B^A$, $h_b^2: B^A \to A^A$.

A RASP is said to be *location independent* if for all $b \in B$, h_b is location independent.

We shall say that h_b (and also h_b^2) is *finitely conditioned* if the set of all images of h_b^2, called the *exits* of h_b (and also of h_b^2), regarded as a mapping from K_o into A^A, is finite; i.e., the set $h_b^2(K_o)$ is finite. Referring again to 3.1 we see the instruction indicated is finitely conditioned. Indeed, the set of all images of h_b^2 is in this case $\{H, \bar{H}\}$ if we regard a as variable and the a_i, $0 \leq i \leq j$, as fixed. Thus, $H, \bar{H}$

are the *exits* of h_b. More generally, if the set of images of h_b^2 consists of m elements (m must of course be greater than zero), the instruction 3.1 would involve an m-valued (rather than 2-valued) relation in place of R and m functions H_0, $H_1, \cdots, H_{m-1}$. The m-valued relation may, however, be replaced by m mutually exclusive, exhaustive 2-valued relations, which is what we do below. If $m = 1$, the instruction may be said to be *unconditioned.*[3]

The analogue of 3.2 for the m-valued case is the conjunction of (a) and (b), below, where we have suppressed the parameters $a_0, \cdots, a_j$.

3.6: (a) $k'(G(k)) = F(k) \wedge (\forall x)[x \neq G(k) \rightarrow k'(x) = k(x)]$.

(b) $(R_0(k) \rightarrow a' = H_0(a)) \wedge (R_1(k) \rightarrow a' = H_1(a)) \wedge \cdots \wedge (R_{m-1}(k) \rightarrow a' = H_{m-1}(a))$.

Here $R_0, R_1, \cdots, R_{m-1}$[4] are supposed to represent mutually exclusive, exhaustive (2-valued) relations; i.e., the states for which R_i holds, $0 \leqq i < m$, constitutes a partition of Σ_o into m subclasses. It may be seen that 3.6(a) represents the mapping h_b^1, while 3.6(b) represents h_b^2 (for a restricted type of instruction).

For particular R_i, H_i, 3.6(b) defines a particular mapping from Σ_o into A. If 3.6(b) defines h_b^2, then 3.6(b) is called a *separated representation* of h_b^2.

3.7: *It is clear that if h_b^2 has a separated representation, then h_b^2 is finitely conditioned.* Conversely,

3.8: PROPOSITION. *If h_b^2 is finitely conditioned, then it possesses a separated representation.*

PROOF. Let the image of $h_b^2: K_o \rightarrow A^A$ be $\{H_0, H_1, \cdots, H_{m-1}\}$. Define $R_i(k)$, $0 \leqq i < m$, to be true iff $h_b^2(k) = H_i$. Then 3.6(b) defines h_b^2. For regarding now h_b^2 as a mapping from $K_o \times A \rightarrow A$, we see that, given k, a, exactly one R_i holds and for that i, $h_b^2(k, a) = H_i(a)$.

Thus, h_b^2 *is finitely conditioned iff it possesses a separated representation.*

The *exits* of a *labeled instruction*, $h_{a,b}: K_o \rightarrow K_o \times A$, are the images of $h_{a,b}^2: K_o \rightarrow A$. A *labeled instruction* is *finitely conditioned* if its set of images is finite.

Our next point is concerned with the range of argument values and function values of k which determine the transition to the next state.

For given b, h_b will be said to be *finitely determined*[5] if for every a there exists a finite sequence $A_{a,b}$ of elements of A, $A_{a,b} = (a_1, \cdots, a_j)$, say, such that for any $b_1, \cdots, b_j$ in B there exist elements a', $a_{i_1}, \cdots, a_{i_l}$ of A and $b_{i_1}, \cdots, b_{i_l}$ of B (l a natural number which also depends on $a, b_1, \cdots, b_j$) for which the following condition is satisfied. If $\sigma = \langle k, a \rangle$ is an element of Σ_o for which $k(a_1) = b_1$, $k(a_2) = b_2, \cdots, k(a_j) = b_j$, then $h_b^2(k, a) = a'$, and $h_b^1(k, a) = k'$ is obtained from k by putting $k'(a_{i_m}) = b_{i_m}$ for $m = 1, \cdots, l$ and $k'(x) = k(x)$ for all other values of the argument. Thus, the control moves from a to a', and $k'(x)$ is ob-

[3] If an instruction is location independent and unconditioned, it may be called *separable*. This means the first component h_b^1 of the instruction h_b is independent of its second argument, while the second component h_b^2 is independent of its first argument.

[4] It is, of course, possible to represent an m-valued relation by $1 + [\log_2 m]$ binary relations (or, in case m is a power of 2, by $\log_2 m$ relations).

[5] In order to appreciate the significance of this definition the reader may find it helpful to refer to the introduction of recursively finitely determined instructions in Section 5.

tained from $k(x)$ by a finite number of changes, depending on a and on the values of $k(x)$ at a finite number of argument places. This condition also is satisfied by 3.1.

h_{b_o} is finitely determined in a trivial sense. If h_b is finitely determined for all $b \in B$, then we shall say that the *RASP is finitely determined.*

This brings us to the important notion of an *instruction schema.* Let J be a given RASP and let l and m be natural numbers. An *instruction schema* ρ is a mapping from the Cartesian product $A^l \times B^m$ into the set of mappings from Σ_o into Σ_o. Denoting the latter set by H_o we have, in symbols, $\rho: A^l \times B^m \rightarrow H_o$. Equivalently, ρ may be regarded as a mapping from $\Sigma_o \times A^l \times B^m$ into Σ_o. We may, therefore, represent an instruction schema by a function $\rho(x, y_1, \cdots, y_l, z_1, \cdots, z_m)$ which takes values in Σ_o as x varies over Σ_o, $y_1, \cdots, y_l$ vary independently over A, and $z_1, \cdots, z_m$ vary independently over B. Or, taking the projections of $\rho(x, y_1, \cdots, z_m)$ into K_o and A, respectively, we may equate ρ to a pair of functions $\langle \rho^1, \rho^2 \rangle$ where $\rho^1 = \rho^1(u, v, y_1, \cdots, y_l, z_1, \cdots, z_m)$ takes values in K_o as u varies over K_o, v varies over A, and the y and z vary over A and B, respectively, as before; and $\rho^2 = \rho^2(u, v, y_1, \cdots, y_l, z_1, \cdots z_m)$ takes values in A for the same range of arguments.

Fixing the *parameters* $y_1, \cdots, y_l, z_1, \cdots, z_m$ in ρ, we obtain a function from Σ_o into Σ_o, e.g., $\rho(x, a_1, \cdots, a_l, b_1, \cdots, b_m)$, which may or may not coincide with some $h_b(x)$. We shall include the possibility $l = m = 0$, for which $\rho(x)$ reduces to a single (and arbitrary) mapping from Σ_o into Σ_o.

Referring back to the beginning of this section for a concrete illustration, we may say that 3.2 is obtained by specialization, i.e., by substituting $a_0, \cdots, a_j$ for $y_0, \cdots, y_j$ in the form which is obtained by replacing "a_i" by "y_i" in 3.2. Here, a_i is regarded as a name of an element of A, while "y_i" is a variable.

Thus, 3.2 yields a mapping from Σ_o into Σ_o for each specialization of its parameters and, in this sense, determines an instruction schema. However, the example involves a feature which is absent from the abstract notion introduced above, since 3.2 actually defines each particular case of the schema by means of a specific "word," i.e., 3.1. This remark leads up to the following definition.

A *designated instruction schema* is an ordered pair $\lambda = \langle \rho, \mu \rangle$ where ρ is an instruction schema as defined above and, for the same l and m, μ is a one-to-one mapping[6] from $A^l \times B^m$ to a subset B_λ of B such that, for all $\sigma \in \Sigma_o$ and $a_1, \cdots, a_l \in A$, $b_1, \cdots, b_m \in B$,

3.9:

$$\rho(\sigma, a_1, \cdots, a_l, b_1, \cdots, b_m) = h(\sigma, \mu(a_1, \cdots, a_l, b_1, \cdots, b_m)).$$

This is equivalent to the statement that the mapping defined by $\rho(x, a_1, \cdots, a_l, b_1, \cdots, b_m)$ coincides with $h_{\mu(a_1, \cdots, a_l, b_1, \cdots, b_m)}(x)$. We take the mapping μ to be one-to-one in order to realize the idea that the elements b' of B_λ are obtained by the substitution of elements $a_1, \cdots, a_l \in A$, $b_1, \cdots, b_m \in B$ for variables which are effectively present. Thus, in 3.2, $l = j + 1$, but $m = 0$.

[6] More generally one may permit μ to be many-to-one. Indeed, in some cases, one may wish to assign a single word to an entire instruction schema.

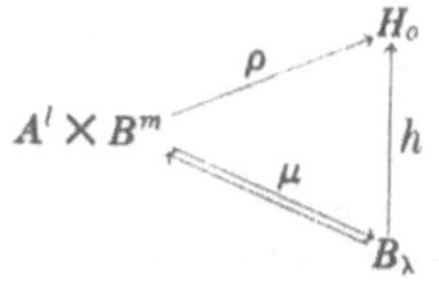

FIG. 1

The relation between ρ, μ and h is described by the commutative diagram given in Figure 1, where we recall that H_o is the set of all mappings from Σ_o into Σ_o. It is evident that ρ is now definable entirely in terms of μ and h.

A designated instruction schema will be called *location independent* if the functions $h_b(x)$ are location independent for all $b \in B_\lambda$, and it will be called *finitely determined* if the $h_b(x)$ are finitely determined for all $b \in B_\lambda$. Actually, we may define these concepts also for undesignated instruction schemata since the definitions of location independence and of finite determination refer entirely to the *mapping defined by* $h_b(x)$, and these definitions can be applied directly to $\rho(x, a_1, \cdots, a_l, b_1, \cdots, b_m)$ for $a_i \in A$, $b_i \in B$.

The notions introduced in this section are fundamental for the sequel. At one point or the other, alternative definitions may well have suggested themselves to the reader. In particular, one might argue that what we have taken to be "labeled instruction" should be called "instruction". However, it seems much more natural to us to regard an instruction as something which is divorced from any location. Three possibilities then suggest themselves for the notion of "instruction": (1) a mapping from Σ_o into Σ_o, (2) a family of mappings from K_o into Σ_o indexed by elements of A, (3) a family of mappings from Σ_o into Σ_o indexed by elements of A. The choice, viz. (1), was made on the basis of simplicity and elegance.

So far we have emphasized the notion of an IMP or RASP as a model for the representation of the processes which go on in a conventional digital computer. It is important to realize that these mathematical models are not restricted only to such interpretations. Thus, we may think of the elements of A as labels which are not tied to any physical storage location. We may also think of these mathematical models as being abstractions of certain types of computational processes carried out by hand or, to be more to the point, computational processes described by "problem-oriented languages."

4. *Computation of Recursive Functions*

In this section we introduce the notion of a function (respectively relation) computable by a RASP. We show first that for an arbitrary RASP, the functions which it computes are closed under composition and that the 2-valued relations are closed under truth functional composition (i.e., closed under Boolean operations). We then show that any particular RASP capable of computing two simple functions and the equality relation is capable of computing all partial recursive functions. This requires that a suitable notion of "program" be introduced as well as the notion of a program computing a function. For the time being we avoid the problem of explicating the intuitive notion of "program" in its full generality, but rather introduce a restricted notion which is sufficiently general for our present purposes. The notion is, however, sufficiently general so that a program may refer to its parts and may modify itself.

Let P be a given RASP and let H be the set of all its instructions. A *program* π is the $(m + 2)$-tuple $\langle p, a_0, e_0, \cdots, e_{m-1}\rangle$, where p is a mapping from a finite subset $\underline{D}\pi$ (also $\underline{D}p$) of A (the *domain* of π and of p) into $H \cup B$, which satisfies 4.1 below;[7] $a_0 \in \underline{D}p$, $p(a_0) \in H$, $i \neq j$ implies $e_i \neq e_j$ and $e_i \in A - \underline{D}p$, $0 \leq i < m$. (The reasons for requiring $e_i \notin \underline{D}p$ will presently become clear.) The element a_0 is called the *entrance* to the program $\langle p, a_0, e_0, \cdots, e_{m-1}\rangle$ and $e_0, \cdots, e_{m-1}$ are the *exits* of this program. For any $a \in \underline{D}p$, if $p(a) \in H$, then $p(a)$ is an *instruction*, a is an *instruction location* and $\langle a, p(a)\rangle$ is a *labeled instruction* of π (and also of p), while if $p(a) \in B$, then $p(a)$ is a *parameter* and a a *parameter location* of π (and also of p). Since H, B are disjoint, so are $p^{-1}H$ and $p^{-1}B$. Let k *holds* π (and also p) mean: "for all $a \in p^{-1}H$, $h_{k(a)} = p(a)$, while for all $a \in p^{-1}B$, $k(a) = p(a)$". If k holds π, $\text{comp}_E(k, a_0)$ is called a computation of π, where $E = \{e_0, \cdots, e_{m-1}\}$. The requirement that a program π is to satisfy may now be given.

4.1: If $\langle\langle k_0, a_0\rangle, \langle k_1, a_1\rangle, \cdots, \langle k_n, a_n\rangle\rangle$ is a successful computation of π then $a_i \in p^{-1}H$ for all $i < n$. If $\langle\langle k_0, a_0\rangle, \langle k_1, a_1\rangle, \cdots, \langle k_i, a_i\rangle, \cdots\rangle$ is an unsuccessful computation of π, then $a_i \in p^{-1}H$ for all i.

Given P, $\pi = \langle p, a, e\rangle$, a program of P, f, a function from a subset of B^r into B^s, $d_0, \cdots, d_{r-1}, v_0, \cdots, v_{s-1}$ a finite sequence of distinct elements of A, then π *is said to compute* f at *datum locations* $d_0, \cdots, d_{r-1}$ and *value locations* $v_0, \cdots, v_{s-1}$ provided that $d_i \notin \underline{D}p$, $0 \leq i < r$, and the following condition holds.

4.2: If k holds π, $k(d_0) = b_0, \cdots, k(d_{r-1}) = b_{r-1}$ and letting $b = \langle b_0, \cdots, b_{r-1}\rangle$, $\sigma = \langle k, a\rangle$, $\sigma_i = \langle k_i, a_i\rangle$, then

(i) if f is defined for b and $f(b) = \langle b_0', \cdots, b_{s-1}'\rangle$, then (a) $\text{comp}_e(\sigma)$ is successful, and (b) if $\text{comp}_e(\sigma) = \langle \sigma, \sigma_1, \cdots, \sigma_n\rangle$, then $k_n(v_i) = b_i'$, $0 \leq i < s$;

(ii) if f is not defined for b, then $\text{comp}_e(\sigma)$ is *not* successful.

Notice that if $k_n(a_n)$ is a passive word, then $\sigma_n = \sigma_{n+1} = \sigma_{n+2} = \cdots$, so that $\text{comp}(\sigma)$ reaches a stable state σ_n. Note, too, that if k_1, k_2 both hold π and agree on d_i, $0 \leq i < r$ and if $\text{comp}_e(k_1, a)$ is successful[8] while $\text{comp}_e(k_2, a)$ is not, then π computes no function at all at datum locations d_i (and any value locations).

Suppose $\langle p_1, a_1, e_1\rangle$ computes f_1 and $\langle p_2, a_2, e_2\rangle$ computes f_2 and suppose we wish to compute $f_2 \circ f_1$ (defined by $(f_2 \circ f_1)(x) = f_2(f_1(x))$) by combining appropriately the two given programs. We immediately encounter the following obstacles: (1) $\underline{D}p_1$ and $\underline{D}p_2$ may overlap, (2) execution of the first program may

[7] We take it for granted that H and B are disjoint. We might have split p into two functions taking values in H and B, respectively, but wish to avoid the ensuing notational complication.

[8] For simplicity we write $\text{comp}_e(k_1, a)$ rather than $\text{comp}_e(\langle k_1, a\rangle)$.

alter the second. This suggests that what is required is a family of programs satisfying certain properties, all of which compute f_i, $i = 1, 2$. This leads to the following definitions.

By the *range of influence* of a program $\pi = \langle p, a, e_0, \cdots, e_{m-1}\rangle$, $RI(\pi)$, we mean that subset of A described by the following requirement.

4.3: $d \in RI(\pi)$ iff there exists $k \in K_o$ such that, k holds p, $\text{comp}_E(k, a)$ is successful and if $\text{comp}_E(k, a)$ terminates in (k_n, a_n), then $k_n(d) \neq k(d)$; $E = \{e_0, \cdots, e_{m-1}\}$.

We shall say that a *RASP P computes a function f* defined on a subset of B^r with values in B^t provided the following conditions hold.

4.4: For every sequence $d_0, \cdots, d_{r-1}, v_0, \cdots, v_{t-1}, a, e$ of distinct elements of A and for all finite subsets A_0, A_1 of A such that $a \notin A_0$ and $A_1 \subseteq A_0$, there exists a program $\pi = \langle p, a, e\rangle$ satisfying:

(1) π computes f at datum locations $d_0, \cdots, d_{r-1}$ and value locations $v_0, \cdots, v_{t-1}$,

(2) $\underline{D}p$ is disjoint from $A_0 \cup \{d_0, \cdots, d_{r-1}, v_0, \cdots, v_{t-1}\}$,

(3) if A_1 is disjoint from $\{v_0, \cdots, v_{t-1}\}$, then $RI(\pi)$ is disjoint from A_1.

It may be observed that if π computes f and $\underline{D}p$ is not empty, then by requirement (2) A must be infinite. It appears that we might have avoided this consequence and yet have obtained Proposition 4.5, but only at the cost of much greater complexity. Thus, allowing A to be infinite promotes simplification.

4.5: PROPOSITION. *If the RASP P computes f_1 and f_2, then it also computes $f = f_2 \circ f_1$.* [We understand f to be defined on $\langle b_0, \cdots, b_{r-1}\rangle$ if and only if f_1 is defined on $\langle b_0, \cdots, b_{r-1}\rangle$ and f_2 is defined on $f_1(b_0, \cdots, b_{r-1})$.] Let f_1 map a subset of B^r into B^s and f_2 a subset of B^s into B^t.

PROOF. Let $d_0, \cdots, d_{r-1}, v_0, \cdots, v_{t-1}, a, e, A_0, A_1$ be given as in 4.4. We are required to find a program $\pi = \langle p, a, e\rangle$ satisfying 4.4 (1), (2), (3). Let $d_0', \cdots, d_{s-1}', a'$ be chosen so that (a) $d_0, \cdots, d_{r-1}, d_0', \cdots, d_{s-1}', v_0, \cdots, v_{t-1}, a, a', e$ are distinct elements, and (b) $\{d_0', \cdots, d_{s-1}'\}$ is disjoint from A_1.

Applying the hypothesis concerning f_2 to

$$d_0', \cdots, d_{s-1}', v_0, \cdots, v_{t-1}, a', e, A_0 \cup \{d_0, \cdots, d_{r-1}\}, A_1,$$

we conclude there exists a program $\pi_2 = \langle p_2, a', e\rangle$ satisfying:

(1_2) π_2 computes f_2 at datum locations $d_0', \cdots, d_{s-1}'$ and value locations $v_0, \cdots, v_{t-1}$,

(2_2) $\underline{D}p_2$ is disjoint from $A_0 \cup \{d_0, \cdots, d_{r-1}, d_0', \cdots, d_{s-1}', v_0, \cdots, v_{t-1}\}$,

(3_2) if A_1 is disjoint from $\{v_0, \cdots, v_{t-1}\}$, then $RI(\pi_2)$ is disjoint from A_1.

Applying the hypothesis concerning f_1 to

$$d_0, \cdots, d_{r-1}, d_0', \cdots, d_{s-1}', a, a', A_0 \cup \{e, v_0, \cdots, v_{t-1}\} \cup \underline{D}p_2, A_1 \cup \underline{D}p_2,$$

we conclude there exists a program $\pi_1 = \langle p_1, a, a'\rangle$ satisfying:

(1_1) π_1 computes f_1 at datum locations $d_0, \cdots, d_{r-1}$ and value locations $d_0', \cdots, d_{s-1}'$,

(2_1) $\underline{D}p_1$ is disjoint from $A_0 \cup \{e, d_0, \cdots, d_{r-1}, d_0', \cdots, d_{s-1}', v_0, \cdots, v_{t-1}\} \cup \underline{D}p_2$,

(3_1) if $A_1 \cup \underline{D}p_2$ is disjoint from $\{d_0', \cdots, d_{s-1}'\}$, then $RI(\pi_1)$ is disjoint from $A_1 \cup \underline{D}p_2$.

We claim that $p = p_1 \cup p_2$ is the sought-for p. From (2_1), (2_2) it follows (2) is satisfied. Now A_1 is disjoint from $\{d_0', \cdots, d_{s-1}'\}$ by (b) and $\underline{D}p_2$ is disjoint from $\{d_0', \cdots, d_{s-1}'\}$ by 2_2; hence by 3_1, $RI(\pi_1)$ is disjoint from $A_1 \cup \underline{D}p_2$. Now consider comp$(k, a)$ where $k(d_i) = b_i$, $0 \leqq i < r$, suppose f is defined on $\langle b_0, \cdots, b_{r-1}\rangle$ and suppose k holds p. Then k holds p_1. Let n be the first instant that comp(k, a) reaches a'. Then $k_n(d_i') = b_i'$, $0 \leqq i < s$, where $f_1(b_0, \cdots, b_{r-1}) = (b_0', \cdots, b_{s-1}')$. Since $RI(\pi_1)$ is disjoint from $\underline{D}p_2$, k_n holds p_2. If q is the first instant that comp(k_n, a') reaches e, then, if $f_2(b_0', \cdots, b_{s-1}') = \langle b_0'', \cdots, b_{t-1}''\rangle$, we have $k_q(v_i) = b_i''$, $0 \leqq i < t$. By 4.1 and (2_1), the first instant that comp(k, a) reaches e is also q. Hence π computes f. Now assume A_1 is disjoint from $\{v_0, \cdots, v_{t-1}\}$; then by (3_1) and (3_2), the range of influence of π is disjoint from A_1. If f is not defined on $\langle b_0, \cdots, b_{r-1}\rangle$, then comp$_e(k, a)$ is not successful.[9] It follows that P computes f.

With the above argument at our disposal, we may now explain the reasons for the exits of a program being excluded from its domain in the definition given. If a' were permitted in the domain of p_1 and $k(a')$ were, say, passive, then when π_1 was combined with π_2, it would have been necessary to change the contents of a'. As a consequence any k holding p would not necessarily hold p_1 and we could not be sure that π_1 would compute as desired.

Let $\pi_1 = \langle p_1, a_1, e_1\rangle$, $\pi_2 = \langle p_2, a_2, e_2\rangle$ be programs (of a given machine) and suppose that (a) $\underline{D}p_1$ is disjoint from $\underline{D}p_2$, (b) $e_1 = a_2$. From 4.1 it follows that

4.6. $\pi_2 \circ \pi_1 \overset{Df}{=} \langle p_1 \cup p_2, a_1, e_2\rangle$ is a program. The *composite* $\pi_2 \circ \pi_1$ of π_1, π_2 is defined (in the case of single exit programs) only when (a) and (b) are both satisfied.

Henceforth, if D is a sequence $\langle d_0, \cdots, d_{r-1}\rangle$, then we let $\hat{D}$ be the set $\{d_0, \cdots, d_{r-1}\}$. By $k(D)$ we shall mean the sequence $\langle k(d_0), \cdots, k(d_{r-1})\rangle$. A portion of the argument of 4.5 also proves

4.7: PROPOSITION. *If π_i computes f_i at datum locations D_i and value locations V_i, $i = 1, 2$, $V_1 = D_2$ and $\underline{D}p_1$, $\underline{D}p_2$, $\hat{D}_1$, $\hat{D}_2$, $\hat{V}_2$ are pairwise disjoint, if $e_1 = a_2$, $e_2 \notin \underline{D}p_1$ and $RI(\pi_1)$ is disjoint from $\underline{D}p_2$, then $\pi_2 \circ \pi_1$ computes $f_2 \circ f_1$ at datum locations D_1 and value locations V_2.*

We turn now to the notion of a program computing a relation. We consider first the 2-valued case. Given a 2-valued relation R defined on a subset B_0 of B^r and given a program $\pi = \langle p, a, e_0, e_1\rangle$, then π is said to *compute R at datum locations $D = \langle d_0, \cdots, d_{r-1}\rangle$* provided that $\hat{D}$ is disjoint from $\underline{D}p$ and the following holds.

4.8: If k holds π then comp(k, a) ultimately reaches an exit of π iff $k(D) \in B_0$. Furthermore,

[9] The verification is left to the reader. In the future we will often fail to mention the case that "f is not defined". The reader should, however, verify the assertion in this case for himself.

(a) If R is true on $k(D)$, the first exit reached by $\text{comp}(k, a)$ is e_0;
(b) if R is false on $k(D)$, the first exit reached by $\text{comp}(k, a)$ is e_1.

Analogously to 4.7, we have the following.

4.9: PROPOSITION. *Let* $\pi = \langle p, a, e_0, e_1\rangle$ *compute the relation* R *at* D *and* $\pi' = \langle p', a', e_0', e_1'\rangle$ *compute the relation* R' *at the same datum locations* D. *Suppose* R, R' *are defined over* B_0, *suppose* $\underline{D}p$ *is disjoint from* $\underline{D}p'$ *and* $RI(\pi)$ *is disjoint from* $\underline{D}p' \cup \hat{D}$. *Then*

(a) $a' = e_0, e_1 = e_1'$ *and* $e_0' \notin \underline{D}p$ *implies* $\langle p \cup p', a, e_0', e_1\rangle$ *computes* $R \wedge R'$ (the conjunction of R and R') *on* B_0 *at* D

(b) $a' = e_1, e_0 = e_0'$ *and* $e_1' \notin \underline{D}p$ *implies* $\langle p \cup p', a, e_0, e_1'\rangle$ *computes* $R \vee R'$ (the disjunction of R and R') *on* B_0 *at* D

(c) $\langle p, a, e_1, e_0\rangle$ *computes* $\sim R$ (the negation of R) *on* B_0 *at* D.

PROOF. (a) Let $k(D) \in B_0$ and suppose that k holds $p \cup p'$. Then k holds p and, by hypothesis, since π computes R, $\text{comp}(k, a)$ will ultimately reach e_0 or e_1. Suppose $R \wedge R'$ is true at $k(D)$, then R is true for the same element of B_0 and so the first exit of π that is reached is e_0. At this instant $\langle k_n, e_0\rangle$ in the computation, $k_n(D) = k(D)$ and k_n holds p', by hypothesis, concerning $RI(\pi)$. Hence, $\text{comp}(k_n, e_0) = \text{comp}(k_n, a')$ will reach e_0' before reaching $e_1 = e_1'$. Recapitulating, $\text{comp}(k, a)$ reaches e_0 before e_1 and then reaches e_0' before e_1. It remains to consider the case that $R \wedge R'$ is false on $k(D)$. Suppose first that R is true on $k(D)$. Then, as before, $\text{comp}(k, a)$ reaches e_0 before e_1. When it reaches state $\langle k_n, e_0\rangle$, k_n holds p' and $k_n(D) = k(D)$. Since R' is false on $k(D) = k_n(D)$, $\text{comp}(k_n, e_0)$ will reach $e_1 = e_1'$ before reaching e_0'. Moreover, e_0' is not reached before instant n since $e_0' \notin \underline{D}p$, by hypothesis, and because p satisfies 4.1. Thus, $\text{comp}(k, a)$ will reach e_1 before e_0'. If R is false on $k(D)$, then $\text{comp}(k, a)$ reaches e_1 before reaching e_0, and using again $e_0' \notin Dp$ and 4.1, we conclude that $\text{comp}(k, a)$ reaches e_1 before e_0'. The observation that $\hat{D}$ is disjoint from $\underline{D}(p \cup p')$ completes the proof of (a).

The proof of (c) is trivial. A proof of (b) may be given analogous to the proof of (a) or we may proceed as follows. Now $\langle p, a, e_1, e_0\rangle$ computes $\sim R$ at D and $\langle p', a', e_1', e_0'\rangle$ computes $\sim R'$ at D; if $a' = e_1, e_0 = e_0'$, then $\langle p \cup p', a, e_1', e_0\rangle$ computes $\sim R \wedge \sim R'$ at D and $\langle p \cup p', a, e_0, e_1'\rangle$ computes $\sim(\sim R \wedge \sim R')$, which is equivalent to $R \vee R'$ at D.

We shall say that a *RASP* P *computes a 2-valued relation* R defined on a subset B_0 of B^r provided the following conditions hold.

4.10: For every sequence $\langle d_0, \cdots, d_{r-1}, a, e_0, e_1\rangle$ of distinct elements of A and for all finite subsets A_0, A_1 of A such that $a \notin A_0$, $A_1 \subseteq A_0$, there exists a program $\pi = \langle p, a, e_0, e_1\rangle$ of P satisfying the following (where $D = \langle d_0, \cdots, d_{r-1}\rangle$).

(1) π computes R at D.
(2) $\underline{D}p$ is disjoint from A_0 (and hence, by (1), disjoint from $A_0 \cup \hat{D}$).
(3) $RI(\pi)$ is disjoint from A_1.

4.11: PROPOSITION. *If the RASP* P *computes* R_1, R_2, *both of which are defined on* $B_0 \subseteq B^r$, *then* P *also computes* $R_1 \wedge R_2$ *on* B_0.

PROOF. Given D, a, e_0, e_1, A_0, A_1, we are required to produce a program π

satisfying 4.10. By hypothesis concerning R_2, there is a program $\pi_2 = \langle p_2, a', e_0, e_1\rangle$, where $a' \notin \hat{D} \cup \{a, e_0, e_1\}$, $a' \in A$, satisfying:

(1_2) π_2 computes R_2 at D,

(2_2) $\underline{D}p_2$ is disjoint from A_0,

(3_2) $RI(\pi_2)$ is disjoint from A_1.

By hypothesis concerning R_1, there is a program $\pi_1 = \langle p_1, a, a', e_1\rangle$ satisfying:

(1_1) π_1 computes R_1 at D,

(2_1) $\underline{D}p_1$ is disjoint from $A_0 \cup \underline{D}p_2 \cup \{e_0\}$,

(3_1) $RI(\pi_1)$ is disjoint from $A_1 \cup \underline{D}p_2 \cup \hat{D}$.

Let $\pi = \langle p_1 \cup p_2, a, e_0, e_1\rangle$ and $R = R_1 \wedge R_2$ on B_0. It follows from 4.9(a) that 4.10(1) holds; (2) and (3) are readily verified.

4.12: COROLLARY. *The class of all 2-valued relations defined over B_0, $B_0 \subseteq B^r$, and computable by a given RASP P is closed under all truth functional combinations.*

PROOF. Using the proposition and the obvious fact that if P computes R, it computes $\sim R$, the result follows.

Let P be a given RASP and suppose $b \in B$ such that $b \neq b_o$. As another test of the suitability of the notions introduced, we consider the question: If a RASP computes a relation, does it compute its characteristic function, and conversely? Given R defined over $B_0 \subseteq B_r$, let $C_R(= C_R{}^b)$ be defined on B_0 and equal to b_o when R is true and equal to b otherwise.

4.13: PROPOSITION. (1) *If P computes the identity function i: $B \to B$ and P computes the relation R, then P computes C_R.* (2) *If P computes the equality relation and the function C_R, then P computes R.*

PROOF (1). Given $d_0, \cdots, d_{r-1}, v, a, e, A_0, A_1, A_1 \subseteq A_0$, it is required to produce a program π satisfying (1), (2), (3) of 4.4.

Let e_0, e_1, d^0, d^1 be distinct elements all of which are outside $\{d_0, \cdots, d_{r-1}, v, a, e\} \cup A_0$. There is a program $\pi_1 = \langle p_1, e_1, e\rangle$ which computes i at datum location d^1 and value location v satisfying:

$\underline{D}p_1$ disjoint from $\{d_0, \cdots, d_{r-1}, v, a, e, e_0, d^0, d^1\} \cup A_0$,

$RI(\pi_1)$ disjoint from A_1, if A_1 does not contain v.

There is a program $\pi_0 = \langle p_0, e_0, e\rangle$ which computes i at datum location d^0 and value location v satisfying:

$\underline{D}p_0$ disjoint from $\{d_0, \cdots, d_{r-1}, v, a, e, e_1, d^0, d^1\} \cup \underline{D}p_1 \cup A_0$,

$RI(\pi_0)$ disjoint from A_1, if A_1 does not contain v.

There is a program $\pi_R = \langle p_R, a, e_0, e_1\rangle$ which computes R at $(d_0, \cdots, d_{r-1})$ which satisfies:

$\underline{D}p_R$ disjoint from $\{d_0, \cdots, d_{r-1}, v, e, e_0, e_1, d^0, d^1\} \cup \underline{D}p_0 \cup \underline{D}p_1 \cup A_0$,

$RI(\pi_R)$ disjoint from $A_1 \cup \underline{D}p_0 \cup \underline{D}p_1 \cup \{d^0, d^1\}$.

Let $p = p_R \cup p_0 \cup p_1 \cup \{\langle d^0, b_o\rangle, \langle d^1, b\rangle\}$ and let $\pi = \langle p, a, e\rangle$, then π is a program since π_R, π_0, π_1 are programs. Let $D = \langle d_0, \cdots, d_{r-1}\rangle$. Suppose k holds p and $k(D) \in B_0$. Assume R is true on $k(D)$. Then e_0 will be the first exit of π_R reached by comp(k, a) and comp(k, a) will not have passed through e up to that instant. At this instant control shifts to π_0. Since we have provided for the contents of $\underline{D}p_0$ to stay fixed, π_0 now computes i at datum location d^0 and value location v. Thus, comp(k, a) reaches e; and, at the first instant in which e is reached, the content of v is b_o (since π_R did not change the content of d^0).

Moreover, it is clear that if $v \notin A_1$, then $RI(\pi)$ is disjoint from A_1. The considerations in the case that R on $k(D)$ is false are similar.

PROOF (2). Given D, a, e_0, e_1, A_0, A_1, it is required to construct a program $\pi = \langle p, a, e_0, e_1\rangle$ which computes R at D and satisfies (2), (3) of 4.10. Let v, e, d be distinct elements of A outside of $\hat{D} \cup \{a, e_0, e_1\} \cup A_0$. By hypothesis there are programs $\pi_= = \langle p_=, e, e_0, e_1\rangle$ and $\pi_c = \langle p_c, a, e\rangle$ satisfying:

$\pi_=$ computes the equality relation at $\langle d, v\rangle$,
$\underline{D}p_=$ is disjoint from $A_0 \cup \hat{D} \cup \{a, e_0, e_1, v, d\}$,
$RI(\pi_=)$ is disjoint from A_1,
π_c computes C_R at datum locations D and value location v,
$\underline{D}p_c$ is disjoint from $A_0 \cup \hat{D} \cup \underline{D}p_= \cup \{e_0, e_1, v, e, d\}$,
$RI(\pi_c)$ is disjoint from $A_1 \cup \underline{D}p_c$ since $v \notin A_1$.

Let $p = p_= \cup p_c \cup \{\langle d, b_o\rangle\}$ and let $\pi = \langle p, a, e_0, e_1\rangle$. It may be readily verified that π is a program that satisfies (1), (2), (3) of 4.10.

We may also introduce the notion of a program $\langle p, a, e_0, \cdots, e_{m-1}\rangle$ computing a given m-valued relation R at datum locations $d_0, \cdots, d_{r-1}$ and the corresponding notion of a RASP computing R. If B has at least m distinct elements, one can then establish 4.13 for the m-valued case. The case $m = 2$ is adequate, however, for our present considerations.

We now introduce a particular RASP P_0 which, we will show, computes all partial recursive functions. We take $A = B$ to be the set of natural numbers N and $b_o = 0$. Thus $P_0 = \langle N, N, 0, K_o, h^1, h^2\rangle$, where h^1, h^2 are defined by three instruction schemata $S(x)$, $D(x, y)$, $C(x, y, z)$, and $K_o = K_f$ or $K_o = K$. These two cases will be treated simultaneously. Let $\bar{S}, \bar{D}, \bar{C}$ be distinct positive integers and let $\#$ be defined by the following requirements:

(a) $\#(\bar{S}, x) = 2^{\bar{S}} \cdot 3^x$, (b) $\#(\bar{D}, x, y) = 2^{\bar{D}} \cdot 3^x \cdot 5^y$, (c) $\#(\bar{C}, x, y, z) = 2^{\bar{C}} \cdot 3^x \cdot 5^y \cdot 7^z$,

(d) the domain of $\#$ is the set of ordered pairs, triples and quadruples indicated.

The active words are $\#(\bar{S}, x)$, $\#(\bar{D}, x, y)$, $\#(\bar{C}, x, y, z)$, where x, y, z vary through all natural numbers. All other words are inactive. We then define: $h_{\#(\bar{S},x)} = S(x)$, $h_{\#(\bar{D},x,y)} = D(x, y)$, $h_{\#(\bar{C},x,y,z)} = C(x, y, z)$. If b is inactive, then h_b is the identity function from states to states of P_0; 0 is inactive. The definitions of S, D, C are indicated below, where $\langle k, a\rangle$ is "present" state and $\langle k', a'\rangle$ is "next" state.

(1) $S(x)$, "add 1 to $k(x)$"; more explicitly:
$$k'(x) = k(x) + 1 \wedge (y \neq x \rightarrow k'(y) = k(y)), a' = a + 1$$

(2) $D(x, y)$, "transfer $k(x)$ to y"; more explicitly:
$$k'(y) = k(x) \wedge (z \neq y \rightarrow k'(z) = k(z)) \wedge a' = a + 1$$

(3) $C(x, y, z)$, "conditional transfer on equality"; more explicitly:
$$k' = k \wedge (k(x) = k(y) \rightarrow a' = z) \wedge (k(x) \neq k(y) \rightarrow a' = a + 1).$$

4.14: LEMMA. *P_0 computes the projection functions* $U_i^r : B^r \rightarrow B$, $U_i^r(b_0, \cdots, b_{r-1}) = b_i$.

PROOF. It is required to show, given the sequence $\langle d_0, \cdots, d_{r-1}, v, a, e\rangle$ of distinct elements of N and for all finite subsets A_0, A_1 of N such that $a \notin A_0$, $A_1 \subseteq A_0$, there exists a program $\pi = \langle p, a, e\rangle$ satisfying (1), (2), (3) of 4.4.

Let m be greater than any of the elements in $\{d_0, \cdots, d_{r-1}, v, a, e\} \cup A_0$. We define the function p by means of a table.

p_i^n:	*Argument*	*Value*
	a	$C(0, 0, m)$
	m	$D(d_i, v)$
	$m + 1$	$C(0, 0, e)$

It is clear that $\pi = \langle p, a, e\rangle$ fulfills our requirements, e.g., $RI(\pi) = \{v\}$. In particular, this shows that the identity function $B \to B$ is computable by P_0.

4.15: LEMMA. *P_0 computes the equality relation (and hence P_0 computes a relation R iff it computes C_R, by 4.13).*

PROOF. Given $d_0, d_1, a, e_0, e_1, A_1, A_0, A_1 \subseteq A_0, a \notin A_0$. Let m be larger than $\max(\{d_0, d_1, a, e_0, e_1\} \cup A_0)$ and let p be defined by the following table.

$p_=$:	*Argument*	*Value*
	a	$C(0, 0, m)$
	m	$C(d_0, d_1, e_0)$
	$m + 1$	$C(0, 0, e_1)$

Then $\pi = \langle p, a, e_0, e_1\rangle$ computes the equality relation at d_0, d_1 and satisfies (2) and (3) of 4.10.

In the cases treated below we will be somewhat less explicit in our argument that P_0 computes the given function.

4.16: The class of partial recursive functions is the smallest class of functions closed under composition and minimalization [2, p. 41] containing:

(1) the constant function c_0, $c_0(x) = 0$,
(2) the successor function S, $S(x) = x + 1$,
(3) the projection functions U_i^n, $U_i^n(x_0, \cdots, x_{n-1}) = x_i$, $0 \leqq i < n$,
(4) addition,
(5) proper subtraction ($x \dot{-} y = 0$ if $x \leqq y$, $x \dot{-} y = x - y$ if $x \geqq y$),
(6) multiplication.

Recall that the operation of minimalization is defined only for total functions. We have shown (4.14) that P_0 computes U_i^n. We know by 4.5 that the class of functions computed by P_0 is closed under composition. We leave it to the reader to verify that P_0 computes c_0 and S. We proceed to show that P_0 computes addition, proper subtraction and multiplication and that the class of functions computable by P_0 is closed under minimalization.

4.17: LEMMA. *Addition of natural numbers is computable by P_0.*

Indeed, it is computed by π_+.

p_+:	*Argument*	*Value*	*Comment*
	a	$C(0, 0, m)$	transfer control to m
	m	$D(d_0, v)$	transfer $k(d_0)$ to v
	$m + 1$	$D(m + 6, \tau_0)$	clear τ_0
	$m + 2$	$C(\tau_0, d_1, e)$	if $k(\tau_0) = k(d_1)$ go to e
	$m + 3$	$S(v)$	add 1 to $k(v)$
	$m + 4$	$S(\tau_0)$	add 1 to $k(\tau_0)$
	$m + 5$	$C(0, 0, m + 2)$	return control to $m + 2$
	$m + 6$	0	

Then $\pi_+ = \langle p_+, a, e\rangle$ computes addition at datum locations d_0, d_1 and value location v. The range of influence of π_+ is $\{v, \tau_0\}$. Each passage of control through the "loop" $m+2$, $m+3$, $m+4$, $m+5$ results in the content of v and τ_0 being augmented by 1. If $k(d_1) = 0$, there are zero passages through the loop and the content of v equals content of d_0 when control reaches e. If $k(d_1) = n$, there are n passages through the loop and at instant $3 + 4n$, $k_{3+4n}(\tau_0) = n = k_{3+4n}(d_1)$ and $k_{3+4n}(v) = k(d_0) + n$.

Remark. If $k(d_0) = k(v)$, then $\langle p_+, m+1, e\rangle$ computes addition at d_0, d_1, v (and contents of a, m may be deleted from p_+).

4.18: LEMMA. *Multiplication of natural numbers is computable by* P_0.

Indeed, it is computed by the program indicated below.

$p_\times$: *Argument*	*Value*	*Comment*
a_1	$C(0, 0, q)$	transfer control to q
q	$D(m + 6, \tau_1)$	clear τ_1
$q + 1$	$D(m + 6, v)$	clear v
$q + 2$	$D(q + 7, e)$	insert code for "return to $q + 3$" in exit of π_+
$q + 3$	$D(v, d_0)$	transfer $k_t(v)$ to d_0
$q + 4$	$C(\tau_1, d_0^1, e_1)$	if $k_t(\tau_1) = k_t(d_0^1)$, exit
$q + 5$	$S(\tau_1)$	add 1 to content of τ_1
$q + 6$	$C(0, 0, m + 1)$	transfer control to π_+
$q + 7$	$\#(\bar{C}, 0, 0, q + 3)$	code for "return to $q + 3$"

Let $p_{\text{prod}} = p_+ \cup p_\times \cup \{\langle e, -\rangle\}$ (a may be deleted from $\underline{D}p_{\text{prod}}$).[10] Then, $\pi_{\text{prod}} = \langle p_{\text{prod}}, a_1, e_1\rangle$ computes multiplication at datum locations d_0^1, d_1 and value location v. The $RI(\pi_{\text{prod}}) = \{\tau_1, v, e, d_0, \tau_0\}$.

PROOF 4.18. Let k hold p_{prod}. Let the number of occurrences of $q + 4$ (as second component) in $\text{comp}_{e_1}(k, a_1)$ be $l \leqq \infty$. Let $j < l$ and let $i(j)$ be the instant in $\text{comp}_{e_1}(k, a_1)$ that $q + 4$ is reached for the jth time. Assume that,

$$*: \quad k_{i(j)}(\tau_1) = j - 1, k_{i(j)}(v) = (j - 1)\cdot k(d_1) = k_{i(j)}(d_0).$$

Then, $a_{i(j)+1} = q + 5$, $a_{i(j)+2} = q + 6$, $a_{i(j)+3} = l + 1$, $a_{i(j)+4} = l + 2, \cdots$. At some later instant (actually $i(j) + 4 + 4\cdot k(d_1)$), the control is again at $l + 2$ and then goes to e, $q + 3$, $q + 4$, since addition is a total function and by the remark. Thus, ($i(j + 1) = i(j) + 7 + 4\cdot k(d_1)$ and)

$$k_{i(j+1)}(v) = k_{i(j)}(d_0) + k(d_1) = j\cdot k(d_1) = k_{i(j+1)}(v)$$

and

$$k_{i(j+1)}(\tau_1) = j.$$

Notice, too, that for $j = 1$, $*$ holds. It follows that $l = 1 + k(d_0^1)$ and $k_{i(l)}(\tau_1) = k(d_0^1)$, $k_{i(l)}(v) = k(d_0^1)\cdot k(d_1) = k_{i(l)}(d_0)$. Now, $a_{1+i(l)} = e_1$ and $k_{i(l)} = k_{1+i(l)}$, since the instruction "$C(\tau_1, d_0^1, e_1)$" does not change the content function k.

That $RI(\pi_{\text{prod}}) \subseteq \{\tau_1, v, e, d_0, \tau_0\}$ is obvious by inspection of program π_{prod}. The reverse inclusion may also readily be verified.

[10] The "$-$" may be replaced by any $b \in B$. For the sake of 4.23 (1), we replace "$-$" by $\#(\bar{C}, 0, 0, q+3)$.

4.19: LEMMA. *The binary relation* $\leqq$ *which holds iff* $k(d_0) \leqq k(d_1)$ *is computable by* P_0.

The program $\pi_{\leqq} = \langle p_{\leqq}, a, e_0, e_1 \rangle$ indicated below computes $\leqq$ at d_0, d_1; $RI(\pi_{\leqq}) = \{\tau\}$.

$p_{\leqq}$:	*Argument*	*Value*	*Comment*
	a	$C(0, 0, n)$	
	n	$D(n+5, \tau)$	clear τ
	$n+1$	$C(d_0, \tau, e_0)$	if $k_t(d_0) = k_t(\tau)$, exit at e_0
	$n+2$	$C(d_1, \tau, e_1)$	if $k_t(d_1) = k_t(\tau)$, exit at e_1
	$n+3$	$S(\tau)$	add 1 to content of τ
	$n+4$	$C(0, 0, n+1)$	
	$n+5$	0	

PROOF. If $k(d_0) \leqq k(d_1)$, then after $k(d_0)$ occurrences of the "loop" $(n+1, n+2, n+3, n+4)$ in the computation, the content of τ is $k(d_0)$ and $\pi_{\leqq}$ exits at e_0. If $k(d_0) > k(d_1)$, then after $k(d_1)$ occurrences of $(n+1, n+2, n+3, n+4)$, *the content of* τ *is* $k(d_1)$ and $\pi_{\leqq}$ exits at e_1.

4.20: LEMMA. P_0 *computes the total function* $x \dot{-} y$, *which is defined as* $x - y$ *when* $x \geqq y$ *and equal to zero otherwise.* The program $\pi_{\dot{-}} = \langle p_{\dot{-}}, a_1, e_0 \rangle$ indicated below computes $\dot{-}$ at d_0, d_1, v; $p_{\dot{-}} = p_{\leqq} \cup p_o \cup \{\langle e_1, - \rangle\}$.[11] The range of influence of $\pi_{\dot{-}}$ is $\{\tau, v, e_1\}$.

p_o:	*Argument*	*Value*	*Comment*
	a_1	$C(0, 0, r)$	
	r	$D(n+5, v)$	clear v
	$r+1$	$D(r+7, e_1)$	insert code for "transfer control to $r+3$" in e_1
	$r+2$	$C(0, 0, n)$	transfer control to $\pi_{\leqq}$
	$r+3$	$S(\tau)$	add 1 to content of τ
	$r+4$	$S(v)$	add 1 to content of v
	$r+5$	$C(\tau, d_0, e_0)$	if content of $\tau = k(d_0)$, exit
	$r+6$	$C(0, 0, r+3)$	return to $r+3$
	$r+7$	$\#(\bar{C}, 0, 0, r+3)$	

PROOF. When control shifts to n, the content of v is zero and the successor control location to e_1 is $r+3$. If $k(d_0) \leqq k(d_1)$, then the control reaches e_0 without changing the content of v so that at this instant t, $k_t(v) = k(d_0) \dot{-} k(d_1)$. If $k(d_0) > k(d_1)$, then the control eventually reaches e_1 and then $r+3$; at this instant the content of τ is $k(d_1)$ (as was observed in the proof of 4.19). Each passage through $(r+3, r+4)$ adds 1 to content of τ and adds 1 to content of v. After $k(d_0) - k(d_1)$ passes through $(r+3, r+4)$, the content of v is $k(d_0) - k(d_1)$, the content of τ is $k(d_0)$. Therefore, the $(k(d_0) - k(d_1))$-th time control reaches $r+5$, $\pi_{\dot{-}}$ exits to e_0.

A program π will be called *self-restoring* if whenever $\langle\langle k_1, a_1\rangle, \cdots, \langle k_n, a_n\rangle\rangle$ is a successful computation of π, then k_1 and k_n agree on $\underline{D}\pi$; π is self-restoring iff $(\underline{D}\pi) \cap RI(\pi)$ is empty.

[11] The "$-$" may be replaced by any $b \in B$. For the sake of 4.23 (1), we replace "$-$" by $\#(\bar{C}, 0, 0, r+3)$.

4.21: LEMMA. *If $f(y, x_1, \cdots, x_n)$ is computable by P_0 by self-restoring programs, then (the partial function) $g(x_1, \cdots, x_n) = [\min_y f(y, x_1, \cdots, x_n) = 0]$ is computable by P_0.*

PROOF. Let $d_1, \cdots, d_n, v, a, e$, distinct elements of A be given and finite subsets A_0, A_1 of A be given such that $a \notin A_0$, $A_1 \subseteq A_0$. It is required to produce a program $\pi = \langle p, a, e\rangle$ satisfying:

(1) π computes g at datum locations $d_1, \cdots, d_n$ and value location v,

(2) Dp is disjoint from $A_0 \cup \{d_1, \cdots, d_n, v\}$,

(3) If A_1 is disjoint from $\{v\}$, then $RI(\pi)$ is disjoint from A_1. Choose $a', e', v', s, s + 1, \cdots, s + 7$ so that they are distinct from each other and distinct from $d_1, \cdots, d_n, v, a, e$ and from the elements of A_0. By hypothesis, there is a program $\pi_f = \langle p_f, a', e'\rangle$ satisfying:

(1_f) π_f computes f at datum locations $v, d_1, \cdots, d_n$ and value location v',

(2_f) Dp_f is disjoint from $A_0 \cup \{v, d_1, \cdots, d_n, v', a, e, e', s, s + 1, \cdots, s + 7\}$,

(3_f) $RI(\pi_f)$ is disjoint from $A_1 \cup \{a, s, s + 1, \cdots, s + 7\} \cup Dp_f$.

p_g :

Argument	*Value*	*Comment*
a	$C(0, 0, s)$	
s	$D(s + 7, e')$	insert code for "transfer control to $s + 3$" in exit of π_f
$s + 1$	$D(s + 6, v)$	clear v
$s + 2$	$C(0, 0, a')$	transfer control to π_f
$s + 3$	$C(v', s + 6, e)$	if value of f is 0, go to exit
$s + 4$	$S(v)$	add 1 to content of v
$s + 5$	$C(0, 0, a')$	transfer control to π_f
$s + 6$	0	
$s + 7$	$\#(\bar{C}, 0, 0, s + 3)$	code for "transfer control to $s + 3$"

Then, π satisfies (2) above, where $p = p_f \cup p_g \cup \{\langle e', -\rangle\}$.[12] Further, $RI(\pi) \subseteq RI(\pi_f) \cup \{v, e'\}$ so that (3) is satisfied. To see that (1) is satisfied, notice that the exit of π can only be reached from control location $s+3$ and will be the successor control location provided at the given instant the content of v' is 0. If, for no y is $f(y, x_1, \cdots, x_n) = 0$, then the computation comp(k, a), for any k holding π and such that $k(d_i) = x_i$, will never reach e (nor will it terminate). If there exists a g such that $f(y, x_1, \cdots, x_n) = 0$ and y_0 is the smallest such, then it is clear that after $1+y_0$ passes through $s+3$, the computation will reach e for the first time.

A *program π will be called fixed* if whenever $\langle \sigma_1, \sigma_2, \cdots, \sigma_n\rangle$ is a successful computation of π, then $k_1, k_2, \cdots, k_n$ all agree on $\underline{D}\pi$, where $\sigma_i = \langle k_i, a_i\rangle$.

4.22: THEOREM. *P_0 computes all partial recursive functions.*

PROOF. 4.16, 4.14, 4.5, 4.17, 4.18, 4.20, 4.21.

Moreover, the constructions show that all partial recursive functions are computable in a "direct" way by programs of P_0, which

4.23: (1) are fixed,

[12] The "$-$" may be replaced by any $b \in B$. For the sake of 4.23 (1), we replace "$-$" by $\#(\bar{C}, 0, 0, s+3)$.

(2) have only a finite range of influence,

(3) have only a finite memory requirement in the following sense which is described somewhat informally.

Let $\pi = \langle p, a, e \rangle$ and $M(\pi)$ be the set of all x such that $x \in \underline{D}p$, $x = e$ or for some k which holds p, the contents of x are read or changed (written on) during the course of the computation comp(k, a) up until the first instant (if ever) that e is reached. The statement that π requires only a finite amount of memory then means that $M(\pi)$ is finite.

We return to these matters in Section 7.

Remark. If π_f has a finite range of influence but $RI(\pi_f)$ intersects $\underline{D}\pi_f$, we may "store" a copy of p_f outside of $RI(\pi_f)$ and construct π in such a way that this copy is not altered and such that before control is shifted to π_f, i.e., between $p_g(s + 4)$, $p_g(s + 5)$, p_f is restored.

Now 4.22 is actually a corollary of the following proposition which, however, is more cumbersome to prove. We shall say that a RASP P *computes a function (resp. relation) by means of programs with a finite RI* provided that 4.4(1), (2), (3) holds together with the requirement that $RI(\pi)$ is finite.

4.24: THEOREM. *If a RASP P, with its set of words B equal to the natural numbers, computes* (a) *the identity function* (i), (b) *the successor function* (S) *in the special sense that datum location and value location are the same,*[13] *and* (c) *the equality relation* ($=$), *by means of self-restoring programs with a finite RI, then P computes all partial recursive functions.*

PROOF. We give only a sketch of the proof of the analogue of 4.21.

Suppose, then, that $f(y, x_1, \cdots, x_n)$ is computable by self-restoring programs of P. Let $d_1, \cdots, d_n, v, a, e, A_0, A_1$ satisfying $A_1 \subseteq A_0$, $a \notin A_0$ be given. It is required to construct a self-restoring $\pi = \langle p, a, e \rangle$ satisfying (1), (2), (3) of 4.21. Let $E_0 = \{d_1, \cdots, d_n, v, a, e\} \cup A_0$. Let a', v', a_6 be distinct from each other and the elements of E_0. Let $E_1 = E_0 \cup \{a', v', a_6\}$, $a_2 \notin E_1$. Suppose $v \notin A_1$.

Choose $\pi_1 = \langle p_1, a, a_2 \rangle$ such that: (a) π_1 computes i at a_6, v, (b) Dp_1 is disjoint from $E_1 \cup \{a_2\}$, (c) $RI(\pi_1)$ is disjoint from $(E_1 \cup \{a_2\}) - \{v\}$ and Dp_1. Let $E_2 = E_1 \cup Dp_1 \cup RI(\pi_1)$, $a_3 \notin E_2$.

Choose $\pi_2 = \langle p_2, a_2, a', a_3 \rangle$ such that (a) π_2 computes $=$ at 0, 0, (b) Dp_2 is disjoint from $E_2 \cup \{a_3\}$, (c) $RI(\pi_2)$ is disjoint from $E_2 \cup \{a_3\}$ and Dp_2. Let $E_3 = E_2 \cup Dp_2 \cup RI(\pi_2)$, $a_4 \notin E_3$.

Choose $\pi_3 = \langle p_3, a_3, e, a_4 \rangle$ such that (a) π_3 computes $=$ at v', a_6, (b) Dp_3 is disjoint from $E_3 \cup \{a_4\}$, (c) $RI(\pi_3)$ is disjoint from $E_3 \cup \{a_4\}$ and Dp_3. Let $E_4 = E_3 \cup Dp_3 \cup RI(\pi_3)$, $a_5 \notin E_4$.

Choose $\pi_4 = \langle p_4, a_4, a_5 \rangle$ such that (a) π_4 computes S at v, v, (b) Dp_4 is disjoint from $E_4 \cup \{a_5\}$, (c) $RI(\pi_4)$ is disjoint from $E_4 \cup \{a_5\}$ and Dp_4. Let $E_5 = E_4 \cup Dp_4 \cup RI(\pi_4)$.

Choose $\pi_5 = \langle p_5, a_5, a', a_6 \rangle$ such that (a) π_5 computes $=$ at 0, 0, (b) Dp_5 is disjoint from E_5, (c) $RI(\pi_5)$ is disjoint from E_5 and Dp_5. Let $E_6 = E_5 \cup Dp_5 \cup RI(\pi_5)$.

[13] This special sense is inessential but convenient to assume here since we are imitating p_g of 4.21.

Choose $\pi_f = \langle p_f, a', a_3 \rangle$ so that (a) π_f computes f at $v, d_1, \cdots, d_n; v'$, (b) Dp_f is disjoint from E_6, (c) $RI(\pi_f)$ is disjoint from $E_6 - \{v'\}$ and Dp_f. Let $p = p_1 \cup p_2 \cup p_3 \cup p_4 \cup p_5 \cup \{\langle a_6, 0 \rangle\} \cup p_f$. As in 4.21 we may argue that $\pi = \langle p, a, e \rangle$ satisfies (1), (2), (3) (and does not modify itself) since the subprograms $\pi_1, \cdots, \pi_5, \pi_f$ have been so chosen so that their ranges of influence are each disjoint from each of their domains.

5. *Recursive RASP Machines*

In Section 4, special RASP's were considered which are capable of carrying out the computations required for the determination of recursive functions. Conversely, if, as is generally accepted, the notion of recursiveness is indeed an adequate explication for the idea of effective calculability, then it will be appropriate to suppose that the operations carried out by an IMP or RASP are, in some definite sense, recursive. We proceed to consider this point. It will be assumed throughout this section that the sets A (addresses) and B (words) are countably infinite.

Suppose, in particular, that both A and B consist of all (finite) strings built up from certain finite alphabets which may well coincide. This corresponds closely to the representation of data or instructions in the memory of a digital computer. It then has a definite meaning (see e.g. [3]) to say that a function whose arguments vary over A or B and which takes values in A or B is computable. Moreover, by the introduction of appropriate artifices we may represent finite sequences of strings by single strings, and this lends a definite meaning to a statement that a specified function from finite sequences of elements of A and B into finite sequences of the same kind is computable. However, for the discussion which follows it will make matters more definite if we consider functions of Gödel numbers of addresses and words rather than of the addresses and words themselves. Still considering sets A and B whose elements are built up from finite alphabets, we may assign to the elements of A and B, separately, natural numbers (or the numerals which represent these numbers) by means of computable functions, as mentioned above. Furthermore, this can be done in such a way that all natural numbers are matched with elements of A and, separately, with elements of B. In this sense, we can obtain effective enumerations of A and of B, leading to an ordering of their elements,[14]

$$\{a_0, a_1, a_2, \cdots\} = A \qquad \{b_0, b_1, b_2, \cdots\} = B.$$

Thus, the Gödel number of the element a_n in the first sequence is precisely n, and we write $N(a_n) = n$, where $N(x)$ is the function which assigns to every string in A its Gödel number. Similarly, we mean by $N(b_n)$ the Gödel number of b_n, so that $N(b_n) = n$. Moreover, it is not difficult to ensure that the empty word has the Gödel number 0 so that our notation remains consistent.

Consider first a finitely supported IMP (Section 2), i.e., an IMP for which K_o consists of the functions of finite support. A function $k(x)$ from A into B

[14] In the case that A or B is the set of natural numbers, we take this order to be the natural order.

will be represented by the number-theoretic function $\bar{k}$ which is defined by

5.1: $\bar{k}(n) = N(k(a_n)), \quad n = 0, 1, 2, \cdots.$

The subscripts of the elements of the support of k then coincide with the set of natural numbers n for which $\bar{k}(n) \neq 0$. $\bar{k}(x)$ is now given by, or may be identified with, a finite set of pairs of natural numbers (n, g), where $g > 0$ such that $g = \bar{k}(n)$, while n does not appear as first element in a pair of the set if $\bar{k}(n) = 0$. Denoting this set of pairs for given k (and hence, for given $\bar{k}$) by G_k, it is now easy to enumerate the G_k (and hence the elements k of K_o) effectively. More particularly, there exists a recursive function of two variables, $f(x, y)$ and a recursive function of one variable, $\gamma(y)$ such that: (i) $x > \gamma(y)$ implies $f(x, y) = 0$, (ii) distinct functions $f(x, m)$ of the variable x are determined by distinct m, and (iii) for every $k(x)$ of finite support there exists a number m for which

5.2: $\bar{k}(n) = f(n, m), \quad n = 0, 1, 2, \cdots.$

We then write $N(k) = m$, and we call m the Gödel number of k. We also write $K_o = \{k_0, k_1, k_2, \cdots\}$ where we have arranged the functions $k(x)$ in the rising order of their Gödel numbers. Then $N(k_n) = n, \quad n = 0, 1, 2, \cdots.$

Let $J = \langle A, B, b_o, K_o, g^1, g^2\rangle$ be a finitely supported IMP. We define the number-theoretic functions $\bar{g}^1(x, y), \bar{g}^2(x, y)$ by

5.3: $\bar{g}^i(n, m) = N(g^i(k_n, a_m)), \quad n, m = 0, 1, 2, \cdots, \quad i = 1, 2.$

J will be called *recursive* if $\bar{g}^1$ and $\bar{g}^2$ are recursive in the ordinary, number-theoretic sense. More generally, if $\bar{g}^1$ and $\bar{g}^2$ are recursive relative to a pre-assigned set of number-theoretic functions, Φ, then we say that J is *recursive in* (*or, relative to*) Φ.

Now let $P = \langle A, B, b_o, K_o, h^1, h^2\rangle$ be a finitely supported RASP. We define the number-theoretic functions $\bar{h}^1(x, y, z), \bar{h}^2(x, y, z)$ by

5.4: $\bar{h}^i(l, n, m) = N(h^i(k_l, a_n, b_m)), \quad l, n, m = 0, 1, 2, \cdots, \quad i = 1, 2.$

P will be called recursive (or recursive relative to a set Φ) if $\bar{h}^1$ and $\bar{h}^2$ are recursive (or recursive relative to Φ).

If an IMP J is recursive relative to the set of functions $\{\bar{h}^1, \bar{h}^2\}$ where $\bar{h}^1, \bar{h}^2$ are given by a finitely supported RASP P, then J will be called *recursive relative to* P.

5.5: PROPOSITION. *Let P be a finitely supported RASP and let J be the IMP which is generated by P. Then J is recursive relative to P.*

PROOF. The relation between the functions g^1 and g^2 which belong to J and the functions h^1 and h^2 which belong to P is given by 3.5. It follows that for $i = 1, 2$,

$$\bar{g}^i(m, n) = N(g^i(k_m(a_n)) = N(h^i(k_m, a_n, k_m(a_n)))$$
$$= N(\bar{h}^i(m, n, N(k_m(a_n)))).$$

But $N(k_m(a_n)) = \bar{k}_m(a_n) = f(n, m)$, by 5.2, and so

5.6: $\bar{g}^i(m, n) = \bar{h}^i(m, n, f(n, m)).$

This shows that $\bar{g}^1$ is recursive in $\bar{h}^1$ and $\bar{g}^2$ is recursive in $\bar{h}^2$ and proves our assertion. As an immediate consequence, we have

5.7: COROLLARY. *The IMP generated by a recursive RASP is recursive.* More generally, if a RASP is recursive in a set Φ, then the IMP generated by it is recursive in Φ.

Now let ρ be an instruction schema in a finitely supported RASP, P, so that $\rho = \langle \rho^1, \rho^2 \rangle$, $\rho^i = \rho^i(u, v, y_1, \cdots, y_l, z_1, \cdots, z_m)$, as explained in Section 3. Then we define the number-theoretic functions $\tilde{\rho}^i = \tilde{\rho}^i(u, v, y_1, \cdots, y_l, z_1, \cdots, z_m)$ by

5.8: $\tilde{\rho}^i(m, n, p_1, \cdots, p_l, g_1, \cdots, g_m)$
$= N(\rho^i(k_m, a_n, a_{p_1}, \cdots, a_{p_l}, b_{g_1}, \cdots, b_{g_m})), \qquad i = 1, 2$

for arbitrary natural numbers $m, n, p_1, \cdots, p_l, g_1, \cdots, g_m$. The instruction schema ρ will be called *recursive* (or *recursive in* a set Φ) if ρ^1 and ρ^2 are recursive (recursive in Φ). This definition is inspired by the idea that the state transitions which are determined by the parameters of a given instruction schema should be given by a process which is, in some sense, uniform.

Now let $\lambda = \langle \rho, \mu \rangle$ be a designated instruction schema in a finitely supported RASP, P. In correspondence with μ we define the number-theoretic function $\tilde{\mu}(y_1, \cdots, y_l, z_1, \cdots, z_m)$ by

5.9: $\tilde{\mu}(p_1, \cdots, p_l, g_1, \cdots, g_m) = N(\mu(a_{p_1}, \cdots, a_{p_l}, b_{g_1}, \cdots, b_{g_m}))$ for arbitrary natural numbers p_i, g_i.

Let β_λ be the range of the function $\tilde{\mu}$; then β_λ is the set of Gödel numbers of elements of B_λ.

We say that λ is *recursive* (in a set Φ) if ρ is recursive (in Φ) and if, at the same time, the function $\tilde{\mu}$ and the set β_λ are recursive (in Φ).

A *RASP P will be said to be generated by the set of designated instruction schemata* $R = \{\lambda_i\}$, $\lambda_i = \langle \rho_i, \mu_i \rangle$, with parameters $y_1, \cdots, y_{l_i}, z_1, \cdots, z_{m_i}$, $l_i \geqq 0$, $m_i \geqq 0$, if all active words of P belong to the union of all B_{λ_i} and such that for all i, $b \in B_{\lambda_i}$ entails $\rho_i(\mu_i^{-1} b) = h_b$. (We recall that a word b is active if $h_b(\sigma) \neq \sigma$ for some state σ.)

5.10: Theorem. *If a finitely supported RASP P is generated by a finite set* $R = \{\lambda_i\}$, $i = 1, \cdots, r$ *of recursive designated instruction schemata, then P is recursive.*

A corresponding result holds for recursiveness relative to a set Φ.

Proof. We have to provide an effective procedure for computing $\tilde{h}^1(l, n, m)$, $\tilde{h}^2(l, n, m)$ for all natural numbers l, n, m. For given l, n and m we first check to see whether m belongs to $\beta_{\lambda_1}, \cdots, \beta_{\lambda_r}$. This can be done effectively since the β_{λ_i} are supposed to be recursive. If m does not belong to any of these sets, then $\tilde{h}^1(l, n, m) = l$, $\tilde{h}^2(l, n, m) = n$. In the alternative case we take the first i for which $m \in \beta_{\lambda_i}$. Then there exist natural numbers $p_1, \cdots, p_{l_i}, g_1, \cdots, g_{m_i}$ such that

5.11: $m = \tilde{\mu}(p_1, \cdots, p_{l_i}, g_1, \cdots, g_{m_i})$.

By the definition of a designated instruction schema, the p_j, g_j are determined uniquely by m, so we can find them by ranging all $(l_i + m_i)$-tuples of natural numbers effectively in a simply infinite sequence and by going through the elements of the sequence computing the right-hand side of 5.11 and seeing whether 5.11 is satisfied. The answer to this question will be positive after a finite number of steps. Having found the appropriate $p_1, \cdots, p_{l_i}, g_1, \cdots, g_{m_i}$, we then compute $\tilde{h}^1(l, n, m)$, $\tilde{h}^2(l, n, m)$ by means of the equations

5.12: $\tilde{h}^j(l, n, m) = \tilde{\rho}^j(l, n, p_1, \cdots, p_{l_i}, g_1, \cdots, g_{m_i}), \qquad j = 1, 2.$

This completes the computation and proves 5.10.

We have considered the possibility that a RASP is generated by a finite number of recursive (designated) instruction schemata because this appears to be the correct abstraction from the actual situation. It is certainly realized in all the special cases considered in the present paper. In particular,

COROLLARY. *P_0, with $K_o = K_f$, (cf. Section 4), is recursive.*

Next, we consider the notion of recursiveness relative to a RASP P which is finitely determined but not necessarily finitely supported (see Section 3). We shall make use of some arbitrary but fixed recursive enumeration of the set of finite sequences of natural numbers (beginning with the empty sequence τ_0) so: $\tau_0, \tau_1, \tau_2, \cdots$. Writing $n = N(\tau_n)$ and calling n the Gödel number of the sequence, we then have a recursive procedure in order to recover from n both the length of τ_n and its elements.

P will be called *recursively finitely determined* if there exist recursive functions $\phi(x, y)$, $\psi(x, y, z)$, $\eta(x, y, z)$, $\zeta(x, y, z)$, which satisfy the following conditions.

5.13: If n is the Gödel number of b and l is the Gödel number of a, then $\phi(n, l)$ is the Gödel number of the sequence obtained by replacing $a_1, \cdots, a_j$ in $A_{a,b}$ by their Gödel numbers; if n is the Gödel number of b, l is the Gödel number of a, and m is the Gödel number of the sequence of Gödel numbers of $b_1, \cdots, b_j$, then $\psi(n, l, m)$ is the Gödel number of a', $\eta(n, l, m)$ is the Gödel number of the sequence of Gödel numbers of $a_{i_1}, \cdots, a_{i_l}$ and $\zeta(n, l, m)$ is the Gödel number of the sequence of Gödel numbers of $b_{i_1}, \cdots, b_{i_l}$. (See the definition of a finitely determined RASP for the notation used here. In this notation, a_i, b_i do not necessarily have Gödel number i.)

Now let J be the IMP generated by the RASP P, and let

5.14: $\sigma^{(0)}, \sigma^{(1)}, \sigma^{(2)}, \cdots$ be a computation in J, $\sigma^{(j)} = \langle k^{(j)}(x), a^{(j)} \rangle$, $j = 0, 1, 2, \cdots$. Since P is not necesarily finitely supported, we may not be able to enumerate the elements $k(x)$ of K_o. However, for every $k(x) \in K$, we can still define a corresponding number-theoretic function $\bar{k}(x)$ by 5.1.

5.15: PROPOSITION. *Let J be the IMP generated by a recursively finitely determined RASP P, and let 5.14 be a computation in J, for an arbitrary but fixed state $\sigma^{(0)}$. Then $N(a^{(j)})$, regarded as a function of j, is recursive in $\bar{k}^{(0)}(x)$. Also, $\bar{k}^{(j)}(x)$, regarded as a function of x and of j, is recursive in $\bar{k}^{(0)}(x)$.*

PROOF. Let j be an arbitrary positive integer. Assuming that we know $N(a^{(0)})$ as well as the functional value of $\bar{k}^{(0)}(x)$ for any given value of x, we have to show that we can compute $N(a^{(j)})$, as well as the functional value of $\bar{k}^{(j)}(x)$ for any given x. Beginning with $\sigma^{(0)}$, we determine $b^{(0)} = k^{(0)}(a^{(0)})$. Evidently, $N(b^{(0)}) = \bar{k}^{(0)}(N(a^{(0)}))$ is computable (from $\bar{k}^{(0)}(x)$). With $n = N(b^{(0)})$, we then make use of the functions ϕ, ψ, η, ζ, whose properties are given by 5.13, in order to compute the appropriate sequences $N(a_1), \cdots, N(a_j)$; $\bar{k}^{(0)}(N(a_1)), \cdots, \bar{k}^{(0)}(N(a_j))$; $N(a_{i_1}), \cdots, N(a_{i_l})$; $N(b_{i_1}), \cdots, N(b_{i_l})$; and also $N(a^{(1)})$. This enables us to compute $\bar{k}^{(1)}(p)$ for any given natural number p. For this purpose we check first to see whether p is the Gödel number of an address whose contents are changed on passing from $k^{(0)}$ to $k^{(1)}$. If it is not, then $\bar{k}^{(1)}(p) = \bar{k}^{(0)}(p)$. If it is, so that $p = N(a_{i_g})$, then the new contents are equal to b_{i_g}; then $\bar{k}^{(1)}(p) = N(b_{i_g})$, and this number has been computed already.

The passage from $N(a^{(1)})$ and $k^{(1)}(x)$ to $N(a^{(2)})$ and $k^{(2)}(x)$ is slightly more complicated. Thus, in order to compute $N(b^{(1)}) = \bar{k}^{(1)}(N(a^{(1)}))$, we now have to check first whether $N(a^{(1)})$ is the Gödel number of an address whose content is changed on passing from $k^{(0)}$ to $k^{(1)}$. Having found $N(b^{(1)})$ and then $\phi(N(b^{(1)}))$, we have to carry out similar checks in order to compute $\bar{k}^{(1)}(N(a_1)), \cdots, \bar{k}^1(N(a_j))$ (where $a_1, \cdots, a_j$ have now taken on the meaning appropriate to $b^{(1)}, a^{(1)}$), and so forth. In general, on passing from $N(a^{(g)})$, $\bar{k}^{(g)}(x)$ to $N(a^{(g+1)})$, $\bar{k}^{(g+1)}(x)$, we have to take into account all changes of content which have taken place up to, and during, the gth step. However, as indicated, this can be done effectively at each stage, enabling us to determine $N(a^{(j)})$ and, for any given x, $\bar{k}^{(j)}(x)$, as asserted. A survey of this procedure also establishes

5.16: Corollary. *If in* $\sigma^{(0)} = \langle k^{(0)}, a^{(0)} \rangle$ *we keep* $k^{(0)}$ *constant but vary* $a^{(0)}$, *then* $N(a^{(j)})$, *regarded as a function of* j *and of* $N(a^{(0)})$, *is recursive in* $\bar{k}^{(0)}(x)$. *And* $\bar{k}^{(j)}(x)$, *regarded as a function of* x *and of* j *and of* $N(a^{(0)})$, *is recursive in* $\bar{k}^{(0)}(x)$.

Now let $\lambda = \langle \rho, \mu \rangle$ be a designated instruction schema in a RASP P. We shall say that λ is recursively finitely determined if λ is finitely determined and if, in addition, it satisfies the following conditions.

5.17: Defining $\bar{\mu}$ as in 5.9, both $\bar{\mu}$ and its range, β_λ, are recursive.

5.18: There exist recursive functions $\phi(y_1, \cdots, y_l, z_1, \cdots, z_m, y)$, $\psi(y_1, \cdots, y_l, z_1, \cdots, z_m, y, z)$, $\eta(y_1, \cdots, y_l, z_1, \cdots, z_m, y, z)$, $\zeta(y_1, \cdots, y_l, z_1, \cdots, z_m, y, z)$, such that if $y_1, \cdots, y_l, z_1, \cdots, z_m$ take as values the Gödel numbers of the parameters of the instruction schema, and y and z are the Gödel numbers of a and of the sequence of Gödel numbers of $b_1, \cdots, b_j$, then ϕ is the Gödel number of the sequence of Gödel numbers of $a_1, \cdots, a_j$, ψ is the Gödel number of a', and η and ζ are the Gödel numbers of the sequences of Gödel numbers of $a_{i_1}, \cdots, a_{i_l}$ and $b_{i_1}, \cdots, b_{i_l}$, respectively. Here again the symbols a_i, b_i are used as in the definition of a finitely determined instruction and have no connection with the Gödel numbering of A and B or with the parameters of the instruction schema.

By using the method of 5.10, we prove without difficulty

5.19: Theorem. *If a RASP P is generated by a finite set of recursively finitely determined (designated) instruction schemata,* $R = \{\lambda_i\}$, $i = 1, \cdots, r$, *then P is recursively finitely determined. In particular,* P_0, *with* $K_o = K$, (cf. 4), *is recursively finitely determined.*

It has been suggested by some that any realistic theory of computers should concern itself with finite systems only. Evidently, for an individual finite system (e.g., in our case, for finite A and B) a discussion of recursiveness such as that given in the present section becomes pointless. By contrast we believe that it is methodologically appropriate to represent a system which in actual fact may be expected to be *very large* by an *infinite model.* (Note that there is ample precedent for this procedure in the more established branches of the Empirical Sciences.) However, it must be said that this general principle is too vague to permit us to decide in each particular case precisely which components of a given system are to remain finite in the idealization and which are to become infinite. Thus, within our framework it is reasonable to suppose that a given computation is produced by the action of a small number of initial instructions on a small number of initial data. This leads to the idea of a finitely supported RASP. But it is also reasonable to look upon the action of the machine at any given stage as a "local" (finite) transformation of an unlimited supply of data, and this leads to the idea of a finitely determined RASP. In either case, we may expect the action of the machine to be effective in the accepted sense of the word, and this leads to the two classes of recursiveness considered in the present section.

It is evident that, at any given stage, a finitely determined RASP will transform a function $k(x)$ of finite support into a function of finite support. It follows that we may transform any given finitely determined RASP into a finitely supported RASP, which is still finitely determined, by merely restricting the set K_o to content-functions of finite support. Moreover, we have the following connection between the two kinds of recursiveness introduced above.

5.20: PROPOSITION. *Let P be a finitely supported RASP which is recursively finitely determined. Then P is recursive.*

PROOF. On the assumptions of 5.20 it is indeed evident that we can compute the Gödel number of $a' = h^2(k, a, b)$ from the Gödel numbers of k, of a, and of b. For a knowledge of the Gödel number of k enables us to compute the Gödel numbers of $b_1, \cdots, b_j$ (in the notation used above for finitely determined RASP's) and, hence, to compute a'. It also enables us to compute $a_{i_1}, \cdots, a_{i_l}$ and $b_{i_1}, \cdots\ b_{i_l}$ and, hence, the Gödel number of $k' = h^1(k, a, b)$. This proves 5.20.

In discussing the recursiveness of finitely determined RASP's, we have made extensive use of the Gödelization of finite sequences. While this artifice is extremely convenient for a theoretical discussion, its use does not, in general, mirror actual or recommended practice. Thus, in relating our procedures to the real situation, we may well think of the direct computation of the length and elements of certain finite sequences in terms of the length and elements of given sequences (see Section 7) rather than of the corresponding computations with Gödel numbers.

Finally, we observe that all our definitions and results can be adapted so as to refer to recursiveness relative to a pre-assigned set of functions Φ.

5.21: THEOREM. *If a program π of a recursively finitely determined RASP computes a (partial) function f (at datum locations D and value locations V), then f is partial recursive.*

PROOF. Let k hold π and let $k(x) = b_o$ for all $x \notin \hat{D} \cup \underline{D}\pi$. By 4.2, if f is defined at $k(D)$, then comp$_e(k, a)$ will be successful where $\pi = \langle p, a, e\rangle$, while if f is not defined, comp$_e(k, a)$ will not be successful. We shall show how we may, in effect, effectively compute comp$_e(k, a)$. To this end, we make the following definition.

Let g be any function from a finite subset of A into B. Let $\bar{g}(x)$ equal $g(x)$ if $x \in \underline{D}g$; otherwise $\bar{g}(x) = b_o$. Let $a \in A$. Define $\nu_b(g, a) = \langle g', a'\rangle$ as follows. By means of ϕ (cf. 5.13), we may effectively compute the finite sequence $\langle a_1, \cdots, a_j\rangle$. Let $b_i,\ 1 \leq i \leq j$, be $g(a_i)$ if defined; otherwise $b_i = b_o$. By means of ψ, η, ζ (cf. 5.13) we may now effectively compute a', $\langle a_{i_1}, \cdots, a_{i_l}\rangle$, $\langle b_{i_1}, \cdots,$

$b_{i_l}\rangle$, respectively. Let $g'(a_{i_r}) = b_{i_r}$, $1 \leqq r \leqq l$; $g'(x) = g(x)$ if x is defined under g and $x \neq a_{i_r}$ for some r; $g'(x)$ is undefined otherwise.

Now let k_0 be any content function of finite support. Let g_0 be any finite function such that $\bar{g}_0 = k_0$ and let $\langle g_{i+1}, a_{i+1}\rangle = \nu_b(g_i, a_i)$, where $b = g_i(a_i)$ if $g_i(a_i)$ is defined; otherwise $b = b_o$. Then $\text{comp}(k_0, a_0) = \langle\langle\bar{g}_0, a_0\rangle, \langle\bar{g}_1, a_1\rangle, \langle\bar{g}_2, a_2\rangle, \cdots\rangle$. In particular, if $k_0 = k$, $a_0 = a$ and g_0 is k restricted to $(\hat{D} \cup \underline{D}\pi)$, then $\text{comp}(k, a)$ may be effectively computed so that if $\text{comp}_e(k, a)$ is successful, the value of f at $k(D)$ may effectively be found.

In a similar way we may prove:

5.22: Proposition. *If a program π of a recursive finitely supported RASP computes a (partial) function f (at datum locations D and value locations V), then f is partial recursive.*

6. *Definitional Extension*

Let $P = \langle A, B, b_o, K_o, h^1, h^2\rangle$ and $\bar{P} = \langle\bar{A}, \bar{B}, \bar{b}_o, \bar{K}_o, \bar{h}^1, \bar{h}^2\rangle$ be two RASP's. $\bar{P}$ will be called a *definitional extension* of P if $A = \bar{A}$, $B = \bar{B}$, $b_o = \bar{b}_o$ and if for every *active* word b in B, and for every $a \in A$, $k \in K_o$,

$$h^1(k, a, b) = \bar{h}^1(k, a, b) \text{ and } h^2(k, a, b) = \bar{h}^2(k, a, b).$$

Thus, $\bar{P}$ is obtained from P simply by "activating" some of the passive words of P.

A definitional extension $\bar{P}$ of P will be said to be *generated* by a set of designated instruction schemata $R = \{\lambda_i\}$ in $\bar{P}$ if every word $b \in B$ which is active in $\bar{P}$ but passive in P belongs to at least one B_{λ_i} (where $\lambda_i = \langle\rho_i, \mu_i\rangle$ and B_{λ_i} is the range of μ_i, as before). $\bar{P}$ is a *finite* (or *finitely generated*) extension of P if there exists such a set $R = \{\lambda_i\}$ which is finite, $i = 1, \cdots, r$.

6.1: Proposition. *Let P and $\bar{P}$ be two finitely supported RASP's such that $\bar{P}$ is a finite definitional extension of P which is generated by the set of designated instruction schemata $R = \{\lambda_i\}$, $i = 1, \cdots, r$. If P and the λ_i are recursive (in a set Φ), then $\bar{P}$ is recursive (in Φ).*

Observe that any finitely supported RASP P is recursive in the set $\{\bar{h}^1, \bar{h}^2\}$. It follows that, for $\Phi = \{\bar{h}^1, \bar{h}^2\}$, the last sentence of 6.1 may be replaced by the conclusion that $\bar{P}$ is recursive in P if the λ_i are recursive in $\{\bar{h}^1, \bar{h}^2\}$.

Proof of 6.1. Let $\bar{\bar{h}}^1(x, y, z)$ and $\bar{\bar{h}}^2(x, y, z)$ be the number-theoretic functions which correspond to $\bar{h}^1$ and $\bar{h}^2$ as in 5.4. Then we have to show how to compute $\bar{\bar{h}}^1(l, n, m)$ and $\bar{\bar{h}}^2(l, n, m)$ for arbitrary natural numbers l, n, m. For this purpose we first check whether m belongs to one of the sets β_{λ_i} (which correspond to B_{λ_i}). If it does, we take the first i for which this is true, and we then compute $\bar{\bar{h}}^i(l, n, m)$, $i = 1, 2$, as in the proof of 5.10. If m does not belong to any β_{λ_i}, then we simply put $\bar{\bar{h}}^i(l, n, m) = \bar{h}^i(l, n, m)$, $i = 1, 2$. For, by the assumption of our theorem, this identity must then be true, irrespective of whether the word b with Gödel number m be active or passive. This completes the proof of 6.1. Notice that we did not require any assumption concerning the recursive character of the set of active words in P or $\bar{P}$.

A similar argument proves

6.2: PROPOSITION. *Let P be a recursively finitely determined RASP. Let* $\bar{P}$ *be a finite definitional extension of P by means of the set* $R = \{\lambda_i\}, \quad i = 1, \cdots, r$ *of recursively finitely determined designated instruction schemata. Then* $\bar{P}$ *is recursively finitely determined.*

Now, let P be a given RASP and let ρ be any *undesignated* instruction schema in P. Still supposing that A and B are countably infinite, we may ask whether there exists a definitional extension $\bar{P}$ of P such that ρ is the first element of a designated instruction schema λ in $\bar{P}$, $\lambda = \langle\rho, \mu\rangle$. In other words, we are asking whether we may find words to designate the transitions defined by fixing the parameters of ρ. We consider the case $l + m > 0$ (see Section 3).

If P contains an infinite number of passive words, then the answer to this question is, trivially, in the affirmative. For in this case, we have only to proceed as follows. Select an infinite set B' of passive words in B, excluding b_o, and choose a one-to-one mapping μ from $A^l \times B^m$ onto B'. Such a mapping exists since both sets are countable. We now define the functions $\bar{h}^1(x, y, z)$ and $\bar{h}^2(x, y, z)$ for $\bar{P}$ by

6.3: $\bar{h}^1(k, a, b) = h^1(k, a, b)$, $\bar{h}^2(k, a, b) = h^2(k, a, b)$, for $k \in K_o$, $a \in A$, and for all b in $B - B'$; while for $b \in B'$, $b = \mu(a_1, \cdots, a_l, b_1, \cdots, b_m)$, we put

6.4: $\bar{h}^1(k, a, b) = \rho^1(k, a, a_1, \cdots, a_l, b_1, \cdots, b_m)$, $\bar{h}^2(k, a, b) = \rho^2(k, a, a_1, \cdots, a_l, b_1, \cdots, b_m)$, for all $k \in K_o$, $a \in A$.

The case $l + m = 0$ is even simpler. In this case the instruction schema reduces to a single instruction, which we may designate by any passive b in B other than b_o, provided such a word exists. In either case our choice of B' and μ is arbitrary to such an extent that we can scarcely claim that the elements of B' are in any sense natural names or designations for the transitions with which they are connected.

Now, suppose that P is finitary and recursive, and let ρ be a *recursive* undesignated instruction schema in P. Disregarding the case $l + m = 0$, we now suppose that B includes an infinite subset of passive words, B', such that the set β of Gödel numbers of B' is recursive. We may suppose, if necessary after a preliminary adjustment, that B' does not include b_o. We now choose a *recursive* one-to-one mapping μ from $A^l \times B^m$ to B' (i.e., μ is such that the corresponding number-theoretic function is recursive), and we define $\bar{P}$ again by 6.3 and 6.4. The designated instruction schema $\lambda = \langle\rho, \mu\rangle$ is recursive and so the corresponding $\bar{P}$ is recursive, by 6.1. Similar conclusions apply if we wish to "*designate*" a finite number of undesignated instruction schemata, either simultaneously or successively. If $m + l = 0$, $\bar{P}$ is recursive if P is recursive (provided a passive $b \neq b_o$ is available at all, as explained above).

Next, let P be recursively finitely determined, but not necessarily finitely supported. Let ρ be an *undesignated* instruction schema which is *recursively finitely determined*. That is to say, ρ satisfies the condition 5.18. Taking the case $l + m > 0$ and supposing again that B includes an infinite subset of passive words such that β is recursive, we define μ and then $\bar{h}^1$ and $\bar{h}^2$ as before (by 6.3 and 6.4). The designated instruction schema $\lambda = \langle\rho, \mu\rangle$ obtained in this way is finitely recursively determined and so is the RASP $\bar{P}$, by 6.2. The same conclusion applies for $l + m = 0$, for an arbitrary choice of $b \neq b_o$.

Since the connection between the elements of B' and the corresponding instructions (transitions) $\bar{h}_b(x)$ is now effective, according to one of the two types of recursiveness or the other, we may say that our constructions adequately attribute a *meaning* to these words. We need read no more into the meaning of "the meaning of a word" than our ability to interpret it effectively for the purpose in hand.

There is another kind of extension which is more elementary from the mathematical point of view. This is based on the enlargement of the fundamental "vocabulary," i.e., of the sets A and B. It is conceivable that one might find it convenient to widen these sets before carrying out a definitional extension as described above. We shall not find it necessary to consider this type of extension for the purposes of the present paper.

We have explained previously that we regard the assignment of meaning, in the form of instructions, to words which were not previously endowed with such meaning as an essential feature of the enrichment of a language such as occurs on passing from a ML to a POL. This enrichment may take place in several superimposed stages, i.e., in our terminology, we may have to carry out several definitional extensions of a given RASP P successively.

7. *Sequential Functions*

If one uses RASP machines only to compute recursive functions (in the sense of 4), it is quite clear that the full capacity of the machine is not being utilized in the sense that 4.23(1), (2), (3) hold. Now, digital computers may well be called upon to compute a function, say, Σ which takes a finite sequence of numbers of "arbitrary" length into their sum or a sequence of n^2 (for any $n > 0$) numbers (regarded as a matrix) into the sequence of n^2 numbers which represents its inverse (when it exists). These are examples of what we shall call *sequential functions*.

A *sequential function over* B is a mapping f from a subset of $B_\infty = \bigcup_{i=1}^{\infty} B^i$ into B_∞, where B^i is the set of all i-tuples of elements of B. To distinguish this class of functions from the subclass of functions previously considered which take a subset of B^r into B^s, for some r, s, we refer to this latter class as *ordinary functions*. We wish to consider natural senses in which sequential functions may be said to be computed by a RASP or a RASP program.

We consider first the notion of a program $\pi = \langle p, a, e \rangle$ of a given RASP computing a sequential function at the infinite number of datum locations d_0, d_1, d_2, $\cdots$ and infinite number of value locations v_0, v_1, v_2, $\cdots$. Given a $u \in B^i$, we would like to represent it in the machine by the infinite sequence $\langle k(d_0), k(d_1), \cdots \rangle$. It is not unreasonable to require of the mapping that

$$*: \text{ if } b_1, \cdots, b_i \text{ goes into } b_1', \cdots, b_i', \cdots, \text{ then } b_j = b_j' \text{ for all } j \leqq i.$$

We restrict ourselves to finite sequences over $B - \{b_o\}$ and represent them by infinite sequences satisfying $*$, and the requirement

$$**: \quad j > i \text{ implies } b_j' = b_o .$$

[A digital computer may have two representations of zero, both a "positive zero" and a "negative zero". Either of these zero representations is a natural candidate for the role of b_o.]

It may appear that our restriction to $(B - \{b_o\})_\infty$ has been too severe. Indeed, the unique mapping of $(B - \{b_o\})_\infty$ into B^∞ satisfying $*$ and $**$ may be uniquely extended to a mapping, satisfying $*$ and $**$, of the set B_∞^- of elements of B_∞ which do not terminate with b_o. For our purposes this is not satisfactory for it may be verified (by an argument similar to that of 7.4) that no program $\pi = \langle p, a, e_0, e_1\rangle$ of a finitely determined RASP, (e.g., P_0), can compute the binary relation which holds if and only if $u \in B_\infty^-$ has, say, an even number of occurrences of b_o.

The above discussion is intended to motivate the following definition.

7.1: A *program* $\pi = \langle p, a, e\rangle$ *of a RASP* P *is said to compute a sequential function* over $B' = B - \{b_o\}$ *at* (the infinite sequence of distinct) *datum locations* $d_0, d_1, \cdots$ *and* (the infinite sequence of distinct) *value locations* $v_0, v_1, \cdots$, $v_i \neq d_j$, provided that $d_i \notin \underline{D}p$ for $i \geqq 0$ and the following condition holds.

7.2: If k holds p, $k(d_i) = b_i \in B'$ for $0 \leqq i < r$, $k(d_i) = b_o$ for $i \geqq r$ and (i) if f is defined for $b = \langle b_0, \cdots, b_{r-1}\rangle$, $f(b) = \langle b_0', \cdots, b_{s-1}'\rangle$, $k(v_i) = b_o$ for $i \geqq s$, then

(a) $\text{comp}_e(k, a)$ is successful

(b) If $\text{comp}_e(k, a)$ terminates with $\langle k_n, a_n\rangle$, then $(a_n = e)$ and $k_n(v_i) = b_i'$, $0 \leqq i < s$ and $k_n(v_i) = b_o$ for $i \geqq s$ (cf. 4.2). (ii) if f is not defined for b, then $\text{comp}_e(k, a)$ is not successful. A function from $(B')^r$ into $(B')^s$ is a sequential function over B' and we have then two senses (4.2 and 7.1) in which it may be computed by a RASP. Under appropriate hypotheses it may be expected that these two notions coincide. We omit, however, further consideration of this point in this paper.

7.3: We shall say that a RASP P computes a sequential function f under the following condition. For all a, e, A_0, A_1 such that $a \notin A_0$, $A_1 \subseteq A_0$, there exists a program $\pi = \langle p, a, e\rangle$ of P and two infinite sequences of distinct elements, $D = \langle d_0, d_1, \cdots\rangle$ and $V = \langle v_0, v_1, \cdots\rangle$ such that $\hat{D}$ is disjoint from $\hat{V}$ and

(a) π computes f at datum locations D and value locations V,

(b) $\underline{D}p$ is disjoint from $A_0 \cup \hat{D} \cup \hat{V}$,

(c) If A_1 is disjoint from $\hat{V}$, then $RI(\pi)$ is disjoint from A_1.

Notice that this definition is *not* a direct analogue of 4.4. It may be seen that the direct analogue is inappropriate since it would lead to the conclusion that no sequential function is computable by a RASP. Thus, there are two quite distinct senses in which an ordinary function may be computable by a RASP. In most cases the context will make clear what sense is intended; in other cases the sense intended may be made explicit.

7.4: THEOREM. *If*

(a) *the program* $\pi = \langle p, a, e\rangle$ *computes the sequential function* f (in the sense of 7.1) *at* $d_0, d_1, \cdots$ *and* $v_0, v_1, \cdots$,

(b) *the instructions of* π *are finitely determined and*

(c) π *is fixed* (cf. 4.23, 1), *then there exists* r *such that if* $b, b' \in B_\infty'$, $b = \langle b_0, b_1, \cdots, b_s \rangle$, $s \geqq r - 1$, $b' = \langle b_0', b_1', \cdots, b_{s'}' \rangle$, $s' \geqq r - 1$, $b_i = b_i'$ *when* $0 \leqq i < r$, *then* $f(b) = f(b')$.[15]

PROOF. Let $h_{b_1}, \cdots, h_{b_n}$ be the instructions of the given program and suppose $p(a_i) = b_i$. By (b) there is a finite sequence A_{a_i, b_i} (cf. definition of finitely determined instruction) on which h_{b_i}, $1 \leqq i \leqq n$, "depends". Making use of (c) we conclude: there exists a finite set, $A' = \bigcup_i \hat{A}_{a_i, b_i}$, which satisfies the following property.

Suppose k_0, k_0' both hold π, $k_0(v_j) = b_o = k_0'(v_j)$ for all j, and also agree on A'. Let S be that subset of A consisting of those elements x such that $k_0(x) = k_0'(x)$, so that S contains A'. Let $\text{comp}(\sigma_0) = \langle \sigma_0, \sigma_1, \cdots \rangle$, $\text{comp}(\sigma_0') = \langle \sigma_0', \sigma_1', \cdots \rangle$, where $\sigma_i = \langle k_i, a_i \rangle$, $\sigma_i' = \langle k_i', a_i' \rangle$. It follows immediately by induction on i that

$*$: if $a_0 = a = a_0'$, then (1) $a_i = a_i'$ for all i and (2) k_i and k_i' agree on S for all i.

Now let $r = 1 + \max(\{i \mid d_i \in A'\} \cup \{0\})$. Let $k_0(d_i) = b_i$ for all i. Then, if $f(b)$ is defined and equal to $\langle c_0, \ldots, c_t \rangle$, $\text{comp}_e(k_0, a)$ will be successful; and if $\langle k_n, e \rangle$ is the terminal state of $\text{comp}_e(k_0, a)$, then $k_n(v_i) = c_i$ for each $i \leqq t$. Similarly, letting $k_0'(d_i) = b_i'$ for all i, we may "find" $f(b')$. Inasmuch as $v_i \in S$ for all i, it follows by $*$ that $k_n(v_i) = k_n'(v_i)$ for all i, so that $f(b) = f(b')$.

We now wish to show, by contrast, that every partial recursive sequential function over the *positive* integers can be computed by P_0, provided that we do not make the restriction 7.4 (c). To this end we require some definitions.

Let G be the mapping from all finite sequences of *positive* integers into positive integers defined by:

$$G(\langle b_1, b_2, \cdots, b_n \rangle) = 2^{b_1} 3^{b_2} \cdots (Pr(n))^{b_n}.$$

It is clear that G is a one-to-one mapping *into* the positive integers, that its range is recursive, and that both G and G^{-1} are effectively computable. We shall call a *sequential function* f *over the positive integers partial recursive* if the corresponding ordinary number-theoretic function $\bar{f}$ is partial recursive, where $\bar{f}$ is defined by the requirement: $\bar{f}(x) = y$ iff x, y are in the range of G and $f(G^{-1}(x)) = G^{-1}(y)$.

COROLLARY TO 7.4: *There are recursive sequential functions which are not computable by fixed programs of* P_0. *Indeed, the function* Σ, *which takes* $\langle x_1, \cdots, x_n \rangle$ *into* $\sum_{i=1}^{n} x_i$ *is one such, where* $x_i > 0$.

7.5: THEOREM. *All partial recursive sequential functions over the positive integers are computable by* P_0.

PROOF. Let f be a partial recursive sequential function. The "flowchart" (Figure 2) indicates a program which computes f at datum locations $\langle d, d + 2, d + 4, \cdots \rangle$ and value locations $\langle d + 1, d + 3, d + 5, \cdots \rangle$. In this flowchart, box 2 represents a program which computes the successor function at datum

[15] This entails that not all recursive sequential functions (defined below) are computable by fixed programs of finitely determined RASP's.

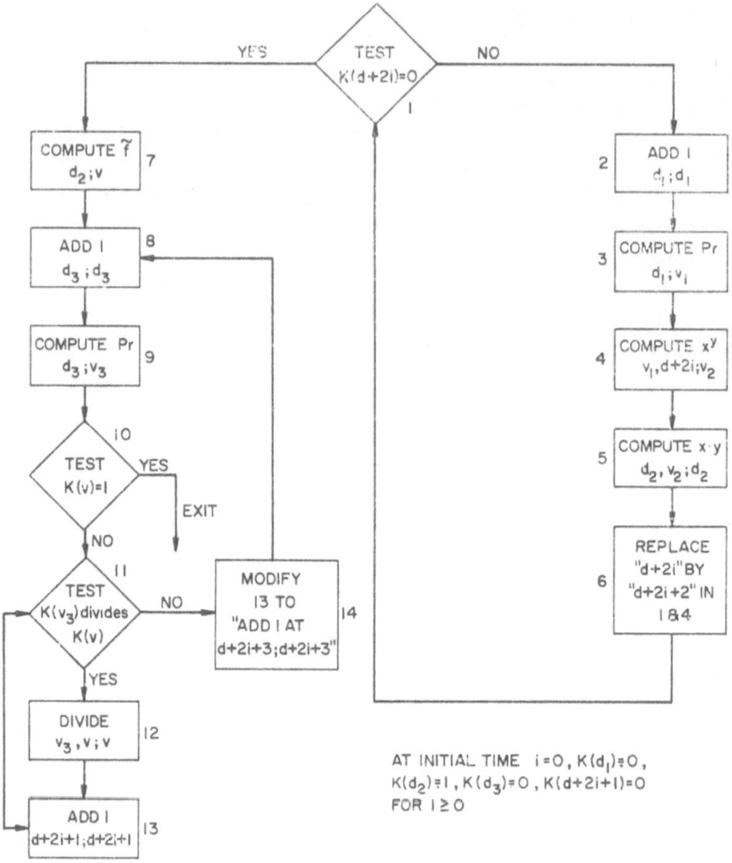

FIG. 2. Flowchart for Theorem 7.5

location d_1 and value location d_1 (cf. 4.24, footnote 13). "Pr" represents the function which takes i into the ith prime and box 3 represents a program which computes Pr at datum location d_1 and value location v_1. The exit of program 2 is the entrance to program 3, the exit of program 3 is the entrance to program 4, etc. Program 12 divides $k(v_3)$ into $k(v)$ and puts the quotient in v. Program 1 computes the equality relation at $d + 2i$ and another location which is not identified but which is supposed to contain zero. One of the exits of program 1 is the entrance to program 7 and the other the entrance to program 2. The label "$d + 2i$" is a variable one. The entire program is such that $d + 2i$ successively assumes the values d, $d + 2$, $d + 4$, etc. until a zero content is reached. We are assured by 4.23 that the programs indicated exist except for 6 and 4 which require further comment.

Program 6, for example, computes the function, say, θ which takes $\#(\bar{C}, x, y, z)$ into $\#(\bar{C}, x + 2, y, z)$ and any number which is not of the form $\#(\bar{C}, x, y, z)$

into, say, 0. Then θ is recursive by virtue of the definition of $\divideontimes$. Program 1 may be taken to consist of a single instruction located, say, at a. Then θ is computed by program 6 at a, a.

The right-hand "loop" of the flowchart (consisting of programs 1, 2, 3, 4, 5, 6, 1) is used to encode the given finite sequence of positive integers into a positive integer. When the exit "yes" of program 1 is reached, the encoding computation is completed. Program 7, as indicated, computes $\bar{f}$. Programs 8 through 14 are used to decode the value of $\bar{f}$ obtained—a positive integer—into a finite sequence of positive integers.

One final remark in connection with this argument. The cardinality of each program indicated by the flowchart is supposed to be independent of the values of the parameters d, d_1, v_1, d_2, v_2, etc. This property of P_0 is involved in finding the space for the program as well as the infinite number of data and value locations.

Essentially, the same argument as for 5.21 yields:

7.6, 7.7: THEOREM. *Both* 5.21 *and* 5.22 *hold in the case that f is a partial sequential function.*

COROLLARY. *If a program of P_0 computes a sequential function f (at datum locations D and value locations V), then f is a partial recursive sequential function.*

7.8: DISCUSSION.

1. *Instruction Modification.* The combined results 7.4 and 7.5 give some insight into the role of instruction modification in digital computers. As has been noted, it is an immediate consequence of 7.4 that no finitely determined machine can compute all recursive sequential functions by fixed programs, i.e., by means of programs whose instructions are never altered or "modified". On the other hand, 7.5 points out that the finitely determined machine P_0 can compute all partial recursive sequential functions. In particular, P_0 can compute Σ, which takes $\langle x_1, \cdots, x_n\rangle$ into $\sum_{i=1}^{n} x_i$, $n \geqq 1$, by programs whose instructions are altered; but P_0 cannot compute Σ by programs whose instructions are not altered. Moreover, finitely determined RASP's appear to be sufficiently adequate mathematical models of the central processing unit of digital computers before the advent of index registers for one to conclude that 7.4 applies to such computers. It is easy to convince oneself that 7.5 applies to the proper idealization of any general purpose computer which was designed, say, since the Princeton Institute for Advanced Study one.

2. *Index Registers.* It appears rather clear that a RASP $P_0' = \langle N, N, \cdots\rangle$ generated by the following instruction schemata is capable of computing all partial recursive sequential functions by means of *fixed* programs. Of course, P_0' is not finitely determined.

(1) $S'(x_1, x_2)$, "add 1 to $k(x_1 + k(x_2))$", more explicitly:

$$k'(x_1 + k(x_2)) = k(x_1 + k(x_2)) + 1 \wedge (y \neq x_1 + k(x_2) \rightarrow k'(y) = k(y)) \wedge a' = a + 1.$$

(2) $D'(x_1, x_2, y_1, y_2)$, "transfer $k(x_1 + k(x_2))$ to $y_1 + k(y_2)$", more explicitly:

$$k'(y_1 + k(y_2)) = k(x_1 + k(x_2)) \wedge (z \neq y_1 + k(y_2) \rightarrow k'(z) = k(z)) \wedge a' = a + 1.$$

(3) $C(x_1, x_2, y_1, y_2, z_1, z_2)$, "conditional transfer on equality", more explicitly:

$$k' = k \wedge (k(x_1 + k(x_2)) = k(y_1 + k(y_2)) \rightarrow a' = z_1 + k(z_2)) \wedge (k(x_1 + k(x_2)) \neq k(y_1 + k(y_2)) \rightarrow a' = a + 1).$$

3. *Routines.* We have discussed (essentially) but one sense in which RASP machines compute sequential functions. There is, however, at least one other sense which is quite important simply on pragmatic grounds. Roughly, and tentatively, speaking we mean by *routine* a program whose parameters are permitted to vary. Let f be a sequential function and let f_r be the restriction of f to the domain of r-tuples. We consider the special case that for each r there exists s such that the range of f_r is included in the set of all s-tuples. A *routine may be said to compute* f provided that for each r, there is a specialization π of the given routine's parameters such that π computes f_r in the sense of an ordinary function (at certain data and value locations which depend on the parameters). It is anticipated that this notion will play an important role in the later development of this inquiry. It may be noted now, however, that sequential functions computable by routines still fall under the scope of 7.4. In particular, the function Σ is not computable by any *routine* of any finitely determined machine if instruction alteration is not permitted.

8. *Point of View and Direction*

The compiler problem for a RASP P is, roughly speaking, the problem of finding an algorithm which will operate on any program schema of an extension of P (Section 6) and produce a program schema of P *which performs the same task* as the given program schema. For this reason we have concerned ourselves in some detail (in Sections 4 and 7) with the question of what these tasks are and in precisely which senses programs perform these tasks.

The notion of a RASP machine and its associated notions provide a basis for endowing programming languages with semantics. Thus, we have begun an investigation of the relations between different (poorer or richer, machine- or problem-oriented) languages by associating them with different RASP's, whose possibilities are determined by their transformation functions, $h: \Sigma_o \times B \rightarrow \Sigma_o$. This implies that each one of the languages has a "static" character in that the scope of its instructions is given once and for all. The idea of widening the scope of a language by the *introduction* of new definitions is here expressed by *passage* to a richer language, i.e., to a language with a larger supply of instructions. However, in order to be able to survey the possibilities of enrichment by new definitions simultaneously, it will be useful to introduce a *single* language which is sufficiently flexible to permit the introduction of new instructions or instruction schemata by definition. It is anticipated that this language will be discussed at a more syntactical level, complementing the semantic approach adopted so far. Within this framework one may then discuss the question, "What are the tasks that computers are called upon to perform?" This task goes beyond the computation of ordinary functions and relations (as in Section 4) and should at least include sequential functions and relations (cf. Section 7).

Acknowledgements. The authors gratefully acknowledge many helpful comments of Dr. J. D. Rutledge. The authors thank J. W. Thatcher for calling some errata to their attention.

RECEIVED JANUARY, 1964

REFERENCES

1. NAUR, P. (ED.) Report on the algorithmic language ALGOL 60. *Comm. ACM 3*, 5 (May 1960), 299–314; and *Numer. Math. 2*, (1960), 106–136.
2. DAVIS, M. *Computability and Unsolvability*. McGraw Hill, New York, 1958.
3. HERMES, H. *Aufzählbarkeit, Entscheidbarkeit, Berechenbarkeit, Einführung in die Theorie der Recursiven Funktionen*. Berlin, Springer, 1961.
4. GOLDSTINE, H. H., AND VONNEUMANN, J. Planning and coding of problems for an electronic computing instrument. Rep. 1947, Inst. Adv. Study, Princeton.
5. ASSER, G. Turing-Maschinen und Markowsche Algorithmen. *Z. Math. Log. Grundlag. Math. 5* (1959), 346–365.
6. KAPHENGST, H. Eine abstrakte programmgesteuerte Rechenmaschine. *Z. Math. Log. Grundlag. Math. 5* (1959), 366–379.
7. WANG, H. A variant to Turing's theory of computing machines. *J. ACM 4*, 1 (Jan. 1957), 63–92.
8. GILMORE, P. An abstract computer with a LISP-like machine language without a label operator. In *Computer Programming and Formal Systems*, P. Braffort (Ed.), North Holland, 1963.
9. SHEPHERDSON, J. C. AND STURGIS, H. E. Computability of recursive functions. *J. ACM 10*, 2 (April 1963), 217–255.

Reprinted from *Programming Languages* (edited by F. Genuys), London, Academic Press Inc. (1968), 1–42.

Abstract Algorithms and Diagram Closure

CALVIN C. ELGOT

1. INTRODUCTION

A programming language L presumably should be designed, at least in part, to express algorithms of a given class M . Given an algorithm in the class M , there should be a name in L which expresses it. In addition, L should possess names for the "tasks performable by algorithms" in M . A theory of the language L should presumably include statements about algorithms in M and their tasks.

Unfortunately, the notion of algorithm is vague (although the notion of "function computable by algorithm" is precise if one accepts Church's thesis). The notion "task performable by algorithm" is perhaps still vaguer.

There is no attempt here to explicate these notions in their full generality. On the contrary, an attempt is made to delineate a class of "algorithms" which permits maximum flexibility in the sequencing of "atomic acts", which, from the point of view of "accessing units of data", is simple although restrictive, which permits the set of atomic acts to be arbitrary and which, under suitable specialization and interpretation may be made to yield exactly the class of partial recursive functions. This point of view has the consequence that familiar "algorithms" which operate on natural numbers, in contradistinction to their representations (numerals), may be faithfully represented by our "diagram algorithms" while, because of the restrictive accessing capability of these algorithms, we cannot (in general) faithfully represent familiar algorithms which operate on numerals. The point of view that "algorithms" are to operate on numbers rather than numerals is a reversal of the usual one. Both points of view are present in [SS] where, apparently, the "number point of view" was first introduced. The treatment of the "diagram algorithm" notion is essentially mathematical although linguistic notions are introduced for motivational purposes. Specifically, no deductive framework is introduced in the main text. For more ready comparison with deductive treatments, however, an appendix is included which outlines a procedure for casting the computation notion in the form of rules operating on strings of symbols to produce strings of symbols. A linguistic notion of "concrete diagram" is introduced, as well as a mathematical notion of "abstract diagram", in order to express algorithms in this class. The notion "concrete diagram" is similar to that of "flow chart" which was introduced in [GN] and is used in the preparation of programs for digital computers. What is vaguely referred to above as the "task performable by algorithm" takes on the precise meaning in our specialized context of "direction, diagram-computable by algorithm". In order to promote the perspicuity of diagrams as well as to keep them physically manageable in size, a device is introduced for assigning names to the directions computable by algorithms in this class.

What we call "algorithm", however, is more properly called "relative algorithm" for in the case that the set (B below) of units of data is infinite, what we take as "atomic act" is not truly atomic on the ground that no individual and no finite machine can discriminate among an infinite number of objects "all at once". Similarly, on the same grounds, the usual statements of the "Euclidean Algorithm" in which the representation of the numbers do not figure are statements of <u>relative</u> rather than <u>ab-initio</u> algorithms. On the other hand, the familiar algorithms for, say, the addition of two positive integers given in decimal notation, or, more simply, determining the successor of a positive integer given in decimal notation are ab-initio. In this case, the acts taken as atomic involve only a finite (indeed "small") number of discriminations.

It is natural in the treatment of relative algorithms to take the set of units of data as arbitrary. This set may be finite or not. The possibility that the set is nondenumerably infinite is not ruled out although there is some question whether so-called "algorithms" operating on such sets of units of data are even properly "relative algorithms".

This point of view is similar to that in [ER]; however, their universality is, so to speak, built in, while here it is not. The specialization to "fixed programs" makes the algorithm notion particularly closely related to the notions in [SS]. Here, however, the set of data units and "operations" are arbitrary which permits then the introduction, for example, of "logical equivalence" as indicated below. It is an important part of our point of view that a "box" of a "flow chart" is given a meaning which is independent of the flow chart in which it occurs. Thus a diagram may be construed as a complex imperative sentence constructed from simpler imperative sentences in a standardized way. This point is discussed in greater detail at the end of Section 4.

While the discussion below involves direction symbols which are interpreted, one may also consider "uninterpreted diagrams". An uninterpreted diagram (which we may appropriately call a <u>diagram form</u> or diagram schematum) would become interpreted by specifying a set B and for each direction symbol of rank $\langle n, p, q \rangle$, a direction over B of rank $\langle n, p, q \rangle$.

A statement concerning diagram forms is logically true if it is true under all interpretations. The reader will note that some of the simpler statements below when understood as statements concerning uninterpreted diagrams are actually logically true; e. g., the statements embodied by Figs. 3, 4, 5, 6. We regard "diagram" as analogous to "(declarative) sentence" and diagram form" as analagous to "sentential form".

2. FUNDAMENTAL MATHEMATICAL AND LINGUISTIC NOTIONS TERMINOLOGY

In this section we follow in a general way the point of view and terminology of Alonzo Church as expounded especially in the Introduction of [AC]. Specifically, the underlined words of this section which are not otherwise defined or explained, are understood as in [AC]. In particular we take a declarative sentence to be a linguistic object which expresses a proposition and denotes a truth value. We further follow Church (and Frege) in taking a declarative sentence to be a name, although different from other names in that it may be used assertively to express a proposition or, as other names, as a constituent of a sentence.

We wish to understand the word mapping in such a way that the following holds. If μ is a mapping, there is a set $\mathcal{A}\mu$ called the range of arguments (or source) of μ, there is a set $\mathcal{V}\mu$ called the range of values (or target) of μ, there is a set graph μ called the graph of μ and there is a notion of application of μ to an element x in $\mathcal{A}\mu$; application of μ to x produces at most one value $\mu(x)$. The graph of μ is a set of ordered pairs; the first member of an ordered pair comes from $\mathcal{A}\mu$ and the second member comes from $\mathcal{V}\mu$. More briefly, graph $\mu \subseteq \mathcal{A}\mu \times \mathcal{V}\mu$. The set dom μ of first members of graph μ, we call the domain of μ while the set cod μ of second members of graph μ, we call the codomain of μ. Thus dom $\mu \subseteq \mathcal{A}\mu$ and cod $\mu \subseteq \mathcal{V}\mu$. The graph of μ is required to satisfy:

$$(1) \qquad (\langle x, y\rangle \in \text{graph } \mu \wedge \langle x, z\rangle \in \text{graph } \mu) \Longrightarrow y = z .$$

If $x \in$ dom μ and $\langle x, y\rangle \in$ graph μ, then $\mu(x) = y$. If $x \in \mathcal{A}\mu \sim$ dom μ (set-theoretic subtraction), then "$\mu(x)$" lacks a denotation (and we say it is undefined) although we regard it as having a sense. If $x \notin \mathcal{D}_\mu$, then "$\mu(x)$" lacks even a sense. For our present purposes at least, it suffices to take as foundational point of view a set theory such as von Neumann-Bernays-Gödel and define a mapping to be a set μ such that there exist unique sets X, Y, R satisfying:

(a) $R \subseteq X \times Y$

(b) $\langle x, y\rangle \in R \wedge \langle x, z\rangle \in R \Longrightarrow y = z$,

(c) $\mu = \langle X, R, Y\rangle$.

If μ is a mapping satisfying (a), (b), (c), then $X = \mathcal{D}_\mu$, $Y = \mathcal{V}_\mu$, $R = \text{graph}\ \mu$, $R = \{\langle x, \mu(x)\rangle \mid x \in \text{dom}\ \mu\}$ and we write "$\mu : X \to Y$". If "x" and "y" are names then the sentence "x = y" will be understood in such a way that it always has a denotation even though "x" or "y" may lack denotations. Specifically, we take "x = y" to denote truth if either "x" and "y" both lack denotations or both "x" and "y" have denotations and denote the same element; otherwise, "x = y" denotes falsehood (see [SCK], p. 328).

Let $\mu : X \to Y$, $\nu : X \to Y$. Then the following assertions are equivalent.

(2) $\mu = \nu$

(3) $\forall x \forall y\, (\mu(x) = y \Longleftrightarrow \nu(x) = y)$

(4) $\forall x\, (\mu(x) = \nu(x))$

where x, y are <u>variables</u> which respectively <u>range</u> over X, Y .

We call a mapping μ <u>total</u> if dom $\mu = \mathcal{a}\mu$, <u>surjective</u> if cod $\mu = \mathcal{V}\mu$, <u>injective</u> if $(x \in \text{dom}\, \mu \wedge y \in \text{dom}\, \mu \wedge \mu(x) = \mu(y)) \Longrightarrow x = y$ and <u>bijective</u> if it is total, injective and surjective.

By a <u>relation</u> we mean a mapping whose range of values is the set of <u>truth values</u> consisting of <u>truth</u> (T) and <u>falsehood</u> (F) . We do not demand that a relation be total. A relation, however, is different from other mappings in that application of a relation to an argument may be used to express a propo sition as well as to produce a value. We also find it convenient to define a <u>semirelation</u> to be a mapping whose range of values is the singleton consisting of T or F alone. In particular, if E is a subset of a set B we associate with it a <u>characteristic semirelation</u> $E_B^{\frac{1}{2}}$ and a <u>characteristic relation</u> $E_B^{\frac{1}{2}+\frac{1}{2}}$ both of which have B as its range of arguments and which satisfy:

$$\mathrm{dom}\ E_B^{\frac{1}{2}} = E$$

$$\mathrm{dom}\ E_B^{\frac{1}{2}+\frac{1}{2}} = B$$

$$E_B^{\frac{1}{2}}(x) = T \iff x \in E \iff E_B^{\frac{1}{2}+\frac{1}{2}}(x) = T .$$

We analagously define the characteristic relation and semirelation of subsets of B^n, $n > 0$. In certain contexts we may drop the subscript B without risk of confusion.

A function over B will be understood as a mapping f such that there exist (unique) positive integers n, p satisfying

$$\mathcal{A}f = B^n , \quad \mathcal{V}f = B^p .$$

The numbers n, p are respectively called the argument and value ranks of f. Similarly, for relation over B, semirelation over B.

Because we do not wish to restrict ourselves to two-valued relations, we employ mappings $\mu : B^n \to \{0, 1, \ldots, q-1\}$ $q > 0$, $n > 0$ which we call decisions over B. The rank of such a decision is $\langle n, q \rangle$. (Hereafter, we identify q with $\{0, 1, \ldots, q-1\}$.) It is convenient also to have a special term, "direction", for mappings which are a kind of combination of functions and decisions over B .

A direction over B is a mapping $d: B^n \to B^p \times q$, where $n \geqq 0$, $p \geqq 0$, $q > 0$. The triple $\langle n, p, q \rangle$ is the rank of d . In the case that n, p are positive, we define $d^{(1)}: B^n \to B^p$, $d^{(2)}: B^n \to q$ by the requirement

$$\text{(5)} \qquad \forall x (d(x) = \langle d^{(1)}(x), d^{(2)}(x) \rangle)$$

where x ranges through B^n . We note that $\mathrm{dom}\, d^{(1)} = \mathrm{dom}\, d^{(2)}$ since both equal dom d .

If the rank of d is $\langle n, 0, q \rangle$, $n > 0$, $q > 0$, we identify d with the decision $d^{(2)}$ of rank $\langle n, q \rangle$. †

Admitting the case $n = 0$ is a convenience, which enables us to speak of an element of $B^p \times q$, $p > 0$, $q > 0$, as a total direction with rank $\langle 0, p, q \rangle$; in the case $p = 0$, one may identify an element of q with a direction of rank $\langle 0, 0, q \rangle$. For the same reason, we

---

† We take the point of view here that B^0 is not meaningful and that, therefore, it is not à priori meaningful to speak of a mapping $d: B^n \to B^0 \times q$; in this case, we assign meaning to "$d: B^n \to B^0 \times q$" by saying it means "$d: B^n \to q$". Alternatively, one may take the point of view that $B^0 = \{\phi\} = 1$, that "$d: B^n \to B^0 \times q$" and "$d: B^n \to q$" have distinct meanings and that we are deliberately confounding the distinction (as in the paragraph preceding 3. 2.)

now admit the possibility of argument rank 0 to functions and decisions over B. We note that there is exactly one total direction over B of rank $\langle 0, 0, 1 \rangle$ and exactly one direction of rank $\langle 0, p, q \rangle$, $q > 0$, which is not total.

The value produced by application of a direction to an argument is an ordered pair (if $p > 0$); we call the first member the <u>result</u> of the direction and the second member the <u>outcome</u>. Similarly, if $\langle n, p, q \rangle$ is the rank of a direction, n is the <u>argument rank</u>, p, the <u>result rank</u> and q, the <u>outcome rank</u>.

Additional notions and terminology will be introduced as required.

3. COMMANDS

Ab-initio algorithms may be regarded as involving <u>symbol spaces</u>, each of which is capable of having written in it one of a finite number of symbols. In order to accommodate the "relative algorithm" concept, we extend the idea of symbol space to that of "cell". A <u>cell</u> is capable of accommodating a certain set of "symbols" which may be of arbitrary cardinality. We use the word "address" for the name of a cell. In our formal treatment of diagram algorithms, however, we do not need to distinguish between "address" and "cell". In fact, in the formal development, an <u>address</u> or cell is a linguistic entity which has associated with it (like a variable) a certain set called its range of values. We shall assume an infinite set of cells is given and fixed throughout our discussion. This is analogous to the infinite list of variables which are often parts of formal calculi. Each cell shall have as its range of values a certain, nonempty set, B . This is analogous to the set over which a variable ranges in an interpretation of a formal calculus. One may also stretch the meaning of "linguistic entity" to permit a nondenumerable set of cells. This too has a counterpart in a current practice of model theory.

We think of a cell as containing at any given instant of time an element of its range of values and at different instants of time as containing possibly different values. If a_0 is a cell, we use the notation "$\bar{a}_0$" to mean "the present content of cell a_0". In certain contexts one may wish to use the notation "$\bar{a}_0(t)$" to mean "the content of cell a_0 at time t". Our present purposes, however, do not require it.

By a command, in general, we mean the sense of an imperative sentence. We require, however, only imperative sentences of a very special form and shall mean by "simple imperative sentence" only an imperative sentence of the form described below. The simple imperative sentences we consider are instances of the schematic linguistic entity indicated below. †

3.1

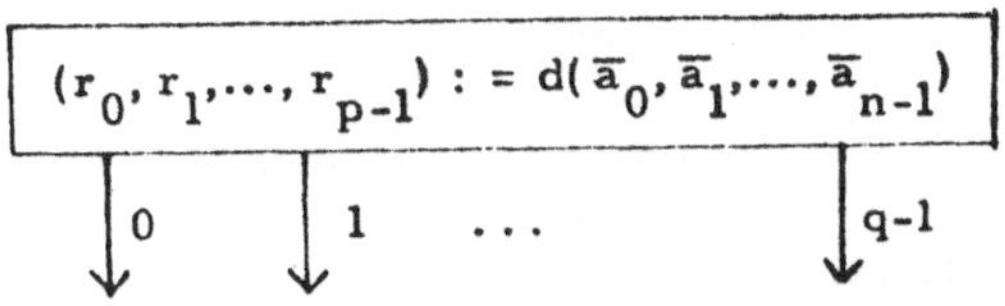

Here d is a direction over B of rank $\langle n, p, q\rangle$, $(a_0, a_1, \dots, a_{n-1})$ is an n-tuple of distinct cells called the <u>argument cells</u> of the command and $(r_0, r_1, \dots, r_{p-1})$ is a p-tuple of distinct cells called the <u>result cells</u> of the command. (It is not ruled out, however, that $a_i = r_j$ for some i, j.)

More specifically, an instance of the displayed entity becomes a sentence if specific values of n, p, q are chosen, "d" is replaced by a name of a direction over B of rank n, p, q, "a_0", "a_1",..., "a_{n-1}" are replaced by distinct addresses (which we haven't displayed) and "r_0", "r_1",..., "r_{p-1}" are replaced by distinct addresses.

The sense of 3.1 is taken to be the same as the sense of the following sentence.

---

† The rectangle itself, the edges (arrows) emanating from the rectangle, the labels attached to the edges and the inscription inside the rectangle are all parts of the imperative sentence.

"Place $d^{(1)}(\bar{a}_0, \bar{a}_1, \ldots, \bar{a}_{n-1})$ respectively into cells $r_0, r_1, \ldots, r_{p-1}$ (displacing $\bar{r}_0, \bar{r}_1, \ldots, \bar{r}_{p-1}$) respectively and select the edge labelled $d^{(2)}(\bar{a}_0, \bar{a}_1, \ldots, \bar{a}_{n-1})$; make no other changes in cell content". (While the sense of this sentence does not permit removing the requirement that the r_i's be distinct, it does not prohibit removing the distinctness requirement from the a_i's. This would have the effect, however, of "building in" closure under "identification of cells" for a given set of commands, which we prefer not to do. When, however, this kind of closure obtains, the requirement that the a_i's be distinct may be removed and regarded as a kind of abbreviation.)

Let A be the set of cells (each of which has B as its range of values). Let K be the set of total mappings $k: A \to B$. We call an element of K a <u>content function</u>. We take the denotation of the command 3.1 at, say, location c to be the mapping $\Delta_c: K \to K \times q$ defined by the following requirements.

(1) $k \in \text{dom}\, \Delta_c$ iff $\langle k(a_0), k(a_1), \ldots, k(a_{n-1})\rangle \in \text{dom}\, d$

(2) if $k \in \text{dom}\, \Delta_c$ and $\Delta_c(k) = \langle k', j\rangle$, then $d^{(2)}(k(a_0), \ldots, k(a_{n-1})) = j$,

$d^{(1)}(k(a_0),\ldots,k(a_{n-1})) = \langle k'(r_0),\ldots,k'(r_{p-1})\rangle$ and if $a \neq r_0,\ldots,a \neq r_{p-1}$, then $k'(a) = k(a)$. In particular, if $p = 0$, $k' = k$. In the case $n = 0$, we take $\mathrm{dom}\,\Delta_c = K$ if d is total; we take $\mathrm{dom}\,\Delta_c = \phi$ if d is not total; (1) is ignored.

It is often convenient to confound the distinction between a direction d of rank $\langle n, p, 1\rangle$ and the function $d^{(1)}$ of rank $\langle n, p\rangle$ which is canonically associated with it.† This we deliberately do when confusion does not threaten. Taking B to be the natural numbers (ω), $\underline{S}$, the successor function, $\underline{\mathrm{const}}_0$, the total function with argument rank 0 whose (sole) value is 0, $\underline{\mathrm{eq}}:\omega^2 \to 2$, $\mathrm{eq}(x, y) = 0 \iff x = y$ and taking the symbols $\underline{a}_0$, $\underline{a}_1$ as part of the set of symbols A, we give three specific examples of sentences.

† If B is infinite, then $\mu: B \to B$ is a function over B but not a decision over B. If, however, $B \in \omega$, $B > 0$, then μ is both a function over B and a decision over B. This latter case has the consequence, for example, that a μ-command would have a single outcome by virtue of being a function, and B outcomes by virtue of being a decision which, if $B > 1$, would be contradictory. We therefore attain the desired convenience "par abus de language".

3.2

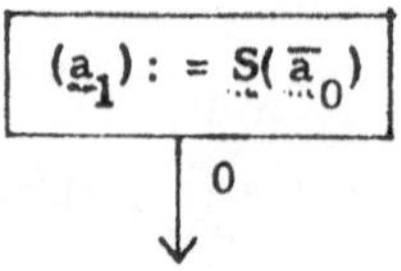

$$(\underline{a}_1) := \underline{S}(\overline{\underline{a}}_0)$$

0

3.3

$$(\underline{a}_0) := \underline{const}_0(\ \ )$$

0

3.4

$$(\ \ ) := \underline{eq}(\overline{\underline{a}}_0,\ \overline{\underline{a}}_1)$$

$0 \qquad 1$

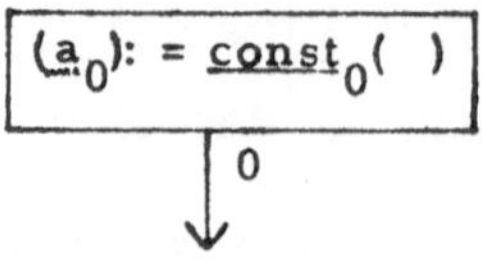

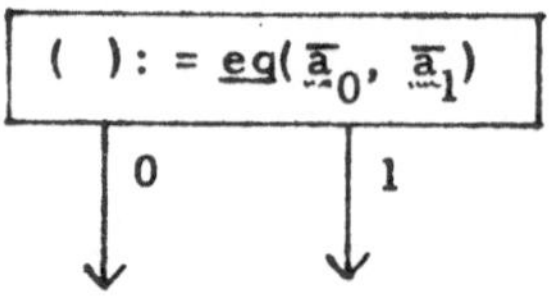

As an "abbreviation" of the last sentence we permit

3.5

$$\overline{\underline{a}}_0 = \overline{\underline{a}}_1$$

$T \qquad F$

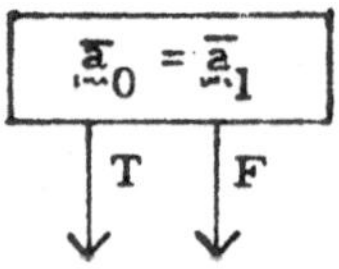

We note that execution of 3.2 results in the following declarative sentence being true:

3.6
$$\overline{\underline{a}}_1 = \underline{S}(\overline{\underline{a}}_0)$$

Note too how 3.2 and 3.6 differ syntactically.

While the formal sentences we have described do involve bars, there often is no confusion in deleting them because of the very special forms of the simple imperative sentences.

4. FINITE CELL DIAGRAM ALGORITHMS

Informally speaking, the main ingredients of a (finite cell diagram) computing algorithm δ are the following.

(1) A finite collection o of occurrences[†] (or locations) of commands. (We speak of "occurrences of commands" rather than "commands" because a given collection may contain more than one occurrence of the same command.)

(2) A mapping θ, called the occurrence-command mapping, which associates with each command occurrence c, $c \epsilon o$, the command $\theta(c)$ which occurs at that occurrence.

(3) Each command occurrence $c \epsilon o$ has associated with it a (finite) set of possible successors. A given execution of the command at c selects a unique element of this set, $\nu(c, i)$, $\nu(c, i) \epsilon o$ where i is an element of the outcome rank of $\theta(c)$. We call ν the direct transition function.

[†] More strictly, we should speak of occurrences of imperative sentences.

(4) An element c_I, $c_I \in o$ is distinguished as the initial command occurrence or entrance.

(5) A finite nonempty, sequence of distinct elements of o is distinguished. The codomain of this sequence is the set of exits. The first time (if ever) during the course of an execution of an algorithm, one "reaches" an exit, one "reads" the result and outcome of the execution. The outcome of the execution is i, if it is the ith exit which is first reached.

(6) A finite sequence of distinct cells, the distinguished argument cells, for recording the data on which the algorithm is to operate.

(7) A finite sequence of distinct cells, the distinguished result cells, at which one finds the result of an execution at the time an exit is first reached.

Concerning the contents of the cells other than the argument cells at the start of an algorithm execution, there appear to be two natural points of view from which to choose. One point of view is that the content is arbitrary and the other, that the content of each of these cells is some distinguished element $b_0 \in B$. While the latter point of view would result in every computing algorithm computing some direction while the former does not, in other ways the former seems simpler, more elegant and richer. It is the former point of view which is selected. We define later a broad class of computing algorithms which we call normal. Normal algorithms always compute directions. Moreover, in a computation of a normal algorithm, one may regard cells as "coming into existence as required".

To a lesser extent we shall deal with (finite cell diagram) generating algorithms. These are much like computing algorithms in that the ingredients (1), (2), (3), (4), (7) are present (but we use the word "value" instead of "result") but (6) is absent and, as far as (5) is concerned, there is in this case but one exit, which we call an ejector to suggest the role it plays. During the course of execution of a generating algorithm, each time the ejector is reach, the contents of the value cells are "ejected" to form a subset of B^p, $p > 0$, if there are p value cells. For the time being we restrict attention to computing algorithms.

We conceive of a concrete computing diagram to be a linguistic object, in fact a complex kind of imperative sentence, the sense of which is a (diagram) computing algorithm. For purposes of communicating a computing algorithm we may display a particular diagram. Of course, other linguistic objects may be exhibited to express the same sense and any of these may be expected to suffer from some ad hoc features. We have chosen one which appears minimally ad hoc. When we talk about diagrams, in contradistinction to using them, however, it is more convenient to have a notion available which, in effect, abstracts the main ingredients of any linguistic method for expressing the same class of algorithms. We define an abstract finite cell computing diagram as follows. (cf. (1)-(7) at beginning of section.)

Definition. An abstract computing (finite cell) diagram δ with (command) locations in C, cells in A and directions in D is a septuple $\langle o, \theta, \nu, c_I, e, \xi, \eta \rangle$, where

(1) o is a finite subset of C;

(2) θ is a mapping with source C, domain o and target the set of ordered triples $\langle d, a, r \rangle$ where $d \in D$ and if the argument and result ranks of d are n, p respectively, then $\underline{a}$ (resp. r) is a finite sequence of distinct elements in A of length n

(resp. p); these ordered triples are called abstract commands and are correlated with the commands 3.1 in an obvious way;

(3) ν is a total mapping with domain the set of abstract edges of δ ; these are ordered pairs $\langle c, i \rangle$ where $c \in o$ and where $i \in q$, q being the outcome rank of $\theta(c)$; an abstract edge of δ is correlated with a concrete edge of a concrete diagram which δ abstracts, in an obvious way; the target of ν is o ;

(4) c_I is an element of o ;

(5) e is a finite sequence[†] of distinct elements (called the exits of δ) of o ;

(6) ξ is a finite sequence of distinct elements of A (called the distinguished argument cells);

[†] A mapping $\mu : \omega \to M$ such that $\text{dom}\, \mu = \omega$ or $\text{dom}\, \mu \in \omega$ is a sequence (of elements of M) of length $\text{dom}\, \mu$. If $\text{dom}\, \mu = \omega$, then μ is a (simple) infinite sequence while if $\text{dom}\, \mu \in \omega$ then μ is a finite sequence.

(7) η is a finite sequence of distinct elements of A (called the distinguished result cells):

subject to the axioms

axiom 1: if c is an exit then $\theta(c)$ is the total $\langle 0, 0, 1\rangle$-rank command ;

axiom 2: if $\langle c, 0\rangle$ is the unique edge of an exit c , then $\nu(c, 0) = c$.

The rank of δ is $\langle n, p, q\rangle$, where n (resp. p, q) is the length of ξ (resp. η, e).

As an abbreviation in concrete diagrams, we shall omit recording $\theta(c)$ when c is an occurrence of an exit and we shall omit the concrete edge emanating from an exit. By virtue of axioms 1 and 2, this abbreviation does not produce ambiguity.

A concrete diagram δ consists of a finite set of (concrete) command occurrences each edge of which is extended to meet some command occurrence in the set. This much of the description of δ specifies o , θ , ν . Each extended edge may be interpreted as expressing, "If this edge is selected, execute next the command at the location indicated by following the sense of the arrow-head." The manner of specifying the initial command occurrence, the exits, the distinguished argument and result cells, is quite ad hoc. It is indicated below where diagrams are used. Another linguistic method to express the kind of algorithm expressed by a concrete diagram may be based on the method used by Post in 1936. In this method, distinct names are introduced for each command occurrence, say in the form of a list with a name on the left and a command occurrence on the right. The set o is then the set of names and θ assigns to an element of o an occurrence of an imperative sentence. The direct transition function may be represented, say, by a (finite) table completely separate from the list. Alternatively, the direct transition function information may be incorporated into the list.

A computation of the diagram[†] $\delta = \langle o, \theta, \nu, c_I, e, \xi, \eta \rangle$ (or execution of the algorithm which the diagram expresses)

[†] More strictly (consult discussion of Euclidean algorithm at end of this section) this notion should be associated with the machine which executes the diagram algorithm rather than the algorithm itself. If our main considerations demanded that we make the distinction, we would define a computation of the algorithm to be a sequence: $\langle k_0, c_0, n_0 \rangle, \langle k_1, c_1, n_1 \rangle, \ldots, \langle k_i, c_i, n_i \rangle, \langle k_{i+1}, c_{i+1}, n_{i+1} \rangle, \ldots$ such that for all i

(1) $\Delta^{(2)}_{c_i}(k_i) = n_i$

(2) $\nu(c_i, n_i) = c_{i+1}$

(3) $\Delta^{(1)}_{c_i}(k_i) = k_{i+1}$.

is a sequence (finite or infinite) $\sigma_0, \sigma_1, \ldots, \sigma_i, \sigma_{i+1}, \ldots$ such that each σ_i is an ordered pair $\langle k_i, c_i \rangle$, where $k_i \in K$, $c_i \in o$ and

1) if $\Delta_{c_i}(k_i) = \langle k', j \rangle$ and $\nu(c_i, j) = c'$, then $k_{i+1} = k'$ and $c' = c_{i+1}$;

2) $c_0 = c_I$.

The sequence $k_0, k_1, \ldots, k_i, k_{i+1}, \ldots$ will be called the track of the above computation, while the sequence $c_0, c_1, \ldots, c_{i+1}, \ldots$ is called its trace. A computing diagram computation is called successful iff the codomain of its trace contains an exit; i. e., if there exists an i such that c_i is an exit where $c_0, c_1, \ldots$ is the trace of the computation. If c_i is an exit, then it follows from axioms 1, 2 that $\sigma_i = \sigma_{i+1} = \sigma_{i+2} = \ldots$.

Given any $k \in K$, there is exactly one computation, call it comp k, of maximum length such that the initial member of its track is k. If the directions which are constituents of commands of δ are all total, then the length of comp k, for all k, is ω. If the directions are not necessarily total, it is still the case that if comp k is successful then the length of comp k is ω.

If comp k is not successful, then its length may be finite or infinite. Assuming δ has q exits, let $\Delta_\delta : K \to K \times q$ be defined for $k \in K$ iff comp k is successful and $\Delta_\delta(k) = \langle k', j\rangle$ iff $\langle k', e_j\rangle$ is in the codomain of comp k (i.e., appears in the sequence comp k), where e_j is the jth exit of δ.

A diagram $\delta = \langle o, \theta, \nu, c_I, e, \xi, \eta\rangle$ computes the command 3.1 if $\xi = (a_0, \ldots, a_{n-1})$, $\eta = (r_0, \ldots, r_{p-1})$, if the length of e is q and if the following holds: if

$k(a_i) = x_i \in B$, $0 \leq i < n$ then

(1) comp k is successful iff $\langle x_0, \ldots, x_{n-1}\rangle \in \text{dom } d$

(2) if $d^{(1)}(x_0, \ldots, x_{n-1}) = \langle y_0, \ldots, y_{p-1}\rangle$ and if $\Delta_\delta(k) = \langle k', j\rangle$, then

(a) $k'(r_i) = y_i$, for $0 \leq i < p$

(b) $k'(a_i) = k(a_i)$ for those i such that $0 \leq i < n$ and $a_i \neq r_m$ for all m satisfying $0 \leq m < p$

(c) $d^{(2)}(x_0, \ldots, x_{n-1}) = j$.

We give an example of a concrete diagram which defines, via the Euclidean algorithm, the greatest common divisor function (gcd).

If δ computes 3.1, we say <u>δ computes d at argument cells $(a_0, \ldots, a_{n-1})$ and result cells $(r_0, \ldots, r_{p-1})$</u>. If δ computes 3.1 (for some $\underline{a}$, r) we also say <u>δ defines d</u>.

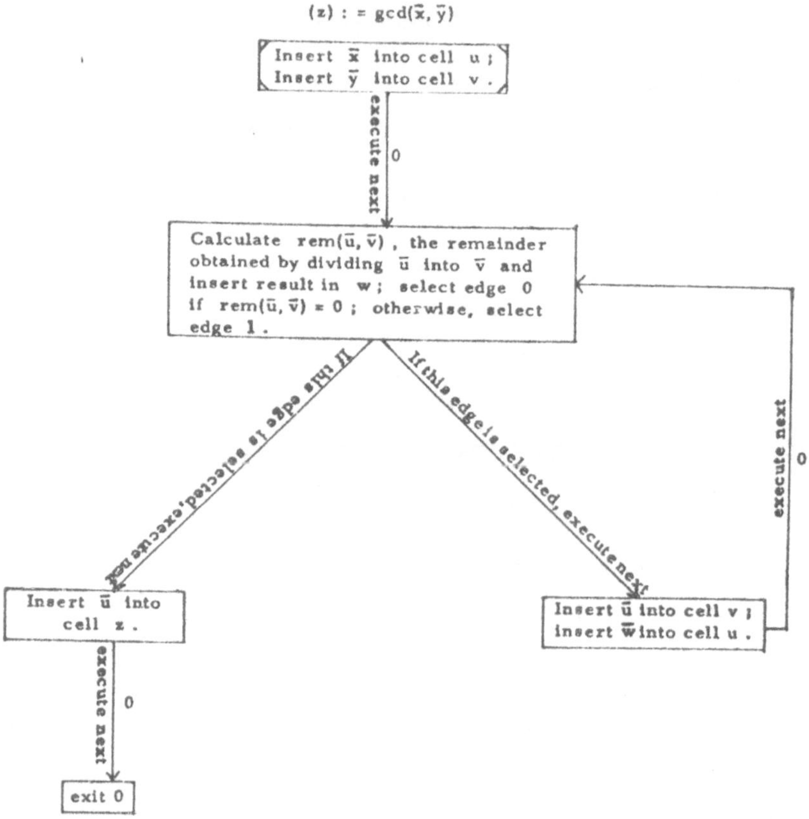

The above diagram may be abbreviated as follows.

$(z) := \gcd(\bar{x}, \bar{y})$

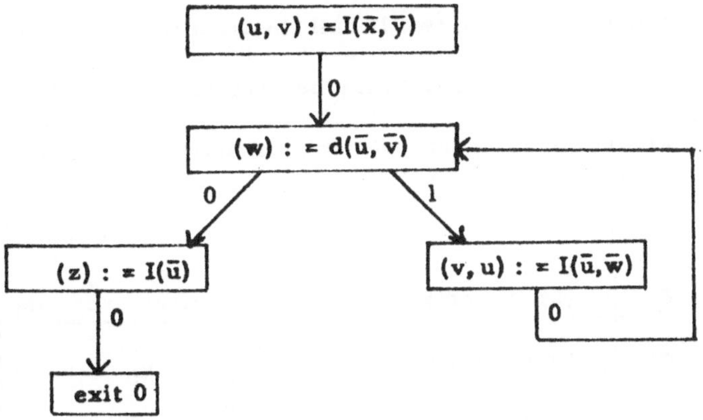

where $d: P^2 \to P \times 2$, and where P is the set of positive integers, $d^{(1)} = \text{rem}$ and $d^{(2)}(p, q) = 0$ if p divides q, $d^{(2)}(p, q) = 1$ otherwise. We use I ambiguously for the identity function $I: P \to P$ and the identity function $I: P^2 \to P^2$; the context is adequate to distinguish the two uses.

In the cases where a rectangle has but one edge emanating from it, we have inscribed "execute next" although the longer form "if this edge is selected, execute next" would have been equally appropriate. It would perhaps have been clearer (but obvious considerations forbade it) to inscribe "if this edge is selected, execute next the command at the location indicated by following the sense of the arrowhead."

The linguistic methods based upon Post's 1936 paper alluded to above (we'll call the first one to be discussed the "two-table method") may be used to express the Euclidean algorithm for greatest common divisor as follows.

θ	$(z) := \gcd(\overline{x}, \overline{y})$
c_0	$(u, v) := I(\overline{x}, \overline{y})$
c_1	$(w) := d(\overline{u}, \overline{v})$
c_2	$(z) := I(\overline{u})$
c_3	$(v, u) := I(\overline{u}, \overline{w})$
c_4	exit

ν	after executing	execute
Start	c_0	c_1
	c_1	c_2 if the outcome of c_1 is 0; c_3 if the outcome of c_1 is 1
	c_2	c_4
	c_3	c_1
	c_4	

The information given in the θ-table, below the horizontal line, is exactly the information provided by the unconnected set of command occurrences in the concrete diagram above. The information given in the ν-table is exactly the information provided by "connecting" the command occurrences together and indicating the initial command occurrence. It is, of course, possible to combine the information provided by the two tables into a single table. The "one-table method" is now illustrated.

		$(z) := gcd(\bar{x}, \bar{y})$
Start	c_0	$(u, v) := I(\bar{x}, \bar{y})$; execute c_1 next
	c_1	$(w) := d(\bar{u}, \bar{v})$; if outcome is 0 execute c_2 next; if outcome is 1 execute c_3 next.
	c_2	$(z) := I(\bar{u})$; execute c_4 next.
	c_3	$(v, u) := I(\bar{u}, \bar{w})$; execute c_1 next.
	c_4	exit

The pivotal question in connection with the one-table method is this: when one says, for example, "execute c_1 next", what does "c_1" mean? If one takes the position that "c_1" expresses the sense of the imperative sentence which occurs to the right of it (in the table immediately above) then it is clear that the sense of "c_1" depends upon the senses of "c_2" and "c_3" and the sense of "c_3" depends upon the sense of "c_1". Hence "c_1" is a complicated reflexive sentence (i. e., it is a complicated sentence which refers to itself) and the algorithm, at least ostensibly, is not analyzable into "simple atomic" sentences.

On the other hand, concerning the same table, one may attribute to "c_i" the sense which is expressed by the imperative sentence (phrase) which occurs to the right of "c_i" up to the first semicolon. In this case, the single table expresses exactly the same thing as is expressed by the pair of tables above or by the concrete diagrams above and the algorithm expressed is analyzable into "atomic commands" which are areflexive.

We believe that the distinction between the areflexive and the reflexive interpretations of the one-table method is worth considering. In fact, we suggest that, on a superficial level at least, there is a distinction between an algorithm and "the" machine which executes that algorithm and that, roughly speaking, the algorithm corresponds to the areflexive interpretation, while the machine corresponds to the reflexive interpretation. (Cf. [CCE], Section 4, where "machine species" is discussed.) To be more explicit, we describe a machine which executes the Euclidean algorithm.

The machine consists of the pair $\langle \tau, c_0 \rangle$ where

$\tau : K \times \{c_0, c_1, c_2, c_3, c_4\} \rightarrow K \times \{c_0, c_1, c_2, c_3, c_4\}$ and

for all $k \in K$,

$\tau(k, c_0) = \langle \Delta^{(1)}_{c_0}(k), c_1 \rangle$;

$\tau(k, c_1) = \langle \Delta^{(1)}_{c_1}(k), c \rangle$, where $c = c_2$ if $\Delta^{(2)}_{c_1}(k) = 0$;

$c = c_3$ if $\Delta^{(2)}_{c_1}(k) = 1$;

$\tau(k, c_2) = \langle \Delta^{(1)}_{c_2}(k), c_4 \rangle$;

$\tau(k, c_3) = \langle \Delta^{(1)}_{c_3}(k), c_1 \rangle$;

$\tau(k, c_4) = \langle k, c_4 \rangle$. (Recall axioms 1 and 2.)

The use of the machine for computing gcd: $P^2 \to P$ involves specifying:

(1) how the arguments are to be represented (encoded) in the "memory" of the machine (more generally if the set of states of the machine is not a product of memory and control states, how to represent the arguments in the state of the machine); in this case, "the" representation may be given by an arbitrary injection $i : P^2 \to K$ such that $i(p, q) = k$, where $k(x) = p$, $k(y) = q$;

(2) when and where the result is to be found; in this case, the "when" may be answered by specifying c_4 as "exit"; i.e., by "reading" the result when c_4 first occurs as control state and the "where" may be answered by specifying the "decoding" function $\pi : K \to P$ such that $\pi(k) = k(z)$; thus if k is the memory state at the time that c_4 first occurs in a computation of the machine, then the result is $\pi(k)$.

Moreover, defining π_1, π_2 by the requirement $\pi_1\langle k, c\rangle = k$, $\pi_2(k, c) = c$ we have: $\gcd(p, q) = \pi\pi_1\tau^r(i(p, q), c_0)$ where, $r = \min s[\pi_2\tau^s(i)p, q), c_0) = c_4]$.

The reader may note that the first concrete diagram used to express the Euclidean algorithm involves the reflexive phrase, "If this edge is selected, execute next." In this connection, observe first that this one reflexive phrase occurs in all concrete diagrams (either explicitly or implicitly by convention) and no other; therefore, by convention, we may omit the phrase (as was done in the second concrete diagram expressing this algorithm). Secondly, by way of explanation, note that in the diagrams above, command occurrences are used autonomously (i.e., an occurrence of a command is used to denote an occurrence of that command) and that reference to these occurrences are achieved, in effect, by pointing; backward pointing to the "old" command occurrence and forward pointing (the sense of the arrowhead) to the "new" command occurrence. Thirdly, note that reflexivity in the interpretation of the diagram may be completely eliminated by introducing names for the command occurrences. This is illustrated by the Euclidean algorithm expressed by the "diagram" below.

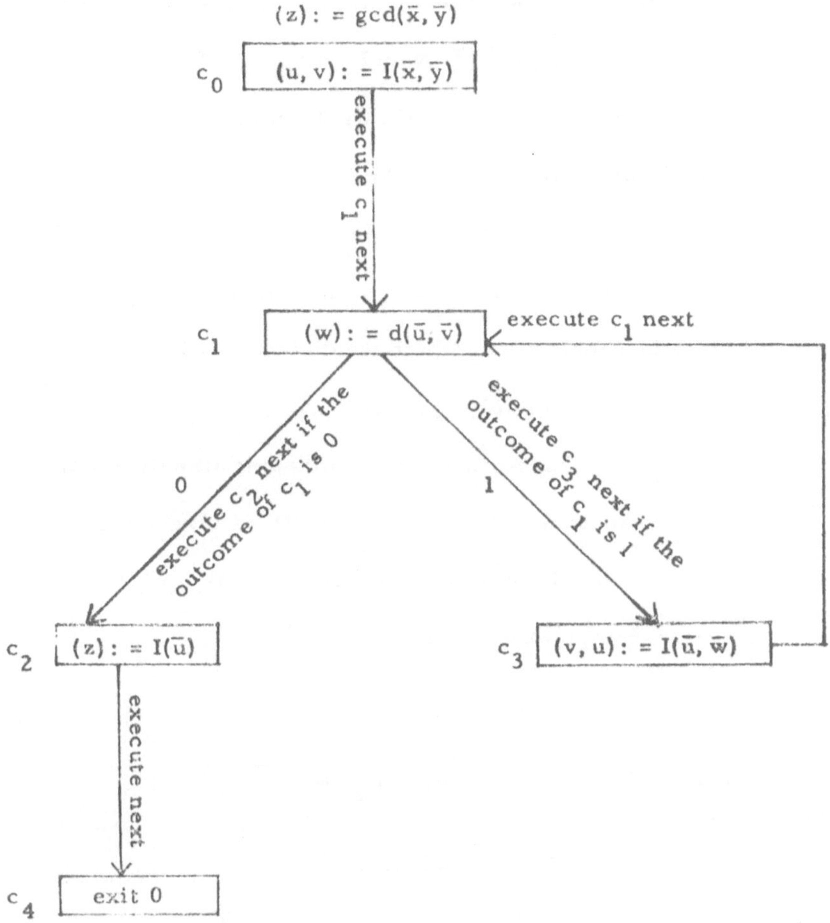

5. DIAGRAM CLOSURE

A direction $d: B^n \to B^p \times q$ is <u>diagram-computable</u> from D, where D is a set of directions over B (which includes the $\langle 0, 0, 1\rangle$ direction), if for every finite sequence ξ (resp. η) of length n (resp. p) of distinct elements of A, there exists a diagram $\langle o, \theta, \nu, c_I, e, \xi, \eta \rangle$ with directions from† D which computes the command $\langle d, \xi, \eta \rangle$. From the form of this definition, there is an à priori possibility that whether or not d is diagram-computable from D depends on the sets A and C. It is easy to see that the requirement that A and C be infinite (a requirement which we do impose) implies that this dependence is only apparent.

Thus we may introduce a mapping (the diagram-closure mapping) which takes a set D of directions over B into the set $\bar{D}$ of directions d (over B) which are diagram-computable from D.

Theorem 1: (1) $D_1 \subseteq D_2 \Longrightarrow \bar{D}_1 \subseteq \bar{D}_2$;

(2) $D \subseteq \bar{D}$;

(3) $\bar{\bar{D}} = \bar{D}$.

The proof of (1) is immediate. The proof of (2), while virtually immediate, is worth indicating if only to illus-

---

†Specifically, where $c \in o$, if $\theta(c) = \langle d', \xi', \eta' \rangle$, then $d' \in D$. The assumption that the $\langle 0, 0, 1 \rangle$ direction is in D is one of convenience. Notice that any $\langle 0, 0, q \rangle$ direction, $q > 0$, is computable from D by a diagram of rank $\langle 0, 0, q \rangle$ whose entrance coincides with an appropriate exit.

trate some linguistic notions. After a short digression, we return to consider (3).

Let $d \in D$, $d: B^n \to B^p \times q$, then the concrete diagram of Fig. 1 defines d.

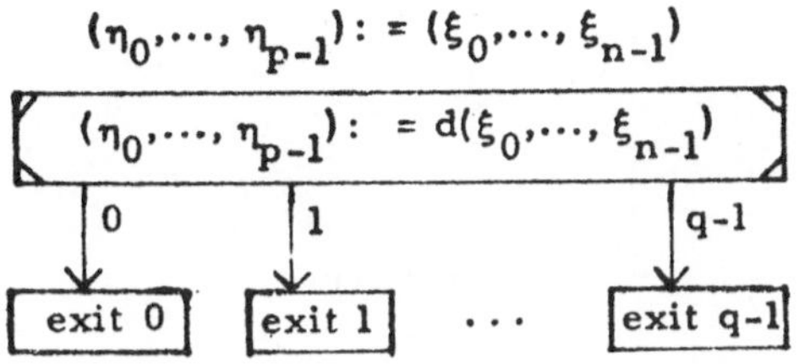

Figure 1

In Fig. 1, the initial command location is indicated by slash marks and the distinguished argument and result cells are indicated above this location (respectively to the right and left of the mark :=). For any ξ, η of the appropriate lengths, this diagram computes the command $\langle d, \xi, \eta \rangle$. Hence $d \in \overline{D}$. The following figure which incorporates the diagram of Fig. 1 may be called a <u>naming</u> or <u>definitional diagram</u> for it is supposed to give a name, vix. "d'" to the direction defined by Fig. 1. Thus, a naming diagram simultaneously defines a direction and introduces a name for it. Here, the symbol "d'" is supposed to be devoid of meaning (and even rank) before the naming diagram is introduced.

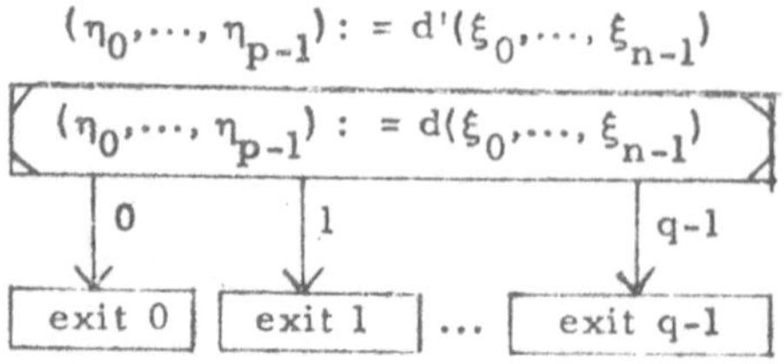

Figure 2

Thus if "d'" is defined by Fig. 2, then $d' = d$.

Fig. 3 is an example of a <u>declarative</u> (or <u>assertive</u>) <u>diagram</u>.

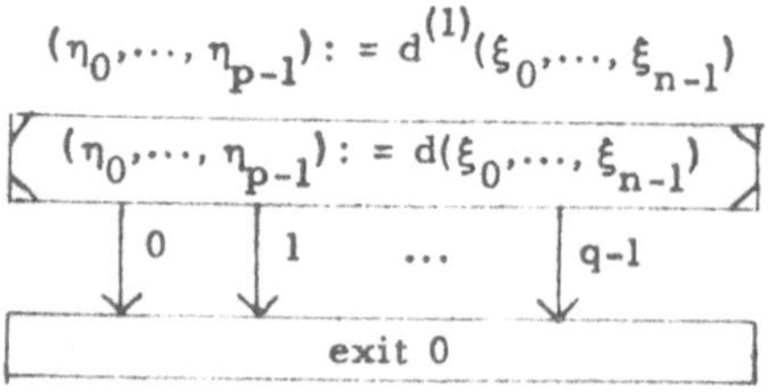

Figure 3

Fig. 3 is a declarative sentence whose sense is: "The direction defined by Fig. 1 is d ." We further illustrate the notion "declarative diagram" by using them to make assertions (Figs. 4, 5, 6).

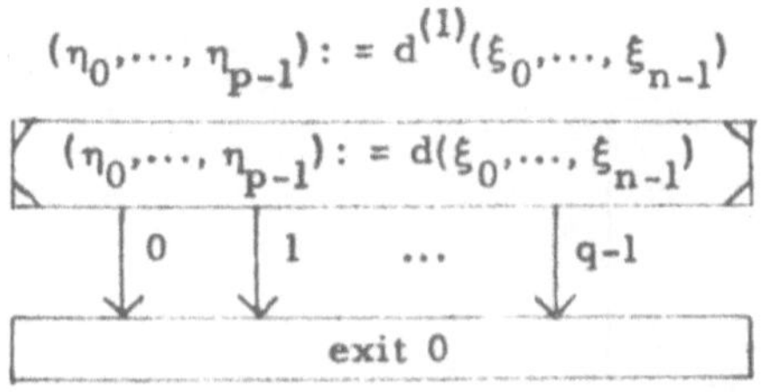

Figure 4

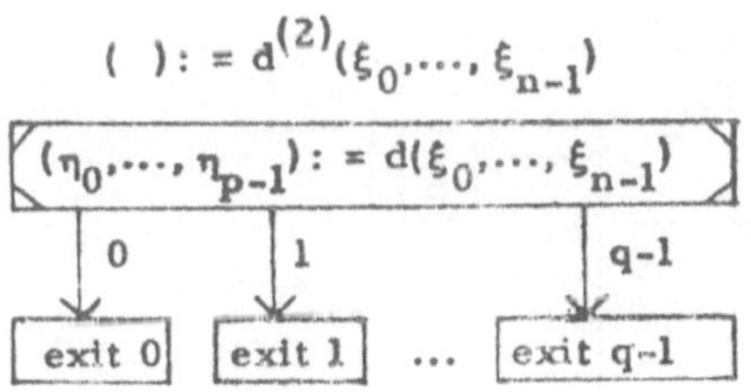

Figure 5

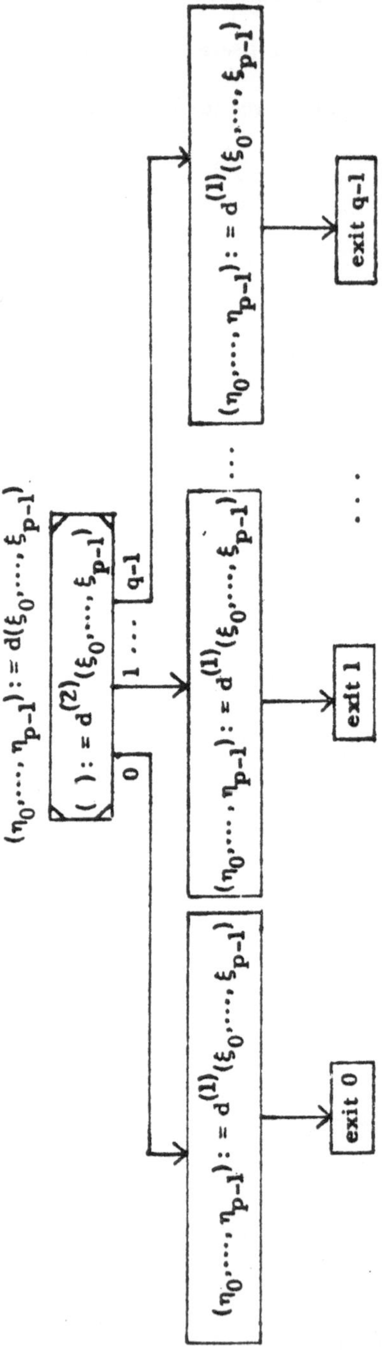

$(\eta_0,\ldots,\eta_{p-1}) := d(\xi_0,\ldots,\xi_{p-1})$
$(\ ) := d^{(2)}(\xi_0,\ldots,\xi_{p-1})$
0
1 . . .
q-1
$(\eta_0,\ldots,\eta_{p-1}) := d^{(1)}(\xi_0,\ldots,\xi_{p-1})$
$(\eta_0,\ldots,\eta_{p-1}) := d^{(1)}(\xi_0,\ldots,\xi_{p-1})$
. . .
$(\eta_0,\ldots,\eta_{p-1}) := d^{(1)}(\xi_0,\ldots,\xi_{p-1})$
exit 0
exit 1
. . .
exit q-1

Figure 6

We return now to Theorem 1 (3) which is fundamental and outline an argument. Let A_δ mean the set of cells whose names appear in δ. This includes the distinguished argument and result cells. Let $d \in \overline{\overline{D}}$. Then d is defined by a diagram δ with directions[†] in $\overline{D}$. Suppose "d'" occurs in δ and $d' \in \overline{D} \sim D$. Fix on some occurrence, say c, (not an exit) of some command, say $\langle d', \xi, \eta \rangle$, which employs d'. Then there is a δ' with directions in D which computes $\langle d', \xi, \eta \rangle$. In fact, because the set A is infinite, we may choose δ' in such a way that its set of <u>auxiliary cells</u> (the set of cells in A_δ which are not distinguished argument or result cells of δ') is disjoint from A_δ; i.e., $A_\delta \cap (A_{\delta'} \sim (\text{cod } \xi \cup \text{cod } \eta)) = \phi$. Similarly, we may assume $o_\delta \cap o_{\delta'} = \phi$. One may construct a diagram δ'' which defines d (as δ does) by <u>substituting</u> δ' for c in δ. The resulting diagram δ'' has fewer occurrences of commands whose directional part is in $\overline{D} \sim D$. Therefore, iteration of this construction ultimately yields a diagram defining d whose directions are all in D. We indicate the construction of δ''.

1. $o_{\delta''}$ is obtained by taking the union of o_δ and $o_{\delta'}$ and deleting c and the exits of δ'.

[†] We may assume that the initial command location of δ is distinct from the exits of δ.

2. $\theta_{\delta''}(x) = \theta_{\delta}(x)$ if $x \in o_{\delta}$; $\theta_{\delta''}(x) = \theta_{\delta'}(x)$ if $x \in o_{\delta'}$.

3. If x is an edge in δ but not an edge of c and if (a) $\nu_{\delta}(x) \neq c$, then $\nu_{\delta''}(x) = \nu_{\delta}(x)$, (b) $\nu_{\delta}(x) = c$, then $\nu_{\delta''}(x) = c'$ where c' is the initial command location of δ' . If x is an edge in δ' but not an exit edge and if (a) $\nu_{\delta'}(x)$ is not an exit, then $\nu_{\delta''}(x) = \nu_{\delta'}(x)$, (b) $\nu_{\delta'}(x)$ is the jth exit, then $\nu_{\delta''}(x) = \nu_{\delta}(c, j)$.

4. If c_I is the initial command location of δ and $c_I \neq c$, then the initial command location c'' of δ'' is c_I ; otherwise, $c'' = c'$.

5, 6, 7. The exit, argument and result sequences of δ'' are the same as those of δ .

By virtue of Theorem 1 holding, we may refer to $\overline{D}$ as the diagram-closure of D and call a set D diagram-closed if $D = \overline{D}$.

Even though δ' computes the command $\langle d', \xi, \eta \rangle$, we cannot assign to δ' the same denotation as $\langle d', \xi, \eta \rangle$ for, in general, one can find a diagram δ whose auxiliary cells are not disjoint from the auxiliary cells of δ' and which contains an occurrence of $\langle d', \xi, \eta \rangle$ such that replacement of an occurrence of $\langle d', \xi, \eta \rangle$ in δ by an occurrence of δ' would produce a diagram δ'' which does not compute the same command as δ . But "denotation" has the property that if two names have the same denotation, then replacement of one by the other in a third name produces a fourth name with the same denotation as the third (cf. [AC]).

To return for a moment to the programming language motivation, consider a finite sequence $\delta_0, \delta_1, \ldots, \delta_n$ of definitional diagrams, where δ_i defines d_i , with the property that "d_j" does not occur in δ_i for $0 \leq i \leq j \leq n$. More specifically, the directional constituents of commands in δ_i are drawn from the set $D \cup \{d_0, \ldots, d_{i-1}\}$. Then, in a natural way, the finite sequence of diagrams defines d_n from D . Moreover, the finite sequence of diagrams may be regarded, again in a natural way, as a partial specification of how d_n is to be computed.

6. FIRST PROPERTIES OF DIAGRAM-CLOSURE AND DIAGRAM-CLOSED SETS

(1) Proposition: If $d \in \overline{D}$ and d' is obtained from d by permuting arguments or by permuting results or by permuting outcomes, then d' is in $\overline{D}$. More precisely, if π, π', π'' are respectively permutations of n, p, q where $\langle n, p, q\rangle$ is the rank of d , and if

$$d'(x_0, \ldots, x_{n-1}) = \langle\langle y_0, \ldots, y_{p-1}\rangle, j\rangle \iff$$

$$d(x_{\pi(0)}, \ldots, x_{\pi(n-1)}) = \langle\langle y_{\pi'(0)}, \ldots, y_{\pi'(p-1)}\rangle, \pi''(j)\rangle$$

then $d \in \overline{D} \implies d' \in \overline{D}$. Indeed $d' \in \{\overline{d}\}$.

(2) Proposition: If d' is obtained from d by the addition or suppression of "dummy variables" then $d \in \overline{D} \implies d' \in \overline{D}$. In fact, $d' \in \{\overline{d}\}$.

(3) Proposition: If $f: B^n \to B^2$ (more strictly, $f: B^n \to B^2 \times 1$) is, say, total and if $f_1: B^n \to B$, $f_2: B^n \to B$ are defined by the requirement that $f(x) = \langle f_1(x), f_2(x)\rangle$ for all $x \in B^n$, then $f_1, f_2 \in \{\overline{f}\}$. For example,

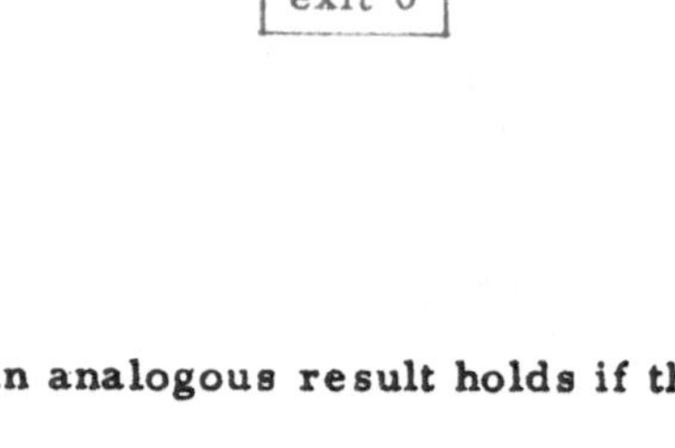

Of course, an analogous result holds if the result rank is greater than 2. It is, however, not always the case that $f \in \{\overline{f_1, f_2}\}$ as shown by the following example.

(4) <u>Example</u>: Let $B = \omega$. Let $f(x, y) = \langle 2xy, 3xy \rangle$ where juxtaposition indicates ordinary multiplication. Then $f_1(x, y) = 2xy$ and $f_2(x, y) = 3xy$. There is no problem in computing the command $\langle f, \xi, \eta \rangle$ where $\xi = \langle \xi_0, \xi_1 \rangle$, $\eta = \langle \eta_0, \eta_1 \rangle$ provided either η_0 or η_1 is distinct from both ξ_0, ξ_1. The reader may verify, though, that the command $\langle f, \xi, \xi \rangle$ is not diagram-computable from $\{f_1, f_2\}$. Thus $f \notin \overline{\{f_1, f_2\}}$.

(5) In view of (4), it is natural to inquire: If the command $\langle f, \xi, \xi \rangle$ is computable from D, does it follow that for all η, the command $\langle f, \xi, \eta \rangle$ is computable from D? This also fails. <u>For example,</u> if D consists solely of the $\langle 0, 0, 1 \rangle$ direction and <u>iden</u> is the total function which maps B identically onto B, then the command $\langle$<u>iden</u>$, \xi, \xi \rangle$ is clearly computable but, if $\eta \neq \xi$, then the command $\langle$<u>iden</u>$, \xi, \eta \rangle$ is not.

(6) On the other hand, it is quite clear that

<u>Proposition</u>: <u>If</u> <u>iden</u> $\in \overline{D}$ <u>and if</u> $\langle d, \xi, \eta \rangle$ <u>is computable from D for some</u> ξ, η, <u>then</u> $\langle d, \xi, \eta \rangle$ <u>is computable from D for all</u> ξ, η.

(7) A somewhat similar situation prevails with respect to identification of arguments.

Proposition: If $iden \in \bar{D}$, then $\bar{D}$ is closed under identification of variables; i.e., if $d \in \bar{D}$ and d' is defined by the requirement

$$d'(x_0,\ldots,x_i,\ldots,x_{j-1},x_{j+1},\ldots,x_{n-1}) = d(x_0,\ldots,x_i,\ldots,x_{j-1},x_i,x_{j+1},\ldots,x_{n-1})$$

then $d' \in \bar{D}$. One can, however, find sets D such that $\bar{D}$ is not closed under identification of arguments. For example, if $f:\omega^2 \to \omega$ is ordinary multiplication and $D = \{f\}$, then $\bar{D}$ is not closed under identification of variables because it does not contain the squaring function. To see this one need only keep in mind that the commands of any diagram with directions in D involves two distinct argument cells. It is natural to inquire whether the assumption that D is closed under identification of variables (in which case, we may as well permit, for the sake of simplicity, commands whose argument cells are not necessarily distinct) implies that $\bar{D}$ is. The matter is resolved only in part by the following proposition.

(8) Proposition: If D is closed under identification of variables, $d: B^n \to q$, $n > 1$, is in $\bar{D}$ and if

$d' : B^{n-1} \to q$ is obtained from d by identification of variables, then $d' \in \bar{D}$. The proof of this proposition follows readily from the Lemma in Section 7 which is needed in another connection as well.

(9) Proposition: For any D, the set of functions of result rank 1 in $\bar{D}$ is closed under substitution (more precisely, under the substitution schema in [SCK]). For example, let $h : B^3 \to B$ be defined from g, f_0, f_1 as follows:

$$h(x_0, x_1, x_2) = g(f_0(x_0, x_1, x_2), f_1(x_0, x_1, x_2))$$

Then for any cell η_0 and any sequence ξ_0, ξ_1, ξ_2 of length 3 of distinct elements:

$$(\eta_0) : = h(\bar{\xi}_0, \bar{\xi}_1, \bar{\xi}_2)$$

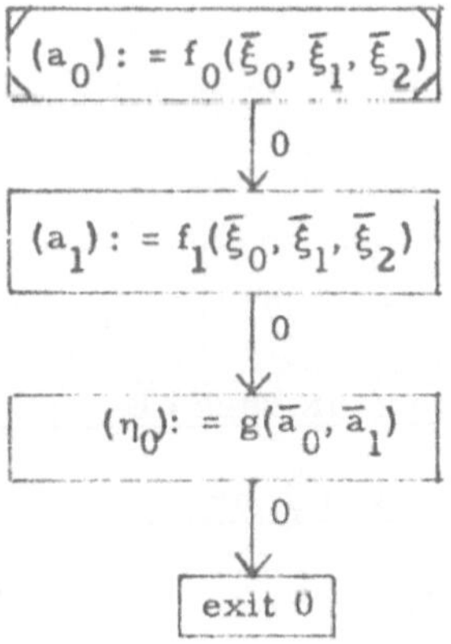

provided cells a_0, a_1 are distinct from each other as well as from ξ_0, ξ_1, ξ_2 .

(10) Proposition: (a) Every direction with an empty graph is in D . (b) If D consists of functions alone, if $f \in \overline{D}$, $f: B^n \to B$ and graph $f \neq \phi$, then there is a term T such that $f(x_0, \ldots, x_{n-1}) \equiv T$, where "$T_1 \equiv T_2$" means all simultaneous valuations of the left- and right-hand members are equal and where T is constructed from "atomic terms", $d(y_0, \ldots, y_{m-1})$, where $d \in D$ and $y_0, \ldots, y_{m-1}$ are distinct variables chosen from a given infinite list of variables, by repeated substitution. If D contains functions of argument rank n or less, then T may be chosen so that the set of variables which appear in it is a subset of $\{x_0, \ldots, x_{n-1}\}$.

(11) It follows from (9), (7), (2), (1) that

Proposition: If $\text{iden} \in \overline{D}$, then the set of functions in D of result rank 1 is closed under explicit definability.

(12) Let D consist solely of total directions of result rank 0 and outcome rank 2 so that D may be identified with a set of total (two-valued) relations. (Not necessarily total relations are considered in Section 8.)

<u>Proposition:</u> <u>If $R: B^n \to \{T, F\}$, $R \in \overline{D}$, is total then $R(x_0, \ldots, x_{n-1})$ is equivalent to a truth functional combination of terms $R_i(y_0, \ldots, y_{m-1})$, where $R_i \in D$</u> .

(13) <u>Proposition:</u> If $B = \omega$ and if $D = \{S, \underline{const}_0, \underline{eq}\}$, (where $\underline{S}$ is the successor function, $\underline{const}_0$ is the total function of argument rank 0 whose value is 0 and $\underline{eq}$ is the rank $\langle 2, 0, 2\rangle$ direction which corresponds to equality between natural numbers, i.e., $d^{(2)}(x, y) = 0 \iff x = y$, where $x, y \in \omega$, then

(1) if $f: \omega^n \to \omega$, then $f \in \overline{D} \iff f$ is partial recursive

(2) if $R: \omega^n \to \{T, F\}$ is total, then $R \in \overline{D} \iff R$ is recursive.

In fact, if f is the total function defined in terms of the total function h , by the following requirements

(a) $f(0) = m$

(b) $f(S(x)) = h(x, f(x))$

then

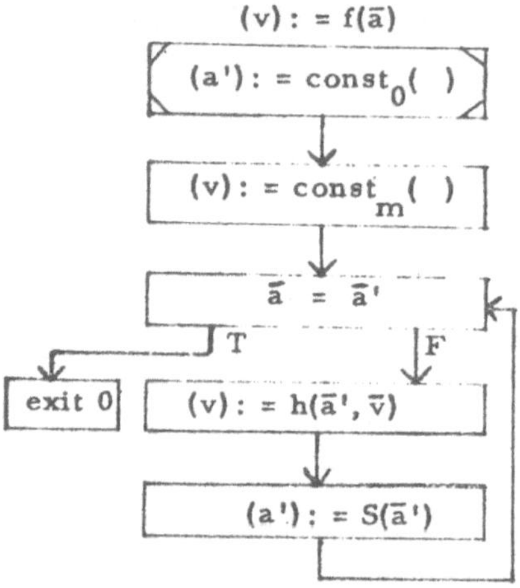

so that $f \in \{S, \underline{const}_0, \underline{eq}, h\}$. Here we have used

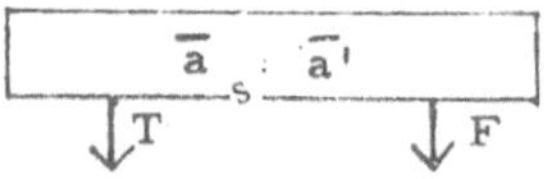

as an "abbreviation" for

Also, if R is total and f is defined by

f(x) = $\mu y R(y, x)$ = the least y such that R(y, x)

then

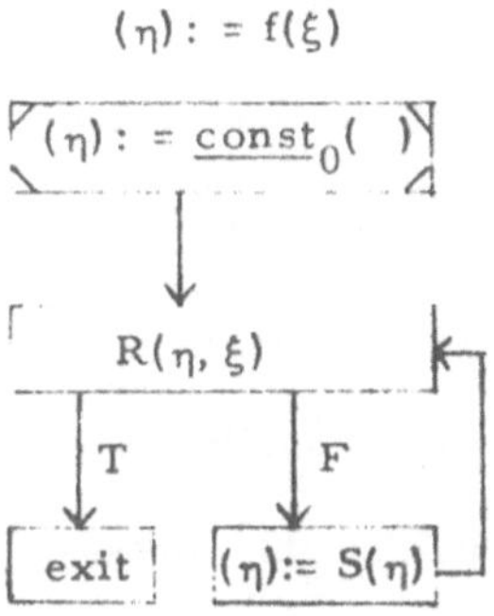

so that $f \in \overline{\{S, \underline{const}_0, \underline{eq}, R\}}$.

(14) We remark further that the set $D = \{S, \underline{const}_0, \underline{eq}\}$ is minimal in that if D' is a proper subset of D then $\overline{D}'$ is a proper subset of $\overline{D}$. The set D is a finite <u>diagram-base</u> for the set of partial recursive functions. One readily supplies others.

(15) If f_{eq} is the total direction of rank $\langle 2, 1, 1\rangle$ such that $f^{(1)}_{eq}(x, y) = 0$ if $x = y$ and $f^{(1)}_{eq}(x, y) = 1$ if $x \neq y$, then $\{S, const_0, f_{eq}\}$ is <u>not</u> a diagram-base for the set of partial recursive functions. (cf. (10)(b))

7. A HELPFUL LEMMA

Suppose δ computes the command $\langle d, \xi, \eta \rangle$ and suppose $x_0, \ldots, x_{m-1}$ is a list (without repetitions) of all the argument cells of this command which are not also result cells. Then we know from the definition of this concept that at the end of any (finite) successful computation of δ the content of $x_0, \ldots, x_{m-1}$ is the same as the content of these cells at the beginning of the computation. This, of course, does not rule out the possibility that during the course of such a computation the contents of these cells undergo some change. If, say, the content of x_i undergoes change during the course of a computation, then x_i must appear as a result cell in some command of δ. Equivalently, if x_i does not appear as a result cell in any command of δ, then during the course of any computation of δ the content of x_i remains fixed. It is, however, possible that while x_i appears as a result cell in some command of δ, that command is never employed in any computation of δ; if this is true for each command in which x_i occurs as a result cell, then, of course, it is again the case that the content of x_i remains fixed during the course of every computation of δ. Let $x'_0, \ldots, x'_{m-1}$ be distinct elements of A such that $\{x'_0, \ldots, x'_{m-1}\} \cap A_\delta = \phi$.

Lemma. If $\langle d, \xi, \eta \rangle$ is computed by δ then there exists a δ' utilizing the same directions as δ and which also computes $\langle d, \xi, \eta \rangle$ but has the further property: for all i satisfying $0 \leq i < m$, if x_i is an argument cell but not a result cell of the command $\langle d, \xi, \eta \rangle$ then in no command of δ' does x_i appear as a result cell.

We do not give a detailed proof but indicate the construction. The idea of the construction is to introduce additional command locations which have the effect of "remembering" when in the course of a computation of δ, a command has x_i as a result cell, modifying such a command the first time it is employed in the computation by replacing any x_i which appears as a result cell by x_i' and on subsequent uses of any command employing x_i, replacing both the occurrences of x_i as argument and result cells by x_i'. The desired diagram δ' is obtained from the diagram δ'' briefly indicated below, by proper identification of exits. Let m be as above and let S be the set of all subsets of m. Then we take

(1) $\quad o_{\delta''} = o_\delta \times S$

(2) $\quad$ If $c \in o_\delta$ is not an exit of δ, $s \in S$, then we

define $\theta''(c, s)$ to be the command obtained from $\theta(c)$ by

(a) replacing any occurrence of x_i, $0 \leq i < m$, as result cell in $\theta(c)$ by x_i'

(b) if $j \in s$ and x_j occurs as an argument cell of $\theta(c)$, then replace x_j by x_j'.

(3) Of ν'' we require: $\nu''(\langle c, s\rangle, i) = \langle \nu(c, i), s'\rangle$ where s' is the union of s and the set of indices j such that x_j is a result cell of $\theta(c)$.

(4) We take $c_I'' = \langle c_I, \phi \rangle$.

(5) The exits of δ'' are the elements of $E \times S$ where E is the set of exits of δ; their order is irrelevant. δ' is obtained from δ'' merely by identifying all the exits of δ'' with the same first component and ordering the exits as follows: $\{e_0\} \times S$, $\{e_1\} \times S, \ldots$

8. DIAGRAM-COMPUTABLE AND -SEMICOMPUTABLE SETS

We say a subset E of B^n, $n > 0$ is <u>diagram-computable</u> (<u>resp. diagram-semicomputable</u>) from D

iff the characteristic relation (resp. semirelation) of E is computable.

Theorem. If D is a set of total directions over B, or more generally, if there exists a set of total directions D' over B such that $\bar{D} = \bar{D}'$, then a subset E of B^n is diagram-computable from D iff E and $B^n \sim E$ are both diagram-semicomputable from D.

For the proof, the notion of the consensus of two commands is introduced. Let $\langle d, \xi, \eta \rangle$, $\langle d', \xi', \eta' \rangle$ be two commands. The consensus is defined iff the sets of result cells of the two commands is disjoint; i.e., iff $\mathrm{cod}\ \eta \cap \mathrm{cod}\ \eta' = \phi$. In this case the consensus is $\langle d'', \xi'', \eta'' \rangle$ where η'' is the sequence obtained by following (concatenating) the sequence η with the sequence η'; ξ'' is obtained by following ξ by the elements in $\mathrm{cod}\ \xi' \sim \mathrm{cod}\ \xi$ in the order in which they appear in $\mathrm{cod}\ \xi'$ and where the set q'' of outcomes of d'' is the ordinary numerical product $q'' = qq'$ of the sets of outcomes of d and d'. It will facilitate the exposition to assume some definite bijective correspondence is given between q'' and the set of ordered pairs $q \times q'$ and use as labels of command edges, elements of $q \times q'$ rather than elements of q''. The direction d'' is then defined as follows:

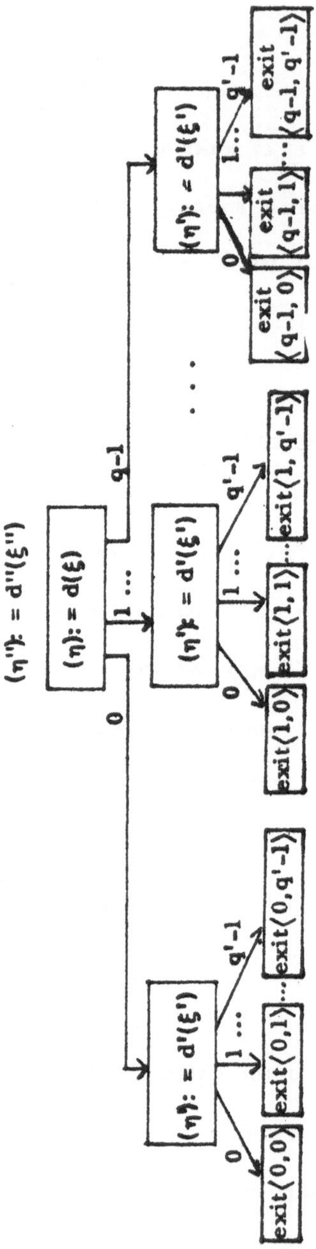
(η''): = d''(ξ'')
(η): = d(ξ)
0
1 ...
q-1
(η'): = d'(ξ')
0
1 ...
q'-1
exit⟨0,0⟩
exit⟨0,1⟩
exit⟨0,q'-1⟩
(η'): = d'(ξ')
0
1 ...
q'-1
exit⟨1,0⟩
exit⟨1,1⟩
exit⟨1, q'-1⟩
(η'): = d'(ξ')
0
1...
q'-1
exit ⟨q-1, 0⟩
exit ⟨q-1, 1⟩
exit ⟨q-1, q'-1⟩

We note that if each of two commands is diagram-computable from D' , then so is their consensus (cf. Section 6). Roughly speaking, the effect of executing the consensus of two commands is the effect of simultaneously executing each of the two commands. Suppose δ and δ' respectively diagram-compute (from D') $E^{\frac{1}{2}}$ and $(B^n \sim E)^{\frac{1}{2}}$ at argument cells $\underline{a} \in A^n$ We wish to construct a diagram which diagram-computes (from D') $E^{\frac{1}{2}+\frac{1}{2}}$. We may assume the sets of auxiliary cells of δ and δ' are disjoint and that (from the Lemma of the previous section) no a_i , $0 \leq i < n$, appears as a result cell in any command of either δ or δ' . It thus follows that the set of result cells of δ is disjoint from that of δ' .

Let

$\delta = \langle o, \theta, \nu, c_I, e, \underline{a}, r \rangle$, where dom $r = \phi$,

$\delta' = \langle o', \theta', \nu', c'_I, e', \underline{a}, r \rangle$

$\delta'' = \langle o'', \theta'', \nu'', c''_I, e'', \underline{a}, r \rangle$, where e and e' are sequences of length one (which we identify with the elements themselves) and where, letting $c \in o \sim \{e\}$, $c' \in o' \sim \{e'\}$,

(1) $o'' = o \times o'$;

(2) $\theta''(c, c')$ is the consensus of $\theta(c)$ and $\theta(c')$; this

is defined since the result cells of $\theta(c)$ and $\theta(c')$ are disjoint;

(3) $\nu''(\langle c, c'\rangle, \langle j, j'\rangle) = \langle \nu(c,j), \nu'(c', j')\rangle$;

(4) $c''_I = \langle c_I, c'_I\rangle$;

(5) cod $e'' = (\{e\} \times o') \cup (o \times \{e'\})$; (the ordering of cod e'' is irrelevant).

Let $k \in K$ be arbitrary. Let $\sigma'' = \text{comp}_{\delta''} k = (\sigma''_0, \sigma''_1, \ldots)$. Since the elements of D' are total, the length of σ'' is ω. Let $\sigma = \text{comp}_\delta k$, $\sigma' = \text{comp}_{\delta'} k$. As usual we let $\sigma_i = \langle k_i, c_i\rangle$, $\sigma'_i = \langle k'_i, c'_i\rangle$, $\sigma''_i = \langle k''_i, c''_i\rangle$. One readily verifies that for each i : if $x \in A_\delta$, then $k''_i(x) = k_i(x)$; if $x \in A_{\delta'}$, then $k''_i(x) = k'_i(x)$; $c''_i = \langle c_i, c'_i\rangle$.

Thus $\text{comp}_{\delta''} k$ is successful iff either $\text{comp}_\delta k$ or $\text{comp}_{\delta'} k'$ is successful. A diagram δ'' which computes the characteristic relation $E^{\frac{1}{2}+\frac{1}{2}}$ of E is obtained from δ'' by identifying the elements of the set $\{e\} \times o'$ and identifying the elements of the set $o \times \{e'\}$ and appropriately labelling them. Thus $E^{\frac{1}{2}+\frac{1}{2}}$ is diagram-computable from D' (also from D). It is easy to see that $E^{\frac{1}{2}}$ (resp. $(B^n \sim E)^{\frac{1}{2}}$) is diagram-computable from $E^{\frac{1}{2}+\frac{1}{2}}$, so that if $E^{\frac{1}{2}+\frac{1}{2}}$ is diagram-computable from D so is $E^{\frac{1}{2}}$ (resp. $(B^n \sim E)^{\frac{1}{2}}$) for example:

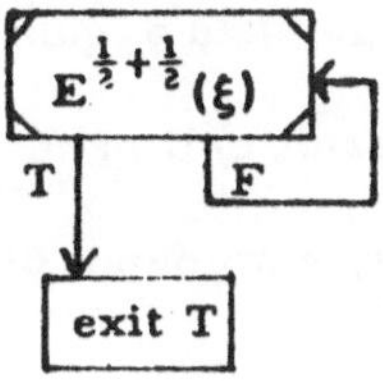

In a similar way, one may prove, for example:

Theorem. If R and R' are diagram-computable from a set of total directions, then so is their strong disjunction (cf. [SCK]).

9. NORMAL DIAGRAMS

It has been pointed out earlier that a diagram doesn't always compute a direction. The restriction to normality or to finite-automaton-normality is a not unreasonable one and results in every diagram computing a direction.

A diagram δ is normal if the trace $c = (c_0, c_1, \ldots)$ of any computation of δ has the property:

(1) for every j, if x is an argument cell of $\theta(c_j)$, then either x is a distinguished argument cell of δ or there exists $i < j$ such that x is a result cell of $\theta(c_i)$ (where θ is the occurrence-command mapping of δ);

(2) if the computation is successful and

(a) if x is a distinguished argument cell but not a distinguished result cell of δ then x is not a result cell of any command $\theta(c_j)$;

(b) if y is a distinguished result cell which is not a distinguished argument cell, then for some i, y is a result cell of $\theta(c_i)$.

Proposition: Let δ be a normal diagram with distinguished argument cells $(\xi_0, \ldots, \xi_{n-1})$. Let $\underset{\sim}{x}_0, \ldots, \underset{\sim}{x}_{n-1}$ be variables which range over B and let $c = (c_0, c_1, \ldots)$ be the trace of some computation of δ. Let $k \in K$ be any content function such that $k(\xi_i) = \underset{\sim}{x}_i$, $0 \leq i < n$, and suppose the trace of comp k is c. Then for any $\xi \in A$ and any instant $t \in \omega$, there exists a term $T(\underset{\sim}{x}_0, \ldots, \underset{\sim}{x}_{n-1})$ built from $\underset{\sim}{x}_0, \ldots, \underset{\sim}{x}_{n-1}$ and names of the functional parts of directions (including directions with argument rank zero) which expresses the content of ξ at time t; i.e., such that the content of ξ at t is $T(\underset{\sim}{x}_0,\ldots,\underset{\sim}{x}_{n-1})$. The proposition is readily established by induction on the length of a computation.

Corollary: Every normal diagram (of rank $\langle n, p, q\rangle$) computes a direction (of rank $\langle n, p, q\rangle$).

In contradistinction to the notion "normal diagram" the notion "finite-automaton-normal diagram" to be intro duced is purely syntactic and readily recognizable. First we define "path". A (finite automaton) path in a diagram δ is a sequence $(\langle c_0, i_0\rangle, \langle c_1, i_1\rangle, \ldots, \langle c_m, i_m\rangle)$ such that c_0 is the initial command occurrence of δ and for each j, $\langle c_j, i_j\rangle$ is an occurrence-edge of δ and for each j such that $j+1$ is in the domain of the sequence, $\nu(c_j, i_j) = c_{j+1}$ where ν is the direct transition function of δ. The path is successful if $\nu(c_m, i_m)$ is an exit of δ.

A finite-automaton-normal diagram δ is one which satisfies: for every path $(\langle c_0, i_0\rangle, \langle c_1, i_1\rangle, \ldots)$ in δ:

(1) for every j, if x is an argument cell of $\theta(c_j)$, then either x is a distinguished argument cell of δ or there exists $j' < j$ such that x is a result cell of $\theta(c_{j'})$ (where θ is the occurrence-command mapping of δ) and

(2) if the path is successful and

(a) if x is a distinguished argument cell but not a distinguished result cell of δ then x is not a result cell of any command $\theta(c_j)$;

(b) if y is a distinguished result cell which is not a distinguished argument cell then, for some i, y is a result cell of $\theta(c_i)$.

It is easy to see that if a diagram is <u>finite-automaton-normal, then it is normal.</u> Under reasonable conditions if $d \in \overline{D}$, then there is a normal diagram with directional constituents in D which computes d . For example:

<u>Proposition</u>: If there exists $b \in B$ such that $\underline{\text{const}}_b \in \overline{D}$, ($\underline{\text{const}}_b : B^0 \to B$, $\underline{\text{const}}_b(\ ) = b$) then for every $d \in \overline{D}$ there exists a diagram with directional constituents in D which computes d .

<u>Proof</u>: Let δ compute d and let the cells of δ other than the distinguished argument cells be $a_0, a_1, \ldots, a_{r-1}$. By the Lemma of Section 7 we may assume that δ satisfies condition (2) above. Modify δ to δ' by prefixing

the initial command occurrence by the following sequence of commands:

$\boxed{(a_0) := \underline{const}_b()}$ ↓ , $\boxed{(a_1) := \underline{const}_b()}$ ↓ , ... , $\boxed{(a_{n-1}) := \underline{const}_b()}$ ↓

10. GENERATING ALGORITHMS

The notion "(finite cell diagram) generating algorithm" has already been mentioned in Section 4. It is related to "generating diagram" as "computing algorithm" is related to "computing diagram". Generating diagrams do not have any distinguished argument cells.

An abstract generating (finite cell) diagram δ with command locations in C , cells in A and directions in D is a sextuple $\langle o, \theta, \nu, c_I, e, \eta \rangle$, where $o, \theta, \nu, c_I, e, \eta$ are as in Section 4, subject to the axioms:

<u>axiom 1</u> (a) e has length 1 ; (however, we shall identify e with e(0)) ;

(b) η has length greater than or equal to 1 ;

<u>axiom 2</u> $\theta(e)$ is the unique command of rank $\langle 0, 0, 1 \rangle$. The rank of δ is dom η (i. e., the length of η) .

Let $E \subseteq B^p$. A generating diagram <u>generates</u> E if it has rank p and if for any $k \in K$, if $\text{comp}\, k = (\langle k_0, c_0\rangle, \langle k_1, c_1\rangle, \ldots)$ then

$$\langle x_0, \ldots, x_{p-1}\rangle \in E \iff \exists i [c_i = e \wedge k_i(\eta_0) = x_0 \wedge \ldots \wedge k_i(\eta_{p-1}) = x_{p-1}] .$$

A (generating) diagram generates at most one set E. Conditions (1) of the normality definitions in Section 9, give <u>normality definitions for generating diagrams</u>. We note that if the diagram is finite-automaton-normal, the directional constituent of the initial command occurrence of such a diagram has argument rank zero. Moreover, a normal diagram always generates some set.

<u>A set E, $E \subseteq B^n$, is diagram-generable from D</u> if there exists a diagram whose directional constituents are in D, which generates it.

<u>Theorem.</u> If E is generable from D' and each $d \in D'$ is generable from D, then E is generable from D.

The proof is similar to the proof that "$\bar{\bar{D}} = \bar{D}$" indicated in Section 5.

<u>Proposition:</u> Let $E \subseteq B$. Suppose E is generable from D. Suppose further that <u>eq</u> is computable from D. Then E is semicomputable from D.

Proof indication: Let δ generate E at η and let $\xi \in A \sim A_\delta$. A diagram δ' which computes $E_B^{\frac{1}{2}}$ at ξ is obtained from δ by introducing an exit (i. e., introducing an additional occurrence which is designated as an exit), replacing the ejector of δ by an occurrence of the command

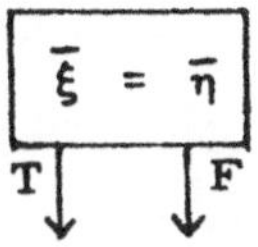

and extending the T-edge of this command occurrence to meet it. The F-edge of this occurrence goes to the same location as the ejector edge of δ. A detailed proof that this construction does what is claimed calls the previous theorem into play.

Under more powerful hypotheses semicomputability implies generability.

The same proposition holds for $E \subseteq B^n$, $n > 1$.

We also note that the notions "definitional diagram" and "assertive diagram" extend to generating diagrams in a natural way. The "assertive" notion is illustrated by schematically indicating the modification of δ to δ' immediately below.

If $\qquad (\eta) :\in E$

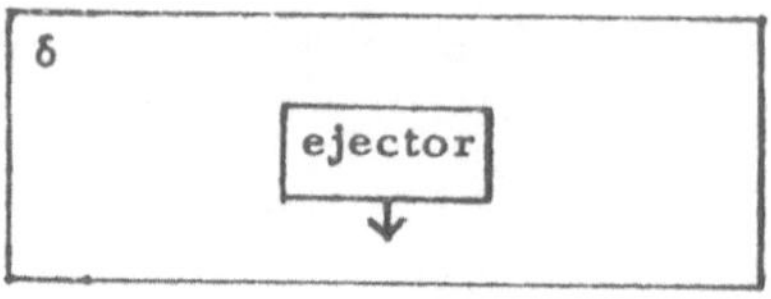

then $\qquad (\ ) := E^{\frac{1}{2}}(\bar{\xi})$

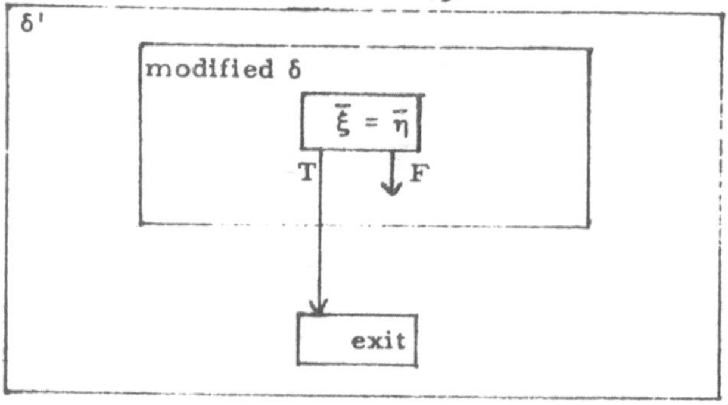

In the case of generating diagrams, the symbol " $:\in$ " is used in place of " $:=$ ", as exemplified above.

Proposition: If $E^n \subseteq B^n$, $n > 0$, is generable from D then so is E.

11. FUNCTIONS AND THEIR GRAPHS

Proposition: Suppose eq is computable from D and $f : B \to B$ then

(1) (a) if f is computable from D and B is

generable from D, then graph f is semicomputable from D;

(b) if f is total as well, then graph f is computable from D;

(2) if graph f is generable from D, then f is computable from D.

The proof is indicated below.

(1) (a) If $(\xi) :\in B$

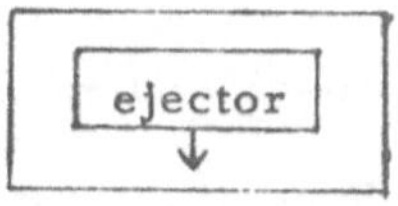

then $(\ ) := (\text{graph } f)^{\frac{1}{2}}(\bar{\xi}', \bar{\eta})$

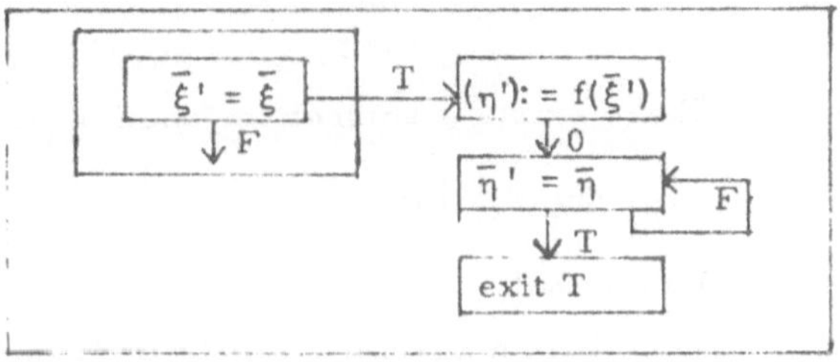

(1) (b) Modify the diagram immediately above by introducing an additional exit marked "F" and connecting the F-edge of $\langle eq, \eta', \eta \rangle$ to this new exit rather than to itself.

(2) If $(\xi, \eta) : \in (\text{graph } f)$

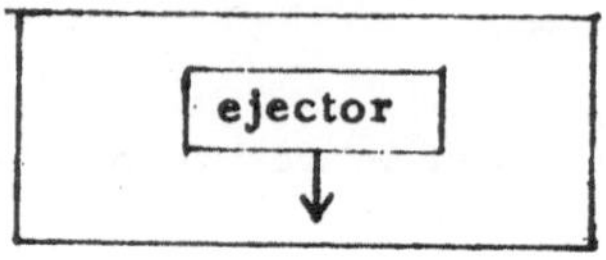

then $(\eta) : = f(\bar{\xi}')$

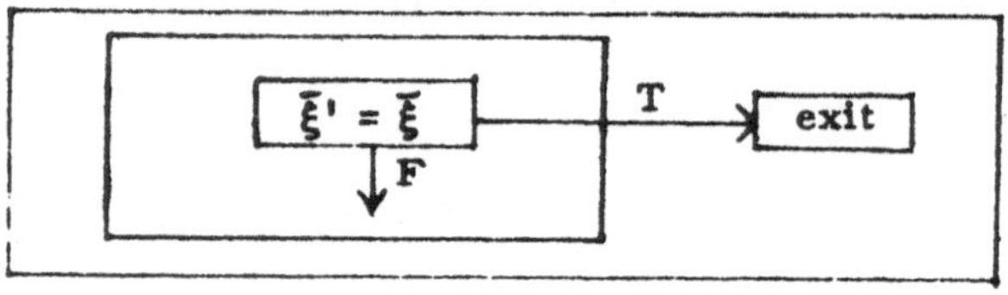

12. CONDITIONS FOR UNIVERSALITY

Let D be a diagram-closed set of directions over B which contains a base of total directions. In what follows, we shall require the following <u>main hypothesis</u>.

H(D) : there exists a surjective function $\psi : B \to F(B)$, where F(B) is the set of all finite subsets of B , (so that B is infinite) such that

1. dom ψ is diagram-computable from D ,

2. $\text{const}_{b_0} \in D$, where b_0 is some element satisfying $\psi(b_0) = \phi$,

3. $f_u \in D$, where $f_u : B^2 \to B$, and if $b \in \text{dom } \psi$ then $\psi(f_u(b, b')) = \psi(b) \cup \{b'\}$,

4. $\underline{EQ} \in D$, where $\underline{EQ}: B^2 \to \{T, F\}$ and if $b, b' \in \mathrm{dom}\ \psi$ then $\underline{EQ}(b, b')$ is defined and $\underline{EQ}(b, b') \iff \psi(b) = \psi(b')$.

We require too

$H'(D)$: B is generable from D .

Note that $H'(D)$ implies that B is at most denumerable. Thus $H(D)$ and $H'(D)$ together imply that B is denumerably infinite.

Proposition 1: Assume $H(D)$. Then $\underline{eq}$ and the relation $\underline{EL}$ are diagram-computable from D where $\underline{EL}$ is defined by the requirement

$$\underline{EL}(b, b') \iff b' \in \psi(b) .$$

The figure below indicates the argument that $\underline{eq} \in D$.

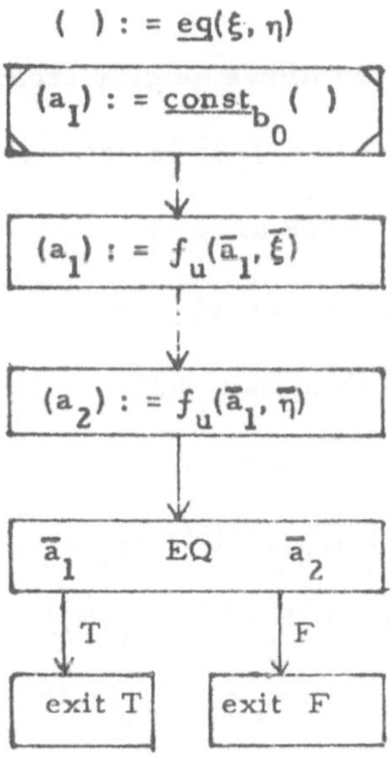

Proposition 2: Assume $H'(D)$. Then, for every $b \in B$, $const_b \in D$.

Proposition 3: Assume $H(D)$, $H'(D)$. Then $iden \in D$. There is a function $f_{\sim} \in D$ which satisfies for $b \in dom\ \psi$:

$$\psi(f_{\sim}(b, b')) = \psi(b) \sim \{b'\} .$$

Proposition 4: Assume $H(D)$, $H'(D)$. The function $J : B^2 \to B$ defined by

$$J(x, y) = f_u(f_u(b_0, f_u(b_0, x)), f_u(f_u(b_0, x), y))$$

is total, injective, is in D, $J^{-1} \in D$ and dom J^{-1} is a subset of dom ψ and is computable from D.

Proof indication. It follows quickly from 3. of $H(D)$ that J is total. The proof that J is injective, which exploits 3. of $H(D)$, is similar to the argument that $\{\{x\}, \{x, y\}\}$ is a "good" definition for $\langle x, y \rangle$. Namely, suppose $J(x, y) = J(x', y')$. Note that $\psi J(x, y)$ has at most two elements. In fact, we have

$\psi J(x, y) = \psi(f_u(b_0, f_u(b_0, x))) \cup \{f_u(f_u(b_0, x), y)\}$ and

$\psi(f_u(b_0, f_u(b_0, x))) = \psi(b_0) \cup \{f_u(b_0, x)\} = \{f_u(b_0, x)\}$.

Let $b_1 = f_u(b_0, x)$ and let $b_2 = f_u(f_u(b_0, x), y)$ so that

$\psi J(x, y) = \{b_1, b_2\}$. Suppose $J(x, y) = J(x', y')$. Then $\psi J(x, y) = \{b_1, b_2\} = \psi J(x', y')$ where $\psi(b_1) = \{x\}$ and $\psi(b_2) = \{x, y\}$ and either

(1) $\psi(b_1) = \{x'\}$ and $\psi(b_2) = \{x', y'\}$, in which case, $x = x'$ and $y = y'$ or

(2) $\psi(b_1) = \{x', y'\}$ and $\psi(b_2) = \{x'\}$, in which case $x = x' = y'$ and $x' = x = y$ so that $y = y'$ as well.

The function $J \in D$ because $f_u \in D$, $\text{const}_{b_0} \in D$, and D is closed under substitution.

The rest of the argument is now indicated. Assume $H(D)$ and $H'(D)$. There exist functions $f_1, f_2 \in D$ with domains which are diagram-computable from D and which are subsets of dom ψ such that $f_1 : B \to B$, $f_2 : B \to B^2$, $b_1 \in \text{dom } f_1 \implies \psi(b_1) = \{f_1(b_1)\}$ and $b_2 \in \text{dom } f_2 \wedge f_2(b_2) = \langle x, y\rangle \implies \psi(b_2) = \{x, y\}$. Then dom $J^{-1} \iff \exists b_1, b_2, x, y[\psi(b) = \{b_1, b_2\} \wedge [(\psi(b_1) = \{x\} \wedge \psi(b_2) = \{x, y\}) \vee (\psi(b_2) = \{x\} \wedge \psi(b_1) = \{x, y\})]]$. It is convenient to refer to an element b in the domain of ψ as a <u>code</u> (or Gödel number) for $\psi(b)$. Then, every finite subset of B has at least one code. Making use of J as well as ψ, one can also introduce "codes" for all finite subsets of B^2. Let ψ_J be the mapping with source B and target the set of all finite subsets of B^2 defined as follows:

$$x \in \operatorname{dom} \psi_J \iff x \in \operatorname{dom} \psi \wedge \forall y (y \in \psi(x) \implies y \in \operatorname{dom} J^{-1})$$

$$x \in \operatorname{dom} \psi_J \implies \psi_J(x) = \{J^{-1} y \mid y \in \psi(x)\} .$$

Since J is total, and ψ is surjective, it follows ψ_J is surjective. If $b \in \operatorname{dom} \psi_J$ then b is a <u>code</u> for $\psi_J(b)$. In case $\psi_J(b)$ is the graph of a function $f: B \to B$ with graph equal to $\psi_J(b)$, then we also say b is a code for f. A function <u>APP</u>: $B^2 \to B$ depending on ψ, J is defined by the following requirements:

(1) $\langle b, x \rangle \in \operatorname{dom} \underline{APP} \iff b \in \operatorname{dom} \psi_J$, $\psi_J(b)$ is the graph of a function f, $x \in \operatorname{dom} f$

(2) $(\langle b, x \rangle \in \operatorname{dom} \underline{APP} \wedge \langle x, y \rangle \in \psi_J(b)) \implies \underline{APP}(b, x) = y$.

Proposition 5: Assume H(D) , H'(D) . Then **APP** ϵ D . Moreover, dom **APP** is computable from D .

Proof indication.

$(\eta) : \epsilon$ B

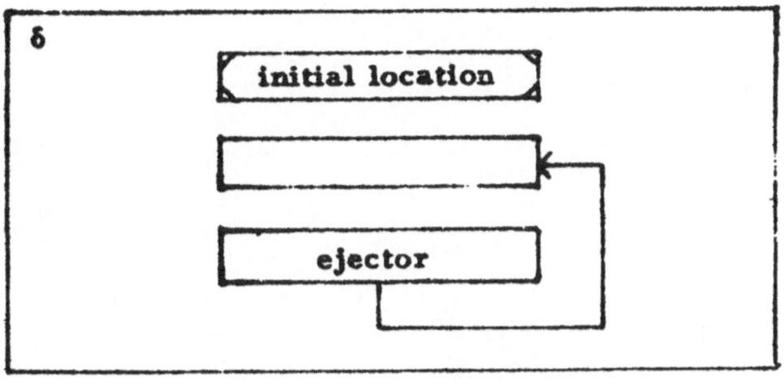

where the location to which the ejector goes need not be distinct from the initial location nor need it be distinct from the ejector.

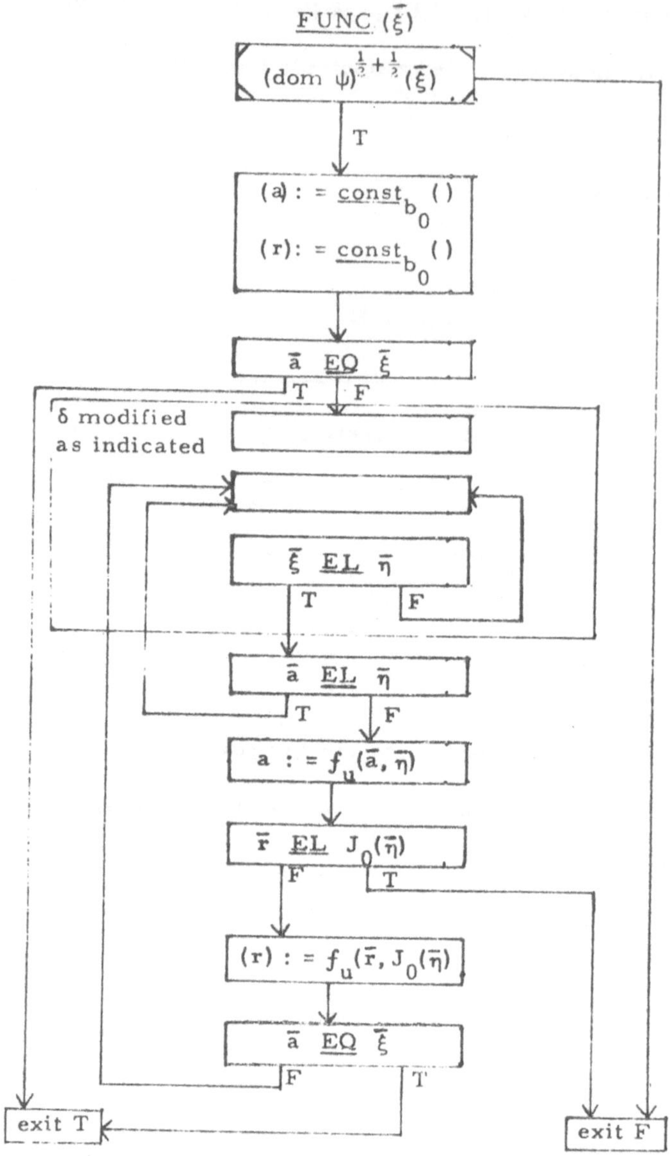
FUNC ($\bar{\xi}$)
(dom ψ)$^{\frac{1}{2}+\frac{1}{2}}$($\bar{\xi}$)
T
(a) : = const$_{b_0}$ ( )
(r): = const$_{b_0}$ ( )
$\bar{a}$ EQ $\bar{\xi}$
T
F
δ modified as indicated
$\bar{\xi}$ EL $\bar{\eta}$
T
F
$\bar{a}$ EL $\bar{\eta}$
T
F
a : = f_u($\bar{a}$, $\bar{\eta}$)
$\bar{r}$ EL J_0($\bar{\eta}$)
F
T
(r) : = f_u($\bar{r}$, J_0($\bar{\eta}$))
$\bar{a}$ EQ $\bar{\xi}$
F
T
exit T
exit F

Then: FUNC(b) $\Longleftrightarrow$ $\psi_J(b)$ is the graph of a function.

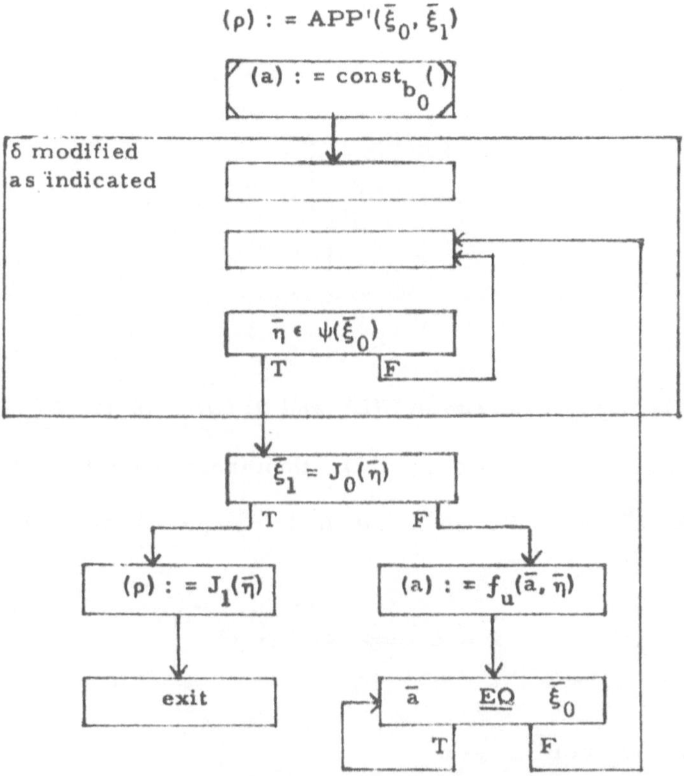

Then, APP' is an extension of APP and

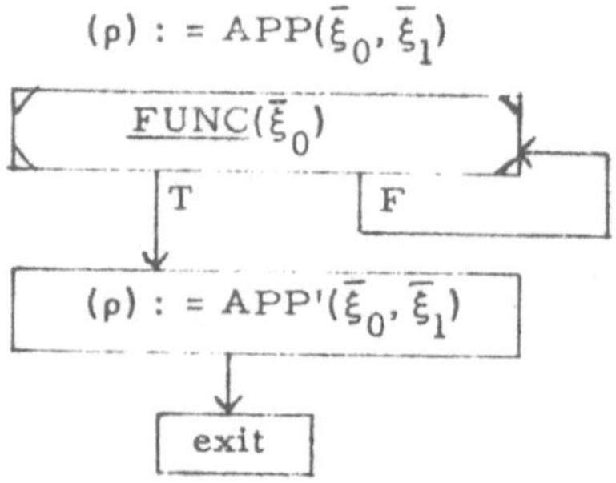

<u>Theorem 7</u>. Assume H(D) and H'(D). If E, $E \subseteq B^n$, is semicomputable from D, then there is a total relation $R : B^{n+1} \to \{T, F\}$ such that R is computable from D and

$$x \in E \iff \exists y\, R(x, y)$$

where $x \in B^n$, $y \in B$.

A proof may be given based upon the following observations.

(1) If $\sigma = (\sigma_0, \sigma_1, \ldots, \sigma_{r-1})$ is a successful computation of, say, δ of minimal length, then $i < j < r \Rightarrow \sigma_i \neq \sigma_j$. Let $\hat{\sigma} = \{\sigma_0, \sigma_1, \ldots, \sigma_{r-1}\}$.

(2) If σ and σ' are successful computations of minimal length and $\hat{\sigma} = \hat{\sigma}'$ then $\sigma = \sigma'$.

Moreover, (1) and (2) hold if $\sigma_i = \langle k_i, c_i \rangle$ is replaced by $\langle k_i \restriction A_\delta, c_i \rangle$. Let us call a sequence $(\langle k_0 \restriction A_\delta, c_0 \rangle, \langle k_1 \restriction A_\delta, c_1 \rangle, \dots)$ where $(\langle k_0, c_0 \rangle, \langle k_1, c_1 \rangle, \dots)$ is a computation of δ, a <u>computation[†] of δ (over A_δ)</u>. We <u>assume</u>, without loss of generality, that $A \subseteq B$, $C \subseteq B$. Then we may assign codes to successful computations over A_δ as follows. Let σ be a successful computation over A_δ. Then b is a code for σ provided that

(a) $b \in \operatorname{dom} \psi_J$

(b) $\langle x, y \rangle \in \psi_J(b) \Longrightarrow \psi_J(x)$ is the graph of a function f_x

(c) $x \in \operatorname{dom} \psi_J \wedge \langle x, y \rangle \in \psi_J(b) \Longrightarrow \langle f_x, y \rangle \in \hat{\sigma}$.

Let δ generate B from D at, say, η. Assuming $\text{const}_b \in D$ we may, without loss of generality, assume δ is normal so that the <u>order</u> in which elements of B are ejected is independent of the initial content func-

---

[†]More generally, we may speak of computations of δ over A', where $A_\delta \subseteq A' \subseteq A$. Then "a computation of δ over A" is the same thing as "a computation of δ".

tion. We may define a total function $S_\delta : B \rightarrow B$ as follows. Let comp $k = (\langle k_0, c_0\rangle, \langle k_1, c_1\rangle, \ldots)$. Then $S_\delta(x) = y$ provided that if i is the smallest number such that both $k_i(\eta) = x$ and c_i is the ejector (of δ) and if j is the smallest number such that both $k_j(\eta) = y$ and c_j is the ejector, then $i < j$; moreoever, for all m, if $i < m < j$ and c_m is the ejector, then there exists n, $n < i$, $(k_m(\eta) = k_n(\eta))$ and c_n is the ejector.

Let b_1 be the first element which is ejected by δ. Then the structures $\langle B, S_\delta, b_1\rangle$ and $\langle \omega, S, \theta\rangle$ are isomorphic.

Without indication of proof, we state some further results.

Theorem 8. Assume $H(D)$, $H'(D)$. Let δ be a normal diagram which generates B. Then $S_\delta \in D$. Also the following statements are equivalent, where $E \subseteq B^n$:

(1) E is semicomputable from D;

(2) E is generable from D;

(3) $E = \phi$ or there exists a total function f, $f \in D$, such that cod $f = E$:

(4) There exists a total relation $R : B^{n+1} \to \{T, F\}$ such that R is computable from D and $x \in E \iff \exists y\, R(x, y)$ where $x \in B^n$, $y \in B$.

Let δ be as above. Let R be a relation with domain B^{n+1}. Then we define a function $f: B^n \to B$ as follows:

$x \in \operatorname{dom} f \iff \exists y\, R(x, y)$

$x \in \operatorname{dom} f \implies f(x) = \mu_\delta\, y\, R(x, y)$

where $x \in B^n$, $y \in B$ and $\mu_\delta y$ means "the least y such that" in the order of B induced by δ.

<u>Proposition 9</u>: Assume $H(D)$, $H'(D)$. If R, δ, f are as above, then $f \in D$.

One may also prove the following analogue of the Kleene Normal Form theorem.

<u>Theorem 10.</u> Assume $H(D)$, $H'(D)$ and assume further that D has a finite base. For every n, p, q, $n \geq 0$, $p \geq 0$, $q > 0$, there exists a total direction $U : B \to B^p \times q$ depending only on p and q and there exists a total relation $R : B^{n+2} \to \{T, F\}$ depending only on n such that

$U \in D$, $R \in D$ and for every $d : B^n \to B^p \times q$, $d \in D$, there exists $b \in B$ satisfying

$$d(x) = U\mu_\delta y\ R(b, x, y) .$$

<u>Corollary</u>: With the assumptions of Theorem 10 and R as in Theorem 10: for every set E semicomputable from D , there exists $b \in B$ satisfying

$$x \in E \iff \exists y\, R(b, x, y)$$

Proof: Take $p = 0$, $q = 1$.

APPENDIX

This section is devoted to indicating how the diagram-computation notion may be cast into a syntactical framework analagous to a deductive one. While this aim involves making various ad hoc decisions concerning the representation of some semantical notions, it, presumably, results in facilitating comparisons with syntactic approaches to a theory of computability or of algorithms. This discussion may also be of interest, possibly, from the point of view of using diagrams as constituents of programming languages.

In (1) below, we indicate by means of an example, how diagrams, with $B = \omega$, $D = \{S, eq, const_0\}$, may be cast into this form. "Proof" is defined in a familiar way and, of course, it turns out that proofs and computations (over a subset of A) correspond in a way which will be made clear. In (2), below, diagram forms are considered. With any interpretation over a denumerable domain, a set of equations is associated and a notion of "deduction from a set of equations" (cf. [SCK] p. 264 ff.) is introduced which corresponds to computations of the interpreted diagram form. (The restriction to denumerable interpretations comes about as a result of the decision to use only finite alphabets in this discussion.) In (1), the syntactic system obtained depends upon the diagram considered and, similarly, in (2) the system obtained depends upon the diagram form under consideration. In (3), below, we very briefly indicate how a single deductive system may be defined which is adequate to "imitate" all diagrams with $B = \omega$ and $D = \{S, \underline{eq}, \underline{const}_0\}$. Our discussion terminates without discussing a system which bears the relation to (2) that (3) bears to (1).

(1) **A system will be constructed to imitate the following diagram δ .**

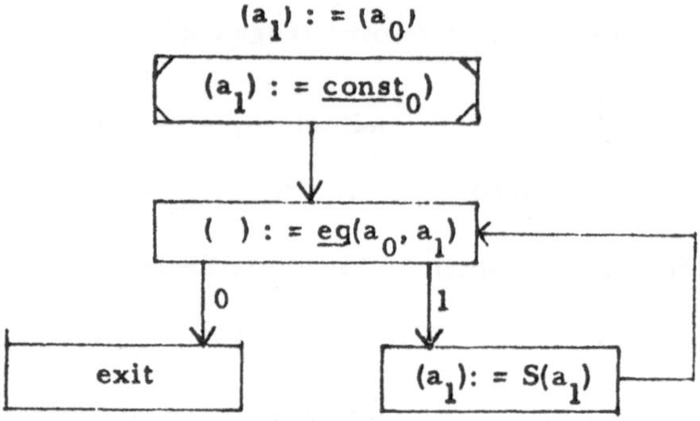

The distinguished argument and result cells are not involved in the formulation of the rules given below, but are involved in stating in what sense the system defines the <u>iden</u> function (from ω to ω). The exact form the rules take depend upon the correlation between the four command occurrences above and four elements of C as indicated by the configuration below.

$$c_0$$

$$c_1$$

$$c_3 \qquad\qquad c_2$$

<u>Alphabet</u> $\quad$ a, b, c, 1

<u>Wffs</u> $\quad$ (Well formed formulas)

$c1^{(m)}ab1^{(n)}a1b1^{(p)}$; $m = 0, 1, 2,\ n \geq 0$,

$p \geq 0$, $1^{(m)} = \overbrace{11\ldots1}^{m}$

Rules

$$cabl^{(n)}albl^{(p)} \vdash clabl^{(n)}alb$$

$$clabl^{(n)}albl^{(p)} \vdash cllabl^{(n)}albl^{(p)}\text{, if } n \neq p$$

$$clabl^{(n)}albl^{(n)} \vdash clllabl^{(n)}albl^{(n)}$$

$$cllabl^{(n)}albl^{(p)} \vdash clabl^{(n)}albl^{(p+1)}$$

$$clllabl^{(n)}albl^{(p)} \vdash clllabl^{(n)}albl^{(p)}$$

Proof: A proof is a finite sequence $w_1, w_2, \ldots, w_r$ of wffs such that $w_i \vdash w_{i+1}$ for each i satisfying $1 \leq i < r$.

Interpretation

$A_\delta = \{a, al\}$

$bl^{(n)} = n$

$o = \{c, cl, cll, clll\}$

$\{k \upharpoonright A_\delta \mid k \in K\} = \{abl^{(n)}albl^{(p)} \mid n \geq 0,\ p \geq 0\}$ = content functions over A_δ

A wff $cl^{(m)}abl^{(n)}albl^{(p)}$ may be interpreted as a state $\langle abl^{(n)}albl^{(p)}, cl^{(m)}\rangle$ of a machine capable of executing the algorithm δ.

A proof which begins with $cabl^{(n)}albl^{(p)}$, $n \geq 0$, $p \geq 0$, then, may be interpreted as a computation of δ and all finite computations of δ over A_δ are correlated (in 1 - 1 fashion) with proofs by means of this interpretation.

(2) A system will be constructed to imitate the following diagram form δ.

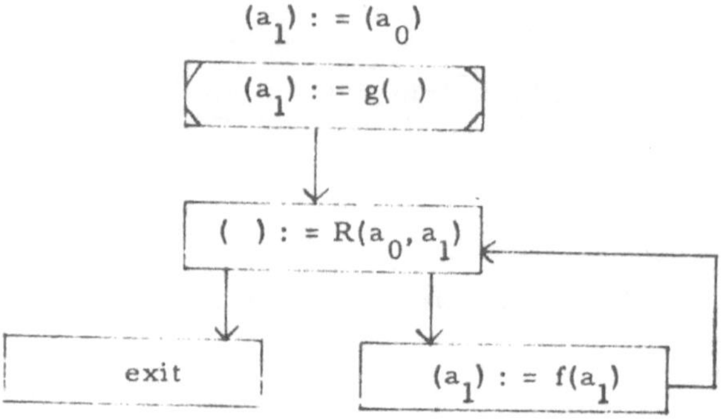

(The set of values of R will be correlated with $\{T, T1\}$.)

Alphabet   a, b, c, 1

f, R, g, T

Wffs   A state-wff is a wff of (1).

An equation-wff is a string of one of the following forms:

$fb1^{(r)}b1^{(s)}$, $r \geq 0$, $s \geq 0$

$Rb1^{(r)}b1^{(s)}T^{(q)}$, $r \geq 0$, $s \geq 0$, $q = 0, 1$

$gb1^{(s)}$

A permissible set $\mathcal{E}$ of equation-wffs is a set of equation-wffs which satisfies:

if $\mathrm{fbl}^{(r)}\mathrm{bl}^{(s)},\ \mathrm{fbl}^{(r)}\mathrm{bl}^{(t)} \in \mathcal{E}$, then $s = t$;

if $\mathrm{Rbl}^{(r)}\mathrm{bl}^{(s)}\mathrm{T}^{(q)},\ \mathrm{Rbl}^{(r)}\mathrm{bl}^{(s)}\mathrm{T}^{(p)} \in \mathcal{E}$, then $p = q$;

at most, one wff begins with g .

Rules for deduction from a set $\mathcal{E}$ of equation-wffs

$$\mathrm{cabl}^{(n)}\mathrm{albl}^{(p)} \vdash \mathrm{cabl}^{(n)}\mathrm{albl}^{(s)} \quad \text{if} \quad \mathrm{gbl}^{(s)} \in \mathcal{E}$$

$$\mathrm{clabl}^{(n)}\mathrm{albl}^{(p)} \vdash \mathrm{cllabl}^{(n)}\mathrm{albl}^{(p)} \quad \text{if} \quad \mathrm{Rbl}^{(n)}\mathrm{bl}^{(p)}\mathrm{Tl} \in \mathcal{E}$$

$$\mathrm{clabl}^{(n)}\mathrm{albl}^{(p)} \vdash \mathrm{clllabl}^{(n)}\mathrm{albl}^{(p)} \quad \text{if} \quad \mathrm{Rbl}^{(n)}\mathrm{bl}^{(p)}\mathrm{T} \in \mathcal{E}$$

$$\mathrm{cllabl}^{(n)}\mathrm{albl}^{(p)} \vdash \mathrm{clabl}^{(n)}\mathrm{albl}^{(s)} \quad \text{if} \quad \mathrm{fbl}^{(p)}\mathrm{bl}^{(s)} \in \mathcal{E}$$

$$\mathrm{clllabl}^{(n)}\mathrm{albl}^{(p)} \vdash \mathrm{cllabl}^{(n)}\mathrm{albl}^{(p)} .$$

Since we are restricting ourselves to interpretations of the diagram form δ over denumerable sets B , there is a bijective correspondence ψ from $\{\mathrm{bl}^{(n)} \mid n \geq 0\}$ to B . With respect to ψ, then, a permissible set $\mathcal{E}$ of equations may, in an obvious way, be correlated with an interpretation of δ . As in (1), a proof is a finite sequence $w_1, w_2, \ldots, w_r$ of state-wffs such that $w_i \vdash w_{i+1}$, $1 \leq i < r$. Again, as in (1), proofs which begin with a state-wff of the form $\mathrm{cabl}^{(n)}\mathrm{albl}^{(p)}$, $n \geq 0$, $p \geq 0$, may be correlated with computations of the interpreted diagram form δ over $A_\delta = \{\mathrm{a}, \mathrm{al}\}$.

(3) Alphabet a, b, c, l, *, e

Z, E, S

The last three symbols will be interpreted as $\underline{\text{const}}_0$, eq and S, respectively. The asterisk will be used as a punctuation mark and e will be used to denote the total. $\langle 0, 0, 1\rangle$ direction over w_1. One may introduce a notion of diagram-string to correspond to "diagram with directions $\{S, eq, const_0\}$" in such a way that the string obtained by concatenating the following strings (in the order indicated)

cZalcl clEaalTcl1lTlcll cllSalalcl clllecll l

may be correlated with the diagram of (1). Each of the four strings is intended to correspond to a command location together with the command at that location. We assume that the notion of an address-occurrence in a diagram-string is clear. If u is a diagram-string, let A_u be the set of all addresses which occur in u. We assume too that the general (in contradistinction to the special notion of state-string in (2)) notion of state-string is clear. If v is a state string, let A_v be the set of all addresses which occur in v. Similarly,

C_u, C_v , the set of command locations in u, v respectively, is assumed understood. Then, a string of the form

$$u * v$$

where u is a diagram-string, v is a state-string, $A_u \subseteq A_v$ and $C_v \subseteq C_u$, will be called a <u>wff</u> (in the sense of this subsection). If the above is clear, it should also be clear that one may introduce a rather complicated rule

$$w \vdash w'$$

such that if $w = u*v$ is a <u>wff</u> then w' is a wff of the form $u'*v'$ with $u = u'$; further, if u is the string indicated above which corresponds to δ of (1), if v is a wff of (1) and if $u*v_1, u*v_2, \ldots, u*v_r$ is a proof in the sense of (3), then

$$v_1, v_2, \ldots, v_r$$

is a proof in the sense of (1)

The reader may also note that the asterisk is only a convenience. No ambiguity results from concatenating diagram-strings and state-strings to form "wffs".

ACKNOWLEDGEMENT

The author thanks J. C. Shepherdson for calling his attention to some misstatements in an earlier version. Thanks are also tendered J.B. Wright, J. W. Thatcher, and J. D. Rutledge for suggestions on improving the manuscript, to Madame Claudine Donio for her help in preparing the earlier French version, and to M. Schutzenberger for making the author's stay in Paris (where this study was carried out) possible.

REFERENCES

[AC] Alonzo Church, "Introduction to Mathematical Logic," Princeton Univ. Press, 1956.

[ER] C. C. Elgot, Abraham Robinson, "Random-Access, Stored-Program Machines ...," J. of Assoc. for Computing Machinery, Vol. 11, pp. 365-399.

[SCK] S. C. Kleene, "Introduction to Metamathematics," Van Nostrand, 1952.

[SS] J. C. Shepherdson and H. E. Sturgis, "Computability of recursive functions," J. of Assoc. for Computing Machinery, April 1963.

[GN] H. H. Goldstine and von Neumann, "Planning and coding of problems for an electronic computing instrument," Rep. 1947, dist. for Advanced Study, Princeton.

[ELP] E. L. Post, "Finite Combinatory Processes-Formulation I," J. Symbolic Logic, 1936.

[CCE] C. C. Elgot, "Direction and Instruction-Controlled Machines," Proceedings of the Symposium on System Theory, Polytechnic Institute of Brooklyn, 1965.

Reprinted from Lecture Notes in Mathematics 188:
Symposium on Semantics of Algorithmic Languages (edited by E. Engeler),
Springer-Verlag Berlin (1971), 71–88.

ALGEBRAIC THEORIES AND PROGRAM SCHEMES

by

Calvin C. Elgot

1. Introduction

The objective of this paper is to point out that the "semantics" of cycle-free program schemes (of the kind intuitively described in Section 2) may be explicated by existing algebraic notions. More explicitly, equivalence classes of schemes correspond to morphisms in a free algebraic theory (cf., Lawvere (1963), and Eilenberg and Wright (1967)), and an interpretation of the collection of morphisms is a coproduct-preserving functor into an appropriate target category.

While some attempt has been made to be reasonably self-contained, the interested reader may wish to consult Elgot (1970), and Eilenberg and Wright (1967).

2. Program Schemes--Intuitive

The intuitive notion "program scheme" we have in mind is close in spirit to that of Ianov (1958), as generalized by Rutledge (1964). The "corresponding" precise notion will be called "machine scheme", described in section 4.

Roughly speaking, a program scheme consists of an indexed family of "instruction schemes", together with a specification of the a priori possible sequencing of instructions (i.e., interpreted instruction schemes). The indices are sometimes called "instruction labels" and sometimes "states". In addition, provision must be made for how the sequencing of instructions comes to a halt. In this context, an instruction may be understood as a function $X \to X \times [n]$, $n \geq 1$, where $[n] = \{1,2,\dots,n\}$ and X is a set. The set X may be regarded as the set of possible states of "external memory" or, some may prefer to say, "storage". If external memory is in state $x \in X$ at some instant of time, if $f: X \to X \times [n]$ is executed at this instant of time, and if $f(x) = (x', i)$, $i \in [n]$, then after execution of f , external memory is in state x' , and the next instruction to be executed is determined by the number i . Thus execution of f simultaneously performs a test on X (with n possible outcomes) and an operation on X .

To illustrate the notion of "program", let $f_i: X \to X \times [2]$, $1 \leq i \leq 3$, be functions. The following "table" describes a program P .

→ 1. f_1;2,3

2. f_2;3,4

3. f_2;5,5

4. f_3;3,5

5. exit

Assuming instruction #1 is the "beginning" instruction, and given an "input" $x_0 \in X$, execution of the program procceds as follows: f_1 is executed; if the outcome is 1, the next instruction to be executed is #2, the first of the two indices following the semicolon on the top line; if the outcome is 2, the next instruction to be executed is #3. Suppose $f_1(x_0) = (x_1, 1)$, $f_2(x_1) = (x_2, 2)$, $f_3(x_2) = (x_3, 1)$; then the sequence of executions is: #1, #2, #3, #4, #5. When instruction #5 is reached, the execution sequence terminates. In this particular case, which is "cycle-free", execution sequences are all finite. If $f_2(x_3) = (x_4, i)$, $i = 1,2$, then the "output" of the execution is x_4 . The "external behavior" of P (cf., Elgot (1970) is a function $|P|: X \to X$ with $|P|(x_0) = x_4$.

The information in the table above may also be presented as follows.

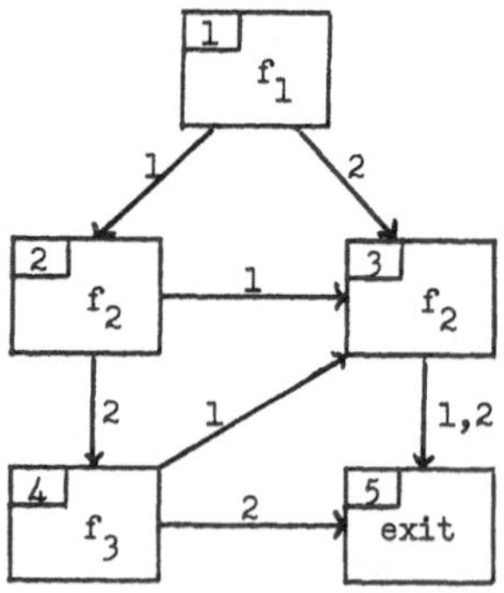

In this section, we will not bother to distinguish a program from the "isomorphism class" of programs which it determines.

The composite QR (cf., Elgot (1970)) of the following two programs

<u>Q</u>	<u>R</u>
$\rightarrow$ 1. f_1;2,3	$\overset{1}{\rightarrow}$ 2. f_2;2,3
2. exit 1	$\overset{2}{\rightarrow}$ 3. f_2;3,4
3. exit 2	4. f_3;3,5
	5. exit

is the program P. The external behavior of Q, which has two exits, is $|Q| = f_1: X \rightarrow X \times [2]$, while the external behavior of R which has two beginning states is $|R|: X \times [2] \rightarrow X$. If $|R|(x,i) = y$, $i = 1,2$, then execution of program R operating on input x, starting with the ith beginning state, yields output y. Moreover, $|P| = |Q||R| = f_1|R|$.

As a further illustration, let T be the following program:

$\overset{1}{\rightarrow}$ 2. exit 2

$\overset{2}{\rightarrow}$ 3. exit 1.

Then QT is the following:

$\rightarrow$ 1. f_1;2,3

2. exit 2

3. exit 1,

while TR is the following:

$\overset{2}{\rightarrow}$ 2. f_2;2,3

$\overset{1}{\rightarrow}$ 3. f_2;3,4

4. f_3;3,5

5. exit.

It should also be noted that $|T|: X \times [2] \rightarrow X \times [2]$ satisfies $|T|(x,1) = (x,2)$, $|T|(x,2) = (x,1)$ for every $x \in X$. Of course, $|QT| = |Q||T|$ and $|TR| = |T||R|$.

3. <u>Algebraic Theories</u>

Categories (cf., MacLane and Birkhoff (1967) and Eilenberg and Wright (1967)) arise naturally in our considerations. A <u>category</u> $\mathcal{A}$ consists of a class Obj $\mathcal{A}$

whose elements $A,B,\ldots$ are called objects, a family of sets $\mathcal{A}(A,B)$ whose elements are called morphisms from A to B, and an operation called composition, required to be associative, which determines for each $\varphi \in \mathcal{A}(A,B)$ and $\psi \in \mathcal{A}(B,C)$ a unique element $\varphi\psi \in \mathcal{A}(A,C)$; in addition, it is assumed, for each B, there exists a (necessarily unique) element $1_B \in \mathcal{A}(B,B)$ such that $\varphi 1_B = \varphi$, $1_B\psi = \psi$. We write B in place of 1_B and $\varphi\colon A \to B$ or $A \xrightarrow{\varphi} B$ in place of $\varphi \in \mathcal{A}(A,B)$.

Let X be a set. We form a category $\mathcal{A} = (X)$ be taking as objects X^n, $n = 0,1,2,\ldots$ and taking as $\mathcal{A}(X^p,X^n)$ the set of all functions with domain X^p and image a subset of X^n.

For each $i \in [n] = \{1,2,\ldots,n\}$, the morphism $\pi_i\colon X^n \to X$ defined by $\pi_i(x_1,x_2,\ldots,x_n) = x_i$ is distinguished and the following basic property noted:

(3.1) for each family $\varphi_i\colon X^p \to X$, $i \in [n]$, of morphisms in (X), there is a unique morphism $\varphi\colon X^p \to X^n$ such that $\varphi_i = \varphi\pi_i$, i.e.,

$$\varphi_i\colon X^p \xrightarrow{\varphi} X^n \xrightarrow{\pi_i} X$$

for each $i \in [n]$.

Let $f\colon [n] \to [p]$, $g\colon [p] \to [q]$ be functions. Define $f^{\#}\colon X^p \to X^n$ by $f^{\#}(x_1,x_2,\ldots,x_p) = (x_{f(1)},x_{f(2)},\ldots,x_{f(n)})$. The reader may readily verify

(3.2) $(fg)^{\#} = g^{\#}f^{\#}$;

(3.3) if X has at least two elements, the function $f \mapsto f^{\#}$ preserves distinctness.

Now let $\mathcal{S}_F$ be the category whose objects are $[n]$, $n \geq 0$, and where morphisms $[n] \to [p]$ are the functions with domain $[n]$ and values in $[p]$. Let $\mathcal{S}_F^{op}$ be the category obtained from $\mathcal{S}_F$ by "reversing arrows". Properties (3.2) and (3.3) may then be reformulated:

(3.4) if X has at least two elements, the function $\mathcal{S}_F^{op} \to (X)$ which takes f into $f^{\#}$ is an embedding.

For $i \in [n]$, we write $i\colon [1] \to [n]$ for the function with value i.

The above remarks are intended to motivate the following notion. A (non-degenerate) algebraic theory T is a category whose objects are $[n]$, $n \geq 0$, which satisfies

(3.4') $\mathcal{S}_F$ is a subcategory of T ;

(3.1') (a) if $\varphi, \psi: [n] \to [p]$ are in T , and if $i\varphi = i\psi$ for each $i \in [n]$, then $\varphi = \psi$;

(b) if $\varphi_i: [1] \to [p]$ is in T , $i \in [n]$, then there exists $\varphi: [n] \to [p]$ such that $i\varphi = \varphi_i$ for each $i \in [n]$.

Let $\Omega = \{\Omega_n\}$, $n \geq 0$, be a family of pairwise disjoint sets which are also disjoint from the set $\{x_1, x_2, x_3, \ldots\}$ of "variables". A revealing "concrete" example of an algebraic theory may be obtained by treating elements of Ω_n as "function symbols". A <u>term</u> is an element of the smallest class U of words such that

(3.5) $x_i \in U$, $i \geq 1$;

(3.6) if $u_1, \ldots, u_n \in U$ and $\omega \in \Omega_n$, then $\omega(u_1, \ldots, u_n)$ is in U .

The "words" indicated above employ parentheses and commas; as is well known, they may be suppressed without introducing ambiguity. We define a category T be taking as objects $[n]$, $n \geq 0$; taking $\varphi: [n] \to [p]$ in T to be $(u_1, \ldots, u_n)$, where u_i is a term in at most the variables $x_1, \ldots, x_p$; and defining $\varphi\psi$, where $\psi: [p] \to [q]$, to be the result of simultaneously substituting ψ_j for x_j in φ for each $j \in [p]$, where ψ_j is the jth term comprising ψ . By identifying $f: [n] \to [p]$ with $(x_{f(1)}, x_{f(2)}, \ldots, x_{f(n)})$ and identifying $\omega \in \Omega_n$ with the term $\omega(x_1, \ldots, x_n)$, we obtain an algebraic theory $T = \mathcal{S}_F[\Omega]$ in which $\Omega_n \subset T([1],[n])$ for all $n \geq 0$; briefly, $\Omega \subset T$.

Let T and T' be algebraic theories. By a <u>theory-morphism</u> or <u>morphism between theories</u>, we mean a function which takes $\varphi: [n] \to [p]$ in T into $\varphi': [n] \to [p]$ in T' in such a way that

(3.7) if $j: [1] \to [p]$, $j \in [p]$, then $j' = j$;

(3.8) $(\varphi\psi)' = \varphi'\psi'$ where $\psi: [p] \to [q]$.

Clearly, a theory-morphism keeps $\mathcal{S}_F$ elementwise fixed.

It is not difficult to show

<u>Proposition 3.1</u>. Every family of functions

$$\Omega_n \to T([1],[n])$$

admits a unique extension to a theory-morphism

$$\mathcal{S}_F[\Omega] \to T' .$$

The theory $\mathcal{S}_F[\Omega]$ is freely generated by Ω .

Let X be a set. We construct a category $[X]_F$ with objects $[n]$, $n \geq 0$, as follows. The set $[X]_F([n],[p])$ of morphisms is the set of all functions $f: X \times [n] \to X \times [p]$. If $f: [n] \to [p]$ is in $\mathcal{S}_F$, we denote by $X \times f: X \times [n] \to X \times [p]$, the function which takes (x,i) into $(x,f(i))$. Note $X \times fg = (X \times f)(X \times g)$. If $X \neq \emptyset$, $\mathcal{S}_F$ may be embedded in $[X]_F$. In this way, $[X]_F$ becomes an algebraic theory.

By replacing functions by partial function or relations, we obtain algebraic theories $[X]_P$, $[X]$ respectively. Note that $[X]_F \subset [X]_P \subset [X]$. It may also be noted that $[X]_F$ fails to have morphisms $[1] \to [0]$, while $[X]_P$ and $[X]$ have exactly one.

Let $\mathcal{S}_P$ be the algebraic theory whose set $\mathcal{S}_P([n],[p])$ of morphisms is the set of all partial functions from $[n]$ into $[p]$. It is easy to see that $\mathcal{S}_P = \mathcal{S}_F[\Omega]$, where Ω_0 consists of a single element and $\Omega_i = \emptyset$ for $i \neq 0$. We also note that every theory-morphism $\mathcal{S}_P \to T$ preserves distinctness.

4. Ω-Machine Schemes (Total)

Let $\Omega = (\Omega_1, \Omega_2, \ldots)$ be a sequence of pairwise disjoint sets. A total Ω-machine scheme M from n to p , $n \geq 0$, $p \geq 0$, consists of

(1) a set S , whose elements are called (internal) states;

(2) a function $b: [n] \to S$; $b(i)$ is the ith beginning state;

(3) an injective function $e: [p] \to S$; $e(j)$ is the jth exit;

(4) a function τ which associates with each non-exit $s \in S$, a pair whose first member is $\omega \in \Omega_m$ and whose second member is a finite sequence $(s_1, s_2, \ldots, s_m)$, $s_j \in S$, $1 \leq j \leq m$, where m is some positive integer. The scheme M is finite if S is finite.

Thus, for example, if $\omega_j \in \Omega_2$, $1 \leq j \leq 3$, the following table describes an Ω-machine scheme:

→ 1. ω_1;2,3

2. ω_2;3,4

3. ω_2;5,5

4. ω_3;3,5

5. exit .

Here $n=1=p$, $S=[5]$. The assignment $\omega_j \mapsto f_j$ determines the program P described earlier.

If $\tau(s)=(\omega,(s_1,s_2,\ldots,s_m))$, we say s is <u>ν-related</u> to s_j for each j , $1 \leq j \leq m$, and write $(s,s_j)\in\nu$. Let

$$\nu^+ = \nu \cup \nu^2 \cup \nu^3 \cup \ldots$$

be the transitive closure of ν . A machine scheme M is <u>cycle-free</u> if its <u>next-state relation</u> ν is areflexive, i.e., $(s,s)\notin\nu^+$ for every $s\in S$. The machine scheme described above is cycle-free.

Let M' be another total Ω-machine scheme from n to p consisting of S' , ν' , e' , τ' . An <u>isomorphism</u> θ: M→M' is a bijection $\theta: S\to S'$ subject to the conditions

(a) $b': [n] \xrightarrow{b} S \xrightarrow{\theta} S'$;

(b) $e': [p] \xrightarrow{e} S \xrightarrow{\theta} S'$;

(c) for each non-exit $s\in S$, if $s'=\theta(s)$ and if

$$\tau(s)=(\omega,(s_1,\ldots,s_m))$$

then

$$\tau'(s')=(\omega,(s_1',\ldots,s_m')) .$$

Let $\bar{M}$ be the isomorphism class of machine schemes n to p determined by the machine scheme M . If N is a machine scheme from p to q , we define the composite $\overline{MN}=\overline{M'N'}$; where $M'\in\bar{M}$, $N'\in\bar{N}$, the sets of states of M' and N' are disjoint and M'N' is defined as in Elgot (1970). If M" is isomorphic to M', and N" is isomorphic to N' , and the sets of states of M" and N" are disjoint, then M'N' is isomorphic to M"N" . Thus $\overline{MN}$ is well defined. The operation of composition is associative.

At this point, we recall that the identity function on $[p]$ is also denoted by $[p]$.

For each function $f: [n] \to [p]$, we define a machine scheme Q_f whose set S_f of states is $[p]$ and such that $e_f = [p]$, $b_f = f$. Then

$$\bar{M}\bar{Q}_{[p]} = \bar{M}$$

and

$$\bar{Q}_{[p]}\bar{N} = \bar{N} .$$

We define the category $C_t[\Omega]$ of Ω-machine schemes as follows. The objects of $C_t[\Omega]$ are $[n]$, $n \geq 0$. A morphism $\varphi: [n] \to [p]$ of $C_t[\Omega]$ is an isomorphism class of finite machine schemes from n to p . Composition of these morphisms has been defined above.

(Actually $C_t[\Omega]$ is not a proper construct in the sense of Gödel-Bernays-von Neumann set theory. The difficulty may be removed by the ad hoc requirement that a "state" be a non-negative integer.)

Restriction of the morphisms of $C_t[\Omega]$ to isomorphism classes of cycle-free machines leads to the subcategory $C_{ct}[\Omega]$ of $C_t[\Omega]$ of <u>cycle-free Ω-machine schemes</u>.

Let $f: [n] \to [p]$ and $g: [p] \to [q]$ be functions. It is easy to see that $\bar{Q}_{fg} = \bar{Q}_f\bar{Q}_g$. Moreover, Q_f is cycle-free. It is convenient to embed $\mathcal{S}_F$ in $C_{ct}[\Omega]$ via the injective functor $f \mapsto Q_f$. Thus $\mathcal{S}_F \subset C_{ct}[\Omega]$.

Each $\omega \in \Omega_p$ determines a morphism $\omega: [1] \to [p]$ in $C_{ct}[\Omega]$ as follows. This morphism is the isomorphism class of the atomic machine

→ 0. ω;1,2,...,p
1. exit 1
2. exit 2
⋮
p. exit p

determined by ω . Clearly distinct elements of Ω_p go into distinct morphisms. We will identify $\omega \in \Omega_p$ with the <u>atomic</u> morphism $\omega: [1] \to [p]$ in $C_{ct}[\Omega]$ which it determines.

A given machine scheme M from n to p may possess non-exit states s which are inaccessible (from any beginning state), i.e., $(b(i),s) \notin \nu^+$ for all $i \in [n]$, where ν is the next-state relation of M. We turn our attention to this phenomenon.

Call a morphism $\gamma: [n] \to [p]$ in $C_t[\Omega]$ f-like, where $f: [n] \to [p]$ is a function, if $\gamma = \bar{M}$ and $f(i) = j$ implies $b_M(i) = e_M(j)$. Every non-exit state of M is inaccessible since an exit of a machine is ν_M-related to no state.

If M is a machine from n to p, its accessible part is the submachine M^{acc} from n to p whose set of states consist of the accessible states of M together with M's exits. The beginning and exit functions of M^{acc} and M agree. For convenience, assume $[s]$ is the set of states of M^{acc}; thus $[s] \subset S_M$.

By the inaccessible part M^{ina} of M, we mean the f-like machine from n to s whose set of states is S_M; whose beginning function is $b: [n] \xrightarrow{f} [s] \xrightarrow{I} S_M$, where f is the beginning function of M^{acc} and I is inclusion; whose exit function is I (so that the exits of M^{ina} are the states of M^{acc}); and whose τ-function is inherited from M.

Let M' be the machine from s to p with set $[s]$ of states whose τ- and exit-functions agree with M^{acc}, but whose beginning function is the identity $[s]$. Let $\psi = \bar{M}'$; then $f\psi = \overline{M^{acc}}$.

Proposition 4.1. Every morphism $\varphi: [n] \to [p]$ in $C_t[\Omega]$ admits a decomposition

$$\varphi: [n] \xrightarrow{\gamma} [s] \xrightarrow{\psi} [p]$$

where $\gamma = \overline{M^{ina}}$ and ψ is defined above.

From the intuitive discussion of section 2, it is clear that every "interpretation" of an f-like morphism, $f: [n] \to [p]$, over a set X should be the function $X \times f: X \times [n] \to X \times [p]$. This consideration motivates the following definition:

A (machine) interpretation, $F: C_{ct}[\Omega] \to T$, where T is an algebraic theory, assigns to each morphism, $\varphi: [n] \to [p]$, a morphism, $F(\varphi): [n] \to [p]$, in such a way that

(a) $F(\varphi\psi) = F(\varphi)F(\psi)$ where $\psi: [p] \to [q]$;

(b) if $\gamma: [n] \to [p]$ is f-like, then $F(\gamma) = f$.

The following is an immediate consequence of Proposition 4.1.

Corollary 4.2. If $\varphi: [n] \to [p]$ is in $C_{ct}[\Omega]$, if F is an interpretation, and if $\psi: [n] \to [p]$ is the accessible part of φ, then $F(\varphi) = F(\psi)$.

Our main assertion concerning $C_{ct}[\Omega]$ states its free nature. Let T be an algebraic theory.

Theorem 4.3. Each family of functions

$$\Omega_n \to T([1],[n]) \ , \quad n > 0 \ ,$$

has a unique extension to an interpretation

$$F: C_{ct}[\Omega] \to T \ .$$

The uniqueness part of the argument essentially depends upon the following points: Let $\varphi: [n] \to [p]$ be in $C_{ct}[\Omega]$. Then $F(i\varphi) = F(i)F(\varphi) = iF(\varphi)$. Since $F(\varphi)$ is in an algebraic theory, it is determined by the n values $F(i\varphi)$, $i \in [n]$. If $n = 1$ and if φ is not f-like for some f, then $\varphi: [1] \to [p]$ admits a unique factorization

$$[1] \xrightarrow{\omega} [m] \xrightarrow{\psi} [p]$$

where $\omega \in \Omega_m$. Cf., also Theorem 15.1 of Elgot (1970).

5. Congruences and Quotients

Let $\mathcal{A}$ and $\mathcal{B}$ be categories with the same class of objects. A functor $F: \mathcal{A} \to \mathcal{B}$ which is the identity on the class of objects will be called strict. The functors of section 4 are strict. A theory-morphism may be described as a strict functor which keeps $\mathcal{S}_F$ pointwise fixed. Strict functors behave much like homomorphisms between general algebras.

By the kernel $\mathcal{K}(F)$ of F, we shall mean the family of relations $K = \{K_{A,B}\}$, where A,B are objects, defined by

$$(\varphi,\varphi') \in K_{A,B} \Leftrightarrow F(\varphi) = F(\varphi') \ ,$$

where $\varphi,\varphi': A \to B$ are morphisms in $\mathcal{A}$. The kernel of F is compatible with composition in the sense:

(5.1) $(\varphi,\varphi') \in K_{A,B} \wedge (\psi,\psi') \in K_{B,C} \Rightarrow (\varphi\psi,\varphi'\psi') \in K_{A,C}$.

Moreover, where $\mathcal{A}(A,B)$ is the set of morphisms in $\mathcal{A}$ from A to B , clearly:

(5.2) each component $K_{A,B}$ of K is an equivalence relation on $\mathcal{A}(A,B)$.

Call a family $K = \{K_{A,B}\}$, $K_{A,B} \subset \mathcal{A}(A,B) \times \mathcal{A}(A,B)$, a <u>congruence</u> on $\mathcal{A}$ if it satisfies (5.1) and (5.2). In analogy with the general algebraic situation, we may construct a quotient category $\mathcal{A}/K$. There is then a natural strict functor

$$N: \mathcal{A} \to \mathcal{A}/K$$

with the property

(5.3) for every strict functor $F: \mathcal{A} \to \mathcal{B}$ with the property $\varphi,\varphi' \in K_{A,B} \Rightarrow F(\varphi) = F(\varphi')$, there is a unique strict functor $G: \mathcal{A}/K \to \mathcal{B}$ such that

$$F: \mathcal{A} \overset{N}{\to} \mathcal{A}/K \overset{G}{\to} \mathcal{B} ,$$

i.e., $F = NG$.

Now let $\mathcal{A} = C_t[\Omega]$, and let K be the smallest congruence on $\mathcal{A}$ containing the family of relations $R = \{R_{[n],[p]}\}$, where

$$R_{[n],[p]} = \{(\gamma,f) \mid \gamma,f: [n] \to [p] , f \in \mathcal{S}_F , \gamma \text{ is f-like}\} ,$$

K is the componentwise intersection of all congruences which componentwise contain R .

<u>Proposition 5.1</u>. Let $\varphi,\psi: [n] \to [p]$ be in $C_t[\Omega]$. We have

$$(\varphi,\psi) \in K_{[n],[p]} \text{ iff } \varphi^{acc} = \psi^{acc} .$$

The proof of the porposition depends upon Proposition 4.1 and the following considerations:

(a) Let $M = M_1 M_2$ where M_1 is a machine scheme from n to 2 and M_2 is a machine scheme from 2 to p . Suppose exit 1 of M_1 is accessible, but exit 2 of M_1 is not. Then the set of states of M^{acc} is the union of the set of non-exit states of M_1 which are accessible from the beginning states of M_1 , the set of states of M_2 which are accessible from the first beginning state of M_2 and the set of exits of M_2 .

(b) If M_2^{acc} is isomorphic to M_3^{acc} , then $(1M_2)^{acc}$ is isomorphic to $(1M_3)^{acc}$, where 1 is the machine from 1 to 2 consisting of two exits and whose beginning state is exit 1 .

We call the quotient $\mathcal{M}_t[\Omega] = C_t[\Omega]/K$, the <u>category of accessible Ω-machine schemes</u>. We denote by $\mathcal{M}_{ct}[\Omega] = C_{ct}[\Omega]/K$ the subcategory of $\mathcal{M}_t[\Omega]$ consisting of morphisms which have cycle-free representatives, i.e., the image of the composite functor IN :

$$C_{ct}[\Omega] \xrightarrow{I} C_t[\Omega] \xrightarrow{N} \mathcal{M}_t[\Omega] ,$$

where I is the inclusion functor.

<u>Corollary 5.2</u>. $\mathcal{S}_F$ is a subcategory of $\mathcal{M}_{ct}[\Omega]$.

From (5.3) and Theorem 4.3, we obtain

<u>Theorem 5.3</u>. Each family of functions

$$\Omega_n \to T([1],[n]) \ , \quad n > 0 ,$$

has a unique extension to a strict functor

$$\mathcal{M}_{ct}[\Omega] \to T$$

which keeps $\mathcal{S}_F$ pointwise fixed.

The category $\mathcal{A} = \mathcal{M}_t[\Omega]$ or $\mathcal{A} = \mathcal{M}_{ct}[\Omega]$ comes close to being an algebraic theory. Not only does it contain $\mathcal{S}_F$ as a subcategory, but it has the property:

(5.4) for every family of morphisms $\varphi_i: [1] \to [p]$, $i \in [n]$, in $\mathcal{A}$, there is a morphism $\varphi: [n] \to [p]$ in $\mathcal{A}$ such that $i\varphi = \varphi_i$ for every $i \in [n]$.

Note, though, if $G: \mathcal{A} \to T$ is a strict functor which keeps $\mathcal{S}_F$ elementwise fixed, and if the hypothesis of (5.5) is satisfied, then

$$iG(\varphi) = G(i)G(\varphi) = G(i\varphi) = G(i\psi) = G(i)F(\psi) = iG(\psi) .$$

Since (5.5) is satisfied in T , and since $iG(\varphi) = iG(\psi)$ for every $i \in [n]$, we conclude that $G(\varphi) = G(\psi)$. This motivates the introduction of the congruence L .

Let L be the smallest congruence on $\mathcal{M}_t[\Omega]$ which satisfies for $\varphi,\psi\colon [n] \to [p]$: if $i\varphi = i\psi$ for each $i \in [n]$ then $(\varphi,\psi) \in L_{[n],[p]}$. It is clear that $\mathcal{M}_t[\Omega]/L$ is an algebraic theory--the <u>theory of Ω-machine schemes</u>. We fix attention on the subtheory $\mathcal{M}_{ct}[\Omega]/L$ of cycle-free machine schemes.

<u>Theorem 5.4</u>. The theory of cycle-free total Ω-machine schemes is (theory-isomorphic to) $\mathcal{S}_F[\Omega]$, the algebraic theory freely generated by Ω.

<u>Proof</u>. From (5.3) and Theorem 5.3, we have: each family of functions

$$\Omega_n \to T([1],[n]) \ , \quad n > 0 \ ,$$

has a unique extension to a strict functor

$$\mathcal{M}_{ct}[\Omega]/L \to T$$

which keeps $\mathcal{S}_F$ pointwise fixed. But this property characterizes $\mathcal{S}_F[\Omega]$.

We have seen that every interpretation $G\colon \mathcal{C}_{ct}[\Omega] \to T$ admits a factorization

$$G\colon \mathcal{C}_{ct}[\Omega] \overset{N_1}{\to} \mathcal{M}_{ct}[\Omega] \overset{N_2}{\to} \mathcal{S}_F[\Omega] \overset{H}{\to} T \ ,$$

where H is a theory-morphism. It will be noted later (section 8) that there is a theory-morphism $J\colon \mathcal{S}_F[\Omega] \to [X]_P$, for an appropriate set X, which is <u>injective</u>, i.e., preserves distinctness of morphisms. It follows, therefore, $\bigcap_G \mathcal{K}(G)$ as G runs through all interpretations $G\colon \mathcal{C}_{ct}[\Omega] \to [X]_P$ is exactly $\mathcal{K}(N_1N_2)$.

6. <u>0-Object Theories</u>

By a 0-object theory, we shall mean an algebraic theory T such that the set $T([1],[0])$ consists of a single morphism $0\colon [1] \to [0]$. It follows that each set $T([n],[0])$ consists of a single morphism $0\colon [n] \to [0]$. In any theory T, $T([0],[p])$ consists of a single morphism $0\colon [0] \to [p]$. Thus $[0]$ is a 0-object in the usual categorical sense. For each n, p there is a unique morphism $0\colon [n] \to [p]$ in a 0-object theory which factors through $[0]$, i.e., $0\colon [n] \to [0] \to [p]$.

The theories $[X]$ and $[X]_P$, but not $[X]_F$, are 0-object theories. We wish to describe and show the existence of free 0-object theories, for they will be related to machine schemes.

Let $\Omega = \{\Omega_n\}$, $n > 0$, and let $\Omega' = \{\Omega_n\}$, $n \geq 0$. The following assertion is evident.

(6.1) Each family of functions

$$\Omega_n \to T([1],[n]) \ , \quad n > 0 \ ,$$

where T is a 0-object theory, has a unique extension to a theory-morphism

$$\mathfrak{S}_F[\Omega'] \to T \ .$$

If K is the kernel of the theory morphism and $\varphi, \varphi' \colon [n] \to [p]$, factor through $[0]$, then $(\varphi, \varphi') \in K_{[n],[p]}$.

Let $Z = \{Z_{[n],[p]}\}$ be the family of relations defined by: $(\varphi, \varphi') \in Z_{[n],[p]}$ iff φ and φ' both factor through $[0]$. Then Z is a theory-congruence (cf., Eilenberg and Wright (1967)), and $\mathfrak{S}_F[\Omega']/Z$ is an algebraic theory. If $\Omega_0 \neq \emptyset$, then this quotient is a 0-object theory $\mathfrak{S}_P[\Omega]$ (which clearly does not depend upon the cardinality of Ω_0). From (6.1), we infer

(6.2) each family of functions

$$\Omega_n \to T([1],[n]) \ , \quad n > 0 \ ,$$

where T is a 0-object theory, has a unique extension to a theory-morphism

$$\mathfrak{S}_P[\Omega] \to T \ .$$

The theory $\mathfrak{S}_P[\Omega]$ is <u>freely generated</u> as a 0-object theory by Ω . If $\varphi \colon [n] \to [p]$ is in the subtheory $\mathfrak{S}_F[\Omega]$ of $\mathfrak{S}_F[\Omega']$, it does not factor through $[0]$. Thus $\mathfrak{S}_F[\Omega]$ is a subtheory of $\mathfrak{S}_P[\Omega]$.

We shall require the following assertions:

<u>Theorem 6.1</u>. Let T be a 0-object theory, and let $\Omega \subset T$. Suppose Ω generates T as a 0-object theory, and suppose bijective maps $\Gamma_n \to \Omega_n$, $n > 0$, are given. If T also satisfies (6.3) below, then the unique extension of these maps to a theory-morphism $\mathfrak{S}_P[\Gamma] \to T$ is bijective.

(6.3) If $\varphi \in T([1],[p])$, $\varphi \neq 0$, admits factorizations

$$\varphi\colon [1] \xrightarrow{\omega} [m] \xrightarrow{\psi} [p]$$
$$\varphi\colon [1] \xrightarrow{\omega'} [n] \xrightarrow{\psi'} [p] ,$$

where $\omega \in \Omega_m$, $\omega' \in \Omega_n$, then $m = n$, $\omega = \omega'$, $\psi = \psi'$.

Theorem 6.2. It is possible to assign to each morphism $\psi\colon [n] \to [p]$ in $\mathcal{S}_p[\Omega]$ a non-negative integer $\deg \psi$ satisfying:

(6.4) $\psi \in \mathcal{S}_p \Leftrightarrow \deg \psi = 0$;

(6.5) $\deg \psi = \Sigma_i \deg i\psi$;

(6.6) if $\psi \neq 0$ and $\omega \in \Omega_n$, then $\deg \omega\psi = 1 + \deg \psi$.

A variant of Theorem 6.2:

Theorem 6.2'. It is possible to assign to each morphism $\psi\colon [n] \to [p]$ in $\mathcal{S}_p[\Omega]$ a non-negative integer $ht(\psi)$ satisfying (6.4), (6.6), and

(6.5') $ht\,\psi = \sup_i ht(i\psi)$.

7. The Embedding Theorem

Let $T = \mathcal{S}_p[\Omega]$. For each $j \in [p]$, define $e_j\colon [p] \to [1]$ by the requirements

(7.1) $je_j = [1]\colon [1] \to [1]$, $je_k = 0\colon [1] \to [1]$ if $j \neq k$.

Let $\Sigma = \{\omega e_j \mid \omega \in \Omega_p , j \in [p]\}$. Now $T([1],[1])$ is a monoid under composition; let M be the smallest submonoid of $T([1],[1])$ containing Σ .

Proposition 7.1. The monoid M is freely generated by Σ .

Proof. Let $x \in M$, and suppose $x \neq 0$. Then, by (7.1), (6.4), (6.5), we have $\deg(e_j x) = \deg x$ so that, by (6.4), $e_j x \neq 0$ and $\deg(\omega e_j x) = 1 + \deg x$ by (6.6) and again by (6.4) $\omega e_j x \notin \mathcal{S}_p$. It follows $0 \notin M$. If $\omega e_j x = \omega' e_k y \neq 0$, we infer by (6.3) that $\omega = \omega'$ and $e_j x = e_k y \neq 0$. Thus $je_k y = je_j x = x \neq 0$. By (7.1), $j = k$ so that $x = y$. It follows that Σ freely generates M .

Let $\hat{M}$ be the monoid whose elements X are subsets of M and whose multiplication is inherited from M . Specifically, if $X, Y \subset M$, then $XY = \{xy \mid x \in X , y \in Y\}$. Let $X + Y \subset M$ be the union of X and Y . We form a category $[\hat{M}]$ whose objects are $[n]$, $n \geq 0$, and whose morphisms from $[n]$ to $[p]$ are $n \times p$ matrices whose entries are elements of $\hat{M}$. Composition of morphisms in $[\hat{M}]$ is taken as ordinary matrix multiplication. The mapping which takes $f: [n] \to [p]$ in $\mathcal{S}_F$ into the $n \times p$ matrix (f_{ij}) defined by

$$f_{ij} = \{1\} \text{ , if } f(i) = j$$
$$f_{ij} = \emptyset \text{ , otherwise ,}$$

embeds $\mathcal{S}_F$ in $[\hat{M}]$. Given row matrices $\varphi_i: [1] \to [p]$, $i \in [n]$, the $n \times p$ matrix φ whose ith row is φ_i is described by the requirement $i\varphi = \varphi_i$ for each $i \in [n]$. Thus $[\hat{M}]$ is an algebraic theory.

<u>Theorem 7.2</u>. The unique theory morphism

$$G: \mathcal{S}_P[\Omega] \to [\hat{M}]$$

which takes $\omega \in \Omega_p$ into the $1 \times p$ matrix

$$[\omega e_1, \omega e_2, \ldots, \omega e_p]$$

is injective.

<u>Proof</u>. The image T of G is a 0-object algebraic theory which is theory-generated by $G(\Omega)$, together with $0: [1] \to [0]$. Let $\varphi \in T([1],[p])$, $\varphi \neq 0$, and suppose

$$\varphi: [1] \xrightarrow{G(\omega)} [m] \xrightarrow{\psi_{ij}} [p]$$
$$\varphi: [1] \xrightarrow{G(\omega')} [n] \xrightarrow{\psi'_{kj}} [p] \text{ .}$$

Since $\varphi \neq 0$, for some j , $\Sigma_i \omega e_i \psi_{ij} = \Sigma_k \omega' e_k \psi'_{kj} \neq \emptyset$. Thus a word in this set has as first "letter" $\omega e_i = \omega' e_k$ for some i,k . It follows $m = n$, $\omega = \omega'$, and $i = k$.

It is now easy to deduce that the matrices (ψ_{ij}) and $\psi'_{kj})$ are equal. By Theorem 6.1, T is isomorphic to $\mathcal{S}_p[\Omega]$.

We note in passing that if $F: \mathcal{S}_P[\Omega] \to [X]_P$ is a theory-morphism, then the sequence

$$[F(\omega e_1), F(\omega e_2), \ldots, F(\omega e_p)]$$

consists of partial functions whose domains are pairwise disjoint.

8. Ω-Machine Schemes

Let $\Omega = \{\Omega_n\}$, $n > 0$ and $\Omega' = \{\Omega_n\}$, $n \geq 0$. There is no difficulty in defining as in section 4 Ω'-machine schemes, even though Ω_0 may be non-empty. They determine a category $C_t[\Omega']$ as in section 4. The kernel of any strict functor of $C_t[\Omega']$ into a 0-object theory T contains the congruence Z which identifies any two morphisms $[n] \to [p]$ which factor through $[0]$. The quotient category $C[\Omega] = C_t[\Omega']/Z$ depends only upon Ω, since $C_t[\Omega']$ possesses morphisms $[1] \to [0]$. The quotient $C_{ct}[\Omega']/Z$ depends not only upon Ω, but also upon whether or not $\Omega_0 = \phi$. If $\Omega_0 = \phi$, then $C_{ct}[\Omega']$ has no morphisms from [1] to [0].

In the case $\Omega_0 = \phi$, the quotient $C_{ct}[\Omega']/Z$ was previously called $C_{ct}[\Omega]$. In case $\Omega_0 \neq \phi$, we call the quotient $C_c[\Omega]$. As before, we define $\mathcal{M}_c[\Omega] = C_c[\Omega]/K$ and, in analogy with Theorem 5.4, we have:

Theorem 8.1. The theory $\mathcal{M}_c[\Omega]/L$ of cycle-free Ω-machine schemes is (theory isomorphic) to $\mathcal{S}_P[\Omega]$, the 0-object theory freely generated by Ω.

We now fill the gap left at the end of section 5. Let W be the underlying set of the monoid M of the previous section.

Theorem 8.2. The unique theory-morphism $\mathcal{S}_P[\Omega] \to [W]_P$, which extends the family of functions

$$\Omega_p \to [W]_P([1],[p])$$

defined by "$\omega \in \Omega_p$ goes into the partial function $\omega e_j x \mapsto (x,j)$, $j \in [p]$", is injective.

The reader may wish to verify this assertion. See Proposition 12.2 of Elgot (1970).

Corollary 8.3. There is an injective theory-morphism $\mathcal{S}_F[\Omega] \to [W]_P$.

9. Concluding Remarks

Let $\varphi\colon [n] \to [p]$ in $C_c[\Omega]$. The morphism φ, together with an assignment $\Omega \to T$ where $T = [X]$ "corresponds" to "cycle-free program". According to the intuitive discussion of section 2, the "external behavior", defined via a notion of "computation", satisfies: (a) the external behavior of a composite of programs is the composite of their external behaviors; (b) the external behavior of an f-like program is f . According to theorem 4.3, an interpretation F is determined by an assignment $\Omega \to T$; by virtue of (a) and (b), the external behavior of the program determined by φ and this assignment is $F(\varphi)$. Thus "external behavior" may be characterized without the intervention of "computation".

Machines with cycles may be treated as in Elgot (1970) by making use of a suitable subtheory of $[\hat{M}]$.

Computing 6, 349—370 (1970)

The Common Algebraic Structure of Exit-Automata and Machines

By

Calvin C. Elgot, Yorktown Heights

(Received February 13, 1970)

Summary — Zusammenfassung

The Common Algebraic Structure of Exit-Automata and Machines. A notion of "exit-automaton" is introduced to play a role analogous to "machine scheme" or "program scheme". Exit-automata and machines are equipped with algebraic structure leading to a simple characterization of their behaviors.

Die allgemeine algebraische Struktur von Ausgangs-Automaten und Maschinen. Ein Begriff des „Ausgangs-Automaten" wird eingeführt, welcher eine analoge Rolle zu den Begriffen „Maschinen-Schema" oder „Programm-Schema" spielen soll. Ausgangs-Automaten und Maschinen sind mit einer algebraischen Struktur versehen, die eine einfache Charakterisierung ihres Verhaltens ermöglicht.

Introduction

Historically speaking, the notion "finite automaton" was, at least in part, inspired by an attempt to abstract the controlled sequencing aspect of a machine (such as Turing machine) or program for a "universal" machine. The studies of finite automaton do not for the most part reflect this origin. A notable exception is [1] J. D. Rutledge's "On Ianov's Program Schemata". (Related papers: [2], [3]).

The notion "exit-automaton" is analagous to "program scheme" as indicated by Proposition 14.2. Isomorphism classes of exit-automata (resp. machines) are endowed with algebraic structure compatible with their behavior in Sections 4—6. This structure is analyzed in Section 7 where the proof of *Theorem 3.3*, a restricted form of the main theorem, is terminated. After introducing additional mathematical notions in Section 9, Section 12, we are prepared to generalize *Theorem 3.3* to *Theorem 13.1* and deal with the main theorem as applied to machines, *Theorem 14.1.* Section 8 on "minimality" is supplementary.

Our discussion is essentially self-contained. Although a little background in automata theory or machine theory would be helpful [4], [5].

Ideas somewhat similar to those discussed here appear in [6], [7] and [8].

1. Relations and Functions

Let X and Y be sets and let f be a relation from X to Y (notation, $f: X \to Y$). If $x \in X$ is f-related to $y \in Y$, we write $x f y$ or $x \overset{f}{\mapsto} y$. If g is a relation from Y to Z, then the composite relation $fg: X \to Z$ is defined by the condition:

$$x\,(f\,g)\,z \Leftrightarrow \exists\, y\,[x f y \wedge y g z], \text{ where } z \in Z.$$

If h is a relation from X to Y we define $f + h: X \to Y$ by the condition:

$$x\,(f + h)\,y \Leftrightarrow [x f y \vee x h y].$$

The *domain of* f, Dom f, is the set of x such that there exists y satisfying $x f y$. If $X = \text{Dom } f$, then f is said to be *total*. If for each $x \in X$ there is at most one y such that $x f y$, then f is a *univalent relation* or a *partial function*. If f is univalent and total, it is a *function*. We note that neither the partial functions nor the functions are closed under $+$.

Let $[n] = \{1, 2, \ldots, n\}$, $n \geqslant 0$; thus $[0] = \emptyset$. For an arbitrary set X, let $X_{n,p}$ be the set of all relations from $X \times [n]$ into $X \times [p]$ where n, p are non-negative integers. We shall neglect to distinguish between $X \times [1]$ and X. Let $[X]$ denote the doubly indexed family $\{X_{n,p}\}_{n,p}$. Let $[X]_P$ denote the doubly indexed family of partial functions and $[X]_F$ the functions. In the case that X consists of exactly one element $X = \{x\}$, we introduce the special notation $\mathfrak{S}_F$ for $[X]_F$, $\mathfrak{S}_P$ for $[X]_P$, and $\mathfrak{S}_R$ for $[X]$. A relation $f: \{x\} \times [n] \to \{x\} \times [p]$ in $\mathfrak{S}_R$ may, in an obvious way, be regarded as a relation from $[n]$ into $[p]$.

Let $f: [n] \to [p]$ be a relation and let X be arbitrary. We denote by $X \times f: X \times [n] \to X \times [p]$ the relation described by:

$$(x, i) \overset{X \times f}{\longmapsto} (y, j) \Leftrightarrow x = y \wedge i f j.$$

Proposition 1.1. *The function* ι *from* $\mathfrak{S}_R$ *into* $[X]$ *which takes* f *into* $X \times f$ *preserves composition;* ι *takes the identity* $[n] \to [n]$ *into the identity* $X \times [n] \to X \times [n]$. *If* $X \neq \emptyset$, ι *is injective (i. e., one-to-one into).*

In the case $X \neq \emptyset$, we embed $\mathfrak{S}_R$ into $[X]$ via ι. Thus $\mathfrak{S}_F \subseteq \mathfrak{S}_P \subseteq \subseteq \mathfrak{S}_R \subseteq [X]$. We now give "another description" of $[X]$.

Let R_X be the set of all relations from X to X and let $[R_X]$ be the set of all (rectangular) matrices whose entries are elements of R_X.

Proposition 1.2. *Let* θ *be the function from* $[X]$ *into* $[R_X]$ *which takes* $f: X\,x\,[n] \to X\,x\,[p]$ *into the* $(n\,x\,p)$ *matrix* (f_{ij}) *defined by*

$$x\,f_{ij}\,y \Leftrightarrow (x, i)\,f\,(y, j).$$

Then θ *is bijective and preserves composition;* $f \in [X]_P$ if each of the n rows of (f_{ij}) consists of p partial functions $X \to Y$ whose domains are pairwise disjoint; $f \in [X]_F$ iff both $f \in [X]_P$ and the domains of each row of (f_{ij}) sum to X. If $f \in \mathfrak{S}_R$, then f_{ij} is the identity on X if $i \overset{f}{\mapsto} j$; otherwise

$f_{ij} = \emptyset$. In particular, if $f: X \to X \times [p]$ is the empty relation, then $f_{1j} = \emptyset$ for each $j \in [p]$.

We embed $\mathfrak{S}_R$ into $[R_X]$ via $\iota\, \theta$.

Corollary 1.3. *Matrices whose rows consist of partial functions with pairwise disjoint domains are closed under matrix multiplication. Ditto, if in addition the domains of the partial functions in a row sum to X.*

Let $j \in [p]$. We shall write $j: [1] \to [p]$ for the function from $[1]$ to $[p]$ whose value is j and $0: [1] \to [p]$ for the empty partial function from $[1]$ to $[p]$. Let $f: [n] \to [p]$ be in $\mathfrak{S}_P$. It is a consequence of this notation that:

$$i \overset{f}{\mapsto} j \Leftrightarrow if = j; \; if = 0 \Leftrightarrow \text{for no } j,\; i f j;$$

if is the composite: $[1] \xrightarrow{i} [n] \xrightarrow{f} [p]$.

We shall use the notation $j^{-1}: [p] \to [1]$ for the partial function from $[p]$ to $[1]$ whose domain is $\{j\}$.

Let $f: X \to X \times [p+1]$ be a relation. Define $f^\dagger: X \to X \times [p]$ by the condition: $x \overset{f^\dagger}{\mapsto} (y, j)$ iff there is a sequence $x_0, x_1, \ldots, x_r, x_i \in X, r \geqslant 0$, such that $x = x_0$, $x_i \overset{f}{\mapsto} (x_{i+1}, p+1)$ for $i < r$ and $x_r \overset{f}{\mapsto} (y, j)$. Let $f_k: X \xrightarrow{f} X x [p+1] \xrightarrow{X x k^{-1}} X$, $1 \leqslant k \leqslant p+1$. The row matrix $\langle f_1, \ldots, f_{p+1}\rangle$ corresponds to f under θ. We note:

$$x_i \overset{f}{\mapsto} (x_{i+1}, p+1) \Leftrightarrow x_i f_{p+1} x_{i+1}$$

$$x_r \overset{f}{\mapsto} (y, j) \Leftrightarrow x_r f_j y.$$

Thus, $x \overset{f^\dagger}{\mapsto} (y, j) \Leftrightarrow x\,(f^r_{p+1} f_j)\, y$ for some $r \geqslant 0$, where $f^r_{p+1} = f_{p+1} f_{p+1} \cdots f_{p+1}$, r times. Hence

$$\begin{aligned} f_j^\dagger &= f_j + f_{p+1} f_j + f^2_{p+1} f_j + \cdots \\ &= (1 + f_{p+1} + f^2_{p+1} + \cdots) f_j \\ &= f^*_{p+1} f_j, \end{aligned}$$

where 1 is the identity from X to X and $f^*_{p+1} = (1 + f_{p+1} + f^2_{p+1} + \cdots)$.

In the case that f is a partial function, $f^\dagger$ is a partial function and for $i \neq k$, the partial functions $f^i_{p+1} f_j$, $f^k_{p+1} f_j$ have disjoint domains. The operation which takes f into $f^\dagger$ is called *iteration*. Iteration (especially as an operation on partial functions) is the central notion of our discussion. Summarizing:

Proposition 1.4. *Let $f: X \to X \times [p+1]$ be a relation. Then $f_j^\dagger = f^*_{p+1} f_j$ for each $j \in [p]$. If f is a partial function, then so is $f^\dagger$ and the partial functions $f^i_{p+1} f_j$ and $f^k_{p+1} f_j$ have disjoint domains if $i \neq k$.*

23*

2. Paths

Let Σ be a non-empty finite set. Let $W = \Sigma^*$ be the set of all words with "letters" in Σ (including the word 1 of length 0). An *(exit-)automaton* A from n to p on Σ consists of the following data:

(2.1) A set S whose elements are called the *(internal) states* of A.

(2.2) A function $b : [n] \to S$; the image b_i of i under b is called the ith *beginning state* of A.

(2.3) An injective function e from $[p]$ into S; the image e_i of i under e is called the ith *exit* of A.

(2.4) A subset E of $S \times \Sigma \times S$ whose elements are called *edges* of A; if $(s, \sigma, s') \in E$, s is the *start* of the edge, σ is its label and s' is the *end* of the edge; a non-exit state which is not the start of any edge is a *0-state*;

subject to the three conditions:

(2.5) (exit) no edge of A starts with an exit of A.

(2.6) (univalence) if $(x, \sigma, y) \in E$ and $(x, \sigma, z) \in E$, then $y = z$.

(2.7) (completeness) for every non-exit state s of A, either s is a 0-state or for every $\sigma \in \Sigma$, there is an $s' \in S$ such that (s, σ, s') is an edge of A.

The automaton A is *finite* if its set of states is finite.

A finite sequence $(s_0, \sigma_1, s_1, \sigma_2, s_2, \ldots, \sigma_r, s_r)$, $r \geqslant 0$, (where $s_i \in S$ for $0 \leqslant i \leqslant r$ and $\sigma_i \in \Sigma$ for $1 \leqslant i \leqslant r$) is a *path* in A from s_0 to s_r if $(s_i, \sigma_{i+1}, s_{i+1})$ is an edge of A for each i, $0 \leqslant i < r$; s_0 is the *start of the path*, s_r is the *end of the path*, the word $\sigma_1 \sigma_2 \ldots \sigma_r$ is the *label of the path* and the non-negative integer r is the *length of the path*. In particular, if $r = 0$, the label of the path is 1; if $r = 1$, the path is an edge and the label of the path is the label of the edge. A path of positive length with the same start and end is a *cycle*. An exit-automaton A is *cycle-free* if there are no cycles in A.

Let s, s' be states in A. We say s' is *accessible from* s if there is a path in A from s to s'; s' *is accessible* if it is accessible from some beginning state of A; A *is accessible* if each non-exit state is accessible.

If $\alpha = (s_0, \ldots, \sigma_r, s_r)$ and $\beta = (s_r, \sigma_{r+1}, \ldots, s_t)$ are paths in A, we define $\alpha\beta = (s_0, \ldots, \sigma_r, s_r, \sigma_{r+1}, \ldots, s_t)$ so that $\alpha\beta$ is a path in A. In particular, if the length of β is 0, i. e., $r = t$, then $\alpha\beta = \alpha$; similarly, if the length of α is 0, i. e., $r = 0$, then $\alpha\beta = \beta$. Let $|\alpha|$ be the label of α. Then, $|\alpha| = \sigma_1 \ldots \sigma_r$, $|\beta| = \sigma_{r+1} \ldots \sigma_t$ and $|\alpha\beta| = |\alpha|\,|\beta| = \sigma_1 \ldots \sigma_r \sigma_{r+1} \ldots \sigma_t$.

Let $s \in S$. The *behavior of* s (in A) is the $(1 \times p)$ matrix whose j^{th} entry, $j \in [p]$, is the set of all $u \in W$ such that $|\alpha| = u$ for some path α in A from s to e_j. The behavior $|A|$ of A is the $(n \times p)$ matrix whose i^{th} row is the behavior of b_i. Thus $u \in |A|_{ij}$ iff there is a path α in A from b_i to e_j with $u = |\alpha|$.

Proposition 2.1. *The behavior of the jth exit, $j \in [p]$, of an exit-automaton is $j : [1] \to [p]$; the behavior of a 0-state is $0 : [1] \to [p]$.*

3. Statement of the Main Theorem (Restricted Form)

Let $\widehat{W}$ be the set of all subsets of W. We embed W into $\widehat{W}$ via the function which takes $u \in W$ into $\{u\} \in \widehat{W}$; thus $W \subseteq \widehat{W}$. Let $[\widehat{W}]$ be the set of all rectangular matrices whose entries are in $\widehat{W}$. Addition in $\widehat{W}$ is defined as union while multiplication in $\widehat{W}$ is inherited from multiplication in W. In particular, $\emptyset + U = U = U + \emptyset$, $\emptyset\, U = \emptyset = U\, \emptyset$, $1\, U = U = U\, 1$ for every $U \subseteq W$. Multiplication (also called composition) of a $(n \times p)$ matrix and a $(p \times q)$ matrix for $p > 0$ is defined in the usual way. In the case $p = 0$, the $(n \times q)$ product matrix is defined to be the *empty matrix*, i. e., the one all of whose entries are $\emptyset$.

Let M_i be a $(1 \times p)$ matrix for each $i \in [n]$, $n \geqslant 0$. The $(n \times p)$ matrix M whose i^{th} row is M_i is said to be obtained from the family $\{M_i\}_i$ by *tupling*.

We embed $\mathfrak{S}_R$ into $[\widehat{W}]$ by means of the injective function which takes $f: [n] \to [p]$ into the $(n \times p)$ matrix (f_{ij}) defined by: $f_{ij} = 1$ if $i f j$; $f_{ij} = \emptyset$ otherwise. Composition is preserved by this function. We note the following.

Proposition 3.1. *Any subclass of* $[\widehat{W}]$ *which contains* $\mathfrak{S}_F$ *and the* (1×0) *matrix and which is closed under composition and tupling contains* $\mathfrak{S}_P$. *If in addition the subclass contains the* (1×2) *matrix* $\langle 1, 1 \rangle$ *whose entries are both* 1, *then the subclass contains* $\mathfrak{S}_R$.

We write $u \leqslant w$ if there exists v such that $u\, v = w$, where $u, v, w \in W$; u is an *initial segment* of w and w is an *extension* of u.

An exit-automaton A will be called *total* if every accessible state of A has non-empty behavior.

Proposition 3.2. *The following conditions are equivalent:*

(1) *The automaton* A *from* n *to* p *is total.*

(2) *For every* $w \in W$, $i \in [n]$, *there exists* $u \in W$, $j \in [p]$ *such that* $(u \leqslant w \vee w \leqslant u)$ *and* u *is the label of a path in* A *from* b_i *to* e_j.

Proof. Let s be an accessible state with empty behavior (so that s is not an exit). Then, by accessibility, there exists $i \in [n]$ and a path in A from b_i to s; let w be the label of the path. Thus, by (2.5), no initial segment u of w is the label of a path from b_i to an exit. Since the behavior of s is empty, no u which extends w is the label of a path from b_i to an exit. Thus (2) $\Rightarrow$ (1).

Now suppose (1) holds so that, in particular, no accessible state of A is a 0-state. Let $w \in W$, $i \in [n]$ and suppose no $u \leqslant w$ is the label of a path from b_i to an exit. By completeness, (2.7), w is the label of a path from b_i to, say, s. Since s has non-empty behavior some extension of w takes b_i into an exit.

Let M be a $(1 \times (p + 1))$ matrix and let M_j, $1 \leqslant j \leqslant p + 1$ be the jth column, $M_j \subseteq W$. We define $M^\dagger$ to be the $(1\, x\, p)$ matrix whose j^{th} column, $(M^{\dagger})_j = M^*_{p+1} M_j$, $1 \leqslant j \leqslant p$, where $M^*_{p+1} = 1 \cup M_{p+1} \cup M_{p+1} M_{p+1} \cup \ldots$ The operation which takes M into $M^{\dagger}$ is called *iteration.*

Let Σ_0 be an enumeration of all the elements of Σ (without repetition) regarded as a row matrix. Let $\mathfrak{B}$ be the set of all matrices which are behaviors of finite exit-automata, let $\mathfrak{B}_c \subseteq \mathfrak{B}$ consist of the behaviors of automata which are also cycle-free and let $\mathfrak{B}_{ct} \subseteq \mathfrak{B}_c$ consist of the behaviors of those automata which are also total.

Theorem 3.3. *The set* $\mathfrak{B}$ *is the smallest subset of* $[\widehat{W}]$ *which contains* Σ_0 *and* $\mathfrak{S}_F$ *and is closed with respect to* (3.1) *composition,* (3.2) *tupling,* (3.3) *iteration. The set* $\mathfrak{B}_{ct}$ *is obtained by deleting* (3.3) *in the above. The set* $\mathfrak{B}_c$ *is the smallest subset which contains* Σ_0, $\mathfrak{S}_P$ *and is closed with respect to* (3.1) *and* (3.2).

The proof of the theorem is given in Sections 4—7.

It is convenient to have available another operation on matrices. Let M_i be $(n_i \times p_i)$ matrices, $i = 1, 2$. The matrix $M = M_1 \oplus M_2$ is $(n \times p)$ where $n = n_1 + n_2$, $p = p_1 + p_2$ and $M_{ij} = (M_1)_{ij}$ for $1 \leqslant i \leqslant n_1$, $1 \leqslant j \leqslant p_1$; $M_{ij} = (M_2)_{(i-n_1),(j-p_1)}$ for $n_1 < i \leqslant n$, $p_1 < j \leqslant p$; $M_{ij} = \emptyset$ otherwise. This operation is associative and so may be extended to several factors.

Let M_i be a $(1 \times p)$ matrix for each $i \in [n]$. Let N be the $(np \times p)$ matrix defined as follows: $N_{j,j} = 1 = N_{p+j,j} = N_{2p+j,j} = \ldots = N_{(n-1)p+j,j}$ for each $j \in [p]$; otherwise $N_{ij} = \emptyset$. The reader may verify

$$(M_1 \oplus M_2 \oplus \ldots \oplus M_n)\, N = (M_1, M_2, \ldots, M_n). \tag{3.4}$$

4. Composition of Automata

Let A, B be automata respectively from n to p and from p to q and suppose $S_A \cap S_B = \emptyset$. We shall define the composite automaton AB. Roughly speaking the composite will be the automaton from n to q obtained from A, B by identifying the jth exit $e_{A,j}$ of A with the jth beginning state $b_{B,j}$ of B. This is slightly more complicated than appears at first glance since it may be the case that $b_{B,j} = b_{B,k}$ for some distinct pair $\{j, k\}$ of indices.

In preparation for the definition of the set S of states of AB, we define a function from $S_A \cup S_B$ into the set of all subsets of $S_A \cup S_B$.

Let $\bar{s} = \{s\}$ if $s \in S_A$ is not an exit of A or if $s \in S_B$ is not a beginning state of B;

$$\begin{aligned} \overline{b_{B,j}} &= \{e_{A,k} \mid b_{B,k} = b_{B,j}\} \cup \{b_{B,j}\}, \\ \overline{e_{A,j}} &= \overline{b_{B,j}}, \quad 1 \leqslant j \leqslant p. \end{aligned} \tag{4.1}$$

Thus, for any $k, j \in [p]$, either $\overline{b_{B,k}} = \overline{b_{B,j}}$ or $\overline{b_{B,k}} \cap \overline{b_{B,j}} = \emptyset$. We observe, moreover, that $\overline{e_{B,k}} = \overline{e_{B,j}} \Rightarrow k = j$.

The composite AB may now be defined:

$$S = \{\bar{s} \mid s \in S_A \cup S_B\}$$

$$b_i = \overline{b_{A,i}} \quad \text{for } 1 \leqslant i \leqslant n$$

$$e_k = \overline{e_{B,k}} \quad \text{for } 1 \leqslant k \leqslant q$$

$$E = \overline{E_A \cup E_B} = \overline{E}_A \cup \overline{E}_B = \{(\bar{x}, \sigma, \bar{y}) \mid (x, \sigma, y) \in E_A \cup E_B\}$$

where $x, y \in S_A \cup S_B$.

By the observation above, the requirement (2.5) is met. The operation which takes $A : n \to p$, $B : p \to q$ into $AB : n \to q$ is called *composition*.

If $\alpha = (s_0, \sigma_1, s_1, \ldots, \sigma_r, s_r)$, $r \geqslant 0$ is a path in A or in B, then $\bar{\alpha} = (\bar{s}_0, \sigma_1, \bar{s}_1, \ldots, \sigma_r, \bar{s}_r)$ is a path in AB. Moreover, $|\alpha| = (\sigma_1 \ldots \sigma_r) = |\bar{\alpha}|$.

Proposition 4.1. *Let $A : n \to p$ and $B : p \to q$ be automata. Then*

$$|AB| = |A|\,|B|.$$

Proof. Let $w \in (|A|\,|B|)_{ik}$, $1 \leqslant i \leqslant n$, $1 \leqslant k \leqslant q$. Then $w \in |A|_{ij} |B|_{jk}$ for some j, $1 \leqslant j \leqslant p$. Hence there exists $u \in |A|_{ij}$, $v \in |B|_{jk}$ such that $w = uv$ and there exist paths α from $b_{A,i}$ to $e_{A,j}$ with label u and β from $b_{B,j}$ to $e_{B,k}$ with label v. Since $\overline{e_{A,j}} = \overline{b_{B,j}}$, $\bar{\alpha}\bar{\beta}$ is defined and is a path in AB from $b_i = \overline{b_{A,i}}$ to $e_k = \overline{e_{B,k}}$ with label w. Thus $w \in |AB|_{ik}$.

Now assume $w \in |AB|_{ik}$. Then there exists a path γ in AB from b_i to e_k with label w. The path γ must pass through or end in $\overline{e_{A,j}} = \overline{b_{B,j}}$ for some j, $1 \leqslant j \leqslant p$. Thus there exist paths α from $b_{A,i}$ to $e_{A,j}$ in A and β from $b_{B,j}$ to $e_{B,k}$ in B such that $\gamma = \bar{\alpha}\bar{\beta}$. Let $u = |\alpha|$, $v = |\beta|$ so that $w = uv$. Since $u \in |A|_{ij}$ and $v \in |B|_{jk}$, we have $w = uv \in (|A|\,|B|)_{ik}$.

Corollary 4.2. *If M is an $(n \times p)$ matrix in $\mathfrak{B}$ (resp. $\mathfrak{B}_c$, $\mathfrak{B}_{ct}$) and N is a $(p \times q)$ matrix in $\mathfrak{B}$ (resp. $\mathfrak{B}_c$, $\mathfrak{B}_{ct}$) then MN is in $\mathfrak{B}$ (resp. $\mathfrak{B}_c$, $\mathfrak{B}_{ct}$).*

Proof. There exist finite automata A, B with disjoint sets of states whose behaviours are respectively M, N. The composite automaton AB is also finite and its behaviour, from the Proposition, is MN. If A, B are cycle-free so is AB; hence $\mathfrak{B}_c$ is closed with respect to composition. If A, B are total so is AB; hence $\mathfrak{B}_{ct}$ is closed with respect to composition.

5. Tupling of Automata

The *degree* of a finite exit-automaton A is the number of non-exit states in A. From the construction of the previous section, $\deg(AB) = \deg A + \deg B$. If $\deg A = 0$, then every state of A is an exit and A is cycle-free and total; the behaviour of A is in $\mathfrak{S}_F$. In fact, $|A|_{ij} = 1$ iff $b_i = \sigma_j$; $|A|_{ij} = \emptyset$ otherwise. Conversaly, if $f \in \mathfrak{S}_F$, then there is a unique automaton, which we may also call f, of degree 0 such that $|f| = f$. The automaton from 1 to 0 of degree 1 with no edges has $0 : [1] \to [0]$ as its behavior and it is cycle-free. Thus

Proposition 5.1. $\mathfrak{S}_F \subseteq \mathfrak{B}_{ct}$, $\mathfrak{S}_P \subseteq \mathfrak{B}_c$.

Now let A_i be an automaton from 1 to p for each $i \in [n]$ and suppose $S_i \cap S_j = \emptyset$ for $i \neq j$. We first define $A = A_1 \oplus A_2 \oplus \ldots \oplus A_n$ from n to $n\,p$:

$$S = S_1 \cup S_2 \cup \ldots \cup S_n$$

b_i is the beginning state of A_i, $1 \leqslant i \leqslant n$

$e_j = e_{1,j}$, $e_{p+j} = e_{2,j}$, $e_{2p+j} = e_{3,j}, \ldots, e_{(n-1)p+j} = e_{n,j}$, for $j \in [p]$

$$E = E_1 \cup E_2 \cup \ldots \cup E_n.$$

Informally speaking, A is obtained from $\{A_i\}_i$ by regarding the sequence $A_1, \ldots, A_n$ as a single automaton.

Let $f\colon [n\,p] \to [p]$ be the automaton of degree 0 defined by: $j = f_j = f_{p+j} = f_{2p+j} = \ldots = f_{(n-1)p+j}$, for each $j \in [p]$, where f_x is the result of applying f to $x \in [n\,p]$. Then we define

$$(A_1, \ldots, A_n) = (A_1 \oplus \ldots \oplus A_n) f.$$

The automaton $(A_1, \ldots, A_n)$ may be briefly described as being obtained from $\{A_i\}_i$ by identifying all the jth exits, $j \in [p]$.

Proposition 5.2. *Let $A_i\colon 1 \to p$, $i \in [n]$ and suppose $S_i \cap S_j = \emptyset$ for $i \neq j$.* Then

$$|A_1 \oplus \ldots \oplus A_n| = |A_1| \oplus \ldots \oplus |A_n|$$
$$|(A_1, \ldots, A_n)| = (|A_1|, \ldots, |A_n|).$$

Proof. Immediate from definitions.

Corollary 5.3. *If M_i is a $(1\,x\,p)$ matrix in $\mathfrak{B}$ (resp. $\mathfrak{B}_c$, $\mathfrak{B}_{ct}$) for each $i \in [n]$, then $(M_1, \ldots, M_n)$ is in $\mathfrak{B}$ (resp. $\mathfrak{B}_c$, $\mathfrak{B}_{ct}$).*

6. Iterate of an Automaton

Let A be an automaton from 1 to $p + 1$, $p \geqslant 0$. Let S be its set of states. The automaton $A^\dagger$ from 1 to p arises from A by identifying $e_{p+1} \in S$ with $b_1 \in S$. (It is not ruled out that $e_{p+1} = b_1$).

More precisely, $A^\dagger$ is defined as follows:

$S^\dagger = \{\bar{x} \mid x \in S\}$, where $\bar{x} = x$ if $x \neq e_{p+1}$ and $x \neq b_1$,
$\bar{x} = \{e_{p+1}, b_1\}$ otherwise, so that $\bar{e}_{p+1} = \bar{b}_1$;

$b_1^\dagger = \bar{b}_1$;

$e_j^\dagger = \bar{e}_j = e_j$ for $1 \leqslant j \leqslant p$;

$E^\dagger = \bar{E} = \{(\bar{x}, \sigma, \bar{y}) \mid (x, \sigma, y) \in E\}$.

Proposition 6.1. *Let A be an automaton from 1 to $p + 1$. Then* $|A^\dagger| = |A^\dagger|$.

Proof. We first show that $\overline{|A^\dagger|_j} \subseteq |A|_{p+1}^* |A|_j$ for $1 \leqslant j \leqslant p$. Let α be a path in $A^\dagger$ from $\bar{b}_1 = \bar{e}_{p+1}$ to $\bar{e}_j = e_j$. Let $s \upharpoonleft \alpha$ be the result of replacing the start of α by s'. Let $\alpha \upharpoonright s'$ be the result of replacing the end of α by s'. If α has positive length, then $(s \dashv \alpha) \vdash s' = s \dashv (\alpha \vdash s')$, so that parentheses may be dropped.

Claim. Either $b_1 \dashv \alpha$ is a path in A or there exists a path α_1 in $A^\dagger$ from $\overline{b}_1$ to $\overline{b}_1$ of positive length together with a path α' in $A^\dagger$ from $\overline{b}_1$ to e_j such that $\alpha = \alpha_1 \alpha'$ and $b_1 \dashv \alpha_1 \vdash e_{p+1}$ is a path in A.

To verify the claim, assume $b_1 \dashv \alpha$ is not a path in A. From this assumption it follows that α admits a decomposition $\alpha = \beta_1 \beta$ where β_1 is the path of smallest positive length in $A^\dagger$ which ends with $\overline{b}_1$. If $b_1 \dashv \beta_1 \vdash e_{p+1}$ is a path in A then set $\beta_1 = \alpha_1$, $\beta = \alpha'$ so that $\alpha = \alpha_1 \alpha'$. In the contrary case it must be that $b_1 \dashv \beta_1 \vdash b_1$ is a path in A. It follows there exists $k > 0$ and $\beta_1, \ldots, \beta_k, \beta$ satisfying: $\alpha = \beta_1 \beta_2 \ldots \beta_k \beta$, where β_i is a path in $A^\dagger$ of positive length from $\overline{b}_1$ to $\overline{b}_1$ for $1 \leqslant i \leqslant k$; $b_1 \dashv \beta_i \vdash b_1$ is a path in A for $i < k$, and $b_1 \dashv \beta_k \vdash e_{p+1}$ is a path in A. Take $\alpha_1 = \beta_1 \beta_2 \ldots \beta_k$, $\alpha' = \beta$ so that α_1 has positive length, $\alpha = \alpha_1 \alpha'$ and the *claim is verified.*

From the claim it follows that α admits a decomposition

$$\alpha = \alpha_1 \alpha_2 \ldots \alpha_l \alpha', \text{ where } l \geqslant 0,$$

satisfying

$$\alpha_i \text{ is a path in } A^\dagger \text{ from } \overline{b}_1 \text{ to } \overline{b}_1,$$

$$b_1 \dashv \alpha_i \vdash e_{p+1} \text{ is a path in } A, \text{ and}$$

$$\alpha_i \text{ has positive length,}$$

for each i, $1 \leqslant i \leqslant l$ and $b_1 \dashv \alpha'$ is a path in A.

Note that the labels of α_i and $b_1 \dashv \alpha_i \vdash e_{p+1}$ are the same. Similarly, $|\alpha'| = |b_1 \dashv \alpha'|$. Moreover, $|\alpha_i| \in |A|_{p+1}$ for $1 \leqslant i \leqslant l$ and $|\alpha'| \in |A|_j$ so that the desired inclusion follows.

To establish the opposite inclusion, let α_i, $1 \leqslant i \leqslant l$, $l \geqslant 0$, be paths in A from b_1 to e_{p+1} and let β be a path in A from b_1 to e_j. Then $\overline{\alpha}_1 \overline{\alpha}_2 \ldots \overline{\alpha} \overline{\beta}$ is a path in $A^\dagger$ from $\overline{b}_1$ to $e_j = \overline{e}_j$. Here $\overline{\beta}$ is the result of replacing each state s_i in β by $\overline{s}_i$. The desired inclusion now follows from the definition of behavior.

Corollary 6.2. *If M is a $(1 \times (p + 1))$ matrix in $\mathfrak{B}$ then $M^\dagger$ is in $\mathfrak{B}$.*

7. Decomposition

Let Σ have m elements, $m > 0$. In fact, let $\Sigma_0 = \{\tau_i\}_{i \in [m]}$. There is a unique automaton Σ_0 of degree 1 whose behavior is Σ_0; the automaton Σ_0 is the *atomic automaton* with behavior Σ_0. This automaton is cycle-free and, because $m > 0$, it is total. Thus:

Lemma. Σ_0 is in $\mathfrak{B}_{ct}$.

We are now prepared to prove the main theorem, Section 3. Let $\mathfrak{C} \subseteq [\widehat{W}]$ be the smallest class which contains Σ_0 and $\mathfrak{S}_F$ and is closed with respect to composition, tupling, and iteration. Closure under iteration implies

$0 : [1] \to [p]$ is in $\mathfrak{C}$ for every p so that $\mathfrak{S}_P \subseteq \mathfrak{C}$. From the lemma and previous corollaries $\mathfrak{B}$ satisfies these properties so that $\mathfrak{C} \subseteq \mathfrak{B}$ and it remains to show that $\mathfrak{B} \subseteq \mathfrak{C}$.

By the *degree of* $M \in \mathfrak{B}$, we mean $\inf \{\deg A \mid M = |A|\}$. We prove $\mathfrak{B} \subseteq \mathfrak{C}$ by induction on the degree of $M \in \mathfrak{B}$. If $\deg M = 0$, then $M \in \mathfrak{S}_F$, so that $M \in \mathfrak{C}$.

Call the beginning state of an automaton an *entrance* if it is not the end of any edge of the automaton. An *entrance-automaton* is an exit-automaton each beginning state of which is an entrance.

Suppose now $\deg M > 0$.

Case 1. Let M be a $(1 \times q)$ matrix, $|C| = M$, $\deg C = \deg M$ and suppose C is an entrance-automaton.

Now b_1 cannot be an exit of C, for if it were $|C|$ would be in $\mathfrak{S}_F$ and so would be of degree 0. If b_1 is a 0-state then $|C|$ is empty, i. e., $|C| = 0 : [1] \to [q]$ and so is in $\mathfrak{S}_P \subseteq \mathfrak{C}$. Assume now that b_1 is not a 0-state so that for each $\sigma \in \Sigma$ there is exactly one edge (b_1, σ, s) of C which starts at b_1 whose label is σ. Let A be the automaton from m to q defined as follows: $S_A = S_C - \{b_1\}$, $b_{A,j} = s_j$, for $1 \leqslant j \leqslant m$, $e_{A,k} = e_{C,k}$ for $1 \leqslant k \leqslant q$, $E_A = E_C - \{(b_1, \tau_1, s_1), (b_1, \tau_1, s_2), \ldots, (b_1, \tau_m, s_m)\}$.

Since b_1 is not an exit, $\deg C = 1 + \deg A$ and by inductive assumption $|A| \in \mathfrak{C}$. Now we have $\Sigma_0 A = C$ (more precisely, $\Sigma_0 A \approx C$) so that $\Sigma_0 |A| = |\Sigma_0| \, |A| = |\Sigma_0 A| = |C| = M$ and $M \in \mathfrak{C}$. This concludes the argument for Case 1.

Case 2. Suppose $\deg M > 0$, M is a $(1 \times q)$ matrix $|B| = M$, $\deg B = \deg M$. We define an automaton C, called the *anti-iterate of* B from 1 to $q + 1$ such that $C^\dagger = B$ (more strictly, $C^\dagger \approx B$) and $\deg C = \deg B$ by adjoining an additional state to S_B to form S_C; this additional state becomes exit $e_{C,q+1}$ and all edges of B which end in b_1 are "redirected" to $e_{C,q+1}$; all other edges of B are edges of C. More precisely, E_C is the union of the following two sets: $\{(x, \sigma, y) \mid (x, \sigma, y) \in E_B,\ y \neq b_{B,1}\}$ and $\{(x, \sigma, e_{C,q+1}) \mid (x, \sigma, b_{B,1}) \in E_B\}$.

Since $\mathfrak{C}$ is closed under iteration and $|C^\dagger| = |C|^\dagger = |B|$, it is sufficient to show $|C| \in \mathfrak{C}$. But C is an entrance automaton so that $|C| \in \mathfrak{C}$ by Case 1.

We note $|C|_{q+1} \neq \emptyset$ iff there is a cycle in B from its beginning state to itself.

For future reference we note: *If A is an entrance automaton, $B = A^\dagger$ and C is the anti-iterate of B, then $A = C$.*

Case 3. Suppose $\deg M > 0$, M is a $(p \times q)$ matrix, $p > 1$, $|A| = M$, $\deg A = \deg M$.

Let M_j be the jth row of M, $1 \leqslant j \leqslant p$. Let A_j be the automaton from 1 to q whose (sole) beginning state is $b_{A,j}$ but is otherwise identical to A. Then $|A_j| = M_j$ and $\deg A_j = \deg A$. We say $\{A_j\}_j$ is obtained from A by *anti-tupling*. To be more economical $A_j = j\,A$ may be replaced by its "accessible part" defined in the next section.

Since $\mathfrak{C}$ is closed under tupling, in order to show $M \in \mathfrak{C}$, it is sufficient to show $M_j \in \mathfrak{C}$ for each j. Thus Case 3 reduces to the earlier cases.

The results for $\mathfrak{B}_c$ and $\mathfrak{B}_{ct}$ are obtained (without the need for Case 2) by observing that the automata obtained by the decompositions are also cycle-free. If the given automata are total, so are the results of decomposition.

From the above argument, we obtain

Theorem 7. *For a finite automaton $A : p \to q$, we have*

$$|A| = |C_1^\dagger, C_2^\dagger, \ldots, C_p^\dagger|, \text{ where } \deg C_j \leqslant \deg A, \tag{7.1}$$

C_j is the anti-iterate of j A and for each j such that $b_{A,j}$ is not an exit of A

$$C_j = D_j E_j; \text{ here } \deg E_j < \deg C_j \tag{7.2}$$

and

(a) $D_j = 0 : 1 \to 0$ *and E_j is obtained from C_j by deleting $b_{A,j}$*

or

(b) $D_j = \Sigma_0 : 1 \to m$ *and E_j is the A described under Case 1.*

8. Minimality

Let $V \subseteq W$ and let $u \in W$. We define: $L_u V = \{v \mid u v \in V\}$. A set V is *recognizable* if the class of sets $\{L_u V \mid u \in W\}$ is finite. It is a familiar (and in any case easily verifiable) fact that if V is recognizable the elements of $\{L_u V \mid u \in W\}$ may be taken as the states of the "minimal" ordinary (i. e.. non-exit) finite automaton which recognizes V. We show below that a similar result holds for exit-automata.

A matrix $M \in [\widehat{W}]$ will be called *recognizable* if each of its entries is recognizable.

An exit-automaton will be called *0-normal* if every state with empty behavior is a 0-state; the automaton is *normal* if it is 0-normal and accessible.

Proposition 8.1. *A normal finite automaton A from n to p is cycle-free if its behavior is finite, i. e., $|A|_{ij}$ is a finite set for each $i \in [n]$, $j \in [p]$.*

Proof. Suppose there is a cycle in A. i. e., a path α of positive length from s to s for some state s of A. Since the labels of edges in A have positive length, the label $|\alpha|$ of α has positive length. By *accessibility* there is a $u \in W$ which is the label of a path from b_i to s for some $i \in [n]$. Since s is not a 0-state, its behavior is non-empty, i. e., some $v \in W$ is the label of a path from s to e_j, for some $j \in [p]$. Thus, by normality, $u |\alpha|^* v \subseteq |A|_{ij}$ and $|A|_{ij}$ is infinite since $|\alpha|$ has positive length.

Assume A is cycle-free. Then there is $r \geqslant 0$ such that every path in A has length at most r. Hence $|A|$ is finite.

Let M be an $(n \times p)$ matrix in $[\widehat{W}]$. We call M *univalent* if it satisfies

$$(u \in M_{ij} \wedge uv \in M_{ik}) \Rightarrow (v = 1 \wedge j = k). \tag{8.1}$$

Thus M is univalent iff each of its rows is univalent.

Proposition 8.2. *The behaviour of an exit-automaton is univalent.*
Let $V = \langle V_1, V_2, \ldots, V_p \rangle$, $V_j \subseteq W$, $j \in [p]$. We define

$$L_u V = \langle L_u V_1, L_u V_2, \ldots, L_u V_p \rangle.$$

Proposition 8.3. *Let $M \in [\widehat{W}]$ be univalent. Then*

$$u \in M_{ij} \Leftrightarrow L_u M_i = j : [1] \to [p]. \tag{8.2}$$

Proof. If $L_u M_i = j$, then $L_u M_{ij} = 1$ so that $u \in M_{ij}$. For the converse, let $u \in M_{ij}$. Then, by (8.1), $v \in L_u M_{ik} \Rightarrow (v = 1 \wedge j = k)$ so that $L_u M_{ij} = 1$ and $L_u M_{ik} = \emptyset$ for $j \neq k$ from which (8.2) follows.

An exit-automaton is *reduced* if no two distinct states have the same behavior; it is minimal if it is normal and reduced.

Proposition 8.4. *Let $M = (M_{ij})$ be a $(n \times p)$ univalent matrix in $[\widehat{W}]$. The following construction, where M_i is the ith row of M, gives a minimal automaton A whose behavior is M.*

S consists of: the row matrices $j : [1] \to [p]$, $j \in [p]$,
the row matrices M_i, for each $i \in [n]$,
$L_u M_i$ where $u \leqslant v$ for some $v \in \Sigma_j M_{ij}$, for each $i \in [n]$.

E consists of the triples $(L_u M_i, \sigma, L_{u\sigma} M_i)$ where $\sigma \in \Sigma$, $L_u M_i \neq \emptyset$, $L_u M_i \neq j$, and $i \in [n]$.

$b_i = M_i$ and $e_j = j$.

Proof. The automaton A is clearly normal. It is reduced because the behavior of state $L_u M_i$ is the $(1 x p)$ matrix $L_u M_i$. In particular, the behavior of $b_i = M_i = L_1 M_i$ is M_i so that $|A| = M$.

Note that if $M_i = \langle \emptyset, \emptyset, \ldots, \emptyset \rangle$ then M_i is a 0-state.

Proposition 8.5. *Let B be a normal automaton with behavior M and let A be the minimal automaton of the previous proposition (with the same behavior). For each state s of B, let $\bar{s}$ be its behavior. Then the function $s \mapsto \bar{s}$ from S_B into S_A is surjective (onto) and $(s, \sigma, s') \in E_B \Rightarrow (\bar{s}, \sigma, \bar{s}') \in E_A$. The function $(s, \sigma, s') \mapsto (\bar{s}, \sigma, \bar{s}')$ from E_B into E_A is also surjective.*

Corollary 8.6. *Any two minimal automata with the same behavior are isomorphic.*

Proposition 8.7. *An $(n \times p)$ matrix M is the behavior of some finite exit-automaton from n to p if M is univalent and recognizable.*

Proof. Let A be a finite automaton from n to p. By Proposition 8.2, $|A|$ is univalent. Given $i \in [n]$, $j \in [p]$, one readily converts A into an ordinary finite automaton with initial state b_i and final set $\{e_j\}$ of states [13] with behavior $|A|_{ij}$. Thus $|A|_{ij}$ is recognizable.

Now suppose M univalent and recognizable. Let A be the automaton of Proposition 8.4 with behavior M. It remains to show A finite. Let $\mathfrak{C}_{ij} = \{L_u M_{ij} | u \in W\}$; $\mathfrak{C}_i = \{L_u M_i | u \in W\}$. Now $\mathfrak{C}_i \subseteq \mathfrak{C}_{i1} \times \mathfrak{C}_{i2} \times \ldots \times \mathfrak{C}_{ip}$. Since $\mathfrak{C}_{ij}$ is finite, it follows A is finite.

Proposition 8.8. *The class of univalent (resp., and recognizable) matrices is closed under composition and iteration.*

Let A be an automaton from n to p, let $S_\emptyset$ be the set of states of A with empty behavior and let $x \in S_\emptyset$. Let B be the automaton from n to p defined as follows: $S_B = (S_A - S_\emptyset) \cup \{x\}$; the edges of B consist of those edges (s, σ, s') where $s, s' \in S_A - S_\emptyset$ together with $\{(s, \sigma, x) | s \in S_A - - S_\emptyset \wedge (s, \sigma, s') \in E_A$ for some $s' \in S_\emptyset\}$; $b_{B,i} = b_{A,i}$ if $b_{A,i} \in S_A - S_\emptyset$; $b_{B,i} = x$ otherwise; $e_{B,i} = e_{A,i}$. In particular, if $S_\emptyset = \emptyset$ then $A = B$. We call B the *0-normal part* of A.

Proposition 8.9. *If B is the 0-normal part of A then B is 0-normal,* $|B| = |A|$, *and* $\deg B \leqslant \deg A$.

Let C from n to p be obtained from A as follows: S_C is the set of accessible states of A together with A' s exits;
$E_C = \{(s, \sigma, s') | (s, \sigma, s') \in E_A \wedge s \in S_C \wedge s' \in S_C\}$; $b_{C,i} = b_{A,i}$, $e_{C,j} = e_{A,j}$. We call C the *accessible part* of A. (The completeness of C follows from the fact that no edge of A starts with an accessible state and ends with an inaccessible one.

Proposition 8.10. *If C is the accessible part of A, then C is accessible,* $|C| = |A|$. *If A is finite,* $\deg C \leqslant \deg A_r$ *if* $C \neq A$, *then* $\deg C < \deg A$. *If A is 0-normal then so is C.*

Proposition 8.11. *For any exit-automaton A there is a minimal exit-automaton B such that* $|A| = |B|$. *If A is finite* $\deg B \leqslant \deg A$; *if A is 0-normal but not minimal then* $\deg B < \deg A$.

Proof. From Propositions 8.10, 8.5.

We recall that "Σ_0" is used to denote both a certain row matrix and the atomic automaton whose behavior is this row matrix.

Proposition 8.12. *If $A = \Sigma_0 B$ where A is a minimal automaton then B is minimal.*

Proof. Any non-exit state s of B is a non-exit state of A so that there is a path from $b_{A,1}$ to s; such a path must pass through a beginning state of B. Thus B is accessible. The behavior of any state in B is the same as its behavior as a state in A; hence B is 0-normal and reduced.

Proposition 8.13. *If $A = B^\dagger$ where A is a minimal automaton from 1 to p and B is an entrance-automaton, then B is minimal. Moreover, B is the anti-iterate of A.*

Proof. Let s be a non-exit state of B. Then s is a non-exit state of A and there is a path α in A from $b_{A,1} = b_{B,1}$ to s; α may be chosen so as not to "pass through" $b_{A,1}$. Then α is also a path in B and B is accessible.

Let B_s be the automaton obtained from B by changing the initial state of B to s so that $|B_s|$ is the behavior of s in B. Since B is an entrance automaton, the simple formula below gives the behavior of s in A:

$$|A_s|_j = |B_s|_{p+1} |B|^*_{p+1} |B|_j + |B_s|_j. \tag{8.3}$$

In particular, if $s = b_{A,1} = b_{B,1}$, we obtain

$$|A|_j = |B|_{p+1} |B|^*_{p+1} |B|_j + |B|_j = |B|^*_{p+1} |B|_j. \tag{8.4}$$

Thus, if $|B_s|$ is empty, we have $|B_s|_{p+1} = \emptyset = |B_s|_j$ so that by (8.3) $|A_s|_j = \emptyset$ and $|A_s|$ is empty. Hence the 0-normality of A implies the 0-normality of B. It also follows from formula (8.3) that two states which have the same behavior in B have the same behavior in A. Since A is reduced so is B.

Since B is an entrance-automaton, it is the anti-iterate of A.

Proposition 8.14. *Let A be a minimal automaton from p to q. Let A_j be the automaton from 1 to q, whose sole beginning state is $b_{A,j}$ but is otherwise identical to A. Let B_j be the accessible part of A_j. Then B_j is minimal.*

The previous three propositions show that if one "starts" with a minimal automaton, the results of the decompositions of Section 7 yield minimal automata (provided Case 3 is modified to produce B_j instead of A_j).

Proposition 8.15. *Let $A = \Sigma_0 B$ have non-empty behavior, where B is a minimal finite automaton from m to p. Then* $\deg|B| < \deg|A| \Leftrightarrow A$ *is minimal.*

Proof. Every $s \in S_A$ with empty behavior in A is in S_B and has empty behavior in B. Thus A is 0-normal so that

$$\deg|A| = \deg A \Leftrightarrow A \text{ is minimal.}$$

Since $\deg A = 1 + \deg B = 1 + \deg|B|$, the result follows.

Example. Let A be an automaton from 1 to $p+1$ and let $B = (1, 2, \ldots, p, A^\dagger)$, where, for each $j \in [p]$, $j: 1 \to p$ is the automaton of degree 0 with behavior $j \in \mathfrak{S}_F$. Then $|AB| = |A^\dagger|$. In particular, if Σ consists of two letters, we have, taking $p = 1$ and $A = \Sigma_0$, $|\Sigma_0 (1, \Sigma_0^\dagger)| = |\Sigma_0^\dagger|$. The automata Σ_0, $\Sigma_0^\dagger$, $(1, \Sigma_0^\dagger)$ have degree 1. The automaton $\Sigma_0 (1, \Sigma_0^\dagger)$ has degree 2 and is not minimal; its behaviour has degree 1.

Example. Let $A: 1 \to p+1$ and let $B = (1, 2, \ldots, p, A)$ where $B: p+1 \to p+1$. Then $|A^\dagger| = |(AB)^\dagger|$.

In particular, if Σ_0 is as above, $|\Sigma_0| = \langle \sigma_1, \sigma_2 \rangle$, then $|\Sigma_0{}^\dagger| = \sigma_2^* \sigma_1 = (\sigma_2 \sigma_2)^* (\sigma_1 + \sigma_2 \sigma_1) = |(\Sigma_0 (1, \Sigma_0))^\dagger|$. In this case, the automaton $AB = \Sigma_0 (1, \Sigma_0)$ is a minimal entrance — automaton from 1 to 2 with behavior $\langle \sigma_1 + \sigma_2 \sigma_1, \sigma_2 \sigma_2 \rangle$ but $(AB)^\dagger$ is *not* minimal.

9. Algebraic Theories

We assume the notion "category" is familiar (cf. [9], [10]). We may now identify $\mathfrak{S}_F$ as a category. The class of objects of $\mathfrak{S}_F$ is the class of sets $[n]$, $n \geqslant 0$. The set of morphisms from $[n]$ to $[p]$ is the set of all functions from $[n]$ to $[p]$.

In any category $\mathfrak{C}$ if $\varphi : A \to B$ and $\psi : B \to C$, the composite morphism from A to C will be written $\varphi\,\psi$. We may now define algebraic theory (cf. [11], [12]).

An *algebraic theory* T is a category which contains $\mathfrak{S}_F$ as a subcategory, which has the same objects as $\mathfrak{S}_F$, and which satisfies:

(9.1) If φ, $\psi : [n] \to [p]$ are in T and if $i\,\varphi = i\,\psi$ for each $i \in [n]$, then $\varphi = \psi$.

(9.2) If $\varphi_i : [1] \to [p]$ is in T for each $i \in [n]$, then there exists $\varphi : [n] \to [p]$ such that $i\,\varphi = \varphi_i$.

The morphism $i\,\varphi$ is the composite $[1] \xrightarrow{i} [n] \xrightarrow{\varphi} [p]$. From (9.1) we infer that the unique function from $[0]$ to $[p]$ is the only morphism from $[0]$ to $[p]$. Let $\varphi_i : [1] \to [p]$ be in T for each $i \in [n]$. The unique morphism $\varphi : [n] \to [p]$ specified by (9.1) and (9.2) is denoted by $(\varphi_1, \ldots, \varphi_n)$. In particular, if $\varphi_i \in \mathfrak{S}_F$ then $\varphi = (\varphi_1, \ldots, \varphi_n) \in \mathfrak{S}_F$ and $i\,\varphi = \varphi_i$.

10. Examples

We may regard $[X]$, $X \neq \emptyset$, as an algebraic theory by treating a relation from $X \times [n]$ into $X \times [p]$ as a morphism from $[n]$ into $[p]$. In this connection, we recall that the function $f \mapsto X \times f$, where f is a function from $[n]$ into $[p]$ is injective and preserves composition. Then $[X]_P$ is a subtheory of $[X]$ and $[X]_F$ is a subtheory of $[X]_P$.

We may regard $[\widehat{W}]$ as an algebraic theory by treating an $(n \times p)$ matrix as a morphism from $[n]$ into $[p]$. Then $\mathfrak{B}_{ct}$ is a subtheory of $\mathfrak{B}_c$ which is a subtheory of $\mathfrak{B}$ which is a subtheory of $[\widehat{W}]$.

11. Description of a Free Theory

Let T and $\overline{T}$ be algebraic theories. By a *functor* (or a morphism between theories) from T into $\overline{T}$ we shall mean a function which takes a morphism $\varphi : [n] \to [p]$ in T into a morphism $\overline{\varphi} : [n] \to [p]$ in $\overline{T}$ subject to the conditions:

(11.1) If $j : [1] \to [p]$, $j \in [p]$, then $\overline{j} = j$.

(11.2) $\overline{\varphi\,\psi} = \overline{\varphi}\,\overline{\psi}$, where $\psi : [p] \to [q]$.

Since $j\,\overline{\varphi} = \overline{j}\,\overline{\varphi} = \overline{j\,\varphi}$ for each $j \in [p]$, it follows that functors preserve tupling and keep $\mathfrak{S}_F$ elementwise fixed.

By Section 8, the degree of $\varphi \in \mathfrak{B}$, $\varphi : [n] \to [p]$, is the degree of the minimal finite automaton from n to p with behavior φ. We note the following properties of degree:

$$\deg \varphi = 0 \text{ iff } \varphi \in \mathfrak{S}_F;$$
$$\text{if } n = 1,\ \varphi = \Sigma_0\, \psi \text{ and } \deg \psi < \deg \varphi, \text{ then } \deg \varphi = 1 + \deg \psi;$$
$$\sup\{\deg i\, \varphi \mid i \in [n]\} \leqslant \deg \varphi \leqslant \Sigma \deg i\, \varphi,\ i \in [n].$$

We note too that if $\varphi : [1] \to [p]$ is in $\mathfrak{B}_{ct}$ and has positive degree, then there exists (a unique) ψ such that $\varphi = \Sigma_0\, \psi$ and $\deg \psi < \deg \varphi$.

Let Ω_r be a set of morphisms in T from $[1]$ to $[r]$ and let $\Omega = \{\Omega_r\}$, $r > 0$. We say T is *freely generated by* Ω iff given any theory $\overline{T}$ and family of morphisms $\overline{\varphi}$ in $\overline{T}$, where $\varphi \in \Omega_r$ and $\overline{\varphi} : [1] \to [r]$ for some $r > 0$, there is a unique functor from T into $\overline{T}$ which takes φ into $\overline{\varphi}$.

Theorem. *$\mathfrak{B}_{ct}$ is freely generated by $\Sigma_0 : [1] \to [m]$.*

Proof. Let $\overline{T}$ be an arbitrary algebraic theory and let $\overline{\Sigma}_0 : [1] \to [m]$ be in $\overline{T}$. For $j : [1] \to [p]$, $j \in [p]$, define $\overline{j} = j$; for $\mu : [0] \to [p]$ define $\overline{\mu} = \mu$. For $\psi : [n] \to [p]$, $n > 1$, define $\overline{\psi}$ by the requirement $i\, \overline{\psi} = \overline{i\, \psi}$, $i \in [n]$. For $\varphi : [1] \to [p]$ of positive degree, φ admits a unique factorization $\varphi = \Sigma_0\, \psi$ where $\deg \psi < \deg \varphi$, $\psi : [m] \to [p]$; define $\overline{\varphi} = \overline{\Sigma}_0\, \overline{\psi}$. Then $\overline{\varphi}$ is unambiguously defined for every φ in $\mathfrak{B}_{ct}$. We now show that the function $\varphi \mapsto \overline{\varphi}$, $\varphi \in \mathfrak{B}_{ct}$, $\overline{\varphi} \in \overline{T}$ is a functor.

We show $\overline{\varphi\, \psi} = \overline{\varphi}\, \overline{\psi}$, $\varphi : [n] \to [p]$, $\psi : [p] \to [q]$ by induction on the degree of φ.

Case 1. $n = 1$, $\deg \varphi > 0$. We have $\varphi = \Sigma_0\, \psi'$, $\deg \varphi' < \deg \varphi$, so that $\overline{\varphi\, \psi} = \overline{(\Sigma_0\, \varphi')\, \psi} = \overline{\Sigma_0\, (\varphi'\, \psi)} = \overline{\Sigma}_0\, \overline{\varphi'\, \psi} = \overline{\Sigma}_0\, (\overline{\varphi'}\, \overline{\psi}) = (\overline{\Sigma}_0\, \overline{\varphi'})\, \overline{\psi} = \overline{\Sigma_0\, \varphi'}\, \overline{\psi} = \overline{\varphi}\, \overline{\psi}$.

Case 2. $n > 1$, $\deg \varphi > 0$. We note that $\overline{\varphi\, \psi} = \overline{\varphi}\, \overline{\psi}$ if for each $i \in [n]$, $i\, \overline{\varphi\, \psi} = i\, (\overline{\varphi}\, \overline{\psi})$. Thus, $i\, \overline{\varphi\, \psi} = \overline{i\, (\varphi\, \psi)} = \overline{(i\, \varphi)\, \psi} = \overline{i\, \varphi}\, \overline{\psi} = (i\, \overline{\varphi})\, \overline{\psi} = i\, (\overline{\varphi}\, \overline{\psi})$. The third equality holds by Case 1 if $\deg i\, \varphi > 0$. If $\deg i\, \varphi = 0$, then $i\, \varphi = j$ for some $j \in [p]$ and $\overline{(i\, \varphi)\, \psi} = \overline{j\, \psi} = j\, \overline{\psi} = \overline{j}\, \overline{\psi}$. The case $\deg \varphi = 0$ should now be clear.

The uniqueness argument is entirely routine.

12. Iterative Functors

Let $\mathfrak{C}$ be a subtheory of $[\widehat{W}]$ which is closed under iteration. Let $\mathfrak{C}'$ be a subtheory of $[\widehat{V}]$ or of $[Y]$ which is closed under iteration, where V is the set of all words on some alphabet and Y is some non-empty set. A functor from $\mathfrak{C}$ into $\mathfrak{C}'$ will be called *iterative* if it preserves iteration.

Proposition 12.1. *Let $\varphi = \Sigma_0 : [1] \to [m]$ and let $\overline{\varphi} : [1] \to [m]$ be an arbitrary morphism in $\mathfrak{C}'$. There is one and only one iterative functor from $\mathfrak{B}$ into $\mathfrak{C}'$ which takes φ into $\overline{\varphi}$. If $\overline{\varphi}$ is univalent then so is $\overline{\psi}$ for every ψ in $\mathfrak{B}$.*

Proof. The uniqueness follows from the main theorem by a routine argument. If $\Sigma_0 = \langle \tau_1, \ldots, \tau_m \rangle$ the desired functor must take $\Sigma_0 j^{-1} = \tau_j$ into $\bar{\varphi} j^{-1}$. We now define a function F from $[\widehat{W}]$ into $\mathfrak{C}'$ by first requiring that F take τ_j into $\bar{\varphi} j^{-1}$ and that F restricted to W be a monoid homomorphism; extend F to $\widehat{W}$ by defining $\overline{U} = \{\Sigma \bar{u} \mid u \in U\}$, where $U \subseteq W$ and finally, for $M \in [\widehat{W}]$, define $((\overline{M})_{ij}) = \overline{(M_{ij})}$. This extends F to all of $[\widehat{W}]$. Again it is routine to verify that F is a "completely additive" functor and from this follows that F is iterative.

Let f be the partial function from W into $W \times [m]$ defined by: $\tau_i u \xrightarrow{f} (u, i)$ so that $f \in [W]_P$.

Proposition 12.2. *The unique functor from* $\mathfrak{B}$ *into* $[W]$ *which takes* Σ_0 *into* f *(defined above) is injective.*

Proof. The functor takes $u \in W$ into $L_u : W \to W$ where $w \overset{L_u}{\longmapsto} v \Leftrightarrow w = uv$ and the functor takes $U \subseteq W$ into L_U where $w \overset{L_U}{\longmapsto} v \Leftrightarrow w = uv$ for some $u \in U$. It is sufficient to observe: $u \in U \Leftrightarrow u \overset{L_U}{\longmapsto} 1$.

13. Automaton Species

We now broadly generalize our preceding considerations, put them in a more natural setting and facilitate making connections with "machines" by introducing the notion "automaton species". At this point *we remove the restriction that* Σ *be finite* and also permit $\Sigma = \emptyset$.

By an *(automaton) species* over Σ, we mean a set Γ of finite sequences of elements of $W = \Sigma^*$. An exit-automaton A of species Γ consists of the data (2.1), (2.2), (2.3), and (2.4) modified so that $E \subseteq S \times W \times S$, subject to condition (2.5) and:

(13.1) for every non-exit $s \in S_A$, there is an enumeration (without repetition)

$$(s, w_1, s_1),\ (s, w_2, s_2),\ \ldots,\ (s, w_r, s_r),\quad r \geqslant 0,$$

of all the edges of A starting with s, such that the finite sequence $(w_1, \ldots, w_r)$ is in Γ.

Condition (13.1) replaces conditions (2.6) and (2.7).

With this new terminology we may identify the automata considered up to now as being of species $\Gamma = \{\Sigma_0, \emptyset\}$, where $\emptyset$ is the empty sequence and where Σ is finite.

Denote by $|\Gamma|$ the class of matrices $|A| \in [\widehat{W}]$ where A is a finite exit-automaton of species Γ. We regard the elements of Γ as row matrices in the obvious way. Thus, if $\Gamma = \{\Sigma_0, \emptyset\}$, $|\Gamma| = \mathfrak{B}$.

Theorem 13.1. *The set* $|\Gamma|$ *is the smallest subset of* $[\widehat{W}]$ *which contains* Γ *and* $\mathfrak{S}_F$ *and is closed with respect to composition, tupling, and iteration. The set* $|\Gamma|_c$ *is the smallest class of matrices which contains* Γ *and* $\mathfrak{S}_F$ *and is closed with respect to composition and tupling. If* $\emptyset \notin \Gamma$, *then* $|\Gamma|_c = |\Gamma|_{ct}$.

The proof is essentially the same as for the special case stated in Section 3.

The minimality discussion of Section 8 is applicable to automata of species Γ where $\emptyset \in \Gamma$ and each member of Γ is a finite sequence of Distinct elements of Σ. We now consider automata of a more special species.

Let $\Omega = \{\Omega_k\}$ where for each $k > 0$, Ω_k is an arbitrary set and $\Omega_k \cap \Omega_l = \emptyset$ for $k \neq l$. We construct an "alphabet" $\Sigma_\Omega = \{(\omega, i) \mid \omega \in \Omega_k, i \in [k]$ for some $k > 0\}$. Let $\Gamma_\Omega = \{((\omega, 1), (\omega, 2), \ldots, (\omega, k)) \mid \omega \in \Omega_k$ for some $k > 0\}$. The result of Section 11 readily generalizes to:

Proposition 13.2. *Setting $\Gamma = \Gamma_\Omega$, $|\Gamma|_c = |\Gamma|_{ct}$ is freely generated by Γ_Ω.*

A similar argument yields the following analogue of Proposition 13.2.

Proposition 13.3. *Let $\Gamma = \Gamma_\Omega \cup \{\emptyset\}$, let $\overline{T}$ be an algebraic theory which possesses a unique morphism $[1] \to [0]$ and let F be a function which takes $\varphi : [1] \to [p]$ in Γ into $\overline{\varphi} : [1] \to [p]$ in $\overline{T}$. Then F admits a unique extension to a functor from $|\Gamma|_c$ into $\overline{T}$.*

14. Machine Species

Let X be a set. By a *machine species* over X we mean a set Γ of finite sequences of relations $f : X \to X$ in R_X. The set Γ is *univalent* if each of its members is a sequence of partial functions $X \to X$ whose domains are pairwise disjoint. An X-machine A of species Γ consists of the data (2.1), (2.2), (2.3), and (2.4) modified so that $E \subseteq S \times R_X \times S$ subject to condition (2.5) and:

(14.1) for every non-exit $s \in S_A$, there is an enumeration (without repetition) (s, f_1, s_1), (s, f_2, s_2), $\ldots$, (s, f_r, s_r), $r \geqslant 0$, of all the edges of A starting with s, such that the finite sequence $(f_1, \ldots, f_r)$ is in Γ.

The behavior $|A|$ of an X-machine A is an element of $[R_X]$. It is defined just as the behavior of automata with multiplication in W (i. e., concatenation) replaced by multiplication in R_X (i. e., composition). The sum of a family of subsets of W is replaced by sum of a family of relations in R_X.

Let $F : W \to R_X$ be a monoid homomorphism; extend F to finite sequences $\{w_i\}_{i \in [r]}$ by setting $\{w_i\}_i \overset{F}{\longmapsto} \{f_i\}_i$ iff $w_i \overset{F}{\longmapsto} f_i$ for each $i \in [r]$. Let Γ' be the image of Γ under F; F induces in an obvious way a function $F : \mathbf{A}_\Gamma \to \mathbf{A}_{\Gamma'}$ (called an *interpretation*) where $\mathbf{A}_\Gamma$ is the class of exit-automata of species Γ and where $\mathbf{A}_{\Gamma'}$ is the class of X-machines of species Γ'. Specifically, the image A' of $A \in \mathbf{A}_\Gamma$ under F is obtained by replacing the edges (x, w, y) of A by (x, f, y) where $w \overset{F}{\longmapsto} f$; A' is the *interpretation* of A under F. We note the commutativity of the diagram below where $F : [\widehat{W}] \to [R_X]$ is uniquely determined by $F : W \to R_X$ as described in Section 12.

$$\begin{array}{ccc} \mathbf{A}_\Gamma & \xrightarrow{A \,\longmapsto\, |A|} & [\hat{W}] \\ \Big\downarrow F & & \Big\downarrow F \\ \mathbf{A}_{\Gamma'} & \xrightarrow{A' \,\longmapsto\, |A'|} & [R_x] \end{array}$$

Let $|\Gamma'|$ be the set of behaviors of finite X-machines of species Γ'. From the commutativity of the diagram and Theorem 13.1 we obtain:

Theorem 14.1. *The set $|\Gamma'|$ is the smallest subset of $[R_X] \approx [X]$ which contains Γ' and $\mathfrak{S}_F$ and is closed with respect to composition, tupling, and iteration. If Γ' is univalent then $|\Gamma'| \subseteq [X]_P$.*

Theorem 14.2. *Let Γ consist of finite sequences of distinct elements of Σ and let $A_i \in \mathbf{A}_\Gamma$, $i = 1,2$. Then $|A_1| = |A_2|$ if for all interpretations F, $|A_1'| = |A_2'|$ where $A_i \overset{F}{\longmapsto} A_i'$. In fact, if $|A_1| \neq |A_2|$ there exists an interpretation F such that Γ' is univalent and $|A_1'| \neq |A_2'|$.*

Proof. Consult Proposition 12.2.

Theorem 14.3. *Theorem 7 holds with "automaton" replaced by "machine".*

Theorem 14.2 is reminiscent of Rutledge's extension [1] of a result of Ianov [2].

Let $f_j \in R_X$, $j \in |p|$. The *atomic machine* from 1 to p determined by $\{f_j\}_j$ is the machine of degree 1 with $p+1$ states $b_1, e_1, \ldots, e_p$ and edges (b_1, f_j, e_j), $j \in [p]$. The state b_1 is the beginning state and e_j is the jth exit.

The machines considered here may be called Mealy *Machines* to distinguish them from the Moore *Machines* we consider elsewhere [12].

15. Turing Machines

The "behavior" of a machine is more usually defined via an appropriate notion of "computation". In order to dispel any fear that a different notion of behavior may be so obtained, we note that from *Theorem 14.3* we have:

Theorem 15.1. *The behavior function $|\ |$ is completely determined by the following four properties:*

(15.1) $|AB| = f|B|$, *where A is the atomic machine determined by $f = \langle f_1, \ldots, f_p \rangle$.*

(15.2) $|A| = |(A_1, A_2, \ldots, A_n)| = |A_1|, |A_2|, \ldots, |A_n|)$ *where $A_i = iA$; $A_1, \ldots, A_n$ is obtained from A by anti-tupling.*

(15.3) $|A^\dagger| = |A|^\dagger$.

(15.4) *If $A : 1 \to p$ and $b_{A,1} = e_{A,j}$, then $|A| = j : [1] \to [p]$. By taking in* (15.1) *$B = [p]$, the identity function on $[p]$, we obtain:*

(15.5) *If A is the atomic machine determined by f, then $|A| = f$.*

We now proceed to give an example.

24*

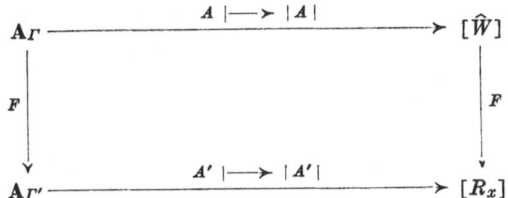

Let $\Sigma_0 = (\tau_0, \tau_1, \ldots, \tau_m)$, $m \geqslant 1$. Let W be the set of all words on Σ; let $\sim$ be the smallest equivalence relation on W such that $w \sim w\,\tau_0$; let $W' = W/\sim$. Alternatively, W' may be viewed as the set of all infinite sequences $\sigma_1\,\sigma_2\,\sigma_3 \ldots, \sigma_i \in \Sigma$, where $\sigma_i = \tau_0$ except for a finite number of indices i. Let $X = W' \times W'$. We call an element of X a TURING *tape*. We define the species $\Gamma = \mathrm{Tur}_1$ as the set consisting of the functions:

$$L : X \to X \text{ defined by} \quad L\,(\tau_i\, u, v) = (u, \tau_i\, v), \tag{1}$$

$$R : X \to X \text{ defined by} \quad R\,(u, \tau_i\, v) = (\tau_i\, u, v), \tag{2}$$

$$p_j : X \to X,\ 0 \leqslant j \leqslant m, \text{ defined by} \quad p_j\,(u, \tau_i\, v) = (u, \tau_i\, v), \tag{3}$$

$$d : X \to X \times [1 + m] \text{ defined by} \quad d\,(u, \tau_i\, v) = (u, \tau_i\, v, 1 + i), \tag{4}$$

where $0 \leqslant i \leqslant m$.

A machine of species Tur_1 is a "particular variety of" TURING *machine*.

Let V be the set of all words on $\{\tau_1, \ldots, \tau_m\}$. Let $v_r : V^r \to X$ be the injection defined by

$$(v_1, \ldots, v_r) \overset{\tau_r}{\longmapsto} (1, v_1\, \tau_0\, v_2\, \tau_0 \ldots \tau_0\, v_r\, \tau_0\, \tau_0\, \tau_0 \ldots)$$

and let $\pi_s : X \to V^s$ be defined by

$$(u, v_1\, \tau_0\, v_2\, \tau_0 \ldots \tau_0\, v_s\, \tau_0\, v_{s+1}\, \tau_0 \ldots) \overset{\pi_s}{\longmapsto} (v_1, v_2, \ldots, v_s).$$

Then for every $\varphi \in |\mathrm{Tur}_1|$, the partial function $v_r\, \varphi\, \pi_s : V^r \to V^s$ is recursive and every recursive partial function is so representable.

Let Tur_2 be the set of functions

$$l_j, r_j : X \to X \times [1 + n],\ 0 \leqslant j \leqslant m, \text{ defined by}$$

$$(\tau_i\, u, v) \overset{l_j}{\longmapsto} (u, \tau_j\, v, 1 + i)$$

$$(u, \tau_i\, v) \overset{r_j}{\longmapsto} (\tau_j\, u, v, 1 + i).$$

We leave it for the reader to verify that $|\mathrm{Tur}_1| = |\mathrm{Tur}_2|$.

16. Appendix

If the automaton species Γ consists of univalent matrices then so does $|\Gamma|$. In particular, the (1×2) matrix $\langle 1, 1\rangle$ whose entries are each the identity $1 \in W$ is not in $|\Gamma|$. In this section we shall see that the assumption that this matrix is in Γ has a dramatic effect on the structure of $|\Gamma|$. We call an automaton species *relational* if $\emptyset \in \Gamma$ and $\langle 1,1\rangle \in \Gamma$.

Let p be a non-negative integer and let x_j be an $(n \times 1)$ matrix in $[\widehat{W}]$ for each $j \in [p]$. We denote by $\langle x_1, \ldots, x_p\rangle$ the $(n \times p)$ matrix whose j^{th}

column is x_j; the operation which produces $\langle x_1, \ldots, x_p\rangle$ from $\{x_j\}_j$ will be called *column-tupling*.

Let Γ be relational. Then it is easy to see that for all $p \geqslant 0$ the $(1\times p)$ matrix $\langle 1, 1, 1, \ldots, 1\rangle \in |\Gamma|$. In fact, under these conditions $\mathfrak{S}_R \subseteq |\Gamma|$.

Proposition 16.1. *Let Γ be a relational automaton species. Then $|\Gamma|$ is closed with respect to column-tupling.*

Proof. Since $|\Gamma|$ is closed with respect to row-tupling, it is sufficient to show $|\Gamma|$ is closed with respect to column-tupling of (1×1) matrices. Let A_j be an automaton from 1 to 1 of species Γ for each $j \in [p]$. Let B be an automaton from 1 to p such that $|B| = \langle 1, 1, \ldots, 1\rangle$. Set $A = B(A_1 \oplus A_2 \oplus \ldots \oplus A_p)$. Then $|A| = \langle |A_1|, \ldots, |A_p|\rangle$.

Proposition 16.2. Given a relational automaton species Γ; let $\Gamma' = \{w_i \mid w = \langle w_1, \ldots, w_r\rangle \in \Gamma,\ i \in [r]\}$. Then $|\Gamma|$ is the smallest subclass $\mathfrak{C}$ of $[\widehat{W}]$ which contains $\mathfrak{S}_F$, Γ' and is closed with respect to composition, row-tupling, column-tupling, and iteration. Moreover, closure under iteration may be replaced by closure under *, the operation which takes the (1×1) matrix M into M^*.

Proof. First note that $\Gamma' \subseteq |\Gamma|$ since $w_i = \langle w_1, \ldots, w_r\rangle\, i^{-1}$, where i^{-1} is the column matrix with 1 in the ith position and otherwise empty. It now follows from the previous proposition that $\mathfrak{C} \subseteq |\Gamma|$. Obviously, since $\mathfrak{C}$ is closed with respect to column-tupling, $\Gamma \subseteq \mathfrak{C}$ so that $|\Gamma| \subseteq \mathfrak{C}$. Thus $|\Gamma| = \mathfrak{C}$.

The formula $\langle 1, U\rangle^{\dagger} = U^*$, where $U \subseteq W$, shows that $|\Gamma|$ is closed with respect to * while the defining formula for † shows that closure under * implies closure under †.

Corollary 16.3. *Let Γ_i be relational automaton species, $i = 1, 2$. If the set of (1×1) matrices in $|\Gamma_1|$ is the same as those in $|\Gamma_2|$, then $|\Gamma_1| = |\Gamma_2|$.*

Corollary 16.4. *Let Γ be relational. If M, N are $(n\times p)$ matrices in $|\Gamma|$ then $M + N$ is in $|\Gamma|$. If P is an $(n\times n)$ matrix in $|\Gamma|$, then P^* is in $|\Gamma|$.*

Proof. We have M_{ij}, $N_{ij} \in |\Gamma|$ and $\langle 1,1\rangle (M_{ij}, N_{ij}) = M_{ij} + N_{ij}$; since $|\Gamma|$ is closed with respect to row- and column-tupling, $M + N \in |\Gamma|$.

Now let A be an exit-automaton of species Γ with behaviour P. To simplify the argument we assume (as we may since, for example, 1 is available as the label of an edge) that A is an entrance automaton. Let B be obtained from A by adjoining new states e'_j, $1 \leqslant j \leqslant 2n$, and edges $(e_i, 1, e'_i)$, $(e_i, 1, e'_{n+i})$ for each $i \in [n]$ and then identifying e'_{n+i} with b_i for each $i \in [n]$; the exits of B are e'_i, $i \in [n]$; $b_{B,i} = b_{A,i} = e'_{n+i}$. Then $|B| = P^*$. Since B is of species Γ, $P^* \in |\Gamma|$.

Corollary 16.5. *Let $\Gamma = \{\Sigma_0, \emptyset, \langle 1,1\rangle\}$, where Σ is a finite set and let M be an $(n\, x\, p)$ matrix. Then $M \in |\Gamma|$ iff M is recognizable; $M \in |\Gamma|_c$ iff M is finite.*

Acknowledgement

The author has had helpful discussions with SAMUEL EILENBERG on much of the material of this paper.

References

[1] RUTLEDGE, J. D.: On IANOV's Program Schemata. Journal of the Association for Computing Machinery **11,** 1 (January, 1964).
[2] IANOV, I. I.: On the Logical Schemata of Algorithms. Problems of Cybernetics **1,** 75–127 (1958).
[3] FELS, E. M.: KALUZHNEN Graphs and IANOV Writs in Logik und Logikkalkül. Munich: Verlag Karl Alber. 1962.
[4] MOORE, E. F., (editor): Sequential Machines. Addison Wesley. 1964.
[5] GINSBURG, S.: An Introduction to Mathematical Machine Theory. Addison Wesley, 1962.
[6] BOHM, C.: On a Family of Turing Machines and the Related Programming Language. ICC Bulletin **3,** 3 (July 1964).
[7] BOHM, C. and G. JACOPINI: Flow Diagrams, Turing Machines and Languages with only Two Formation Rules. Presented at 1964. International Colloquium on Algebraic Linguistics and Automata Theory, Jerusalem, Israel.
[8] GLUSHKOV, V. M. and A. A. LETICHENSKII: Theory of Algorithms and Discrete Processors, in Advances in Information Systems Science. Edited by JULIUS T. TOU.
[9] MACLANE and BIRKHOFF: Algebra. Macmillan. 1967.
[10] EILENBERG and WRIGHT: Automata in General Algebras. Information and Control. **11,** 4 (October, 1967).
[11] LAWVERE, F. W.: Functorial Semantics of Algebraic Theories. Proc. of Nat. Acad. of Sciences **50,** 5, pp. 869–872 (Nov. 1963).
[12] ELGOT, C. C.: The External Behavior of Machines. Proceedings of the Third Hawaii International Conference on System Sciences. 1970.
[13] RABIN and SCOTT: Finite Automata and Their Decision Problems in [3].

Dr. Calvin C. Elgot
IBM Thomas J. Watson Research Center
P. O. Box 218, Yorktown Heights, NY 10598
U.S.A.

Reprinted from Studies in Logic and the Foundations of Mathematics, Vol. 80:
Logic Colloquium '73 (edited by H. E. Rose and J. C. Shepherdson),
North Holland/American Elsevier Publishing Co. (1975), 175–230.

MONADIC COMPUTATION AND ITERATIVE ALGEBRAIC THEORIES*

Calvin C. ELGOT

IBM, Yorktown Heights, N.Y., U.S.A.

and

Columbia University, New York, U.S.A.

Abstract

The notion algebraic theory was introduced by Lawvere in 1963 (cf. S. Eilenberg and J. B. Wright, Automata in general algebras, *Information and Control* **11** (1967) 4) to study equationally definable classes of algebras from a more intrinsic point of view. We make use of it to study Turing machines and machines with a similar kind of control at a level of abstraction which disregards the nature of 'storage' or 'external memory'.

Preliminary remarks

It is important, and not uncommon, to distinguish sharply between an 'algorithm' or 'procedure' and the function (or partial function) it computes. These functions usually take as argument a vector of numbers or generalized numbers and as value a corresponding number or 'undefined'. An algorithm, for computing the function, is usually viewed as taking as argument a representation of the function argument in the form of a string of letters from a 'small' finite alphabet (cf. e.g. [8, 'The Informal Notion of Algorithm', p. 1]), transforming the string according to rather elaborate though well-defined rules, finally obtaining a string which decoded gives the value of the function for the given argument. This point of view finds precise expression, for example, in the various notions of 'Turing machine'.

Our interest here, however, is in the procedure, not at all in the

* The material reported on here evolved from research which was initiated at the University of Bristol and supported by the Science Research Council of Great Britain.

numerical function it may compute. For example, we are interested in a step-by-step record of the external memory state $x \in X$ of a machine as it passes through a sequence of 'instantaneous descriptions'. This information is incorporated in the notion 'track of a machine' defined in Section 5.11 as a certain 'sequacious function' over X determined by the machine.

It should be remarked straightaway that there is nothing 'effectively calculable' about this notion of 'sequacious function'. The operations we define on sequacious functions, however, preserve effectiveness. Enough operations are defined so that, for example, Turing tracks (and, more generally, computational processes) are analyzable into atomic ones by means of three simple operations on Turing tracks (or, more generally, on computational processes). More specifically, each Turing track is expressible from atomic Turing tracks and certain very simple 'base sequacious functions' (corresponding to computations which start at an exit) by means of a finite number of uses of composition, source-tupling (which provides for alternatives to be followed after a test) and conditional iteration (looping, scalar or vector). Thus, in particular, the family of all Turing tracks may be described without recourse to the notion 'Turing machine'. The scalar conditional iteration operation is closely related to Markov's 'reiteration'[6, p. 156] and to the 'whiledo' or 'dowhile' operations used in programming languages[11].

From this point of view, a Turing machine is a 'normal description' (definition in Section 5) for a Turing track. But Turing tracks may also be described by 'sequential descriptions' (definition in Section 5). Indeed, these latter descriptions appear much more perspicuous than the former. Sequential descriptions, in general, thus appear suitable for use by a mathematician while the role of the normal description is, perhaps, storage (in coded form) in a 'universal' machine.

The main notion of this paper is 'iterative algebraic theory' defined in Section 4. This notion is *abstract* in the sense that the elements 'morphisms' of the structure may be entities of any kind. Our main example of an iterative algebraic theory has sequacious functions as its elements.

The main theorem, roughly speaking, asserts that any morphism in the iterative subtheory generated by a family Σ of morphisms (which may be taken to be the family of instructions of a machine) can be described from Σ and two base morphisms by means of a single use of composition, a double use of source-tupling and a single use of *vector* iteration (vector looping) or alternatively, from Σ by means of composition, source-tupling and *scalar* iteration.

Part of the proof of the main theorem indicates an algorithm for converting a sequential description of a morphism into a normal description of the same morphism.

The point of Section 1, which is entirely informal, is to show that the notion 'iterative algebraic theory' arises in a natural way.

Section 2 reviews Lawvere's notion 'algebraic theory'[5] and notes various useful identities.

'Ideal theories' are introduced in Section 3 as a preliminary to the main notion which is introduced in Section 4 where examples are also given. The main theorem is formulated and proved in Section 5. The 'concluding remarks' are intended to tie together some of the 'ends' which we have deliberately left 'loose' in the hope that the many directions for further work along these lines will be clearer.

1. Monadic machine notions—informal

1.1. *External (memory) states*

We are interested in machine notions $\boldsymbol{M}$ (such as 'Turing machine' as, for example, in [6] or [8] or 'register machine' as in [9]) which have a certain common character. Roughly speaking, this 'common character' is the way sequencing of 'atomic acts' is executed by an $\boldsymbol{M}$-machine M.

Each machine notion $\boldsymbol{M}$ of this kind has associated with it a set X whose elements may be called 'external (memory) states'. In the case of a Turing notion $\boldsymbol{M}$ (we presume at least vague familiarity with the machine notion introduced by A. M. Turing [10] or the similar notion introduced by E. L. Post [7]) an element $x \in X$ represents both the position of the reading head of a machine at a given instant of time and the content of the Turing tape at that time-and nothing else.

For the Turing notion discussed in [4], an element of X may be taken as an integer (the position of the reading head) together with a function from the set of integers to the Turing alphabet (the tape content). (Actually the alphabet is pointed and the function essentially finite.) In [1] a candidate for the set X is not conspicuously visible but may readily be provided.

If $x \in X$ is an external state of $\boldsymbol{M}$ and M is an $\boldsymbol{M}$-machine, we also say x is an external state of M.

A remark concerning notation. Let X, Y, Z be sets. Given functions

$$X \xrightarrow{f} Y \xrightarrow{g} Z,$$

we write the composition of f and g as

$$X \xrightarrow{fg} Z$$

or

$$fg : X \longrightarrow Z$$

or

$$X \overset{f}{\text{——}} Y \xrightarrow{g} Z$$

(note the missing arrowhead) or simply

$$fg.$$

The equation $h = fg$ may also be written $h : X \xrightarrow{fg} Z$ or $h : X \overset{f}{\text{——}} Y \xrightarrow{g} Z$.

1.2. *Internal (memory) states*

In addition to an external state, an $\boldsymbol{M}$-machine M, at any instant of time, is in one of a finite number of 'internal (memory) states'. The external and internal state (which together comprise the 'total state') of M at any instant of time, determines its future course. We find it technically desirable to restrict the sets of internal states to finite sets $[m] = \{1, 2, ..., m\}$, where $m \in \mathbf{N}$ is a non-negative integer. Thus $[0] = \emptyset$, $[1] = \{1\}$, $[2] = \{1, 2\}$, etc.

The internal states of a machine M are further partitioned into 'active' and 'quiescent'. The quiescent states are sometimes referred to as 'halt-states' or 'stop-states' or 'exits'. We normalize this partition by specifying the set of internal states as $[s + p]$ and say that M has weight s and p exits. The elements of $[s]$ are active while the elements of $s + [p] = \{s + 1, s + 2, ..., s + p\}$ are quiescent. The element $s + j$ is the jth exit. If X is the set of external states of an $\boldsymbol{M}$-machine M and $[s + p]$ is its set of internal states, we take $X \cdot [s + p] = \{xk; x \in X, k \in [s + p]\}$ to be its set of total states; here $xk = x \cdot k$ is a sequence of length two whose first term is x and second term is k. The set $X \cdot [s + p]$ is in obvious bijective correspondence with $X \times [s + p]$.

In most discussions one of the internal states of a machine is specified as 'initial' or the 'beginning state'. Again, we find it technically desirable to permit several beginning states.

1.3. *Genera*

These machine notions $\boldsymbol{M}$ also have 'atomic acts' associated with them. An atomic act may effect a change in external state, in which case its effect may be specified by a function $X \to X$. (For our purpose the 'nature' of an 'atomic act' is not in itself relevant but only the effect it produces.) An atomic act may also have the effect of 'testing' whether $x \in X$ has a certain property or not; such an atomic act may be specified by a function $X \to [2]$. Tests may also be m-valued, $m \geq 1$, rather than merely 2-valued, yielding a function $X \to [m]$. Atomic acts may also simultaneously transform external memory and test external memory; such a combination act may be described by a function $X \to X \cdot [m]$. We say such a function has m-outcomes.

Thus a machine notion $\boldsymbol{M}$ has associated with it a family $\Gamma = \{\Gamma_m\}_{m \in \mathbf{N}}$, which we call the *genus* of $\boldsymbol{M}$, where Γ_m is a set of functions of type $X \to X \cdot [m]$. Usually the sets Γ_m are finite and, except for a finite number of m's, empty. [The reader may note if $X \neq \emptyset$, then $\Gamma_0 = \emptyset$ (cf. however, Section 1.8, penultimate paragraph).]

The elements of Γ_m are $\boldsymbol{M}$*-generators* with m outcomes.

1.4. *Genus based tables*

The manner in which an $\boldsymbol{M}$-machine M is specified in [4] bears a strong resemblance to the way '(machine language) computer programs' are specified. If M has weight $s \geq 0$ and $p \geq 0$ exits, its 'action' is determined by a 'table' with s 'rows':

$$\begin{array}{llll} 1 & X \xrightarrow{g_1} X \cdot [m_1]; & & [m_1] \xrightarrow{e_1} [s+p] \\ 2 & X \xrightarrow{g_2} X \cdot [m_2]; & & [m_2] \xrightarrow{e_2} [s+p] \\ \vdots & & & \\ s & X \xrightarrow{g_s} X \cdot [m_s]; & & [m_s] \xrightarrow{e_s} [s+p], \end{array}$$

where g_i is an $\boldsymbol{M}$-generator with m_i outcomes and e_i is a function with *source* $[m_i]$ and *target* $[s+p]$ for each $i \in [s]$. The pair comprising a row may be called an 'instruction'. [Post used the word 'direction'.] Assuming M is started with instruction i and given an 'input' $x \in X$, M acts by first evaluating $xg_i = x'i'$; $i'e_i$ specifies the number of the next instruction to be executed; if $i'e_i = s + j, j \in [p]$, then M halts at exit j in external state x'; if $i'e_i = i'' \in [s]$, then a second action is initiated by instruction i'', etc. The determined sequence of acts may be finite or infinite.

The effect of an instruction may be more simply described by appropriately 'combining' its generator part with its 'go to' or 'execute next' part. This we do in Section 1.6.

1.5. *Genus based graphs*

The information in the table of the previous section may be put in graphical form. The associated graph (of weight s and p exits) has $s+p$ nodes and these nodes are 'labelled' $1, 2, \ldots, s+p$. The ith node, $i \in [s]$, is also 'labelled' by $X \xrightarrow{g_i} X \cdot [m_i]$. Emanating from the ith node are m_i directed edges (or arrows) which are labelled $1, 2, \ldots, m_i$, respectively, and the targets of these arrows are specified by e_i. There are no edges emanating from the exit nodes $s+1, s+2, \ldots, s+p$.

1.6. *Sorts*

The finite functions $[n] \xrightarrow{f} [p]$, where $n, p \in \mathbf{N}$ are non-negative integers, play an important role in our considerations. Such functions were used, for example, in Section 1.4 to specify the 'go to' part of an instruction. The family of all such functions (indexed by n, p) forms a category $\mathcal{N}$ under ordinary composition of functions. The objects of $\mathcal{N}$ are the sets $[n]$, $n \in \mathbf{N}$; the set $\mathcal{N}_{n,p}$ is the set of morphisms from $[n]$ to $[p]$. The identities of $\mathcal{N}$ are the identity functions $1_n : [n] \to [n]$.

For X an arbitrary set, let $f_X : X \cdot [n] \to X \cdot [p]$ be the function which satisfies $(xi)f_X = xj$ if $if = j \in [p]$, where $x \in X$, $i \in [n]$. Let $\mathcal{N}_X$ be the family of all f_X, for f in $\mathcal{N}$. Clearly, the mapping $f \mapsto f_X$ of $\mathcal{N}$ into $\mathcal{N}_X$ preserves composition and identities, and if $X \neq \emptyset$, preserves distinctness as well. In fact, if $X \neq \emptyset$, the mapping

$$\mathcal{N} \xrightarrow{f \mapsto f_X} \mathcal{N}_X$$

is an isomorphism.

We are now prepared to combine the 'generator part' of an instruction with its 'go to' part. Let

$$f_i : X \xrightarrow{g_i} X \cdot [m_i] \xrightarrow{e_{i,X}} X \cdot [s+p],$$

i.e.,

$$f_i = g_i e_{i,X} : X \to X \cdot [s+p];$$

here $e_{i,X}$ denotes h_X, where $h = e_i$. The machine action described by the table of Section 1.4 may, according to the explanation given there, be more simply described by the table

$$\begin{array}{ll} 1 & X \xrightarrow{f_1} X \cdot [s+p] \\ 2 & X \xrightarrow{f_2} X \cdot [s+p] \\ \vdots & \\ s & X \xrightarrow{f_s} X \cdot [s+p]. \end{array}$$

Notice, however, that distinct tables of the former kind may well transform into the same table of the latter kind. For example, if $[m_1] \xrightarrow{\pi} [m_1]$ is a permutation, not the identity, if $g_1\pi_X$ is a generator, if row 1 is changed to the row

$$1 \quad X \xrightarrow{g_1\pi_X} X \cdot [m_1]; \qquad [m_1] \xrightarrow{\pi^{-1}e_1} [s+p],$$

then the transformed table remains unchanged.

The information in this latter table may be more succinctly expressed by making use of the function $f = (f_1, \ldots, f_s) : X \cdot [s] \to X \cdot [s+p]$ which satisfies.

$(xi)f = xf_i$. Clearly there is a bijection between the function families $\{f_i : X \to X \cdot [s+p]\}_{i \in [s]}$ and the functions

$$f = (f_1, \ldots, f_s) : X \cdot [s] \to X \cdot [s+p].$$

Indeed, if for each $i \in [s]$, $[1] \xrightarrow{i} [s]$ is the function whose value is i, then

$$i_X f = f_i \quad \text{for all } i \tag{1.6.1}$$

and these equations uniquely determine f. Then

$$(xi)f = x(i_X f). \tag{1.6.2}$$

Thus an $\boldsymbol{M}$-machine M of weight s and p exits gives rise to a function $f : X \cdot [s] \to X \cdot [s+p]$ which we call the *direct transition function* of the machine. If Γ is the genus of M, then the function

$$i_X f \in \Gamma \mathcal{N} \quad \text{for each } i \in [s],$$

where $\Sigma = \Gamma\mathcal{N}$ consists of all compositions ge_X of generators in Γ and functions e in $\mathcal{N}$.

We call $\Sigma = \{\Sigma_m\}_{m \in \mathbf{N}}$ the *sort* of M and note that

$$\Sigma = \Sigma\mathcal{N}. \tag{1.6.3}$$

We now make use of the direct transition function of a machine to express its action more neatly.

1.7. *Machine action redescribed*

Let $f : X \cdot [s] \to X \cdot [s+p]$ be the direct transition function of an $\boldsymbol{M}$-machine M; *let* $\lambda : [p] \to [s+p]$ be the s-translate of inclusion, i.e., $j \in [p] \xrightarrow{\lambda} s + j$; let $(f, \lambda_X) : X \cdot [s+p] \to X \cdot [s+p]$ be the function which agrees with f on xi, $i \in [s]$ and takes $x(s+j)$ into itself for $j \in [p]$. Finally, let

$$f^{1+m} : X \cdot [s] \xrightarrow{f} X \cdot [s+p] \xrightarrow{(f, \lambda_X)^m} X \cdot [s+p].$$

[In the case $p = 0$, f^{1+m} has 'ordinary' meaning.]

Assuming M is started with instruction $i \in [s]$ acting on $x \in X$, either for all m the outcome of $(xi)f^{1+m}$ is in $[s]$ in which case M produces the infinite sequence

$$(xi)f^1, (xi)f^2, (xi)f^3, \ldots$$

or the contrary. In the latter case, let c be the smallest m which fails, then M produces the finite sequence

$$(xi)f^1, (xi)f^2, \ldots, (xi)f^{1+c};$$

$1+c$ is clearly the count of the number of steps required to reach an exit.

If M is started with instruction $s+j, j \in [p]$, then M halts in zero steps in exit j.

1.8. *Terminal behavior*

As indicated earlier, we are not interested here in a notion of a machine computing a numerical function. The most closely related notion which does concern us is 'terminal behavior'.

In order to specify the 'terminal behavior' of a machine it is necessary to specify how the machine is permitted to begin. For a machine M of weight s, p exits and n beginning states (not necessarily distinct) the beginning function is of the type $b : [n] \to [s+p]$.

Let $f : X \cdot [s] \to X \cdot [s+p]$ be the direct transition function of M and let $f^\dagger : X \cdot [s] \to X \cdot [p]$ be the partial function which takes xi into 'undefined' if for all m the outcome of $(xi)f^{1+m}$ is in $[s]$, and which takes xi into $x'j$ if $(xi)f^{1+c} = x'(s+j)$; cf. Section 1.7. Note, incidentally, that $(xi)f^{1+c} = (xi)f^{1+c+m}$ for all $m \in \mathbf{N}$.

There is an obvious partial ordering $\sqsubset$ between partial functions $X \cdot [s] \xrightarrow{\xi} X \cdot [p]$ and we state without proof that $f^\dagger$ may be more succinctly described as the unique minimal (with respect to $\sqsubset$) solution to the equation in ξ (cf. Section 4.3).

$$(1.8.1) \qquad \xi : X \cdot [s] \xrightarrow{f} X \cdot [s+p] \xrightarrow{(\xi, 1_{p, X})} X \cdot [p];$$

i.e., $\xi = f(\xi, 1_{p,X})$. Here $(\xi, 1_{p,X}) : X \cdot [s+p] \to X \cdot [p]$ is the partial function which agrees with ξ on $X \cdot [s]$ and takes $x(s+j)$ into xj for $j \in [p]$.

We record an M-machine M with n beginning states, p exits and weight s by

$$[n] \xrightarrow[s]{M} [p].$$

It is completely specified by a beginning function

$$[n] \xrightarrow{\beta_M} [s+p].$$

and a transition function

$$X \cdot [s] \xrightarrow{\tau_M} X \cdot [s+p].$$

The terminal behavior of M is the partial function

(1.8.2) $$\boldsymbol{MB} : X \cdot [n] \xrightarrow{\beta_{M,X}} X \cdot [s+p] \xrightarrow{(\tau_M^\dagger, 1_{p,X})} X \cdot [p].$$

In particular, if $i\beta_M = s+j$, $i \in [n]$, $j \in [p]$, the ith component of $\boldsymbol{MB}$, i.e., $i_X(\boldsymbol{MB})$ is the function $j_X : X \to X \cdot [p]$.

In order to allow for 'atomic acts' of machines which are the terminal behaviors of other machines, we may permit the transition 'function' of a machine to be partial. In this case, (1.8.1) still has a unique minimal solution and (1.8.2) still defines the terminal behavior. Furthermore, Γ_0 need not be empty (cf. Section 1.3).

The analysis of machine action is facilitated, however, by 'explicitly' taking 'time' into consideration. This we do in the next section.

1.9. *Timed terminal behavior*

We replace the family of partial functions $X \cdot [n] \to X \cdot [p]$ by the family $[X \cdot \mathbf{N}, \text{und.}]^{-}$ of partial functions $X \cdot \mathbf{N} \cdot [n] \xrightarrow{h} X \cdot \mathbf{N} \cdot [p]$ satisfying

(1.9.1) if $x0i \overset{h}{\longmapsto} x'mj$, then $xti \overset{h}{\longmapsto} x'(m+t)j$;

(1.9.2) if h is defined on xti, then it is defined on $x0i$.

Such a partial function $\boldsymbol{MT}$ is the *timed terminal behavior* of a machine

$$[n] \xrightarrow[s]{M} [p]$$

if whenever

$$xi \xrightarrow{\boldsymbol{MB}} x'j$$

in m steps, then

$$x0i \xrightarrow{\boldsymbol{MT}} x'mj.$$

We then interpret

$$xni \xrightarrow{\boldsymbol{MT}} x'(m+n)j$$

as meaning: if machine M 'receives' input $x \in X$ at time $n \in \mathbf{N}$ and is started in its ith beginning state, then M produces $x'j$ at time $m+n$.

Let $X \cdot \mathbf{N} \cdot [s] \overset{f}{\longmapsto} X \cdot \mathbf{N} \cdot [s+p]$ satisfy (1.9.1) and (1.9.2), and be *positive* in the following sense:

(1.9.3) if $x0i \overset{f}{\longmapsto} x'mj$, then $m > 0$.

We show later that the equation (1.9.4) below in

$$X \cdot \mathbf{N} \cdot [s] \overset{\xi}{\longrightarrow} X \cdot \mathbf{N} \cdot [p]$$

has a unique solution (cf. (1.8.1)) among partial functions satisfying (1.9.1) and (1.9.2). (We emphasize 'unique', not 'unique minimal'.)

$$\xi : X \cdot \mathbf{N} \cdot [s] \overset{f}{\longrightarrow} X \cdot \mathbf{N} \cdot [s+p] \xrightarrow{(\xi,\, 1_{p, X \cdot \mathbf{N}})} X \cdot \mathbf{N} \cdot [p] \tag{1.9.4}$$

For any ξ, the partial function $(\xi, 1_{p, X \cdot \mathbf{N}})$ does not satisfy (1.9.3) since $x0(s+j) \mapsto x0j$. Consideration of the solution to (1.9.4) suggests strongly the comment that the presence of the identity

$$1_{p, X \cdot \mathbf{N}} : X \cdot \mathbf{N} \cdot [p] \to X \cdot \mathbf{N} \cdot [p]$$

in our system enables us to express 'stopping'. We denote the unique solution to (1.9.4) by $f^\dagger$ and note $f^\dagger$ is positive.

If $\tau'_M : X \cdot \mathbf{N} \cdot [s] \to X \cdot \mathbf{N} \cdot [s+p]$ satisfies (1.9.1), (1.9.2) and

$$xi \overset{\tau_M}{\longmapsto} x'i' \quad \text{iff} \quad x0i \overset{\tau'_M}{\longmapsto} x'1i', \tag{1.9.5}$$

then we may express MT as follows:

$$MT : X \cdot \mathbf{N} \cdot [n] \xrightarrow{\beta_{M, X \cdot \mathbf{N}}} X \cdot \mathbf{N} \cdot [s+p] \xrightarrow{(\tau'^{\dagger}_M,\, 1_{p, X \cdot \mathbf{N}})} X \cdot \mathbf{N} \cdot [p]; \tag{1.9.6}$$

cf. (1.8.2).

In discussing the action of machines M in a 'timed environment' we permit τ_M to act in any positive amount of time, rather than merely in unit time, in order to provide the possibility of treating the timed behavior of one machine (begun in an active state) as a 'primitive instruction' of another machine.

The notion 'iterative algebraic theory' defined below formulates sufficient salient properties of the timed environment to permit us to prove certain fundamental closure properties of the timed terminal behavior of $\boldsymbol{M}$ machines M of a given sort.

After introducing this notion, the character of the discussion changes from (naive) set-theoretic to axiomatic with the consequent significant change in the domain of applicability of results and the consequent suppression of the irrelevant.

2. Algebraic theories

2.1. *Background*

The notion 'algebraic theory' was introduced by Lawvere in 1963 [5] to study equationally definable classes of algebras from a more intrinsic point of view. According to Lawvere, an algebraic theory is a small category L 'whose objects are the natural numbers $0, 1, 2, \dots$ and in which each object p is the categorical direct product of the object 1 with itself p times'. In greater detail: L is a category whose objects are $0, 1, 2, \dots$ which, in addition, for each $p \in \mathbf{N}$ has distinguished morphisms $\pi_1, \pi_2, \dots, \pi_p : p \to 1$, satisfying

(2.1.1) Given any family of morphisms $f_j : n \to 1, j \in [p]$, there is a unique morphism $f = \langle f_1, f_2, \dots, f_p \rangle : n \to p$ such that $f\pi_j = f_j$ for each j.

Let (X) be the family, indexed by n, p, of sets of functions $X^n \to X^p$. An L-algebra with underlying set X may be described as a functor $L \to (X)$ which takes the morphism $n \xrightarrow{f} p$ in L into the function $X^n \xrightarrow{f^-} X^p$ in (X) subject to the conditions:

(2.1.2) Composition is preserved.

(2.1.3) $\pi_j^- : X^p \to X$ is the projection which picks out the jth component.

We briefly indicate how the theory of a commutative binary operation γ may be presented Lawvere style. There exists an algebraic theory $\gamma\boldsymbol{F}$ which is 'freely generated by $2 \xrightarrow{\gamma} 1$'. The commutativity condition

$$\forall x \, \forall y [\gamma(x, y) = \gamma(y, x)] \tag{2.1.4}$$

is expressed, Lawvere style, in quantifier-free form as

$$\gamma \equiv \langle \pi_2, \pi_1 \rangle \gamma \quad \text{in} \quad \gamma\boldsymbol{F}, \qquad \text{where } \pi_1, \pi_2 : 2 \to 1. \tag{2.1.5}$$

If $\equiv$ is the smallest 'congruence' in $\gamma\boldsymbol{F}$ for which (2.1.5) holds, then $L = [\gamma\boldsymbol{F} / \equiv]$ is an algebraic theory. Given any assignment $\gamma \mapsto \gamma^- : X^2 \to X$, there is a unique extension to a functor $\gamma\boldsymbol{F} \to (X)$ satisfying (2.1.2), (2.1.3); this functor 'factors through L' iff γ^- is commutative. Moreover, there exists a functor $L \to (X)$ satisfying (2.1.2),

(2.1.3) which distinguishes distinct morphisms. Of course, the facts alluded to here are quite general. The allusion to the theory of a commutative binary operation is merely a convenient illustration.

The algebraic theory G of groups may be described as a quotient theory $\{\gamma, \sigma, \rho\}\boldsymbol{F} / \equiv$ for suitable congruence $\equiv$, where $\gamma : 2 \to 1$, $\sigma : 1 \to 1$, $\rho : 0 \to 1$ correspond respectively to (symbols for) group multiplication, inverse, group identity. The theory G may also be described as a quotient $\{\beta, \rho\}\boldsymbol{F} / \sim$ for suitable congruence $\sim$; where $\beta : 2 \to 1$ corresponds to the group operation $(x, y) \to x \cdot y^{-1}$. The more conventional point of view regards 'group' based upon γ, σ, ρ as an 'algebra' of a different species from 'group' based upon β, ρ. From the Lawvere point of view, these two ways of describing 'group' lead to isomorphic algebraic theories.

Let $\phi : [p] \to [n]$ be any function. With ϕ we may associate a morphism $\phi^{\#} : n \to p$ in a Lawvere theory L as follows:

$$\phi^{\#}\pi_j = \pi_{j\phi},$$

i.e.,

$$\phi^{\#} = \langle \pi_{1\phi}, \pi_{2\phi}, \ldots, \pi_{p\phi} \rangle.$$

If $\psi : [q] \to [p]$ is any function, one readily verifies

(2.1.6) $$[\psi\phi]^{\#} = \phi^{\#}\psi^{\#}.$$

If $n = p$ and ϕ is the identity function, then $\phi^{\#}$ is the identity morphism. Thus $\#$ is a contravariant functor from $\mathcal{N}$ into L.

If $\pi_1 \neq \pi_2 : 2 \to 1$, the theory L is non-degenerate. In this case, $\#$ preserves distinctness.

2.2. *Definition*

If L is a Lawvere theory and 'the arrows in L are reversed', we obtain a small category $T = L^{\mathrm{op}}$ 'whose objects are the natural numbers $0, 1, 2, \ldots$, and in which each object n is the categorical coproduct of the object 1 with itself n times'. More concretely: T is a category whose objects are $0, 1, 2, \ldots$ which, in addition, for each $n \in \mathbf{N}$ has *distinguished* morphisms, $1, 2, \ldots, n : 1 \to n$ satisfying:

(2.2.1) Given any family of morphisms $f_i : 1 \to p$, $i \in [n]$ there is a unique morphism $f = (f_1, f_2, \ldots, f_n) : n \to p$ such that $if = f_i$ for each i.

Let $\phi : [n] \to [p]$ be any function. With ϕ we associate a morphism

$\phi' : n \to p$ in T by the requirement

$$i\phi' = j : 1 \to p \quad \text{if} \quad i\phi = j,$$

where $i \in [n]$, $j \in [p]$. If $\psi : [p] \to [q]$ is any function, we have

$$[\phi\psi]' = \phi'\psi'.$$

In the case $n = p$ and ϕ is the identity function, again ϕ' is the identity morphism in T. If T is non-degenerate, the function $\phi \mapsto \phi'$ embeds $\mathcal{N}$ in T.

The notion we use is a slight variant of the notion described above.

By an algebraic (co)theory T we mean a small category whose objects are [0], [1], [2], ... which, in addition, for each $n \in \mathbf{N}$ has *distinguished* morphisms $1, 2, \ldots, n : [1] \to [n]$ satisfying

(2.2.2) given any family of scalar morphisms $f_i : [1] \to [p]$, $i \in [n]$, there is a unique (vector) morphism $(f = f_1, f_2, \ldots, f_n) : [n] \to [p]$ such that $if = f_i$ for each i, i.e., $f_i : [1] \xrightarrow{i} [n] \xrightarrow{f} [p]$.

The operation described by (2.2.2) is called *source-tupling*. It operates on n scalar morphisms to produce one vector morphism whose i^{th} *component* is the ith scalar morphism. In the case $n = 0$, (2.2.2) asserts (for each p) there is a unique morphism $0_p : [0] \to [p]$.

Condition (2.2.2) amounts to assuming the existence of an operation $\{f_i\}_{i \in [n]} \mapsto (f_1, f_2, \ldots, f_n)$ inverse to the operation $f \mapsto \{if\}_{i \in [n]}$. Explicitly, condition (2.2.2) may be replaced by the following two conditions.

(2.2.3) $$i(f_1, f_2, \ldots, f_n) = f_i,$$

(2.2.4) $$f = (1f, 2f, \ldots, nf).$$

If T is non-degenerate, $\mathcal{N}$ may be embedded in T as described above. It is often convenient, in this case, to identify $\mathcal{N}$ with a subtheory of T via this embedding. The identification leads to 'theory' in the sense of [2]. Hereafter, in this paper, we drop the prefix 'co'.

If $\Gamma = \{\Gamma_p\}_{p \in \mathbf{N}}$, where $\Gamma_p \subset T_p$ and T_p is the set of all scalar morphisms in T with target p, we let Γ° be the family of vector morphisms obtained from Γ by source-tupling. Thus if $[n] \xrightarrow{f} [p]$ is in T, we have

$$f \text{ in } \Gamma^\circ \Leftrightarrow if \text{ in } \Gamma \qquad \text{for each } i \in [n],$$

i.e.,

$$f \in \Gamma^\circ_{n,p} \Leftrightarrow if \in \Gamma_p \qquad \text{for each } i \in [n].$$

The distinguished morphisms as well as the morphisms obtained from them by source-tupling are called *base* (alternatively, *ground*) morphisms.

2.3. $[X, Q]$

From the point of view of our applications, the algebraic theory $[X, Q]$ described below is important.

A morphism $[n] \to [p]$ in $[X, Q]$ is a *Q-pointed function* $X \cdot [n] \cup Q \to X \cdot [p] \cup Q$, i.e., a function which keeps the elements of Q pointwise fixed. The distinguished morphism $[1] \xrightarrow{i} [n]$, $i \in [n]$, in $[X, Q]$ is the function $X \cup Q \to X \cdot [n] \cup Q$ which takes x into xi, here we have identified X with $X \cdot [1]$ via the obvious bijection. Composition of morphisms is ordinary composition of functions. It is easy to see that (2.2.2) is satisfied. Indeed, defining f by $(xi)f = xf_i$, we have

$$(xi)f = x(if) \quad \text{for } x \in X,\ i \in [n]. \tag{2.3.1}$$

(Juxtaposition is here used to denote concatenation as in 'xi', application as in '$(xi)f$' and '$x(if)$', composition as in 'if'.) The algebraic theory $[X, Q]$ is degenerate iff $X = \emptyset$.

In the case $Q = \emptyset$, $[X, Q]$ consists of all functions $X \cdot [n] \to X \cdot [p]$ and in case Q has a single element, the morphisms in $[X, Q]$ may be regarded as partial functions $X \cdot [n] \to X \cdot [p]$ by treating the point of Q as 'undefined.'

In Section 4, examples will be given in which Q has many elements.

2.4. *Matrices*

A semiring R is a commutative monoid with respect to the binary operation $+$, a monoid with respect to the binary operation $\cdot$ with $\cdot$ distributing over $+$ and with $0 \cdot x = 0 = x \cdot 0$ for all $x \in R$. In particular, any ring with 1 is a semiring. We now describe an algebraic theory $[R]$. A morphism $[n] \to [p]$ in $[R]$ is an $n \times p$-matrix whose entries are in R. The distinguished morphism $[1] \xrightarrow{i} [n]$, $i \in [n]$, is the $1 \times n$ matrix (a row) whose i^{th} (column) entry is 1 and whose other entries are 0. The verification of (2.2.2) is immediate. We note that if $f_i : [1] \to [p]$, $i \in [n]$, are morphisms in $[R]$, then the morphism $f = (f_1, \ldots, f_n) : [n] \to [p]$ is the $n \times p$ matrix whose i^{th} row is f_i.

We also observe any subcollection of rectangular matrices containing the distinguished rows, closed under matrix multiplication and 'row stacking', i.e., source-tupling, forms in an obvious way an algebraic theory, a *subtheory* of $[R]$.

Clearly the sum of two $n \times p$-matrices in $[R]$ is also in $[R]$. A subtheory, however, may fail to be closed under sum.

The theory $[R]$ is degenerate iff $0 = 1$ in R.

2.5. *Simple properties*

Given morphisms $f_i : [n_i] \to [p]$, $i \in [2]$ in an algebraic theory T, we define the *source-pairing* of $\{f_i\}_{i\in[2]}$ to be the morphism

$$f = (f_1, f_2) : [n_1 + n_2] \to [p]$$

whose i_1^{th} component $i_1 \in [n_1]$ is the i_1^{th} component of f_1 and whose $(n_1 + i_2)^{\text{th}}$ component $i_2 \in [n_2]$ is the i_2^{th} component of f_2. One readily verifies

(2.5.1) $$((f_1, f_2), f_3) = (f_1, (f_2, f_3)), \quad \text{where } f_3 : [n_3] \to [p],$$

(2.5.2) $$(0_p, f) = f = (f, 0_p),$$

(2.5.3) $$(f_1, f_2)g = (f_1 g, f_2 g), \quad \text{where } g : [p] \to [q],$$

(2.5.4) $$0_n f = 0_p, \quad \text{where } n = n_1 + n_2.$$

By virtue of (2.5.1), any meaningful way of parenthesizing $(f_1, f_2, \dots, f_n)$ yields the same morphism and so the source-tupling of scalar morphisms may be extended to source-tupling of vector morphisms.

Let $\kappa : [n_1] \to [n_1 + n_2]$ be the *inclusion* morphism, i.e., the morphism whose i_1^{th} component, $i_1 \in [n_1]$, is the distinguished morphism $[1] \xrightarrow{i_1} [n_1 + n_2]$ and let $\lambda : [n_2] \to [n_1 + n_2]$ be the *translated inclusion* morphism, i.e., the morphism whose i_2^{th} component, $i_2 \in [n_2]$, is the distinguished morphism $[1] \xrightarrow{n_1 + i_2} [n_1 + n_2]$. These two morphisms may more succinctly be described by

(2.5.5) $$i_1 \kappa = i_1 : [1] \to [n_1 + n_2] \quad \text{for each } i_1 \in [n_1],$$

(2.5.6) $$i_2 \lambda = n_1 + i_2 : [1] \to [n_1 + n_2] \quad \text{for each } i_2 \in [n_2].$$

If T is non-degenerate κ, λ may be identified with the inclusion and translated inclusion functions in $\mathcal{N}$.

We further note

(2.5.7) $$\kappa(f_1, f_2) = f_1,$$

(2.6.8) $$\lambda(f_1, f_2) = f_2.$$

Moreover, (f_1, f_2) is uniquely characterized by (2.5.7) and (2.5.8).

There is one other operation which we find useful for perspicacious expression and convenient calculation.

Given morphisms $g_i : [n_i] \to [p_i]$, $i \in [2]$, we define

(2.5.9) $$g_1 \oplus g_2 = (g_1\kappa, g_2\lambda) : [n_1 + n_2] \to [p_1 + p_2],$$

where

$$[p_1] \xrightarrow{\kappa} [p_1 + p_2], \qquad [p_2] \xrightarrow{\lambda} [p_1 + p_2]$$

are again inclusion and translated inclusion.

We can now replace κ and λ with more conveniently indexed equivalents. Recall $1_n : [n] \to [n]$ is the identity.

(2.5.10) $$(1_n \oplus 0_p) = \kappa : [n] \to [n + p].$$

In particular, $1_n \oplus 0_0 = 1_n$.

PROOF.

$$\begin{aligned} 1_n \oplus 0_p &= (1_n\kappa, 0_p\lambda) && (2.5.9)\\ &= (\kappa, 0_{n+p}) && \text{by } (2.5.4)\\ &= \kappa && \text{by } (2.5.2)\end{aligned}$$

(2.5.11) $$(0_n \oplus 1_p) = \lambda : [p] \to [n + p].$$

In particular, $0_0 \oplus 1_p = 1_p$.

PROOF.

$$\begin{aligned} 0_n \oplus 1_p &= (0_n\kappa, 1_p\lambda) && \text{by } (2.5.9)\\ &= (0_{n+p}, \lambda) && \text{by } (2.5.4)\\ &= \lambda && \text{by } (2.5.2)\end{aligned}$$

One may verify

(2.5.12) $$0_n \oplus 0_p = 0_{n+p},$$

(2.5.13) $$0_q \oplus f = f(0_q \oplus 1_p), \qquad f \oplus 0_q = f(1_p \oplus 0_p),$$

(2.5.14) $$(g_1 \oplus g_2) \oplus g_3 = g_1 \oplus (g_2 \oplus g_3),$$

where $g_3 : [n_3] \to [p_3]$,

(2.5.15) $$(g_1 \oplus g_2)(h_1 \oplus h_2) = (g_1h_1 \oplus g_2h_2),$$

where $h_i : [p_i] \to [q_i]$, $i \in [2]$,

(2.5.16) $$(e_1 \oplus e_2)(f_1, f_2) = (e_1 f_1, e_2 f_2),$$

where $e_i : [m_i] \to [n_i], \quad i \in [2]$.

Let $[s] \xrightarrow{f} [s+p], [s] \xrightarrow{g} [s+p]$. The following identity is useful in the specialized contexts which arise later.

(2.5.17) $$(f, 0_s \oplus 1_p)(g, 0_s \oplus 1_p) = (f(g, 0_s \oplus 1_p), 0_s \oplus 1_p).$$

PROOF.

$$\begin{aligned}(f, 0_s \oplus 1_p)(g, 0_s \oplus 1_p) &= (f(g, 0_s \oplus 1_p), (0_s \oplus 1_p)(g, 0_s \oplus 1_p)) && \text{by (2.5.3)}\\ &= (f(g, 0_s \oplus 1_p), (0_s g, 0_s \oplus 1_p)) && \text{by (2.5.16)}\\ &= (f(g, 0_s \oplus 1_p), (0_{s+p}, 0_s \oplus 1_p)) && \text{by (2.5.4)}\\ &= (f(g, 0_s \oplus 1_p), 0_s \oplus 1_p) && \text{by (2.5.2)}\end{aligned}$$

The following notion is required in Section 3.1.

Let T be an algebraic theory with $T_{n,p}$ the set of morphisms $[n] \to [p]$. A *congruence* $\sim$ *on* T is a family of equivalence relations, one on each set $T_{n,p}$, satisfying

(a) $f \sim g : [n] \to [p]$ implies both

$$fh \sim gh, \quad \text{where } h : [p] \to [q],$$

and

$$ef \sim eg, \quad \text{where } e : [m] \to [n];$$

(b) $f, g \in T_{n,p}$, $if \sim ig$ for each $i \in [n]$ implies $f \sim g$.

We also use the following notion where T_1, T_2 are algebraic theories.

A *morphism* $T_1 \xrightarrow{F} T_2$ *between algebraic theories* is a functor which takes $[n]$ into $[n]$ and the distinguished morphism $[i] \xrightarrow{i} [n]$ in T_1 into the distinguished morphism $[i] \xrightarrow{i} [n]$ in T_2. It follows that $\boldsymbol{F}$ preserves source-tupling.

3. Ideal theories

3.1. *Definitions*

By an *ideal theory* I we mean a non-degenerate algebraic theory which satisfies the following right ideal property.

(3.1.1) If $[1] \xrightarrow{f} [p]$ is not distinguished, then neither is fg, where $[p] \xrightarrow{g} [q]$.

Thus, if we call a morphism $[n] \xrightarrow{f} [p]$ in I *ideal* if each of its components is not distinguished, we have

(3.1.2) if f is ideal, then so is fg.

In particular, the morphism 0_p is ideal for each p.

Clearly, ideal morphisms (in an ideal theory) satisfy the left ideal property. Specifically,

(3.1.3) if g is ideal, then so is fg.

The collection $[X \cdot \mathbf{N}, \text{und.}]^-$ introduced in Section 1.9 is isomorphic to a subtheory $[X \cdot \mathbf{N}, \square]^-$ of $[X \cdot \mathbf{N}, Q]$, where Q consists of the single element $\square$ (cf. Section 2.3), in the obvious way. Let $[\![X \cdot \mathbf{N}, \square]\!]$ be the smallest subtheory of $[X \cdot \mathbf{N}, \square]^-$ which contains the positive functions (cf. Section 1.9). While the theories $[X \cdot \mathbf{N}, \square]$ and $[X \cdot \mathbf{N}, \square]^-$ are not ideal, the theory $[\![X \cdot \mathbf{N}, \square]\!]$ is, provided $X \neq \emptyset$.

If we associate the elements of $\mathbf{N}$ with 'time', then the ideal morphisms in $[\![X \cdot \mathbf{N}, \square]\!]$ are the ones which 'operate' in positive time (for all arguments) while the distinguished morphisms operate in zero time (for all arguments).

By an *ideal congruence* $\sim$ on an ideal theory I, we mean a theory congruence satisfying $f \sim g$, f in $\mathcal{N} \Rightarrow f = g$. If $\sim$ is ideal in this sense, then $I/\sim$ is ideal. (Where $\sim$ is a theory congruence; we may have I and $I/\sim$ are ideal but $\sim$ is not.)

It may readily be verified that if T is an algebraic theory freely generated by $\Gamma = \{\Gamma_n\}_{n \in \mathbf{N}}$ (cf. [2, 3]), then T is ideal. In particular, $\mathbf{N}$ is ideal. It may also be verified that the algebraic theory of semigroups is ideal while the algebraic theory of monoids is not. In fact we have for the positive half of the assertion the

PROPOSITION 3.1.1. *If $P = \{P_n\}_{n \in \mathbf{N}}$, where P_n is a set of pairs of ideal* *morphisms* $[1] \to [n]$ *in an ideal theory* I *and* $\mathbf{P}$ *is the smallest congruence which contains* P *(componentwise), then* $I/\mathbf{P}$ *is ideal.*

PROOF. The proof of the Proposition follows from

PROPOSITION 3.1.2. *Let* C_I *be the family (indexed by* n, p*) of binary relations in the ideal theory* I *defined by the requirement, where* $f, f' : [n] \to [p]$, f *is* C_I*-related to* f' *iff*

$$\text{(3.1.4)} \qquad \begin{aligned} &if \text{ is base} \Rightarrow if = if', \\ &if \text{ is ideal} \Rightarrow if' \text{ is ideal.} \end{aligned}$$

Then C_I *is an ideal congruence on* I, *further* C_I *contains every other ideal congruence.*

COROLLARY 3.1.3. *Let* I *and* C_I *be as in Proposition 3.1.2. Then,* $I' = I/C_I$ *is ideal. If* $I_0 \neq \emptyset$, *where* I_0 *is the set of scalar morphisms with target* 0, *then* I' *is completely described by: for each* p *there is exactly one ideal morphism* $[1] \to [p]$. *If* $I_0 = \emptyset$ *but* $I \neq \mathcal{N}$ $(\mathcal{N} \subset I)$, *then* I' *is completely described by: for each positive* p *there is exactly one ideal morphism* $[1] \to [p]$ *and* $I'_0 = \emptyset$.

From Proposition 3.1.1, the observation 'every subtheory of an ideal theory is ideal' and the fact that free theories are ideal, we see ideal theories are in abundant supply.

3.2. *Power-ideal*

We say a morphism $[s] \xrightarrow{f} [s+p]$ in an ideal theory is 'power-ideal' if the morphism

$$\text{(3.2.1)} \qquad f^{1+m} : [s] \xrightarrow{f} [s+p] \xrightarrow{(f,\lambda)^m} [s+p], \quad \text{where } \lambda = 0_s \oplus 1_p,$$

is ideal for some $m \geq 0$ (cf. Section 1.7).

In particular, if f is ideal, then it is power-ideal. We give an example of a non-ideal f which is power-ideal in the case $s = 2$, $p = 1$.

Let $[1] \xrightarrow{g} [1]$ in an ideal theory. Define $[2] \xrightarrow{f} [3]$ by the requirement

$$\begin{aligned} (1_1 \oplus 0_1)f &: [1] \xrightarrow{0_1 \oplus 1_1 \oplus 0_1} [3] \\ (0_1 \oplus 1_1)f &: [1] \xrightarrow{g} [1] \xrightarrow{0_2 \oplus 1_1} [3]; \end{aligned}$$

note, $3 = 0_2 \oplus 1_1$. Then f is not ideal. However, if g is ideal, then so is f^2.

Indeed,

$$\begin{aligned}(1_1 \oplus 0_1)f^2 &= (1_1 \oplus 0_1)f(f, (0_2 \oplus 1_1)) \\ &= (0_1 \oplus 1_1 \oplus 0_1)(f, (0_2 \oplus 1_1) && \text{by definition } f \\ &= ((0_1 \oplus 1_1)f, 0_1(0_2 \oplus 1_1)) && \text{by (2.5.15)} \\ &= ((0_1 \oplus 1_1)f, 0_3) && \text{by (2.5.4)} \\ &= (0_1 \oplus 1_1)f && \text{by (2.5.2)} \\ &= g(0_2 \oplus 1_1) && \text{by definition } f.\end{aligned}$$

$$\begin{aligned}(0_1 \oplus 1_1)f^{(2)} &= (0_1 \oplus 1_1)f(f, (0_2 \oplus 1_1)) \\ &= g(0_2 \oplus 1_1)(f, (0_2 \oplus 1_1)) \\ &= g(0_3, (0_2 \oplus 1_1)) \\ &= g(0_2 \oplus 1_1).\end{aligned}$$

The notion 'power-ideal' may be made more 'definite' via the following

PROPOSITION 3.2.1. $[s] \xrightarrow{f} [s+p]$. *If* $f(f, \lambda)^m$ *is ideal for some* m, *then* $f(f, \lambda)^s$ *is ideal.*

PROOF. We prove the contrapositive. Suppose $if(f, \lambda)^s$, $i \in [s]$ is base. Then each term of the sequence

$$if, if(f, \lambda), \ldots, if(f, \lambda)^s$$

is base. There are $s+1$ terms in the sequence. If these terms are all distinct, then one term is a base morphism $j \in s + [p]$. Then all terms to the right of this one in the infinite sequence have the same value j.

If these $s+1$ terms are not all distinct, then the infinite sequence is ultimately periodic, repeating a piece of the finite sequence above.

3.3. *Matrices of subsets of a monoid*

Let M be a (multiplicative) monoid. The class $M^\wedge$ of all sets $U \subset M$ is a semiring with respect to $+$, $\cdot$, where $+$ is union and $\cdot$ is complex multiplication, i.e., $U \cdot V = \{uv : u \in U, v \in V\}$. The additive identity 0 of $M^\wedge$ is the empty subset of M and the multiplicative identity 1 of $M^\wedge$ is the singleton of the monoid identity. According to Section 2.4, $[M^\wedge]$ is an algebraic theory.

Consider the subcollection $[M^\wedge]^-$ of matrices $\{U_{ij}\}$ satisfying

(3.3.1) if $U_{ij} = 1$, then $U_{ik} = 0$ for $k \neq j$.

The subcollection $[M^\wedge]^-$ is closed under multiplication iff

(3.3.2) the set of elements in M distinct from unity is closed under

It is easy to see that if (3.3.2) holds, $[M^\wedge]^-$ is an ideal theory. There is, however, a smaller subcollection which is of greater interest to us. Let $[\![M^\wedge]\!]$ be the subcollection of matrices $\{U_{ij}\}$ satisfying

(3.3.3) if $1 + U_{ij} = U_{ij}$, then $U_{ij} = 1$ and $U_{ik} = 0$ for $k \neq j$.

Clearly condition (3.3.3) implies condition (3.3.1). Again we have $[\![M^\wedge]\!]$ is closed under multiplication iff M satisfies (3.3.2). If M satisfies (3.3.2), $[\![M^\wedge]\!]$ is an ideal theory.

4. Iterative theories

4.1. *Definition and ready results*

Let $[s] \xrightarrow{f} [s+p]$ be a morphism in an ideal theory I. We say f has a *conditional iterate* (briefly, an *iterate*) in I, if the equation (4.1.1) in $[s] \xrightarrow{\xi} [p]$ has a unique solution:

$$\xi = f(\xi, 1_p). \tag{4.1.1}$$

We denote the unique solution to (4.1.1) by $f^{\dagger}$. (Note that if $f = 1_s \oplus 0_p$, equation (4.1.1) becomes $\xi = \xi$.)

PROPOSITION 4.1.1. *If f is ideal and ξ satisfies* (4.1.1), *then ξ is ideal.*

We say the ideal theory I is *closed with respect to conditional iteration* (briefly, I is an *iterative theory*) iff every ideal morphism f in I has an iterate $f^{\dagger}$ in I.

Alternatively, an iterative theory may be defined as an ideal theory which is enriched by an operation, *conditional iteration*, which takes an ideal morphism $[s] \xrightarrow{f} [s+p]$ into a morphism $[s] \xrightarrow{f^{\dagger}} [p]$ such that

$$f^{\dagger} = f(f^{\dagger}, 1_p), \tag{4.1.2}$$

$$g = f(g, 1_p) \Rightarrow g = f^{\dagger}, \tag{4.1.3}$$

where $[s] \xrightarrow{g} [p]$.

We wish to show that (4.1.2) implies (4.1.7). First we make some related observations of a more general character.

PROPOSITION 4.1.2. *Let $[s] \xrightarrow{f} [s+p]$, $[s] \xrightarrow{\xi} [p]$ be morphisms in an algebraic theory T. We have*

(4.1.4)[1] $$(f, 0_s \oplus 1_p)^{1+m} = (f^{1+m}, 0_s \oplus 1_p),$$

$$(f, 0_s \oplus 1_p)^{1+m}(\xi, 1_p) = (f^{1+m}(\xi, 1_p), 1_p), \tag{4.1.5}$$

$$(\xi, 1_p) = (f, 0_s \oplus 1_p)^{1+m}(\xi, 1_p) \Leftrightarrow \xi = f^{1+m}(\xi, 1_p), \tag{4.1.6}$$

$$\xi = f(\xi, 1_p) \Rightarrow \xi = f^{1+m}(\xi, 1_p). \tag{4.1.7}$$

PROOF OF (4.1.4). Induction on m. If $m = 0$, then (4.1.4) is trivial. Assume now $m > 0$.

[1] This very nice identity was suggested by J. D. Rutledge.

$$\begin{aligned}(f, 0_s \oplus 1_p)^{1+m} &= (f, 0_s \oplus 1_p)^m (f, 0_s \oplus 1_p) \\ &= (f^m, 0_s \oplus 1_p)(f, 0_s \oplus 1_p) && \text{by inductive assumption} \\ &= (f^m(f, 0_s \oplus 1_p), 0_s \oplus 1_p) && \text{by (2.5.17)} \\ &= (f^{1+m}, 0_s \oplus 1_p).\end{aligned}$$

PROOF OF (4.1.5).

$$\begin{aligned}(f, 0_s \oplus 1_p)^{1+m}(\xi, 1_p) &= (f^{1+m}, 0_s \oplus 1_p)(\xi, 1_p) && \text{by (4.1.4)} \\ &= (f^{1+m}(\xi, 1_p), (0_s \oplus 1_p)(\xi, 1_p)) && \text{by (4.5.3)} \\ &= (f^{1+m}(\xi, 1_p), 1_p) && \text{by (2.5.16) and (4.5.2).}\end{aligned}$$

PROOF OF (4.1.6). Immediate from (4.1.5) and $\xi = \eta \Leftrightarrow (\xi, 1_p) = (\eta, 1_p)$.

PROOF OF (4.1.7). Suppose $\xi = f(\xi, 1_p)$. By (4.1.6), $(\xi, 1_p) = (f, 0_s \oplus 1_p)(\xi, 1_p)$. Hence, $(\xi, 1_p) = (f, 0_s \oplus 1_p)^{1+m}(\xi, 1_p)$ and by (4.1.6) $\xi = f^{1+m}(\xi, 1_p)$.

COROLLARY 4.1.3. *Let f be an ideal morphism in an iterative theory. Then*

(4.1.8) $$f^\dagger = f^{1+m}(f^\dagger, 1_p),$$

(4.1.9) $$(f^\dagger, 1_p) = (f, 0_s \oplus 1_p)^{1+m}(f^\dagger, 1_p),$$

(4.1.10) $$f^\dagger = (f^{1+m})^\dagger.$$

PROOF. Equality (4.1.8) follows from (4.1.2) and (4.1.7);
(4.1.9) follows from (4.1.8) and (4.1.6);
(4.1.10) follows from (4.1.8) and (4.1.3).

If we represent the morphism $[s] \xrightarrow{f} [s+p]$ by

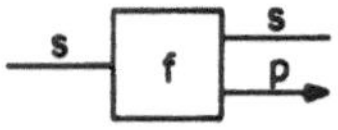

Fig. 1.

and the morphism $[s] \xrightarrow{f^\dagger} [p]$ by

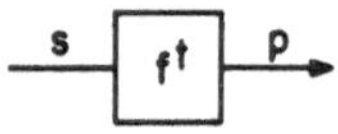

Fig. 2.

then equation (4.1.2) may be given a pictorial interpretation: Fig. 2 and

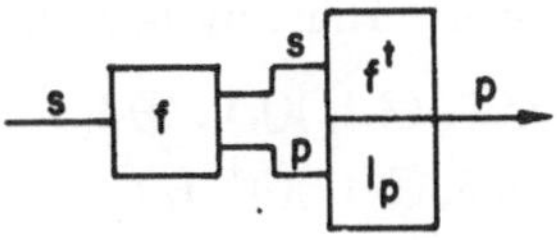

Fig. 3.

represent the same morphism. This suggests the following alternative rendering of $f^\dagger$:

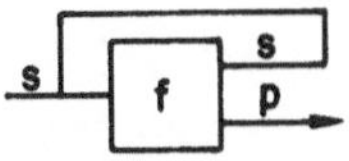

Fig. 4.

4.2. *Iterating power-ideal morphisms*

Let I be an iterative theory. We will show that the iteration operation in I may be extended to power-ideal morphisms in I. Let $[s] \xrightarrow{f} [s+p]$ be an arbitrary morphism in I.

PROPOSITION 4.2.1. *If, for some m, $h = f^{1+m}$ is ideal, then*

(4.2.1) $$h^\dagger = f(h^\dagger, 1_p).$$

Moreover, where $[s] \xrightarrow{g} [p]$,

(4.2.2) $$g = f(g, 1_p) \Rightarrow g = h^\dagger.$$

Briefly: if for some m, f^{1+m} is ideal, then $f^\dagger$ exists and $f^\dagger = h^\dagger$.

PROOF. $h^\dagger = f^{1+m}(h^\dagger, 1_p)$ by (4.1.2).

(4.2.3) $$(h^\dagger, 1_p) = (f, 0_s \oplus 1_p)^{1+m}(h^\dagger, 1_p) \qquad \text{by (4.1.6).}$$

Let $k = f^{2+m}$. Then k is also ideal and so

(4.2.4) $$(k^\dagger, 1_p) = (f, 0_s \oplus 1_p)^{2+m}(k^\dagger, 1_p).$$

By repeatedly substituting (4.2.3) into itself we get

(4.2.5) $$(h^\dagger, 1_p) = (f, 0_s \oplus 1_p)^{(1+m)(2+m)}(h^\dagger, 1_p).$$

In similar fashion,

(4.2.6) $$(k^\dagger, 1_p) = (f, 0_s \oplus 1_p)^{(2+m)(1+m)}(k^\dagger, 1_p).$$

By (4.1.3) and (4.1.6), $k^{\dagger} = h^{\dagger}$. Thus, by (4.2.4),

$$\begin{aligned}(h^{\dagger}, 1_p) &= (f, 0_s \oplus 1_p)(f, 0_s \oplus 1_p)^{1+m}(h^{\dagger}, 1_p) \\ &= (f, 0_s \oplus 1_p)(h^{\dagger}, 1_p) \qquad \text{by (4.2.3).}\end{aligned}$$

Equality (4.2.1) now follows by (4.1.6).

To establish (4.2.2):

$$g = f(g, 1_p) \Rightarrow (g, 1_p) = (f, 0_s \oplus 1_p)(g, 1_p) = (f, 0_s \oplus 1_p)^{1+m}(g, 1_p).$$

Thus: $g = f(g, 1_p) \Rightarrow g = f^{1+m}(g, 1_p)$ and by (4.1.3) the implication (4.2.2) follows.

4.3. $[\![X \cdot \mathbf{N}, \square]\!]$ *is iterative*

We have already noted that $[\![X \cdot \mathbf{N}, \square]\!]$ is an ideal theory when $X \neq \emptyset$. As an algebraic theory, it is quite special in that there is a unique morphism $[1] \xrightarrow{\theta} [0]$ in it, viz. the function with constant value $\square$. The corresponding morphism in $[\![X \cdot \mathbf{N}, \text{und.}]\!]$ is the empty partial function. Let

$$\theta_s = \underbrace{(\theta, \theta, \dots, \theta)}_{s} : [s] \to [0].$$

In this section we make frequent use of $\theta_s \oplus 1_p : [s+p] \to [p]$ We note immediately

(4.3.1) $$(0_s \oplus 1_p)(\theta_s \oplus 1_p) = 1_p.$$

Indeed, $(0_s \oplus 1_p)(\theta_s \oplus 1_p) = 0_0\theta_s \oplus 1_p\, 1_p = 0_0 \oplus 1_p = 1_p$.

We use the abbreviation $e = \theta_s \oplus 1_p$; dropping the indices causes no trouble since they are fixed in the discussion below. Explicitly:

$$xt(s+j) \overset{e}{\longmapsto} xtj, \quad j \in [p], \quad x \in X, \quad t \in \mathbf{N},$$

$$xti \overset{e}{\longmapsto} , \qquad i \in [s].$$

Let

$$f^{(m)} : [s] \xrightarrow{f} [s+p] \xrightarrow{(f, 0_s \oplus 1_p)^m} [s+p] \xrightarrow{e} [p],$$

i.e., $f^{(m)} = f^{1+m}e$. [In the case where f is the (timed) transition function of a monadic machine, $f^{(m)}$ represents the part of the timed behavior contributed by the first $1+m$ steps of computation.]

Notice that $f^{(m)} \subset f^{(1+m)}$ in the sense if $zf^{(m)} \neq \square$, then $zf^{(m)} = zf^{(1+m)}$, where $z \in X \cdot \mathbf{N} \cdot [s+p]$. Thus

(4.3.2) $$f^{(0)} \subset f^{(1)} \subset f^{(2)} \subset \cdots.$$

In order to prove a theorem which goes beyond our immediate requirements (but which takes no additional effort), we define a morphism $[s] \xrightarrow{f} [s+p]$ in $[X \cdot \mathbf{N}, \square]^{-}$ to be *semipositive* iff

$$(x0i)f = x'mi', \; i' \in [s] \Rightarrow m > 0.$$

PROPOSITION 4.3.1. *If* $[s] \xrightarrow{f} [s+p]$ *in* $[X \cdot \mathbf{N}, \square]^{-}$ *is semipositive, then* $f^{\dagger}$ *exists. In fact*

$$f^{\dagger} = \bigcup_m f^{(m)}. \tag{4.3.3}$$

PROOF. Assume first there exists $[s] \xrightarrow{\xi} [p]$ satisfying

$$\xi = f(\xi, 1_p). \tag{4.3.4}$$

Suppose $xti \xmapsto{\xi} y(t+m)j$. According to (4.1.7) we have

$$\xi = f^{1+m}(\xi, 1_p). \tag{4.3.5}$$

Since f is semipositive, we must have

$$\text{if } xti \xmapsto{f^{1+m}} x't'i', \; i' \in [s], \quad \text{then } t' \geq t+1+m;$$

therefore $i' \in s + [p]$ and

$$xti \xmapsto{f^{1+m}(\xi, 1_p)} x't'(i'-s) = y(t+m)j.$$

It follows that

$$xti \xmapsto{f^{(m)}} y(t+m)j,$$

$$\xi \subset \bigcup_m f^{(m)}. \tag{4.3.6}$$

Now suppose

$$xti \xmapsto{f^{(m)}} yuj.$$

Then

$$xti \xmapsto{f^{1+m}} yu(s+j)$$

and by (4.3.5)

$$xti \xrightarrow{\xi} yuj.$$

Thus

$$\bigcup_m f^{(m)} \subset \xi. \tag{4.3.7}$$

Hence, if there exists ξ satisfying (4.3.4), then $\xi = \cup_m f^{(m)}$. It remains to show that $\cup_m f^{(m)} = f(\cup_m f^{(m)}, 1_p)$.

It suffices to show

(4.3.8) $$f^{(1+m)} = f(f^{(m)}, 1_p).$$

We observe first

(4.3.9) $$(f, 0_s \oplus 1_p)e = (fe, 1_p).$$

Indeed,

$$(f, 0_s \oplus 1_p)e = (fe, (0_s \oplus 1_p)e) = (fe, 1_p) \qquad \text{by (4.3.1).}$$

In the case $m = 0$, (4.3.8) asserts

$$f(f, 0_s \oplus 1_p)e = f(fe, 1_p)$$

which follows from (4.3.9). For the case $m > 0$ we have

$$f^{(1+m)} : [s] \xrightarrow{f^{1+m}} [s+p] \xrightarrow{(f, 0_s \oplus 1_p)} [s+p] \xrightarrow{e} [p]$$

$$f^{(1+m)} : [s] \xrightarrow{f^{1+m}} [s+p] \xrightarrow{(fe, 1_p)} [p].$$

Now,

$$\begin{aligned} f(f^{(1+m)}, 1_p) &= f^{2+m}(fe, 1_p) && \text{by (4.1.5) with } \xi = fe \\ &= f^{3+m}e && \text{by (4.3.9)} \\ &= f^{(2+m)} \end{aligned}$$

and (4.3.8) is established.

COROLLARY 4.3.2. *$[\![X \cdot \mathbf{N}, \square]\!]$ is iterative.*

PROOF. Semipositive implies ideal.

4.4. *$[\![M^\wedge]\!]$ is iterative for free and 'other' monoids M*

We have already seen in Section 3.3, that $[\![M^\wedge]\!]$ is ideal. By a (strict) *length monoid M*, we mean a monoid which is equipped with a *length function $l : M \to N$* which satisfies

(4.4.1) $$(xy)l = xl + yl,$$

(4.4.2) $$xl = 0 \Rightarrow x = 1.$$

From (4.4.1) we immediately obtain

(4.4.3) $$1l = 0.$$

Free monoids and free abelian monoids may be equipped with length functions in an obvious way.

From (4.4.1)–(4.4.3) we infer

(4.4.4) $$xy = 1 \Rightarrow x = 1 \ \&\ y = 1.$$

Now, in any algebraic theory $[R]$, where R is a semiring, if $[s] \xrightarrow{f} [s+p]$ is in $[R]$, then f is uniquely determined by the two morphisms

(4.4.5) $$f_1 : [s] \xrightarrow{f} [s+p] \xrightarrow{1_s \oplus \theta_p} [s],$$

(4.4.6) $$f_2 : [s] \xrightarrow{f} [s+p] \xrightarrow{\theta_s \oplus 1_p} [p],$$

where $\theta_p : [p] \to [0]$ is the unique $p \times 0$ matrix. In fact, f_1 consists of the first s columns of f, and f_2 consists of the next p columns. If

$$[s] \xrightarrow{g_1} [q] \quad \text{and} \quad [p] \xrightarrow{g_2} [q]$$

in $[R]$, then clearly

$$fg = [f_1\, f_2] \begin{bmatrix} g_1 \\ g_2 \end{bmatrix} = f_1 g_1 + f_2 g_2,$$

where $[f_1\, f_2]$ is a 1×2 'block' matrix and

$$\begin{bmatrix} g_1 \\ g_2 \end{bmatrix}$$

is a 2×1 block matrix.

In particular, the equation

(4.4.7) $$\xi = f(\xi, 1_p)$$

in the 'unknown' matrix $[s] \xrightarrow{\xi} [p]$ becomes

(4.4.8) $\xi = f_1 \xi + f_2$, where f_1, f_2 are given by (4.4.5), (4.4.6).

The following proposition may be regarded as known, especially when M is a monoid freely generated by some set.

PROPOSITION 4.4.1. *If M is a length monoid and if $[s] \xrightarrow{f} [s+p]$ is ideal in $[\![M^\wedge]\!]$, then there is a unique morphism*

$$\xi = f^\dagger = f_2 + f_1 f_2 + f_1^2 f_2 + \cdots$$

in $[\![M^\wedge]\!]$ which satisfies (4.4.7). Briefly, $[\![M^\wedge]\!]$ is an iterative theory.

4.5. *Sequacious functions – motivation*

If L and M are $\boldsymbol{M}$-machines each with one beginning state and exit, there is an obvious composite machine LM obtained by identifying the exit of L with the beginning state of M. Let X be the set of external states of $\boldsymbol{M}$. Given $x_1 \in X$ as beginning external state of L, there is a unique finite sequence

(4.5.1) $$x_1 x_2 \dots x_l \in X^+$$

or infinite sequence

(4.5.2) $$x_1 x_2 \dots x_k \dots \in X^\infty$$

of external states determined by the action of L. The sequence (4.5.1) or (4.5.2) is the *track of (the computation of) L determined by* x_1. (We are ignoring the 'trace' which 'records' internal states.) If

(4.5.3) $$x_l x_{l+1} \dots \in X^+ \cup X^\infty$$

is the track of M determined by x_l, then

(4.5.4) $$x_1 x_2 \dots x_l x_{l+1} x_{l+2} \dots \in X^+ \cup X^\infty$$

is the track of LM determined by x_1 if (4.5.1) obtains; otherwise (4.5.2) is the track of LM determined by x_1.

For $x \in X$, let $xL' \in X^+ \cup X^\infty$ be the track of L determined by x. It is convenient to extend the function $L' : X \to X^+ \cup X^\infty$ to

$$L^+ : X^+ \cup X^\infty \longrightarrow X^+ \cup X^\infty$$

by the requirement that L^+ keep X^∞ pointwise fixed and

$$(ux)L^+ = u(xL'),$$

where $u \in X^*$ is a finite sequence (possibly null) and $x \in X$. Then

$$x(LM)' = x(LM)^+ = xL^+M^+.$$

The extension of these considerations to machines with several beginning states and exits leads to the notion 'sequacious function.'

4.6. *Sequacious functions – definition*

To facilitate computation we define $ui \leq vj$ (resp. $ui < vj$), where $u \in X^+, i \in [n], ui \in X^+ \cdot [n], v \in X^+, j \in [p], vj \in X^+ \cdot [p]$, to mean u is a prefix of v (resp. u is a proper prefix of v), i.e., $uw = v$ for some (unique) $w \in X^*$ (resp. $w \in X^+$). Thus the relations $\leq$ and $<$ ignore the final terms i, j in the above finite sequences. In the case that $n = 1 = p$, we suppress $i = 1 = j$ as usual and the relations $\leq$ and $<$ take on ordinary meaning. In similar fashion, if $v \in X^\infty$ and $uw = v$, where $u \in X^+$, $w \in X^\infty$, we write $u \leq v$ and $u < v$. Also $v \leq v$ for $v \in X^+ \cdot [p] \cup X^\infty$.

We note that both relations are transitive and

$$ui \leq vj \Rightarrow wui \leq wvj,$$
$$ui < vj \Rightarrow wui < wvj, \quad \text{where } w \in X^*.$$

A morphism $[n] \xrightarrow{f} [p]$ in $[X^+, X^\infty]$ is *sequacious* iff for each $i \in [n]$, $x \in X$, $u \in X^*$,

(4.6.1) $$xi \leq xif,$$

(4.6.2) $$(uxi)f = (ux)if = u(xif).$$

Thus f is completely determined by its values on $X \cdot [n]$.

PROPOSITION 4.6.1. *Suppose* $[n] \xrightarrow{f} [p]$ *in* $[X^+, X^\infty]$ *is sequacious,* $i \in [n]$ *and* $v \in X^+$. *Then*

(4.6.3) $$vi \leq vif$$

(4.6.4) $$(uvi)f = (uv)if = u(vif) \quad \textit{for } u \in X^*.$$

PROOF OF (4.6.3). Let $v = ux$, where $u \in X^*$, $x \in X$. Then $xi \leq xif$ so that

$$vi = uxi \leq uxif = vif.$$

PROOF OF (4.6.4). Let $v = u'x$, where $u' \in X^*$ and $x \in X$. Then

$$(uvi)f = (uv)if = (uu'x)if = uu' \cdot xif = u \cdot u'(xif) = u \cdot (u'x)if = u \cdot vif.$$

PROPOSITION 4.6.2. *A morphism* $[n] \xrightarrow{f} [p]$ *in* $[X^+, X^\infty]$ *is sequacious iff if is sequacious for each* $i \in [n]$.

PROOF. Let $f_i = if$. Suppose first f_i is sequacious for each i. Then $x \leq xf_i = xif$ for each $i \in [n]$, $x \in X$ and so $xi \leq xif$. Verifying (4.6.2):

$$(ux)if = (ux)f_i = u \cdot xf_i = u \cdot xif.$$

The converse is equally immediate.

COROLLARY 4.6.3. *If* $f_i : [1] \to [p]$ *in* $[X^+, X^\infty]$ *is sequacious for each*

$i \in [n]$, then so is

$$(f_1, f_2, \ldots, f_n) : [n] \to [p].$$

4.7. *$[\![X^+, X^\infty]\!]$ is ideal*

Let $[X^+, X^\infty]^-$ be the family, indexed by n, p, of sequacious morphisms $[n] \xrightarrow{f} [p]$ in $[X^+, X^\infty]$.

PROPOSITION 4.7.1. *$[X^+, X^\infty]^-$ is a subtheory of $[X^+, X^\infty]$.*

PROOF. By virtue of Corollary 4.6.3 and the fact that the base morphisms of $[X^+, X^\infty]$ are sequacious, it is only necessary to prove that if $[n] \xrightarrow{f} [p] \xrightarrow{g} [q]$ are sequacious, then so is fg. Verifying (4.6.1):

$$xi \leq xif \leq xifg \qquad \text{by (4.6.3).}$$

Now verifying (4.6.2):

$$(ux)ifg = ((ux)if)g = (u \cdot xif)g = u \cdot xifg \qquad \text{by (4.6.4)}$$

A sequacious morphism $[n] \xrightarrow{f} [p]$ is *positive* iff for each $i \in [n]$, $x \in X$

$$xi < xif. \tag{4.7.1}$$

PROPOSITION 4.7.2. *If $[n] \xrightarrow{f} [p]$ is a positive morphism in $[X^+, X^\infty]^-$ and $v \in X^+$, $i \in [n]$, then*

$$vi < vif. \tag{4.7.2}$$

PROOF. Let $v = ux$, where $u \in X^*$, $x \in X$. Since $xi < xif$, it follows

$$vi = uxi < uxif = vif.$$

We now show that the family of positive morphisms in $[X^+, X^\infty]^-$ satisfies the ideal property.

PROPOSITION 4.7.3. *Let $[n] \xrightarrow{f} [p] \xrightarrow{g} [q]$ be sequacious morphisms. If f is positive, then so is fg.*

PROOF. $xi < xif \leq xifg$ by Proposition 4.6.1, (4.6.3). Thus $xi < xifg$.

Let $[\![X^+, X^\infty]\!]$ be the smallest subtheory of $[X^+, X^\infty]^-$ which contains all positive morphisms. Then

COROLLARY 4.7.4. *$[\![X^+, X^\infty]\!]$ is ideal.*

PROOF. Immediate from Proposition 4.7.3.

It remains to prove that $[\![X^+, X^\infty]\!]$ is closed with respect to conditional iteration, which we do in Section 9.

REMARK. Actually a more general result obtains. Call a morphism $[s] \xrightarrow{f} [s+p]$ in $[X^+, X^\infty]^-$ *semipositive* iff $xif = vi'$ & $i' \in [s] \Rightarrow xi < xif$. Essentially the same proof as given in Section 9 shows that the equation $\xi = f(\xi, 1_p)$ has a unique solution in $[X^+, X^\infty]^-$ when f is semipositive.

4.8. $f^\infty : [s] \to [p]$

Let $[s] \xrightarrow{f} [s+p]$ be an ideal morphism in $[\![X^+, X^\infty]\!]$, i.e., f is sequacious and positive. We wish to define $[s] \xrightarrow{f^\infty} [p]$ as a kind of limit of the morphisms

$$f^{1+m} : [s] \xrightarrow{f} [s+p] \xrightarrow{(f, 0_s \oplus 1_p)^m} [s+p].$$

Let $uif^{1+m} = w_m$, $u \in X^+$, $i \in [n]$. Then exactly one of the following three cases obtain.

Case A. $w_m = v_m i_m$, $v_m \in X^+$, $i_m \in [s]$ for all m. Then by Proposition 4.7.3 and (4.7.2),

$$uif^{1+m} < uif^{2+m} \quad \text{for all } m \tag{4.8.1}$$

and we define $uif^\infty \in X^\infty$ to be the unique solution to the conditions

$$uif^{1+m} < uif^\infty \quad \text{for all } m.$$

Case B. For some m, $w_m = v_m i_m$, where $i_m \in s + [p]$. Let c be the smallest such m. Then

$$w_c = v_c i_c, \qquad w_c = w_{c+1} = w_{c+2} = \cdots \tag{4.8.2}$$

and we define

$$uif^\infty = v_c j, \quad \text{where } i_c = s + j.$$

Case C. For some m, $w_m \in X^\infty$. Let c be the smallest such m. Then

$$w_c = w_{c+1} = w_{c+2} = \cdots \tag{4.8.3}$$

and we define

$$uif^\infty = w_c.$$

PROPOSITION 4.8.1. *f^∞ is sequacious, where f in $[\![X^+, X^\infty]\!]$ is ideal.*

PROOF. Verifying (4.6.1): $xi \leq w_m = xif^{1+m} \leq xif^\infty$. To verify (4.6.2) we consider cases A–C.

If Case A obtains for $x \in X$, $i \in [n]$, then Case A also obtains for $ux \in X^+$, $i \in [n]$ for all $u \in X^*$. Now xif^∞ is characterized by

$$xif^{1+m} < xif^\infty \quad \text{for all } m.$$

Thus,

$$(ux)if^{1+m} = u \cdot xif^{1+m} < u \cdot xif^\infty \quad \text{for all } m.$$

But '$(ux)if^{1+m} < (ux)if^\infty$ for all m' characterizes $(ux)if^\infty$ so that

$$(ux)if^\infty = u \cdot xif^\infty.$$

The remaining two cases are left to the reader.

4.9. $[\![X^+, X^\infty]\!]$ *is iterative*

Let $[s] \xrightarrow{f} [s+p]$ in $[\![X^+, X^\infty]\!]$ be ideal. We claim $\xi = f^\infty$ is a unique solution to (4.1.1). We start with a statement which implies the uniqueness part of the assertion.

PROPOSITION 4.9.1. *If $\xi : [s] \to [p]$ is in $[X^+, X^\infty]^-$ and*

$$\xi = f(\xi, 1_p), \quad \text{where } f : [s] \to [s+p] \text{ is ideal in } [\![X^+, X^\infty]\!]. \tag{4.9.1}$$

Then $\xi = f^\infty$.

PROOF. By (4.1.7), $\xi = f^{1+m}(\xi, 1_p)$ for all m. Suppose, for all m, $xif^{1+m} = v_m i_m$, where $x \in X$, $i \in [s]$, $v_m \in X^+$, $i_m \in [s]$, i.e., Case A obtains. By Proposition 4.7.3 and (4.7.2),

$$xif^{1+m} < xif^{2+m} \leq xif^{2+m}(\xi, 1_p) \quad \text{for all } m$$

so that

$$xif^{1+m} < xi\xi \quad \text{for all } m, \qquad xi\xi = xif^\infty.$$

Suppose now Case B obtains so that

$$xif^{1+c} = v_c i_c, \quad \text{where } i_c = s + j.$$

Then

$$xi\xi = xif^{1+c}(\xi, 1_p) = v_c j = xif^\infty.$$

Finally, if Case C obtains,

$$xif^{1+c} = w_c \in X^\infty$$

so that

$$xi\xi = xif^{1+c}(\xi, 1_p) = w_c = xif^\infty.$$

PROPOSITION 4.9.2. *If* $[s] \xrightarrow{f} [s+p]$ *is ideal in* $[\![X^+, X^\infty]\!]$, *then* $f^\infty = f(f^\infty, 1_p)$.

PROOF. Let $v \in X^+, i \in [s]$. Suppose first that Case A obtains, i.e.,

$$vif^{1+m} = v_m i_m, \quad \text{with } v_m \in X^+ \text{ and } i_m \in [s].$$

Then

$$vif^{1+m} < vif^{2+m} < vif^\infty \quad \text{for all } m.$$

It follows

$$v_0 i_0 f^{1+m} = v_{1+m} i_{1+m}$$

so that Case A obtains again and $vif^\infty = v_0 i_0 f^\infty$. Thus,

$$vif^\infty = v_0 i_0 f^\infty = vif(f^\infty, 1_p).$$

Suppose now $vif = vif^1 = v_0 i_0$, with $v_0 \in X^+$, $i_0 \in [s]$ and that Cases B or C obtain. In the former case, we have, where $i_c = s + j$, that Case B obtains also for $v_0 i_0$ and

$$vif^\infty = v_c j = v_0 i_0 f^\infty = vif(f^\infty, 1_p);$$

in the latter case we have, where $w_c \in X^\infty$, Case C obtains also for $v_0 i_0$ and

$$vif^\infty = w_c = v_0 i_0 f^\infty = vi(f^\infty, 1_p).$$

There are two remaining cases. If $vif = v_0 i_0$ and $i_0 = s + j$, then

$$vif^\infty = v_0 j = v_0 i_0 (f^\infty, 1_p) = vif(f^\infty, 1_p).$$

Finally, if $vif = w_0 \in X^\infty$, then

$$vif^\infty = w_0 = vif(f^\infty, 1_p).$$

From the previous two Propositions we immediately obtain

THEOREM 4.9.3. $[\![X^+, X^\infty]\!]$ *is an iterative theory.*

4.10. *Finite-valued sequacious functions*

The notion of a morphism in $[X^+, X^\infty]$ being sequacious makes use of the multiplicative semigroup structure on X^+ (alternatively, the monoid structure on X^*) and the X^+-module (alternatively, the X^*-module) structure on X^∞ (i.e., for $u, v \in X^+$, $w \in X^\infty$, we have in X^∞, $u(vw) = (uv)w$; alternatively, for $u, v \in X^*$, $1 \in X^*$, we have $u(vw) = (uv)w$ and

$1w = w$). One often wishes to identify all the infinite computations of a machine. This leads to the consideration of the algebraic theory $[X^+, \square]$, where $\square = X^\infty$ is the solitary element of the single point set $\{\square\}$. The singleton $\{\square\}$ becomes an X^+-module (resp. an X^*-module) by defining for $u \in X^+$ (resp. $u \in X^*$) $u\square = \square$. This leads to the transitive relations $\leq$, $<$ on $X^+ \cup \{\square\}$; in particular, $u \leq \square$, $u < \square$ for $u \in X^+$ and $\square \leq \square$. Again, these relations may be extended to relations between $X^+ \cdot [n] \cup \{\square\}$ and $X^+ \cdot [p] \cup \{\square\}$ by ignoring $i \in [n]$ and $j \in [p]$.

Finally, we define a morphism $[n] \xrightarrow{f} [p]$ in $[X^+, \square]$ to be sequacious iff (4.6.1) and (4.6.2) hold.

As before (cf. Section 7), the family $[X^+, \square]^-$ of sequacious morphisms forms a subtheory of $[X^+, \square]$. The morphism f is *positive* iff (4.7.1) holds and the smallest subtheory $[\![X^+, \square]\!]$ of $[X^+, \square]^-$ which contains the positive morphisms is ideal. *Moreover,* $[\![X^+, \square]\!]$ *is closed with respect to conditional iteration.* The proof is similar to but simpler than the proof that $[\![X^+, X^\infty]\!]$ is iterative (cf. Section 9).

The relationship between the infinite and finite 'situations' may be summarized as follows. The unique X^+-module morphism between the X^+-module X^∞ and the X^+-module $\square$ *induces a theory-morphism* C_1 *between the algebraic theories* $[X^+, X^\infty]$ *and* $[X^+, \square]$ satisfying where $[n] \xrightarrow{f} [p]$ in $[X^+, X^\infty]$:

(4.10.1) $$vif \in X^+ \cdot [p] \Rightarrow vi(fC_1) = vif,$$

(4.10.2) $$vif \in X^\infty \Rightarrow vi(fC_1) = \square.$$

Furthermore,

(4.10.3) if f is sequacious, i.e., f in $[X^+, X^\infty]^-$, then fC_1 is sequacious, i.e., fC_1 in $[X^+, \square]^-$ and

(4.10.4) if f is sequacious and positive then so is fC_1.

This leads us to a general comparison of ideal theories.

4.11. *Morphisms between ideal theories.*

Let I and J be ideal theories. A *theory-morphism* $I \xrightarrow{F} J$ *is ideal* if it preserves idealism, i.e., if $[n] \xrightarrow{f} [p]$ an ideal morphism in I implies $[n] \xrightarrow{fF} [p]$ is an ideal morphism in J.

THEOREM 4.11.1. *If* $I \xrightarrow{F} J$ *is an ideal morphism between iterative theories I, J, then* **F** *preserves iteration, i.e.,* $f^{\dagger}\mathbf{F} = (f\mathbf{F})^{\dagger}$ *for all ideal morphisms* $[s] \xrightarrow{f} [s+p]$ *in I.*

PROOF. $f^{\dagger} = f(f^{\dagger}, 1_p) \Rightarrow f^{\dagger}\mathbf{F} = (f\mathbf{F})((f^{\dagger}, 1_p)\mathbf{F}) = (f\mathbf{F})(f^{\dagger}\mathbf{F}, 1_p)$ since theory-morphisms preserve pairing. By (4.1.3) we have $f^{\dagger}\mathbf{F} = (f\mathbf{F})^{\dagger}$.

COROLLARY 4.11.2. *The ideal morphism* $[\![X^+, X^\infty]\!] \xrightarrow{C_1} [\![X^+, \square]\!]$ *preserves conditional iteration.*

Let $[n] \xrightarrow{f} [p]$ be a morphism in $[X^+, \square]^-$, i.e., a finite sequacious function. Define $[X^+, \square]^- \xrightarrow{C_2} [X \cdot \mathbf{N}, \square]^-$ as follows, where x, $x' \in X$, $u \in X^*$, $i \in [n]$, $j \in [p]$, t, $m \in \mathbf{N}$, $m =$ length u.

$$xif = ux'j \Rightarrow xti(fC_2) = x'(t+m)j, \tag{4.11.1}$$

$$xif = \square \Rightarrow (xti)(fC_2) = \square. \tag{4.11.2}$$

PROPOSITION 4.11.3. *The functor* $[X^+, \square]^- \xrightarrow{C_2} [X \cdot \mathbf{N}, \square]^-$ *is a theory morphism which takes positive morphisms in* $[X^+, \square]^-$ *into positive morphisms in* $[X \cdot \mathbf{N}, \square]^-$.

PROOF. The distinguished morphism $[1] \xrightarrow{i} [n]$ in $[X^+, \square]^-$ is the sequacious function satisfying $x \mapsto xi$. The functor C_2 takes this function into the function $xt \mapsto xti$. Now suppose $[p] \xrightarrow{g} [q]$ in $[X^+, \square]^-$ and suppose $x'j \xrightarrow{g} u'x''k$, where $u' \in X^*$, $x'' \in X$, $k \in [q]$. Then

$$xifg = (ux'j)g = u(x'jg) = uu'x''k$$

and if $m' =$ length u' we have

$$xti \xmapsto{(fg)C_2} x''(t+m+m')k$$

and

$$xti \xmapsto{(fC_2)(gC_2)} x''(t+m+m')k.$$

The remainder of the proof is left to the reader.

Thus, the morphism $[X^+, \square]^- \xrightarrow{C_2} [X \cdot \mathbf{N}, \square]^-$ induces an ideal morphism $[\![X^+, \square]\!] \xrightarrow{C_2} [\![X \cdot \mathbf{N}, \square]\!]$ and by Theorem 4.11.1:

COROLLARY 4.11.4. *The ideal morphism* $[\![X^+, \square]\!] \xrightarrow{C_2} [\![X \cdot \mathbf{N}, \square]\!]$ *preserves conditional iteration.*

We also note without proof that the canonical map which takes a function

$X \cdot \mathbf{N} \cdot [n] \xrightarrow{f} X \cdot \mathbf{N} \cdot [p]$ in $[\![X \cdot \mathbf{N}, \square]\!]$ into $X \cdot [n] \xrightarrow{fC_3} X \cdot [p]$,
where

$$x0i \stackrel{f}{\longmapsto} x'mj \Rightarrow xi \stackrel{fC_3}{\longmapsto} x'j,$$

and

$$x0i \stackrel{f}{\longmapsto} \square \Rightarrow xi \stackrel{fC_3}{\longmapsto} \square$$

is a morphism $[\![X \cdot \mathbf{N}, \square]\!] \xrightarrow{C_3} [X, \square]$ between algebraic theories which preserves †, if † is defined in $[X, \square]$ as in Section 1.8, with 'undefined' replaced by '$\square$'.

5. The main theorem

5.1. *Statement of the main theorem*

Let I be an iterative theory. *By an (ideal) normal description*

$$[n] \xrightarrow[s]{D} [p] \text{ over } I \text{ of weight } s,$$

we mean a base morphism $\beta_D : [n] \to [s+p]$ in I and an ideal morphism $\tau_D : [s] \to [s+p]$. We say the normal description

$$[n] \xrightarrow[s]{D} [p]$$

describes the morphism

$$|D| : [n] \xrightarrow{\beta_D} [s+p] \xrightarrow{(\tau_D^\dagger, 1_p)} [p] \text{ in } I.$$

The concept 'normal description' is an abstract counterpart to 'monadic machine'. The new terminology reflects our point of view concerning the role of the concept.

By a *sort* Σ over I we mean a singly indexed family $\Sigma = \{\Sigma_p\}_{p \in N}$, $\Sigma_p \subset I_p$, which is closed with respect to right multiplication by base morphisms, i.e., if $[1] \xrightarrow{f} [p] \xrightarrow{g} [q]$, $f \in \Sigma_p$ and g is base, then $fg \in \Sigma_q$. The condition may be briefly expressed as: $\Sigma\mathcal{N} \subset \Sigma$.

If $\Gamma = \{\Gamma_p\}_{p \in N}$, $\Gamma_p \subset I_p$, is any singly indexed family (a *genus* over I), *then* $\Gamma\mathcal{N}$ is a sort.

In the case of a genus Γ, clearly: Γ^0 (cf. Section 2.2) is closed with respect to source pairing of vector morphisms. In the case of a sort Σ, we have in addition, Σ° is closed under $\oplus$, $\Sigma^\circ\mathcal{N} \subset \Sigma^\circ$, $\mathcal{N}\Sigma^\circ \subset \Sigma^\circ$. Moreover, if $1_1 \in \Sigma_1$, then $\mathcal{N} \subset \Sigma^\circ$. But, of course, Σ° need not be closed under composition. A (*vector*) *morphism* in Σ° is said to be of sort Σ.

By *scalar iteration* in I we mean the iteration operation restricted to scalar morphisms. In order to state succinctly the (abbreviated form of the) main theorem, it is convenient to make the following definitions: $\Sigma^\dagger$ (resp. $\Sigma^{\dagger_1}$) is the smallest subtheory of I which contains Σ and is closed with respect to conditional iteration (resp. scalar iteration); $|\Sigma|$ is the family of all morphisms $|D|$, where D is a normal description and τ_D is an ideal morphism in Σ°.

MAIN THEOREM. *If Σ is a sort over the iterative theory I, then*

$$\Sigma^\dagger \subset |\Sigma| \subset \Sigma^{\dagger_1}.$$

Clearly, $\Sigma^{\dagger_1} \subset \Sigma^{\dagger}$ so that '$\Sigma^{\dagger} = |\Sigma| = \Sigma^{\dagger_1}$' immediately follows from the main theorem. The following is equivalent to the main theorem.

COROLLARY 5.1.1. *If Γ is a genus over iterative theory I, then*

$$\Gamma^{\dagger} \subset |\Gamma\mathcal{N}| \subset \Gamma^{\dagger_1}.$$

The main theorem has more constructive content than is indicated by its abbreviated statement. First, $\Sigma^{\dagger}$ can be described more constructively as follows: each morphism in $\Sigma^{\dagger}$ arises from a finite number of morphisms in Σ and $\mathcal{N}$ by a finite number of applications of pairing, composition, iteration. More explicitly, by a *sequential description* (based upon pairing, composition, iteration (resp. scalar iteration)) of a morphism in $\Sigma^{\dagger}$ (resp. $\Sigma^{\dagger_1}$) we mean a finite sequence $[n_k] \xrightarrow{f_k} [p_k]$, $1 \leq k \leq l$ of morphisms in I satisfying:

$$f_k \text{ is in } \Sigma \text{ or is in } \mathcal{N}$$

or

$$f_k = (f_i, f_j), \quad \text{where } i, j < k, \quad p_i = p_j$$

or

$$f_k = f_i f_j, \quad \text{where } i, j < k, \quad p_i = n_j$$

or

$$f_k = f_i^{\dagger}, \quad \text{where } i < k, \quad n_i \leq p_i, \quad f_i \text{ ideal (resp. } f_i \text{ scalar).}$$

The sequential description is said to describe the last morphism f_l.

The proof that $\Sigma^{\dagger} \subset |\Sigma|$, in effect, provides an algorithm for converting a sequential description of a morphism in $\Sigma^{\dagger}$ to a normal description of the same morphism. The proof that $|\Sigma| \subset \Sigma^{\dagger_1}$, implicitly contains an algorithm for converting a normal description of a morphism into the reversal of a (scalar) sequential description of the same morphism.

To prove that $\Sigma^{\dagger} \subset |\Sigma|$ it is sufficient to show that $|\Sigma|$ contains Σ, $\mathcal{N}$ and that $|\Sigma|$ is closed with respect to source-pairing, composition and conditional iteration. This is done in Sections 5.2–5.6.

The proof that $|\Sigma| \subset \Sigma^{\dagger_1}$ is by induction on the weight of a normal description. It is carried out in Sections 5.7–5.10.

5.2. $\mathcal{N} \subset |\Sigma|$

Let $[n] \xrightarrow{b} [p]$ be in $\mathcal{N}$. Let $[n] \xrightarrow[0]{D} [p]$ be the normal description defined by:

$$\tau_D = 0_p, \qquad \beta_D = b.$$

Then D is a normal description of sort Σ, i.e., τ_D is in Σ^0. Further

$$\tau_D^\dagger = 0_p, \ (\tau_D^\dagger, 1_p) = 1_p, \qquad |D| = \beta_D(\tau_D^\dagger, 1_p) = b.$$

For handy reference we state

PROPOSITION 5.2.1. *For every base morphism b there is a unique normal description D of weight 0 such that* $|D| = \beta_D = b$.

5.3. $\Sigma \subset |\Sigma|$

We note first

PROPOSITION 5.3.1. *In any algebraic theory the equation*

$$\xi : [s] \xrightarrow{h} [p] \xrightarrow{0_s \oplus 1_p} [s+p] \xrightarrow{(\xi, 1_p)} [p]$$

has a unique solution, viz. h. Briefly,

$$(h(0_s \oplus 1_p))^\dagger = h.$$

PROOF. $(0_s \oplus 1_p)(\xi, 1_p) = (0_s\xi, 1_p) = 1_p$.

COROLLARY 5.3.2. $\Sigma \subset |\Sigma|$.

PROOF. We already have from Section 2 that the base morphisms in Σ are in $|\Sigma|$. Let h be an ideal morphism of sort Σ. The argument works as well for vector as for scalar h. Then $h(0_s \oplus 1_p)$ is ideal and of sort Σ. Let $[s] \xrightarrow[s]{D} [p]$ be the normal description defined by $\tau_D = h(0_s \oplus 1_p)$ and $\beta_D = 1_s \oplus 0_p$. Then

$$|D| = \beta_D(\tau_D^\dagger, 1_p) = (1_s \oplus 0_p)(h, 1_p) = (h, 0_p) = h.$$

5.4. $|DE| = |D| \cdot |E|$

Given normal descriptions $[n] \xrightarrow[s]{D} [p] \xrightarrow[t]{E} [q]$ over I. We define $[n] \xrightarrow[s+t]{DE} [q]$ as follows.

$$\beta_{DE} : [n] \xrightarrow{\beta_D} [s+p] \xrightarrow{1_s \oplus \beta_E} [s+t+q], \tag{5.4.1}$$

$$\tau_{DE} : [s+t] \xrightarrow{(\tau_D(1_s \oplus \beta_E),\ \tau_E(0_s \oplus 1_{t+q}))} [s+t+q]. \tag{5.4.2}$$

PROPOSITION 5.4.1. *If D and E are ideal (resp. of sort* Σ *), then so is DE.*

PROPOSITION 5.4.2. *Assuming* τ_D, τ_E *are ideal and* τ_{DE} *given by* (5.4.2), *we have*

$$\tau^{\dagger}_{DE} = (\tau^{\dagger}_{D}\beta_{E}(\tau^{\dagger}_{E}, 1_q), \tau^{\dagger}_{E}) \tag{5.4.3}$$

or equivalently,

$$(1_s \oplus 0_t)\tau^{\dagger}_{DE} : [s] \xrightarrow{\tau^{\dagger}_{D}} [p] \xrightarrow{\beta_E} [t+q] \xrightarrow{(\tau^{\dagger}_{E}, 1_q)} [q], \tag{5.4.4a}$$

$$(0_s \oplus 1_t)\tau^{\dagger}_{DE} = \tau^{\dagger}_{E}. \tag{5.4.4b}$$

PROOF. Since $\tau^{\dagger}_{DE}$ is the unique solution to the equation

$$\xi : [s+t] \xrightarrow{\tau_{DE}} [s+t+q] \xrightarrow{(\xi, 1_q)} [q], \tag{5.4.5}$$

it is sufficient to show that

$$\xi = (\tau^{\dagger}_{D}\beta_{E}(\tau^{\dagger}_{E}, 1_q), \tau^{\dagger}_{E}) = (\tau^{\dagger}_{D} \cdot |E|, \tau^{\dagger}_{E}) \tag{5.4.6}$$

satisfies (5.4.5). In any algebraic theory, the solutions to (5.4.5) are the same as the simultaneous solutions to the pair of equations

$$[1_s \oplus 0_t]\xi : [s] \xrightarrow{1_s \oplus 0_t} [s+t] \xrightarrow{\tau_{DE}} [s+t+q] \xrightarrow{(\xi, 1_q)} [q], \tag{5.4.7a}$$

$$[0_s \oplus 1_t]\xi = [t] \xrightarrow{0_s \oplus 1_t} [s+t] \xrightarrow{\tau_{DE}} [s+t+q] \xrightarrow{(\xi, 1_q)} [q]. \tag{5.4.7b}$$

Verification (a). Substitution of (5.4.6) in the left of (a) yields

$$\tau^{\dagger}_{D}\beta_{E}(\tau^{\dagger}_{E}, 1_q). \tag{5.4.8}$$

Substituting (5.4.6) in the right of (a), making use of $(1_s \oplus 0_t)\tau_{DE} = \tau_D \cdot (1_s \oplus \beta_E)$ and the associativity of source-pairing, we obtain

$$[s] \xrightarrow{\tau_D} [s+p] \xrightarrow{1_s \oplus \beta_E} [s+t+q] \xrightarrow{(\tau^{\dagger}_{D}\beta_E(\tau^{\dagger}_{E}, 1_q), (\tau^{\dagger}_{E}, 1_q))} [q] =$$

$$= [s] \xrightarrow{\tau_D} [s+p] \xrightarrow{(\tau^{\dagger}_{D}\beta_E(\tau^{\dagger}_{E}, 1_q), \beta_E(\tau^{\dagger}_{E}, 1_q))} [q]$$

$$= [s] \xrightarrow{\tau_D} [s+p] \xrightarrow{(\tau^{\dagger}_{D}, 1_p)} [p] \xrightarrow{\beta_E(\tau^{\dagger}_{E}, 1_p)} [q] \qquad \text{by (2.5.3)}$$

$$= [s] \xrightarrow{\tau^{\dagger}_{D}} [p] \xrightarrow{\beta_E(\tau^{\dagger}_{E}, 1_p)} [q]$$

which is (5.4.8) so that equation (5.4.7a) is verified for the substitution (5.4.6).

Verification (b). Substitution of (5.4.6) in the left of (b) yields $\tau^{\dagger}_{E}$.

Substitution of (5.4.6) in the right of (b) and making use of

$$(0_s \oplus 1_t)\tau_{DE} = \tau_E(0_s \oplus 1_{t+q})$$

yields

$$[t] \xrightarrow{\tau_E} [t+q] \xrightarrow{0_s \oplus 1_{t+q}} [s+t+q] \xrightarrow{(\tau_D^\dagger \beta_E(\tau_E^\dagger, 1_q), (\tau_E^\dagger, 1_q))} [q] =$$

$$= [t] \xrightarrow{\tau_E} [t+q] \xrightarrow{(\tau_E^\dagger, 1_q)} [q]$$

$$= [t] \xrightarrow{\tau_E^\dagger} [q],$$

which verifies (b) and with it (5.4.3).

From the calculation verifying (a), we extract

(5.4.9) $$[s+p] \xrightarrow{1_s \oplus \beta_E} [s+t+q] \xrightarrow{(\tau_{DE}^\dagger, 1_q)} [q] =$$

$$= [s+p] \xrightarrow{(\tau_D^\dagger, 1_p)} [p] \xrightarrow{\beta_E(\tau_E^\dagger, 1_q)} [q]$$

$$= (\tau_D^\dagger, 1_p) \cdot |E|.$$

PROPOSITION 5.4.3. *For ideal normal descriptions D, E over I, we have*

$$|DE| = |D| \cdot |E|.$$

PROOF. $|DE| = \beta_{DE}(\tau_{DE}^\dagger, 1_q)$

$$= \beta_D(1_s \oplus \beta_E)(\tau_{DE}^\dagger, 1_q)$$

$$= \beta_D(\tau_D^\dagger, 1_p) \cdot |E| \qquad \text{by (5.4.9)}$$

$$= |D| \cdot |E|.$$

COROLLARY 5.4.4. *$|\Sigma|$ is closed under composition.*

PROOF. Immediate from Propositions 5.4.3 and 5.4.1.

5.5. $|\Sigma|$ *is closed under source-pairing*

Let $[n] \xrightarrow[s]{D} [p]$ be a normal description. Define the normal description $[n+p] \xrightarrow[s]{(D, 1_p)} [p]$ by $\tau_{(D,1_p)} = \tau_D$, $\beta_{(D,1_p)} = (\beta_D, 0_s \oplus 1_p)$. Then

PROPOSITION 5.5.1. $|(D, 1_p)| = (|D|, 1_p)$.

PROOF. $|(D, 1_p)| : [n+p] \xrightarrow{(\beta_D, 0_s \oplus 1_p)} [s+p] \xrightarrow{(\tau_D^\dagger, 1_p)} [p]$.

Thus

$$|(D, 1_p)| = (\beta_D(\tau_D^\dagger, 1_p), (0_s \oplus 1_p)(\tau_D^\dagger, 1_p))$$

$$= (|D|, 1_p).$$

Let $[n+q] \xrightarrow[s]{D \oplus 1_q} [p+q]$ be the normal description defined by composition of normal descriptions as

$$D \oplus 1_q = E(D(1_p \oplus 0_q), 1_{p+q}),$$

where E is the unique normal description of weight 0 with $\beta_E = 1_n \oplus 0_p \oplus 1_q$; cf. Proposition 5.2.1. Then

PROPOSITION 5.5.2. $|D \oplus 1_q| = |D| \oplus 1_q$.

PROOF. $|D \oplus 1_q| = (1_n \oplus 0_p \oplus 1_q)\,|(D(1_p \oplus 0_q), 1_{p+q}|$

$$= (1_n \oplus 0_p \oplus 1_q)(|D|(1_p \oplus 0_q), 1_{p+q})$$

$$= (|D|\,(1_p \oplus 0_q), 0_p \oplus 1_q)$$

$$= |D| \oplus 1_q.$$

Define

$$(1_p, D) : [p+n] \xrightarrow[0]{E} [n+p] \xrightarrow[s]{(D,\, 1_p)} [p], \qquad \beta_E = (0_n \oplus 1_p, 1_n \oplus 0_p).$$

PROPOSITION 5.5.3. $|(1_p, D)| = (1_p, |D|)$.

Given normal descriptions $[n] \xrightarrow[s]{D} [p]$, $[q] \xrightarrow[t]{E} [p]$, we define

$$(D, E) : [n+q] \xrightarrow[s]{D \oplus 1_q} [p+q] \xrightarrow[t]{(1_p,\, E)} [p].$$

PROPOSITION 5.5.4. $|(D, E)| = (|D|, |E|)$.

PROOF. $|(D, E)| = |D \oplus 1_q|\,|1_p, E|$

$$= (|D| \oplus 1_q)(1_p, |E|)$$

$$= (|D|, |E|).$$

5.6. *Closure of $|\Sigma|$ with respect to †*

The proof that $|\Sigma|$ is closed with respect to iteration has been analyzed into several parts. As a possible aid to rapid interpretation of the assertions, we use a 'picture' notation as a supplement to our usual notation. The pictures are identified as 'Fig. 5.6.1', 'Fig. 5.6.2', etc. and below the identification we describe the morphism in our usual notation. The first proposition states that the morphism represented by Figs. 5.6.1 and 5.6.2 (under suitable hypotheses) are the same.

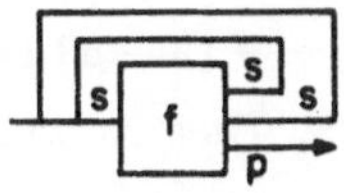

Fig. 5.6.1. $[s] \xrightarrow{f^{\dagger\dagger}} [p]$.

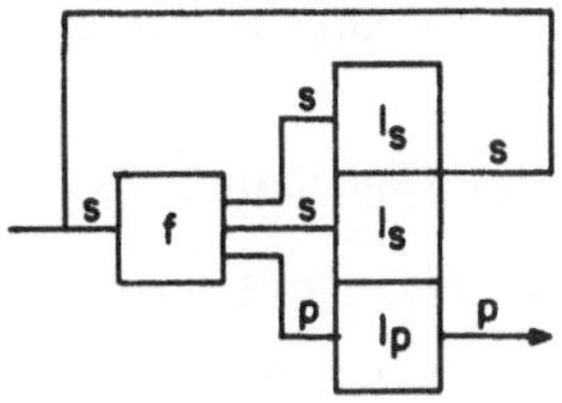

Fig. 5.6.2. $(f((1_s, 1_s) \oplus 1_p))^\dagger$.

PROPOSITION 5.6.1. *Let* $[s] \xrightarrow{f} [2s+p]$ *be an ideal morphism in the iterative theory I. Then* $g : [s] \xrightarrow{f} [2s+p] \xrightarrow{(1_s, 1_s) \oplus 1_p} [s+p]$ *is ideal and* $f^{\dagger\dagger} = g^\dagger$. *If f is of sort* Σ, *then so is g.*

PROOF. $f^{\dagger\dagger} = f^\dagger(f^{\dagger\dagger}, 1_p) = f(f^\dagger, 1_{s+p})(f^{\dagger\dagger}, 1_p)$

$= f(f^\dagger(f^{\dagger\dagger}, 1_p), (f^{\dagger\dagger}, 1_p))$ by (2.5.3)

$= f(f^{\dagger\dagger}, (f^{\dagger\dagger}, 1_p))$

$= f((f^{\dagger\dagger}, f^{\dagger\dagger}), 1_p)$

$= f((1_s, 1_s) \oplus 1_p)(f^{\dagger\dagger}, 1_p)$ by (2.5.16) and (5.5.3)

$= g(f^{\dagger\dagger}, 1_p)$.

Thus, by (4.1.3), $g^\dagger = f^{\dagger\dagger}$.

The following proposition states that the morphism in U represented by Fig. 5.6.3 and the one represented by Fig. 5.6.4 are equal.

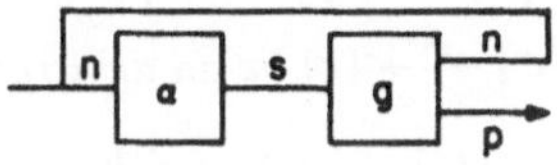

Fig. 5.6.3. $(\alpha g)^\dagger$.

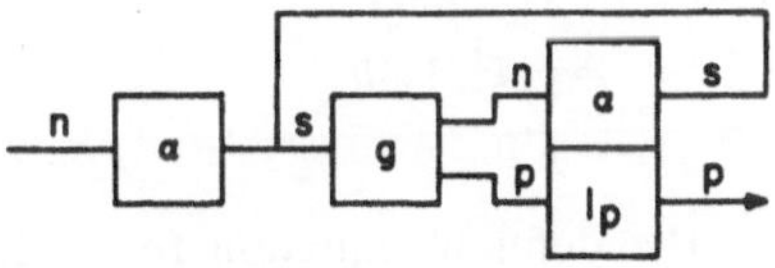

Fig. 5.6.4. $\alpha(g(\alpha \oplus 1_p))^\dagger$.

PROPOSITION 5.6.2. *Let* $[s] \xrightarrow{g} [n+p]$ *be ideal in the iterative theory I and* $[n] \xrightarrow{\alpha} [s]$ *be arbitrary. Then* αg *and* $g[\alpha \oplus 1_p]$ *are ideal and*

$$(\alpha g)^{\dagger} = \alpha(g(\alpha \oplus 1_p))^{\dagger}.$$

Moreover, if α *is in* $\mathcal{N}$, *we have if* g *is of sort* Σ, *then so is* αg *and* $g(\alpha \oplus 1_p)$.

PROOF. Let $k = g(\alpha \oplus 1_p)$. Calculating:

$$\begin{aligned} \alpha g(\alpha k^{\dagger}, 1_p) &= \alpha g(\alpha \oplus 1_p)(k^{\dagger}, 1_p) && \text{by (2.5.16)} \\ &= \alpha k(k^{\dagger}, 1_p) \\ &= \alpha k^{\dagger}. \end{aligned}$$

Since $\alpha k^{\dagger}$ satisfies the defining equation for $[\alpha g]^{\dagger}$, we conclude $[\alpha g]^{\dagger} = \alpha k^{\dagger}$.

The following proposition states that the morphism in I represented by Fig. 5.6.5 and the one represented by Fig. 5.6.6 are equal.

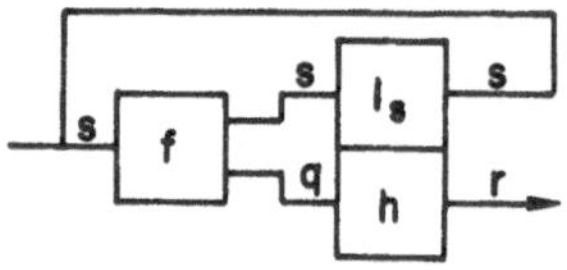

Fig. 5.6.5. $(f(1_s \oplus h))^{\dagger}$.

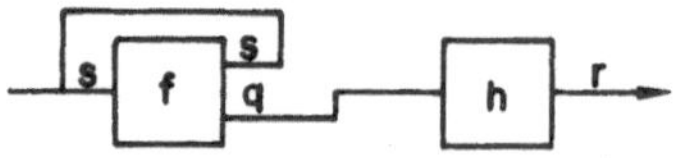

Fig. 5.6.6. $f^{\dagger}h$.

PROPOSITION 5.6.3. *Let* $[s] \xrightarrow{f} [s+q]$ *be an ideal morphism in the iterative theory I and let* $[q] \xrightarrow{h} [r]$ *be an arbitrary morphism in I. Then*

$$(f(1_s \oplus h))^{\dagger} = f^{\dagger}h.$$

PROOF. $$\begin{aligned} f(1_s \oplus h)(f^{\dagger}h, 1_r) &= f(f^{\dagger}h, h) && \text{by (2.5.16)} \\ &= f(f^{\dagger}, 1_q)h && \text{by (2.5.3)} \\ &= f^{\dagger}h. \end{aligned}$$

Since $f^{\dagger}h$ satisfies the defining equation for $(f(1_s \oplus h))^{\dagger}$, the result follows.

COROLLARY 5.6.4. *Let* $[s] \xrightarrow{f} [s+n+p]$ *be an ideal morphism in the iterative theory I and let* $[n] \xrightarrow{\alpha} [s]$ *be arbitrary. Then*

$$(f(1_s \oplus \alpha \oplus 1_p))^\dagger = f^\dagger(\alpha \oplus 1_p).$$

Assume now α *is in* $\mathcal{N}$. *If* f *is of sort* Σ, *then so is* $f(1_s \oplus \alpha \oplus 1_p)$.

PROOF. Take $q = n + p$, $r = s + p$ and $h = \alpha \oplus 1_p$ in Proposition 5.6.3.

The following theorem states that the morphism represented by Fig. 5.6.7 and the one represented by Fig. 5.6.8 are equal.

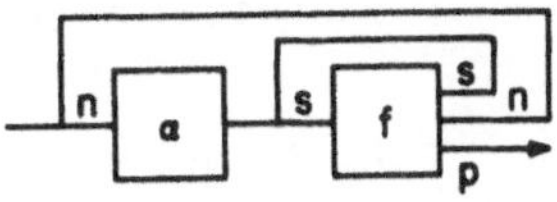

Fig. 5.6.7. $(\alpha f^\dagger)^\dagger$.

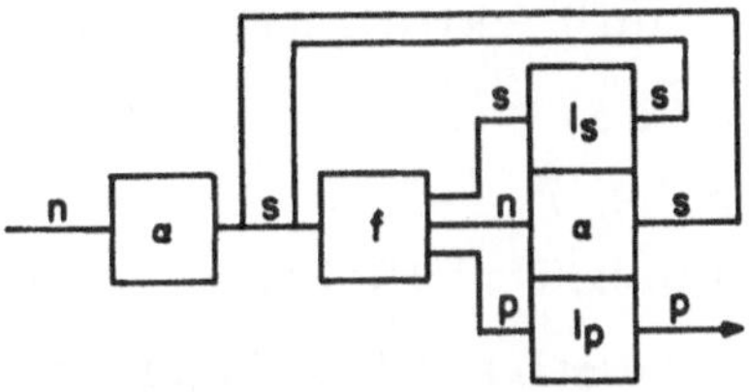

Fig. 5.6.8. $\alpha(f(1_s \oplus \alpha \oplus 1_p))^{\dagger\dagger}$.

THEOREM 5.6.5. *Let* $[s] \xrightarrow{f} [s+n+p]$ *be an ideal morphism in the iterative theory I and let* $[n] \xrightarrow{\alpha} [s]$ *be arbitrary. Then* $f(1_s \oplus \alpha \oplus 1_p)$ *is ideal and*

$$(\alpha f^\dagger)^\dagger = \alpha(f(1_s \oplus \alpha \oplus 1_p))^{\dagger\dagger}.$$

PROOF. By Corollary 5.6.4,

$$(f(1_s \oplus \alpha \oplus 1_p))^\dagger = f^\dagger(\alpha \oplus 1_p).$$

Hence

$$\begin{aligned}(f(1_s \oplus \alpha \oplus 1_p))^{\dagger\dagger} &= (f^\dagger(\alpha \oplus 1_p))^\dagger,\\ \alpha(f(1_s \oplus \alpha \oplus 1_p))^{\dagger\dagger} &= \alpha(f^\dagger(\alpha \oplus 1_p))^\dagger\\ &= (\alpha f^\dagger)^\dagger\end{aligned}$$

by Proposition 5.6.2 with $g = f^\dagger$.

The following theorem states that the morphism described by Fig. 5.6.9 and the one described by Fig. 5.6.7 are equal.

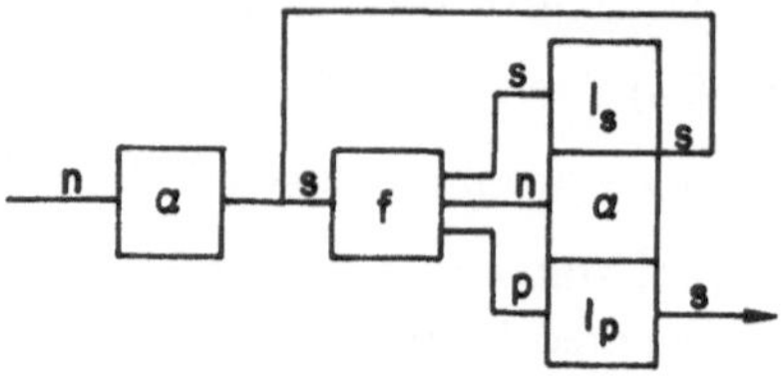

Fig. 5.6.9. $\alpha(f(1_s, \alpha) \oplus 1_p))^\dagger$.

THEOREM 5.6.6. *Let* $[s] \xrightarrow{f} [s+n+p]$ *be an ideal morphism in an iterative theory I and let* $[n] \xrightarrow{\alpha} [s]$ *be arbitrary. Then* $f((1_s, \alpha) \oplus 1_p)$ *is ideal and*

$$(\alpha f^\dagger)^\dagger = \alpha(f((1_s, \alpha) \oplus 1_p)^\dagger.$$

PROOF. By Theorem 5.6.5,

$$\begin{aligned}(\alpha f^\dagger)^\dagger &= \alpha(f(1_s \oplus \alpha \oplus 1_p))^{\dagger\dagger} \\ &= \alpha(f(1_s \oplus \alpha \oplus 1_p)((1_s, 1_s) \oplus 1_p))^\dagger \qquad \text{by Proposition 5.6.1.}\end{aligned}$$

But $(1_s \oplus \alpha \oplus 1_p)((1_s, 1_s) \oplus 1_p) = ((1_s, \alpha) \oplus 1_p)$ and so the result follows.

Let $[n] \xrightarrow[s]{D} [n+p]$ be the normal description consisting of

$$\beta_D : [n] \xrightarrow{\alpha \oplus 0_{n+p}} [s+n+p] \quad \text{and} \quad \tau_D : [s] \xrightarrow{f} [s+n+p].$$

Then $|D| = \beta_D(\tau_D^\dagger, 1_{n+p}) = \alpha f^\dagger$. Thus, if we define $[n] \xrightarrow[s]{D^\dagger} [p]$ by

$$\beta_{D^\dagger} : [n] \xrightarrow{\alpha \oplus 0_p} [s+p], \qquad \tau_{D^\dagger} : [s] \xrightarrow{\tau_D} [s+n+p] \xrightarrow{(1_s, \alpha) \oplus 1_p} [s+p],$$

we have

COROLLARY 5.6.7. *If D is a normal description in the iterative theory I and* $|D|$ *is ideal then* $\beta_D = \alpha \oplus 0_{n+p}$, *where* $[n] \xrightarrow{\alpha} [s]$ *is in* $\mathcal{N}$ *and*

$$|D|^\dagger = |D^\dagger|.$$

Moreover, if τ_D *is of sort* Σ, *then so is* $\tau_{D^\dagger}$. This completes the proof that $\Sigma^\dagger \subset |\Sigma|$.

5.7. *Scalar normal descriptions*

We suppose an iterative theory I given and held fixed throughout this section.

We turn our attention now to proving $|\Sigma| \subset \Sigma^{\dagger_I}$. Inasmuch as $\Sigma^{\dagger_I}$ is closed under source-tupling, it is sufficient to prove that every morphism described by scalar normal description $[1] \xrightarrow[s]{D} [p]$ is in $\Sigma^{\dagger_I}$. If $\beta_D = i : [1] \to [s+p]$, where $i \in s + [p]$, then $|D| = i - s$ is in $\mathcal{N} \subset \Sigma^{\dagger_I}$. We may, therefore, assume $i \in [s]$. Our first task is to show that i may be assumed to be 1.

PROPOSITION 5.7.1. *Let* $[s] \xrightarrow{f} [s+p]$ *be ideal and let* $\pi : [s] \to [s]$ *be a permutation in* $\mathcal{N}$. *Then* $\pi f[\pi^{-1} \oplus 1_p]$ *is ideal and*

$$\pi f^{\dagger} = [\pi f[\pi^{-1} \oplus 1_p]]^{\dagger}. \tag{5.7.1}$$

PROOF. Note that $[\pi^{-1} \oplus 1_p](\pi f^{\dagger}, 1_p) = (f^{\dagger}, 1_p)$ so that

$$\pi f[\pi^{-1} \oplus 1_p](\pi f^{\dagger}, 1_p) = \pi f(f^{\dagger}, 1_p) = \pi f^{\dagger}.$$

Thus, $\pi f^{\dagger}$ satisfies the defining equation for $[\pi f[\pi^{-1} \oplus 1_p]]^{\dagger}$ so that (5.7.1) follows.

COROLLARY 5.7.2. *Let* $[n] \xrightarrow[s]{D} [p]$ *be a normal description with* τ *ideal. Let* $[n] \xrightarrow[s]{E} [p]$ *be the normal description defined by* $\tau_E = \pi\tau_D[\pi^{-1} \oplus 1_p]$, $\beta_E = \beta_D[\pi^{-1} \oplus 1_p]$. *Then* τ_E *is ideal and* $|E| = |D|$.

PROOF.
$$\begin{aligned} |E| &= \beta_E(\tau_E^{\dagger}, 1_p) \\ &= \beta_D[\pi^{-1} \oplus 1_p](\pi\tau_D^{\dagger}, 1_p) && \text{by Proposition 5.7.1} \\ &= \beta_D(\tau_D^{\dagger}, 1_p) = |D|. \end{aligned}$$

COROLLARY 5.7.3. *In Corollary* 5.7.2, *if* $n = 1$, $1\beta_D = i \in [s]$ *and* $1\pi = i$, *we have* β_E *is inclusion and* $|D| = |E|$.

5.8. *Factoring*

We say a morphism $[1+s] \xrightarrow{f} [1+s+p]$ in an algebraic theory T *factors outside of* 1 iff it admits a decomposition

$$f : [1+s] \xrightarrow{g} [s+p] \xrightarrow{0_1 \oplus 1_{s+p}} [1+s+p] \text{ in } T. \tag{5.8.1}$$

The g in (5.8.1) is uniquely determined because

(a) if $s+p\neq 0$, $0_1\oplus 1_{s+p}$ has a right inverse, say $\theta:[1+s+p]\to[s+p]$ in $\mathcal{N}$ so that $f\theta=g$;

(b) if $f:[1]\xrightarrow{g}[0]\xrightarrow{0_1}[1]$, then $g=fg$ since $fg=g0_1g$ and $0_1g:[0]\to[0]$ is the identity.

PROPOSITION 5.8.1. *Let* $[1+s]\xrightarrow{f}[1+s+p]$ *be an ideal morphism in an iterative theory I and suppose f factors outside of* 1. *Then*

$$f^\dagger=(g_1(g_2^\dagger,1_p),g_2^\dagger), \tag{5.8.2}$$

where

$$g_1:[1]\xrightarrow{1_1\oplus 0_s}[1+s]\xrightarrow{g}[s+p], \tag{5.8.3}$$

$$g_2:[s]\xrightarrow{0_1\oplus 1_s}[1+s]\xrightarrow{g}[s+p] \tag{5.8.4}$$

and g is as in (5.8.1).

PROOF. Since $f^\dagger$ is the unique solution to the equation

$$\xi:[1+s]\xrightarrow{g}[s+p]\xrightarrow{0_1\oplus 1_{s+p}}[1+s+p]\xrightarrow{(\xi,1_p)}[p],$$

i.e., the equation $\xi=g(0_1\oplus 1_{s+p})(\xi,1_p)$, it is sufficient to prove for

$$\xi_0=(g_1(g_2^\dagger,1_p),g_2^\dagger)$$

that

$$g_1(g_2^\dagger,1_p)=g_1(0_1\oplus 1_{s+p})(\xi_0,1_p),$$
$$g_2^\dagger=g_2(0_1\oplus 1_{s+p})(\xi_0,1_p).$$

But $(0_1\oplus 1_{s+p})(\xi_0,1_p)=(g_2^\dagger,1_p)$ and the result follows.

COROLLARY 5.8.2. *Let* $[1]\xrightarrow[1+s]{E}[p]$ *be a normal description in an iterative theory I with* $\beta_E:[1]\to[1+s+p]$ *inclusion and* $\tau_D=f:[1+s]\to[1+s+p]$ *ideal. Suppose that f factors outside of* 1 *so that* (5.8.1) *holds. Then*

$$|E|=g_1(g_2^\dagger,1_p).$$

OBSERVATION 5.8.3. Let $[s+p]\xrightarrow[s]{F}[p]$ be the normal description defined by

$$\tau_F=g_2:[s]\to[s+p],\qquad \beta_F=1_{s+p}:[s+p]\to[s+p].$$

Then $|F|=(g_2^\dagger,1_p)$ so that $|E|=g_1\cdot|F|$. If, for each $i\in[s+p]$, $[1]\xrightarrow[s]{iF}[p]$ is the normal description defined by $\tau_{iF}=\tau_F$, $\beta_{iF}=$

$i:[1] \to [s+p]$, then $|E| = g_1(|1F|, |2F|, \dots, |(s+p)F|)$ and the weight of iF is s while the weight of E is $1+s$.

The normal description iF may also have been described as the normal description composition

$$[1] \xrightarrow[0]{i} [s+p] \xrightarrow[s]{F} [p].$$

5.9. *Scalar anti-iteration*

Let $[1] \xrightarrow[1+s]{D} [p]$ be a normal description in an algebraic theory T and let $\beta_D = 1_1 \oplus 0_{s+p}$. We define the normal description $[1] \xrightarrow[1+s]{D^\dagger} [1+p]$, where $\beta_{D^\dagger} = 1_1 \oplus 0_{s+1+p}$ and where $[1+s] \xrightarrow{\tau_{D^\dagger}} [1+s+1+p]$ is given by

$$(5.9.1) \quad \tau_{D^\dagger} : [1+s] \xrightarrow{\tau_D} [1+s+p] \xrightarrow{\pi \oplus 1_p} [s+1+p] \xrightarrow{0_1 \oplus 1_{s+1+p}} [1+s+1+p]$$

and $\pi = (0_s \oplus 1_1, 1_s \oplus 0_1)$. We call $D^\dagger$ the *scalar anti-iterate* of D. For later reference we make the

OBSERVATION 5.9.1. The morphism $\tau_{D^\dagger}$ factors outside of 1.

The main point of this section is, however,

PROPOSITION 5.9.2. *In an algebraic theory T, $D^{\dagger\dagger} = D$.*

PROOF. It suffices to show $\tau_{D^{\dagger\dagger}} = \tau_D$. Thus we postmultiply $\tau_{D^\dagger}$ by $(1_{1+s}, 1_1 \oplus 0_s) \oplus 1_p$; cf. the definition (preceding Corollary 5.6.7) of † applied to normal descriptions. We need only show that the composition

$$(\pi \oplus 1_p) \cdot (0_1 \oplus 1_{s+1} \oplus 1_p) \cdot ((1_{1+s}, 1_1 \oplus 0_s) \oplus 1_p) = 1_{1+s+p}.$$

Thus it is sufficient to show

$$\pi \cdot (0_1 \oplus 1_{s+1}) \cdot (1_{1+s}, 1_1 \oplus 0_s) = 1_{1+s}.$$

Calculating:

$$\begin{aligned}
&\pi \cdot (0_1 \oplus 1_s \oplus 1_1) \cdot (1_{1+s}, 1_1 \oplus 0_s) = \\
&= \pi \cdot (0_1 \oplus 1_s, 1_1 \oplus 0_s) = (0_s \oplus 1_1, 1_s \oplus 0_1) \cdot (0_1 \oplus 1_s, 1_1 \oplus 0_s) \\
&= ((0_s \oplus 1_1)(0_1 \oplus 1_s, 1_1 \oplus 0_s), (1_s \oplus 0_1)(0_1 \oplus 1_s, 1_1 \oplus 0_s)) \\
&= ((0_{1+s}, 1_1 \oplus 0_s), (0_1 \oplus 1_s, 0_{1+s})) \\
&= (1_1 \oplus 0_s, 0_1 \oplus 1_s) = 1_{1+s}.
\end{aligned}$$

5.10. $|\Sigma| \subset \Sigma^{\dagger_1}$

Let $[n] \xrightarrow[1+s]{D} [p]$ be a Σ-normal description in an iterative theory I. *Assume the morphism described by any Σ-normal description of weight s is in $\Sigma^{\dagger_1}$.* Under this assumption, we show $|D|$ is in $\Sigma^{\dagger_1}$. Note that $|iD| = i \cdot |D|$, $i \in [n]$, so that $|D|$ is in $\Sigma^{\dagger_1}$ if $|iD|$ is in $\Sigma^{\dagger_1}$ for each i. If $|iD|$ is not in $\mathcal{N}$, then $i\beta_D \in [1+s]$ and by Corollary 5.7.3 $|iD| = |E|$, where E is a scalar normal description with $\beta_E = 1$.

Thus, *we may assume* $n = 1$ and $\beta_D = 1_1 \oplus 0_s$. Let $E = D^{\downarrow}$. By Observation 5.9.1, E factors outside of 1 so that by Observation 5.8.3,

$$|D\ | = |E| = g_1(|1F|, |2F|, \ldots, |(s+p)F|), |E^\dagger| = |E|^\dagger$$

by Corollary 5.6.7 and $E^\dagger = D$ by Proposition 5.9.2, so that

$$|D| = |E|^\dagger = [g_1(|1F|, |2F|, \ldots, (s+p)F|)]^\dagger. \tag{5.10.1}$$

By inductive assumption, $|iF|$ is in $\Sigma^{\dagger_1}$ for each i. Since D is of sort Σ, g_1 is in Σ and it follows from (5.10.1) that $|D|$ is in $\Sigma^{\dagger_1}$.

If D has weight 0, then $|D|$ is in $\mathcal{N}$ and hence in $\Sigma^{\dagger_1}$.

5.11. *Corollaries to main theorem*

The following corollaries are immediate.

COROLLARY 5.11.1. *Let Γ be a genus over an iterative theory. Then*

$$|\Gamma\mathcal{N}| = \Gamma^\dagger = \Gamma^{\dagger_1}.$$

COROLLARY 5.11.2. *Let Σ be a sort over an iterative theory. Then $|\Sigma|$ is also a sort over that iterative theory. Moreover, $\|\Sigma\| = |\Sigma|$.*

This latter corollary states that the morphisms described by normal descriptions of sort Σ are the same as the morphisms described by normal descriptions of sort $|\Sigma|$. In the case where the iterative theory is $[\![X^+, X^\infty]\!]$ and X is the set of external states of a Turing notion $\boldsymbol{M}$, $\Sigma = \Gamma\mathcal{N}$ and Γ corresponds to the atomic acts of the Turing notion, then a normal description $\cdot$ D corresponds to an $\boldsymbol{M}$-machine M, $|D|$ is the (sequacious function) described by M (*the track of* M) and Corollary 5.11.2 states that the totality of Turing tracks is not enlarged by allowing as an 'atomic act' another Turing track.

It is necessary to explain in greater detail the connection between our informal discussion of machines and iterative theories. At this point the reader may wish to review Section 5.5 and Section 1. Let X be the set of external states of a machine notion $\boldsymbol{M}$ and let $\Gamma = \{\Gamma_q\}_{q\in\mathbf{N}}$, where Γ_q is a set of functions $X \xrightarrow{g} X \cdot [q]$, correspond to the atomic acts associated with the notion $\boldsymbol{M}$. To consider sequacious functions described by $\boldsymbol{M}$-machines M, we correlate with g the unique sequacious function $X^+ \cup X^\infty \xrightarrow{g'} X^+ \cdot [q] \cup X^\infty$ which satisfies

$$x \overset{g}{\longmapsto} x'k \Rightarrow x \overset{g'}{\longmapsto} xx'k.$$

Then g' is positive and $g = g'\boldsymbol{C}_1\boldsymbol{C}_2\boldsymbol{C}_3$ (cf. Section 4.11). Let $\Gamma'_p = \{g' \mid g \in \Gamma_p\}$. Then corresponding to the machine notion $\boldsymbol{M}$ is the genus Γ' over $[\![X^+, X^\infty]\!]$ and corresponding to an $\boldsymbol{M}$-machine $M = (\beta_M; \tau_M)$ is a normal description $D = (\beta_M; \tau'_M)$, where $\tau_D = \tau'_M = ((1\tau_M)', (2\tau_M)', \ldots)$, of sort $\Gamma'\mathcal{N}$. The sequacious function described by M is the morphism $|D|$ in $[\![X^+, X^\infty]\!]$.

For example, if M is started with instruction $i \in [s]$, acting on $x \in X$ and M produces (cf. Section 1.7) the finite sequence

$$(xi)\tau_M^1 \quad (xi)\tau_M^2 \ldots (xi)\tau_M^{1+c} = x_1 i_1 \quad x_2 i_2 \ldots x_{1+c} i_{1+c}$$

and $i_{1+c} = s + j$, then

$$xi \xrightarrow{\tau_D^\dagger} x_1 x_2 \ldots x_{1+c^j}$$

because

$$\begin{aligned} xi\tau_D^1 &= xx_1 i_1, \\ xi\tau_D^2 &= xx_1 i_1 \tau_D = xx_1 x_2 i_2, \\ &\vdots \\ xi\tau_D^{1+c} &= xx_1 x_2 \ldots x_{1+c}(s+j). \end{aligned}$$

5.12. *Kleene regular sets and recursive functions*

We briefly indicate how some familiar ideas arise in this setting.

First, let I be the iterative theory $[\![M^\wedge]\!]$, where $M = A^*$ is the monoid freely generated by a set A. Let Γ_p, $p \geq 0$, be the set of all $1 \times p$ matrices $(a_1\ a_2 \ldots a_p)$, where $a_j \in A$; here we have identified an element $a_j \in A$ with the singleton $\{a_j\}$ subset of A^*. Then $\Sigma = \Gamma\mathcal{N}$ is the family $\{\Sigma_p\}_{p\in\mathbf{N}}$ consisting of rows whose entries are finite subsets of A. In particular, Γ_0 is the set consisting solely of the unique 1×0 matrix $[1] \xrightarrow{\theta} [0]$ and $\theta 0_p \in \Sigma_p$ is the $1 \times p$ matrix all of whose entries are $\emptyset$. It can be shown that, in this case, the morphisms $[1] \rightarrow [1]$ in $|\Sigma| = \Sigma^{\dagger} = \Sigma^{\dagger_1}$ are exactly the regular subsets of non-null words and the set consisting of the null word alone.

To indicate (even without proof) how these considerations are related to recursive functions is a bit more complicated. Let $X = \mathbf{N}^{1+r}$. We shall introduce a genus $\Gamma = \{\Gamma_m\}_{m \in \mathbf{N}}$ over $[\![X^+, X^\infty]\!]$ by treating X as a subset of X^+ and 'cylindrifying' certain functions on $\mathbf{N}$.

Concerning cylindrification, if $\mathbf{N} \xrightarrow{f} \mathbf{N} \cdot [p]$ is a function let $\mathbf{N}^{1+r} \xrightarrow{f_s} \mathbf{N}^{1+r} \cdot [p]$, $0 \leq s \leq r$, be the function satisfying, where $x, x' \in \mathbf{N}$, $y \in \mathbf{N}^s$, $z \in \mathbf{N}^{r-s}$. $x \overset{f}{\longmapsto} x'j \Rightarrow (y, x, z) \overset{f_s}{\longmapsto} (y, x', z)j$. *The function f_s is obtained from f by an s-fold left cylindrification and an $(r$-$s)$-fold right cylindrification.*

Let Γ_1 be the set of all functions (in fact, there are $3(1+r)$ of them) obtained from the following three functions by cylindrification:

$\mathsf{S} : \mathbf{N} \rightarrow \mathbf{N}$, the successor function,

$\mathsf{Z} : \mathbf{N} \rightarrow \mathbf{N}$, the constant function with value 0,

$\mathsf{P} : \mathbf{N} \rightarrow \mathbf{N}$, the predecessor function, i.e., the function satisfying for

$$x \in \mathbf{N},\ x + 1 \overset{\mathsf{P}}{\longmapsto} x,\ 0 \overset{\mathsf{P}}{\longmapsto} 0.$$

Let Γ_2 be the set of $1 + r$ functions obtained from the following function (test on 0) by cylindrification.

$$\mathsf{T} : \mathbf{N} \rightarrow \mathbf{N} \cdot [2], \quad 0 \overset{\mathsf{T}}{\longmapsto} 01, \quad 1 \overset{\mathsf{T}}{\longmapsto} 12, \quad 2 \overset{\mathsf{T}}{\longmapsto} 22, \quad 3 \overset{\mathsf{T}}{\longmapsto} 32,$$

etc. Finally, let $\Gamma_m = \emptyset$ for $m \neq 1$, $m \neq 2$.

The functions in Γ may be treated as functions in $[\![X^+, X^\infty]\!]$ since sequacious functions are determined by their values on X. Let $J = \Gamma^{\dagger}$ be the iterative subtheory of $[\![X^+, X^\infty]\!]$ generated by Γ. If we identify functions $X \cup \{\square\} \rightarrow X \cup \{\square\}$ with partial functions $X \rightarrow X$ in the usual way, then for sufficiently large r the partial recursive functions $\mathbf{N} \rightarrow \mathbf{N}$ are exactly the partial functions which are describable as compositions

$$\mathbf{N} \xrightarrow{x \mapsto (x, 0)} \mathbf{N}^{1+r} \xrightarrow{fC_1C_2C_3} \mathbf{N}^{1+r} \xrightarrow{\pi} \mathbf{N},$$

where $X^+ \cup X^\infty \xrightarrow{f} X^+ \cup X^\infty$

is in J and 0 is a vector of zeros and π is projection onto first component.

6. Concluding remarks

Let $\mathscr{D}_T$ be the family of normal descriptions over an algebraic theory T. We have defined the composition $[n] \xrightarrow[s+t]{DE} [q]$ of two normal descriptions

$$[n] \xrightarrow[s]{D} [p] \xrightarrow[t]{E} [q].$$

It is straightforward but a bit lengthy to show that with respect to this operation, $\mathscr{D}_T$ is a category; $\mathscr{D}_T$ is 'enriched' by a weight function, source-pairing operation, iteration, etc. There is an obvious isomorphism between $\mathcal{N}$ and the family of normal descriptions of weight 0 but $\mathscr{D}_T$ is not an algebraic theory. Indeed, for each s and p there is a morphism $[0] \xrightarrow[s]{} [p]$ in $\mathscr{D}_T$. (Algebraic theories have, for each p, a unique morphism $[0] \rightarrow [p]$.) Nor does (2.5.7) hold if f_2 has positive weight; nor does (4.1.2). Similar remarks hold for 'machines' in place of 'normal descriptions.'

Let X be a set of external states for a machine notion M of sort Σ and let $\mathcal{M}_{X,\Sigma}$ be the (enriched) category of machines. If we call machines in $\mathcal{M}_{X,\Sigma}$ with the same track, *track equivalent* $\overset{\text{track}}{\sim}$, then $\mathcal{M}_{X,\Sigma} / \overset{\text{track}}{\sim}$ is an iterative theory. In fact, if we regard Σ as a sort in $[\![X^+, X^\infty]\!]$, then this iterative theory is $\Sigma^\dagger$.

Perhaps the most natural equivalence to place on $\mathcal{M}_{X,\Sigma}$ is *computational equivalence* viz.: $M_1 \overset{\text{comp.}}{\sim} M_2$ iff M_1 and M_2 describe the same computational process as informally described in Section 1.7 (and $\beta_{M_1} = \beta_{M_2}$). Unfortunately, however, $\mathcal{M}_{X,\Sigma} / \overset{\text{comp.}}{\sim}$ does not satisfy (4.1.2) because, vaguely speaking, 'the second time around the loop $f^\dagger$, the internal states (instruction numbers) will be different'. But $\mathcal{M}_{X,\Sigma} / \overset{\text{comp.}}{\sim}$ is an algebraic theory. These mathematical considerations reinforce our belief that $\mathcal{M}_{X,\Sigma}$ should be regarded as a category of 'names of computational processes'. A machine M in $\mathcal{M}_{X,\Sigma}$ names $M / \overset{\text{comp.}}{\sim}$ or more concretely the pair β_M and an appropriate function which takes $xi \in X \cdot [s]$ into $(X \cdot [s])^* X \cdot [p]$ or $(X \cdot [s])^\infty$.

We have not dealt systematically with genus-based machines. They are clumsier to describe; however they may admit an interesting notion of equivalence properly between $\overset{\text{comp.}}{\sim}$ and $\overset{\text{track}}{\sim}$ which would not make sense for sort-based machines.

We do not know whether the assumption that τ_D and τ_E are power-ideal implies that τ_{DE} and $\tau_{D^\dagger}$ are power ideal (where D, E are normal descriptions). If the answer is in the affirmative then 'power-ideal' can replace 'ideal' in many of our considerations.

Acknowledgements

It is a pleasure to acknowledge the many helpful suggestions of Jesse B. Wright, Joseph D. Rutledge, Stephen L. Bloom and John C. Shepherdson.

References

[1] Martin Davis, *Computability and Unsolvability* (McGraw-Hill, New York, 1958).
[2] Samuel Eilenberg and Jesse B. Wright, Automata in general algebras, *Information and Control* **11** (4) (1967).
[3] Calvin C. Elgot, Algebraic theories and program schemes, in: E. Engeler, ed., *Symposium on Semantics of Algorithmic Languages* (Springer, Berlin, 1971).
[4] Hans Hermes, *Enumerability, Decidability, Computability* (Springer, Berlin, 1965).
[5] F. William Lawvere, Functorial semantics of algebraic theories, *Proceedings of the National Acedemy of Sciences* **50** (5) (1963) 869–872.
[6] A. A. Markov, *Theory of Algorithms* (Academy of Sciences of the USSR, 1954). [English translation in 1961.]
[7] Emil Post, Finite combinatory processes, formulation I, *Journal of Symbolic Logic* **1** (1936) 103–105.
[8] Hartley Rogers, Jr., *Theory of Recursive Functions and Effective Computability* (McGraw-Hill, New York, 1967).
[9] J. C. Shepherdson and H. E. Sturgis, The computability of partial recursive functions, *Journal of the Association of Computing Machinery* **10** (1963) 217–255.
[10] Alan M. Turing, On computable numbers, with an application to the Entscheidungsproblem, *Proceedings of the London Mathematical Society*, Ser. 2 **42** (1936) 230–265; corrections, Ibid., **43** (1937) 544–546.
[11] Niklaus Wirth, On certain basic concepts of programming languages, Technical Report No. CS65, pp. 27–28, Computer Science Department, School of Humanities and Sciences, Stanford University, Stanford, Calif. (May, 1967).

Reprinted from JOURNAL OF COMPUTER AND SYSTEM SCIENCES

Vol. 16, No. 3, June 1978

On the Algebraic Structure of Rooted Trees

CALVIN C. ELGOT

Mathematical Sciences Department, IBM T. J. Watson Research Center, Yorktown Heights, New York

STEPHEN L. BLOOM AND RALPH TINDELL

Department of Mathematics, Stevens Institute of Technology, Hoboken, New Jersey 07030

Received October 13, 1976; revised November 15, 1977

Many kinds of phenomena are studied with the aid of (rooted) digraphs such as those indicated by Figs. 1.1 and 1.2.

Figure 1.1

Figure 1.2

These two digraphs, while different, usually represent the same phenomenon, say, the same "computational process." Our interest in rooted trees stems from the fact that these two digraphs "unfold" into the SAME infinite tree. In some cases at least it is also true that different (i.e. non-isomorphic) trees represent different phenomena (of the same kind). In these cases the unfoldings (i.e. the trees) are surrogates for the phenomena.

1. INTRODUCTION

1.1 *Outside influence*

Many kinds of phenomena are studied with the aid of (rooted) digraphs such as those indicated by Figs. 1.1 and 1.2.

Figure 1.1

Figure 1.2

These two digraphs, while different, usually represent the same phenomenon, say, the

0022-0000/78/0163-0362$02.00/0

same "computational process." Our interest in rooted trees stems from the fact that these two digraphs "unfold" into the SAME infinite tree viz. that of Fig. 2.1.8. In some cases at least it is also true that different (i.e. non-isomorphic) trees represent different phenomena (of the same kind; for example, see Theorem 4.7 [5]). In these cases the unfoldings (i.e. the trees) are surrogates for the phenomena.

We give a specific example. Flowchart schemes, in the sense of [5], are appropriately labelled, rooted, diagraphs. Two such flowchart schemes are "strongly equivalent" (cf. [5]) iff they unfold into the same tree. Moreover, as indicated in [5], two flowchart schemes are strongly equivalent iff they are "semantically equivalent" i.e. equivalent under all interpretations.

It may be remarked that a "flowchart scheme" is a variant of the notion of a "primitive normal description." This latter notion plays a central role in [1, 5], and is exploited in Sections 3 and 4.

The abstract development relies heavily on the notion algebraic theory" introduced by [14] (cf. also [9]). Generalizations of this notion, (including one used here), are discussed in ([8], cf., in particular, p. 113).

1.2. *On Section 2*

The longest section, Section 2, deals, in concrete fashion with rooted trees (locally finite and ordered). We adopt a Peano-like characterization of this kind of tree as definition (and in Appendix 1 relate this characterization to appropriate graph theoretic concepts). Sufficient properties of appropriately labelled trees, on which we define "composition," "source-tupling," and "iteration," are indicated to assert that they form an "iterative algebraic theory" (without, however, using the concept). In this connection we find it convenient to faithfully represent trees by matrices of sets of words and to introduce the notion "profile" of a tree. The final subsection shows that every tree is a component of a (possibly infinite) vector iteration of a "primitive tree," the trees of "finite index" being obtained by finite vector iterations.

1.3 *On Sections 3 and 4*

While an appreciation of Section 2 requires very little background, an appreciation of Sections 3 and 4 requires rather more. These two sections axiomatically characterize a subcollection $\Gamma\, tr$ of the collection $\Gamma\, Tr$ of trees nearly on the basis of tree composition alone. The subcollection $\Gamma\, tr$ is the iterative subtheory of $\Gamma\, Tr$ generated by Γ. This characterization is our main result: $\Gamma\, tr$ is the collection of trees of finite index in $\Gamma\, Tr$ and is the iterative algebraic theory freely generated by Γ. The argument for Theorem 3.4, which is the heart of the proof of the main result, depends upon a new insight concerning trees viz. (3.4) of Theorem 3.4. The setting of the main result involves the "base category" $\mathcal{N}$, (the skeletal category of finite sets). If one replace $\mathcal{N}$ by $\mathcal{S}$, (the category of (all) sets), one obtains an analogue of the main result: the collection of trees called $\Gamma\, Tr(\mathcal{S})$ is the "completely iterative algebraic theory over $\mathcal{S}$ freely generated by Γ." The trees in $\Gamma\, Tr(\mathcal{S})$ differ from those in $\Gamma\, Tr = \Gamma\, Tr(\mathcal{N})$ in that the outdegree (or 'rank") of a vertex is not restricted to be finite and in that the local order is replaced by local indexing.

If, however, the "rank" of each γ in Γ is a finite set, each tree in $\Gamma\ Tr(\mathscr{S})$ is locally finite. In this case we have: the collection of all trees in $\Gamma\ Tr(\mathscr{S})$ whose singly rooted components have finite index may be described as the "(finitely) iterative algebraic theory over $\mathscr{S}$ freely generated by Γ" or the "scalar iterative algebraic theory over $\mathscr{S}$ freely generated by Γ."

1.4 *Historical background*

In the above we have introduced "iterative theory" as a convenient summery for a collection of facts concerning the trees $\Gamma\ Tr$ and "iterative theory freely generated by Γ" as an axiomatic description of $\Gamma\ tr$. Actually the notion "iterative theory" preceded in time [3] the recognition of $\Gamma\ Tr$ as a particular instance of the notion and the existence of free iterative theories [1] preceded in time the recognition of $\Gamma\ tr$ as the iterative theory freely generated by Γ. From a purely mathematical point of view the usefulness of this result stems from the fact that (a) to establish certain assertions, e.g. identities, for all iterative theories, it is sufficient to establish these assertions for free ones and (b) many true assertions concerning iterative theories are transparently true in the tree theory $\Gamma\ tr$.

The suggestion that a suitable collection of trees might provide a concrete description of "free iterative theories" was first made by Goguen *et al.* [11]. A proof of this fact was first offered by Ginali (cf. [13]). In her readable thesis, [13], Ginali characterizes the trees involved as "regular" and relates the material to studies of Mitchell Wand, Erwin Engeler, the above-mentioned authors, and others.

2

2.1 *Unlabelled Rooted Trees*

According to many texts on graph theory a "tree" is an (undirected) graph which is connected and acyclic. Our concern is with certain kinds of "rooted trees," i.e. trees equipped with distinguished vertices, called roots. Furthermore, the rooted trees discussed here have the property that the set of "immediate successors" of any vertex is finite and linearly ordered; we call these trees "locally finite" and "locally ordered."

Before giving a formal definition, we indicate some examples of rooted, locally finite, locally ordered trees. See Figs. 2.1.0–2.1.11.

The singly rooted, locally finite, locally ordered tree (briefly, "tree") represented by Fig. 2.1.0 has three *vertices*; one vertex, the *root*, has *rank* 2; the *first* and *second* successors of the root have rank zero. The fact that we regard the phrases "first successor" and "second successor" as meaningful, suggests that Figs. 2.1.0 and 2.1.2 represent the same tree, and Figs. 2.1.4 and 2.1.5 represent different (i.e. "non-isomorphic") trees. (The phrase "ordered tree" is sometimes used in this connection, but we prefer to say "locally ordered.") The vertices of rank 0 in a tree are called the *leaves* of the tree. Thus, the tree of Fig. 2.1.3 has four leaves, and the tree of Fig. 2.1.7 has no leaves. Figure 2.1.10 indicates a *doubly rooted* tree.

Figure 2.1.0

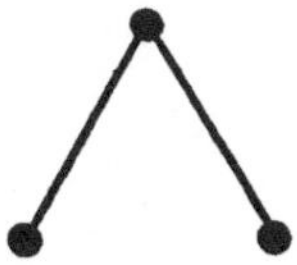

Figure 2.1.1 (B_1)

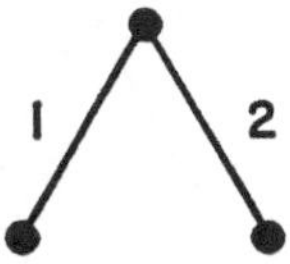

Figure 2.1.2

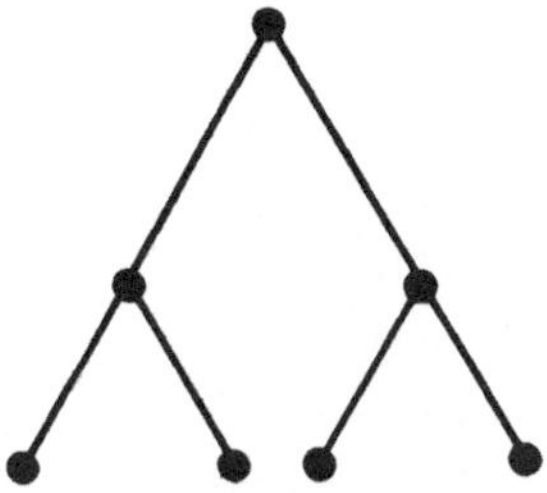

Figure 2.1.3 (B_2)

Figure 2.1.4

Figure 2.1.5

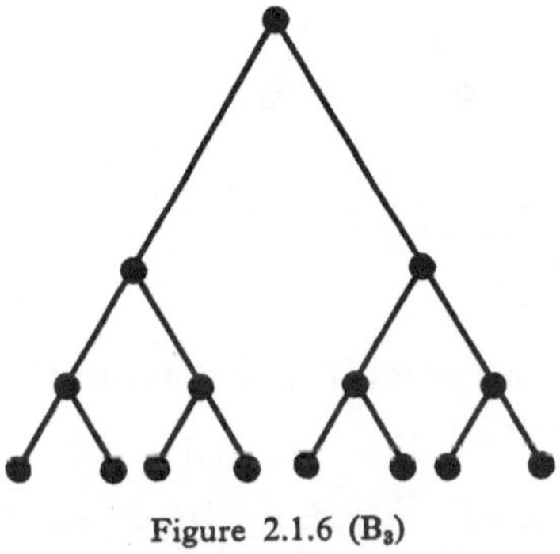

Figure 2.1.6 (B_3)

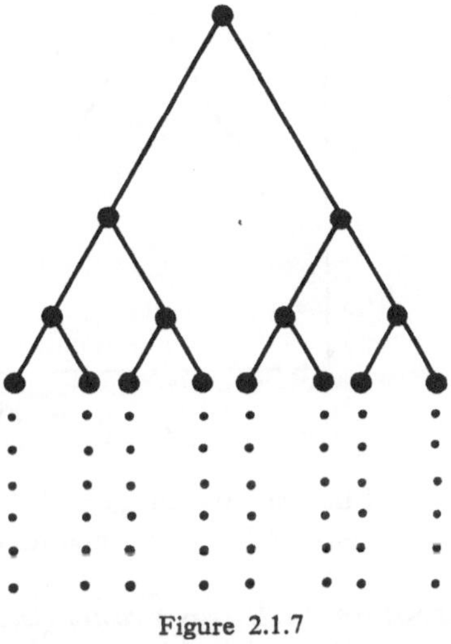

Figure 2.1.7

One might say that the trees represented by Figs. 2.1.0 and 2.1.1 are "isomorphic" or "the same." In our technical discussion we do not identify isomorphism with equality. We now give our formal definition.

Definition 2.1.1. A *ranked set* is a set V together with a function $\rho\colon V \to N$, where N

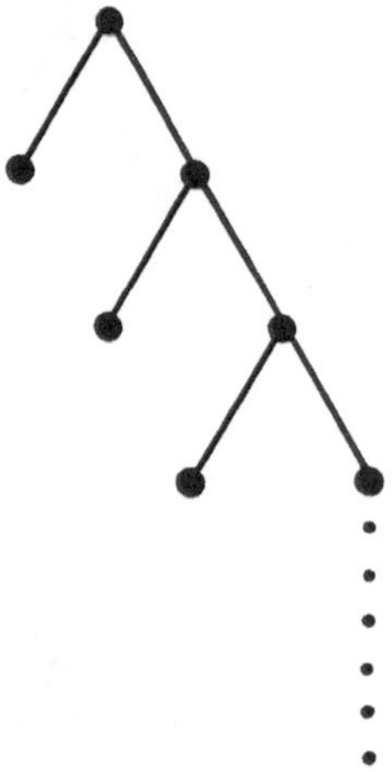

Figure 2.1.8

Figure 2.1.9

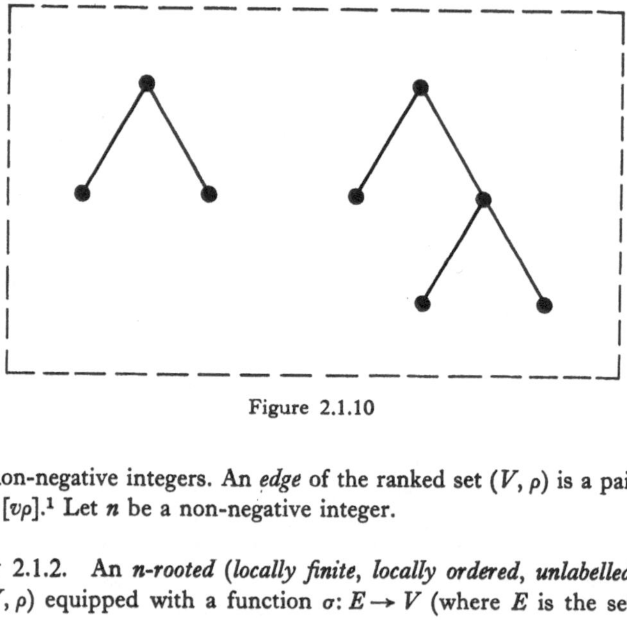

Figure 2.1.10

is the set of non-negative integers. An *edge* of the ranked set (V, ρ) is a pair (v, i) where $v \in V$ and $i \in [v\rho]$.[1] Let n be a non-negative integer.

DEFINITION 2.1.2. An *n-rooted* (*locally finite, locally ordered, unlabelled*) *tree* T is a ranked set (V, ρ) equipped with a function $\sigma: E \to V$ (where E is the set of edges of

[1] For $n \in N$, $[n]$ denotes the set $\{1, 2, ..., n\}$. In particular, $[0]$ is the empty set $\varnothing$, $[1] = \{1\}$, etc. If $f: X \to Y$ is a function and $x \in X$, we write the value of f at x variously as xf of $f(x)$. The composition of $f: X \to Y$ with $g: Y \to Z$ is written $fg: X \to Z$ or $X \xrightarrow{f} Y \xrightarrow{g} Z$ (note the missing arrowhead).

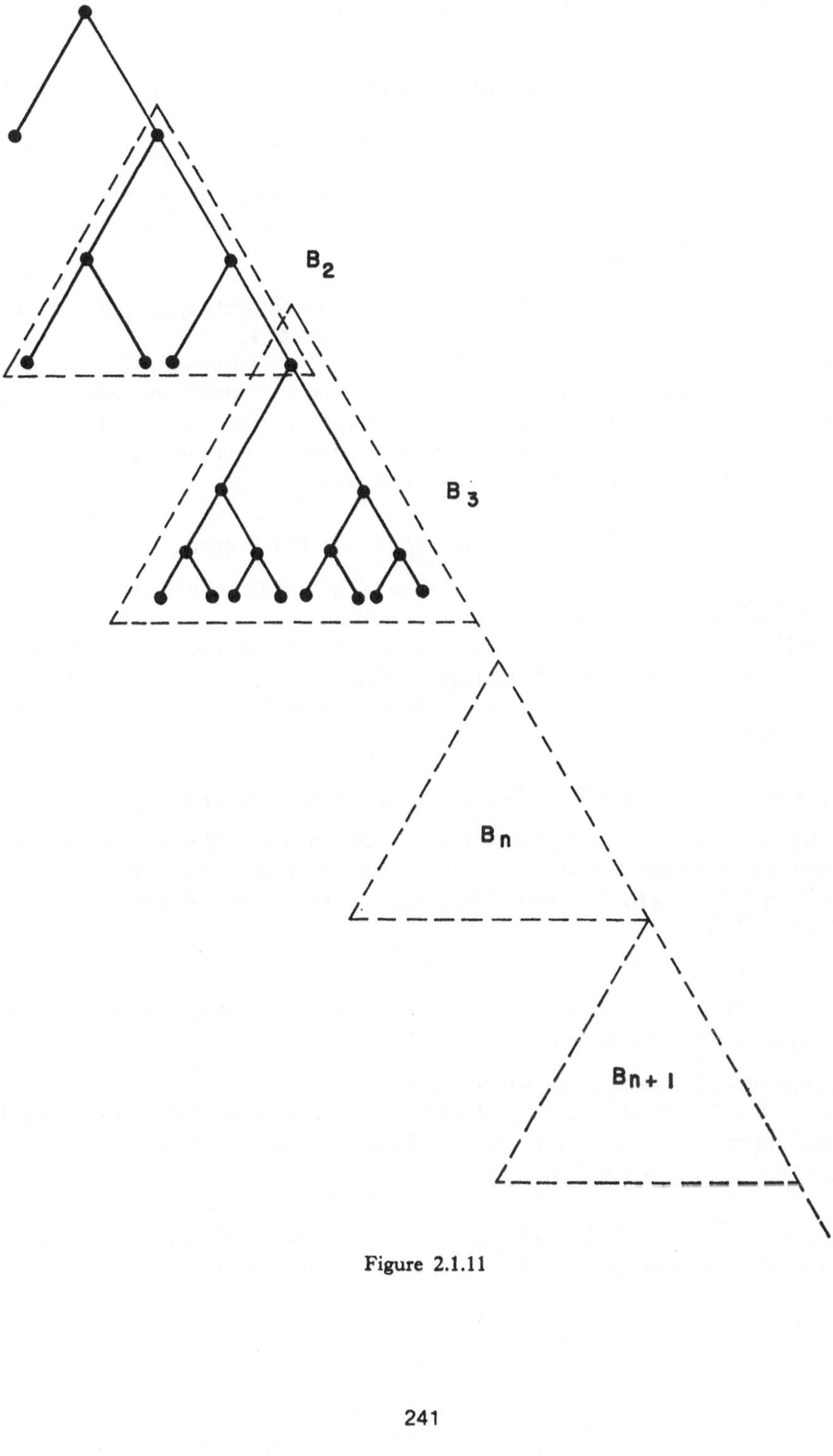

Figure 2.1.11

$(V, \rho))$ and an ordered set of n distinct elements $\epsilon_1, \ldots, \epsilon_n$ of V satisfying the following (Peano-like) conditions:

$$\sigma \text{ is injective; i.e. } \sigma(v, i) = \sigma(v', i') \Rightarrow (v, i) = (v', i'). \tag{2.1.1}$$

$$\text{No element } \epsilon_i,\ i \in [n], \text{ is in the range of } \sigma. \tag{2.1.2}$$

If V' is any subset of V which contains $\epsilon_1, \ldots, \epsilon_n$, and is "closed under σ," then $V' = V$. (V' *is closed under* σ if $\sigma(v, i) \in V'$ whenever $v \in V'$ and $i \in [v\rho]$.) (2.1.3)

The elements of V are called the *vertices* of T. The elements $\epsilon_1, \ldots, \epsilon_n$ are the first, second,..., nth *roots* of T; σ is called the *successor function* of T.

The above definition might be labelled "Proposition" or "Theorem." The terminology reflects our view that this is more appropriately a definition. Certainly the definition is the result of some analysis of the subject. A discussion of the relation between the common notion of tree and the special case of our singly rooted trees is given in Appendix I.

We list some elementary properties of n-rooted trees.

PROPOSITION 2.1.3. *The set V of vertices of a 0-rooted tree is empty.*

Indeed, let $V' = \varnothing$ in the "induction clause" (2.1.3) of Definition 2.1.2.

If $T = ((V, \rho), \sigma, \epsilon_1, \ldots, \epsilon_n)$, abbreviated $T = (V, \rho, \sigma, \epsilon_1, \ldots, \epsilon_n)$, is an n-rooted tree and $v, v' \in V$ we say v' is an *immediate successor* of v if $v' = \sigma(v, i)$, some $i \in [v\rho]$. We say v' is a *descendant* of v if there is a finite sequence $v_1, v_2, \ldots, v_k$, $k \geqslant 1$, of vertices such that $v = v_1$, $v' = v_k$ and v_{i+1} is an immediate successor of v_i, for $1 \leqslant i < k$. In particular, for each v, v is a descendant of v.

PROPOSITION 2.1.4. *Let $T = (V, \rho, \sigma, \epsilon_1, \ldots, \epsilon_n)$ be an n-rooted tree.*

(a) *For any vertex $v \in V$, the collection of all descendants of v, denoted vD_T, is a 1-rooted tree, where the rank function on vD_T is ρ restricted to vD_T, the successor function on vD_T is σ restricted to vD_T, and the root of vD_T is v. (This* tree, *as well as its set of vertices, is denoted vD_T.)*

(b) *If neither $v \in v'D_T$ nor $v' \in vD_T$, then $vD_T \cap v'D_T = \varnothing$.*

(c) *V is the set of all descendants of the roots $\epsilon_1, \ldots, \epsilon_n$. More specifically V is the union $\epsilon_1 D_T \cup \cdots \cup \epsilon_n D_T$ of the disjoint sets $\epsilon_1 D_T, \ldots, \epsilon_n D_T$.*

These facts follow easily from Definition 2.1.2.

We see that D_T, defined in 2.1.4(a) above is a function mapping V into the set of singly rooted "subtrees" of T. For $v \in V$, the tree vD_T is called the "tree of descendants of v," or the "descendency tree of T at v."

DEFINITION 2.1.5. Let $T = (V, \rho, \sigma, \epsilon_1, \ldots, \epsilon_n)$ and $T' = (V', \rho', \sigma', \epsilon_1', \ldots, \epsilon_n')$ be n-rooted trees. An *isomorphism* θ: $T \to T'$ is a bijective function $V \to V'$ such that

(a) $\epsilon_i \theta = \epsilon_i'$, each i, $1 \leqslant i - n$.

(b)   For each $v \in V$, $v\rho = v\theta\rho'$ (i.e. θ preserves rank).

(c)   For each edge (v, i) in T, $\sigma(v, i)\theta = \sigma'(v\theta, i)$.

The conditions (b) and (c) may be expressed by saying the following two diagrams commute,

$$
\begin{array}{ccc}
& E \xrightarrow{\ \sigma\ } V & \\
(v, i) \quad \Big\downarrow & & \Big\downarrow \theta \\
(v\theta, i) \quad E' \xrightarrow[\ \sigma'\ ]{} V'
\end{array}
$$

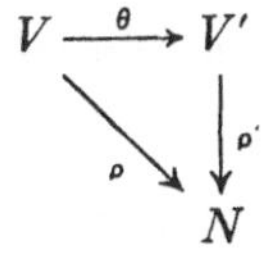

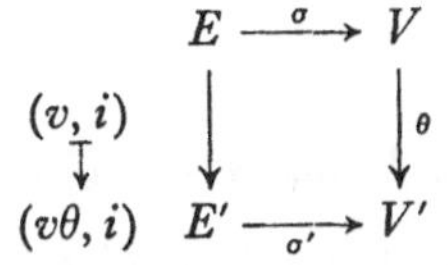

where E and E' are the set of edges of T and T' respectively.

PROPOSITION 2.1.6.   *Let T and T' be n-rooted trees. If θ and θ' are isomorphisms $T \to T'$, then $\theta = \theta'$.*

Proof.   We use the notation of Definition 2.1.5. Let $X \subseteq V$ be the set of vertices v of T such that $v\theta = v\theta'$. Clearly the roots of T belong to X. But if $v \in X$ and $(v, i) \in E$, then $\sigma(v, i)\theta = \sigma'(v\theta, i) = \sigma'(v\theta', i) = \sigma(v, i)\theta'$. Thus X is closed under σ, proving $X = V$.

Thus *if two n-rooted trees are isomorphic, they are "uniquely isomorphic."*

A tree is *m-homomgeneous* if $v\rho = m$, for each vertex v of T. Clearly a singly rooted, 1-homogeneous tree may be identified with the natural numbers N, with $\epsilon_1 = 0$, and $\sigma(n, 1) = n + 1$. The proof of the following proposition is straightforward.

PROPOSITION 2.1.7.   *If T_1 and T_2 are m-homogeneous n-rooted trees, then T_1 is isomorphic to T_2.*

By virtue of 2.1.7, one may speak of "the" m-homogeneous n-rooted tree; for example Fig. 2.1.7 depicts "the" 2-homogeneous singly rooted tree.

Note. m-homogeneous trees are defined for the sake of example only.

At this point, we want to select a "canonical" representative for each isomorphism class of n-rooted trees. First we treat the case $n = 1$. The selection may be done in a number of ways, of course. The following choice[2] appears in the literature. Let $[\omega] = \{1, 2, ...\}$.

---

[2] Knuth [6, p. 558, 15] attributes (essentially) this idea to Francis Galton, "Natural Inheritance" (Macmillan, 1889, p. 249).

Let T be a singly rooted tree. Using (2.1.3), one shows that there is a unique map from V to $[\omega]^*$, $v \in V \Rightarrow \bar{v} \in [\omega]^*$, satisfying

$$\text{the root } \epsilon \text{ goes to the null sequence } \Lambda; \tag{2.1.4}$$

$$\text{if } v \in V,\ i \in [v\rho], \text{ then } \overline{\sigma(v, i)} = \bar{v}i. \text{ Let } \bar{T} \text{ be the image of } T. \tag{2.1.5}$$

A rank and successor function may be defined on $\bar{T}$ in exactly one way to make $T \to \bar{T}$ and isomorphism. Trees of the form $\bar{T}$ are called (singly rooted) *normal* trees. We note the following facts:

PROPOSITION 2.1.8. *If $V \subseteq [\omega]^*$, $\rho: V \to N$ a function, then $T = (V, \rho)$ is a singly rooted normal tree iff*

$$\Lambda \in V \tag{2.1.6}$$

$$\textit{if } v \in V, \textit{ then } vi \in V \textit{ iff } i \in [v\rho]; \tag{2.1.7}$$

$$\textit{if } V' \subseteq V,\ \Lambda \in V', \textit{ and for all } v \in V',\ i \in [v\rho], \textit{ we have } vi \in V', \tag{2.1.8}$$

then $V' = V$.

PROPOSITION 2.1.9. *If (V, ρ) is a normal tree, then $\sigma(v, i) = vi$, and ρ is determined by V. Thus we may identify a single rooted normal tree with its set of vertices.*

PROPOSITION 2.1.10. *If $V_i \subseteq [\omega]^*$, $i = 1, 2$ are isomorphic singly rooted normal trees, then $V_1 = V_2$.*

Now we will define the notion of a normal n rooted tree, $n > 1$. By Proposition 2.1.4(c), the set of vertices of such a tree is the union of the sets of vertices of n disjoint singly rooted trees.

DEFINITION 2.1.11. A subset V of $[n] \times [\omega]^*$ is (the set of vertices of) a *normal n-rooted tree* if for each $i \in [n]$, the set $V_i = \{v \in [\omega]^* \mid (i, v) \in V\}$ is (the set of vertices of) a normal singly rooted tree.

[By identifying $[1] \times [\omega]^*$ with $[\omega]^*$, we may allow $n = 1$ in Definition 2.1.11.]

If $V \subseteq [n] \times [\omega]^*$ is a normal n-rooted tree, its ith root is the vertex (i, Λ); ρ and σ are determined by V. Note too that an edge of a normal n-rooted tree is of the form $((i, v), j)$.

The following is obvious.

PROPOSITION 2.1.12. *Let T be an n-rooted tree. There is a unique n-rooted normal tree $\bar{T}$ isomorphic to T. Thus if T_1 and T_2 are isomorphic normal n-rooted trees, $T_1 = T_2$.*

Remark 2.1.13. The characterization of rooted trees embodied in Definition 2.1.2 permits the formulation of the *principle of (mathematical) tree induction*, the analogue of the principle of mathematical induction. Namely, assume $P(v)$ is a proposition which depends on a vertex v of a rooted tree. The principle asserts: if $P(\epsilon_i)$ for $i \in [n]$ and $P(v) \Rightarrow P(\sigma(v,j))$, $j \in [\rho(v)]$, then $P(v)$ for all v.

2.2. *Trees* $T: n \to p$

A tree $T: n \to p$ (also written $n \to^T p$) consists of an n-rooted tree T' as in Section 2.1 together with a function τ from a subset (called the set of *termini* of T) of the leaves of T' into $[p]$. The function τ is called the *termini* function of T. We say the tree $T = (T', \tau): n \to p$ is normal if T' is normal. By 2.1.11, the normal tree T is fully specified by (V, τ), where V is the set of vertices of T'.

A tree T in the sense of Section 2.1 may be regarded as a tree $T: n \to 1$ by taking $\rho^{-1}(0)$, the set of all leaves of T, as the set of termini, and letting $\tau: \rho^{-1}(0) \to [1]$ be the constant function. The tree T may also be regarded as a tree $T: n \to 0$ by taking the empty set $\varnothing = [0]$ as the set of termini, and taking the unique function $[0] \to [0]$ as τ.

Given trees $T_i: n \to p$, $i = 1, 2$, and a bijection $\theta: V_1 \to V_2$ between their sets of vertices, we say $\theta: T_1 \to T_2$ is an *isomorphism* if θ preserves ρ, σ, the roots, the property of being a terminus, and τ. It should be clear that if T_1, $T_2: n \to p$ are isomorphic, they are uniquely isomorphic, generalizing Proposition 2.1.6.

For each p, and each $j \in [p]$, there is a *root-tree* $\mathbf{j}_p: 1 \to p$ (alternatively $1 \to^j p$) determined up to isomorphism by the following description. The tree $\mathbf{j}_p$ consists of a single vertex ϵ whose rank is 0. The root ϵ is also a terminus, and $\tau(\epsilon) = j$. These root trees play a significant role in our discussion.

We now define an operation of *composition* on trees $n \to p$. Strictly speaking, this "operation" is really an operation "up to isomorphism." Given trees $T: n \to p$, $U: p \to q$, *composition* produces the tree $T \cdot U: n \to q$ defined (up to isomorphism) as follows. Let $U_i = \epsilon_i D_U$, the tree of descendants of the ith root of U, $i \in [p]$. The tree $T \cdot U$ is obtained from T by attaching a copy of U_i to each terminus v of T such that $\tau(v) = i$. For example if $T = \mathbf{j}_p: 1 \to p$, $T \cdot U$ is isomorphic to U_j.

As a second example, let $T: 2 \to 3$ be the tree indicated in Fig. 2.2.1. Let $U: 3 \to 2$ be the tree indicated in Fig. 2.2.2. Then $T \cdot U: 2 \to 2$ is the tree indicated by Fig. 2.2.3. Let $T: n \to p$ be a normal tree. We associate with T an "augmented matrix" $\bar{T} = (A; a)$, where A is an $n \times p$ matrix, and a is a $n \times 1$ matrix. $A_{ij} \subseteq [\omega]^*$, $i \in [n]$, $j \in [p]$, consists of those words $v \in [\omega]^*$ such that (i, v) is a terminus of T and $\tau(i, v) = j$; $a_i \subseteq [\omega]^*$, $i \in [n]$,

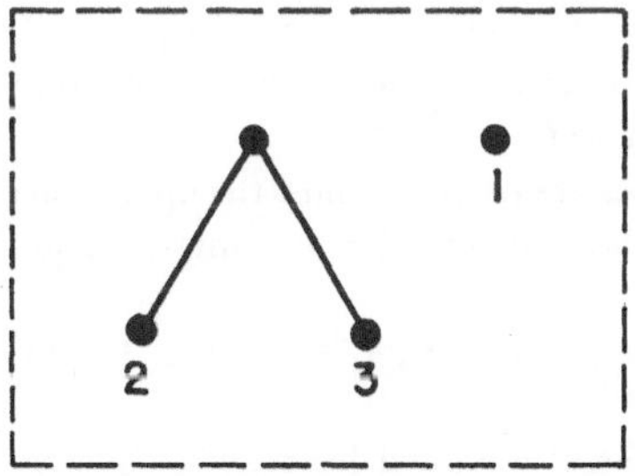

Figure 2.2.1

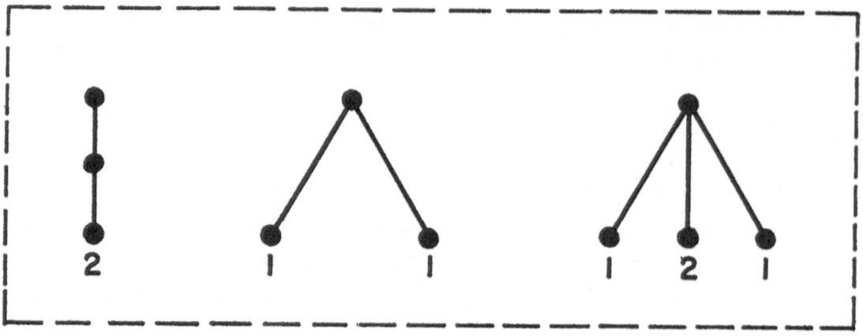

Figure 2.2.2

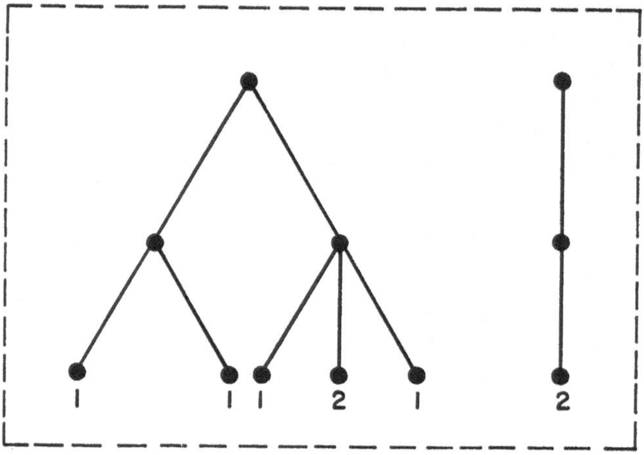

Figure 2.2.3

consists of all $v \in [\omega]^*$ such that (i, v) is a vertex of T which is *not* a terminus. For example, in the case $n = 1$ and $T = \mathbf{j}_p$, $A_{1j} = \{\Lambda\}$, $A_{ik} = \varnothing$, $k \neq j$; $a = a_1 = \varnothing$. [When $n = 1$, we identify $(1, v)$ with v.]

In the case of finite trees $T: n \to p$, a more efficient representation as an augmented $n \times p$ matrix $\bar{T} = (A; a)$ is available (but we will not use this alternative); viz. $a_i \subseteq [\omega]^*$ is taken as the set of words $v \in [\omega]^*$ such that (i, v) is a non-terminus *leaf* of T; A_{ij} is unchanged. Thus this representation takes into account only the "successful paths" i.e. the paths from a root to a leaf.

These augmented matrices are useful to show the operation of composition is associative.

Now let $\bar{U} = (B; b)$ be an augmented $p \times q$ matrix, and define

$$(A; a) \cdot (B; b) = (AB; a + Ab) \tag{2.2.1}$$

where AB is the $n \times q$ matrix obtained by ordinary matrix multiplication (addition of matrix entries being union, and multiplication of matrix entities being (complex) concatenation of sets of words), Ab is the n column vector obtained by multiplying the $n \times p$ matrix A with the $p \times 1$ column vector b; $a + Ab$ is the n column vector whose ith component is $a_i \cup (Ab)_i$.

It is possible to define the augmented matrix $\bar{T} = (A; a)$ for an arbitrary (i.e. not necessarily normal) tree $T: n \to p$. First, define the *label of the path* from the ith root ϵ_i of T to any vertex v in $\epsilon_i D_T$ to be the word $w \in [\omega]^*$, where $\theta(v) = (i, w)$ and where $\theta: T \to T'$ is the isomorphism between T and a normal tree T'. Then define A_{ij} to be the set of all labels of paths from ϵ_i to a terminus v of T such that $\tau(v) = j$; a_i is the set of all labels of paths from ϵ_i to a nonterminus. Clearly, if T itself is normal, this definition of $\bar{T}$ agrees with the previous one.

PROPOSITION 2.2.1. *If $n \to^T p \to^U q$ are trees, then*

$$\overline{T \cdot U} = \bar{T} \cdot \bar{U} \tag{2.2.2}$$

where the multiplication on the right is given by (2.2.1).

Furthermore, if T is not isomorphic to T', $\bar{T} \neq \bar{T}'$.

By an $n \times p$ *surmatrix* we mean an $n \times p$ augmented matrix of the form $\bar{T}$, where $T: n \to p$ is a tree. If T is normal, with the set $V \subseteq [n] \times [\omega]^*$ of vertices, then each set $V_i = \{v \mid (i, v) \in V\}$, $i \in [n]$ is (the set of vertices of) a normal tree $T_i: 1 \to p$. We note that

$$V_i = A_{i1} \cup A_{i2} \cup \cdots \cup A_{in} \cup a_i; \tag{2.2.3}$$

$$\operatorname{dom} \tau_i = A_{i1} \cup A_{i2} \cup \cdots \mathrm{K}\, A_{in}; \tag{2.2.4}$$

$$\tau_i(v) = j \Rightarrow v \in A_{ij}. \tag{2.2.5}$$

Thus $n \times p$ surmatrices serve as simple "surrogates" or representations for normal trees $n \to p$.

The following is immediate from Proposition 2.1.1.

COROLLARY 2.2.2. *The set of surmatrices is closed under multiplication*; *i.e. if $(A; a)$ is an $n \times p$ surmatrix and $(B; b)$ is a $p \times q$ surmatrix then the product $(A; a) \cdot (B; b) = (AB; a + Ab)$ is an $n \times q$ surmatrix.*

Remark. In the notation for surmatrices, a semicolon, rather than a comma, is used to avoid any possible confusion with "source pairing" in algebraic theories.

2.3. *Tr*

Let AUG be the collection of all augmented matrices $(A; a)$ where $A_{ij} \subseteq [\omega]^*$, $a_i \subseteq [\omega]^*$ and let SUR be the subcollection of all surmatrices. If $f_i: 1 \to p$, $i \in [n]$, are augmented "row matrices" then define $(f_1, f_2, \ldots, f_n): n \to p$ to mean the $n \times p$ augmented matrix whose ith augmented row is f_i. We call this operation *source-tupling*. Let $\mathbf{j}_p$ be the surmatrix which is surrogate for the normal root tree $\mathbf{j}_p$ and let $1_p = (\mathbf{1}_p, \mathbf{2}_p, \ldots, \mathbf{p}_p)$ be the $p \times p$ surmatrix $(A; a)$ where A is the identity matrix and a is empty. We have in AUG and in SUR (cf. Corollary 2.2.2) for $n \to^f p \to^g q \to^d r$

$$(f \cdot g) \cdot h = f \cdot (g \cdot h) \tag{2.3.1}$$

$$1_n \cdot f = f = f \cdot 1_p \tag{2.3.2}$$

$$f = (\mathbf{1}_n \cdot f, \mathbf{2}_n \cdot f, ..., \mathbf{n}_n \cdot f) \tag{2.3.3}$$

$$\mathbf{i}_n \cdot (f_1, f_2, ..., f_n) = f_i, \qquad \text{where} \quad f_i: 1 \to p, i \in [n]. \tag{2.3.4}$$

By virtue of satisfying (2.3.1)–(2.3.4), AUG and SUR are "algebraic theories," SUR being a subtheory of AUG (see [3] for the definition of "algebraic theory" as used here). We note the obvious fact

$$\text{if } f = (A; a): 1 \to p \text{ is a surmatrix, } f \neq \mathbf{j}_p, \text{ for } j \in [p], \text{ then } \Lambda \notin \bigcup_j A_{ij}. \tag{2.3.5}$$

From (2.3.5) it follows that

$$\text{in SUR, if } f: 1 \to p \text{ is not } \mathbf{j}_p \text{ for any } j \in [p], \text{ then } f \cdot g \text{ is not } \mathbf{j}_q \text{ for any } g: p \to q, j \in [q]. \tag{2.3.6}$$

By virtue of satisfying (2.3.6) (as well as (2.3.1)–(2.3.4)) SUR is an "ideal theory" [3] The next property of SUR we wish to note concerns solutions to equations in SUR.

Let $f = (A; a): n \to p + n$ be in AUG. We decompose the $n \times (p + n)$ matrix A into "blocks" $A = [BC]$ where B is an $n \times p$ matrix, and C is an $n \times n$ matrix; specifically $B_{ij} = A_{ij}$, $i \in [n], j \in [p]$; $C_{ij} = A_{i(p+j)}$, $i, j \in [n]$.

In AUG, consider the equation in the "unknown" $\xi: n \to p$

$$\xi = f \cdot (1_p, \xi) \tag{2.3.7}$$

where $(1_p, \xi) = (\mathbf{1}, \mathbf{2}_p, ..., \mathbf{p}_p, \xi_1, ..., \xi_n)$ and $\xi_i = \mathbf{i}_n \cdot \xi$. Using

$$f = (A; a) = ([BC]; a) \text{ and}$$
$$\xi = (Y; v),$$

equation (2.3.1) becomes

$$(Y; v) = ([B \quad C]; a)\left(\begin{bmatrix} 1 \\ Y \end{bmatrix}; \begin{bmatrix} 0 \\ v \end{bmatrix}\right) = (B + CY; Cv + a) \tag{2.3.8}$$

where the "1" on the right is the $p \times p$ matrix with $\{\Lambda\}$ along the diagonal and $\varnothing$ elsewhere. The "0" is a $p \times 1$ matrix with $\varnothing$ everywhere. The equation (2.3.8) is equivalent to

$$\begin{aligned} Y &= CY + B, \\ v &= Cv + a. \end{aligned} \tag{2.3.9}$$

By repeated substitution we obtain $Y = CY + B = C(CY + B) + B = C^2Y + CB + B = C^3Y + C^2B + CB + B = \cdots$.

Thus (2.3.9) is equivalent to

$$\begin{aligned} Y &= C^{r+1}Y + \sum_{i=0}^{r} C^iB, \\ v &= C^{r+1}v + \sum_{i=0}^{r} C^ia, \qquad \text{all} \quad r \geqslant 0. \end{aligned} \tag{2.3.10}$$

If we let

$$\begin{aligned} Y &= \sum_{i=0}^{\infty} C^iB, \\ v &= \sum_{i=0}^{\infty} C^ia, \end{aligned} \tag{2.3.11}$$

then a simple calculation shows (2.3.11) satisfies (2.3.7), and hence (2.3.8) and (2.3.9) as well.

Call a matrix *positive* if the union of all entries consists only of words of positive length; i.e. Λ is not in the union. We claim that if C is positive, the solution (2.3.11) to (2.3.7) is *unique*. To establish the uniqueness, it is helpful to introduce the following operations $\mathbf{s}_r$ on $n \times p$ matrices D, where $D_{ij} \subseteq [\omega]^*$.

$$\begin{aligned} &\mathbf{s}_r(D_{ij}) = \{w \in D_{ij} \mid \text{length } w \leqslant r\} \\ &\mathbf{s}_r(D) \text{ is the } n \times p \text{ matrix whose } (i,j)\text{th entry is } \mathbf{s}_r(D_{ij}). \end{aligned} \tag{2.3.12}$$

Now if C is positive, i.e. $\mathbf{s}_0(C) = 0$, then for all $r \geqslant 0$, $\mathbf{s}_r(C^{r+1}) = 0$, and $\mathbf{s}_r(C^{r+1}Y) = 0$, so that from (2.3.10), $\mathbf{s}_r(Y) = \mathbf{s}_r(\sum_{i=0}^{r} C^iB)$, for all $r \geqslant 0$. This uniquely characterizes Y. The argument is identical for ν. In summary, we have

PROPOSITION 2.3.1. *The equation* (2.3.7) *always has a solution in* AUG. *If* $f = ([BC]; a)$ *and* $\xi = (Y; \nu)$, *then* (2.3.11) *is one such solution*; *if* C *is positive, this solution is unique.*

In the case that C is positive and f is in SUR (i.e. f is a surrogate for a normal tree) we wish to establish that the unique solution to (2.3.7) is also in SUR. By Proposition 2.3.1, it is sufficient to show there is some ξ in SUR which satisfies (2.3.7). We will not give this argument, but rely on extrapolation from the particular case where $n = 1$, $p = 1$ and $f: 1 \to 2$ is the surrogate of the tree indicated in Fig. 2.3.1.

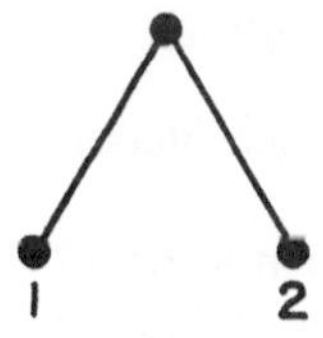

Figure 2.3.1

Figure 2.3.2

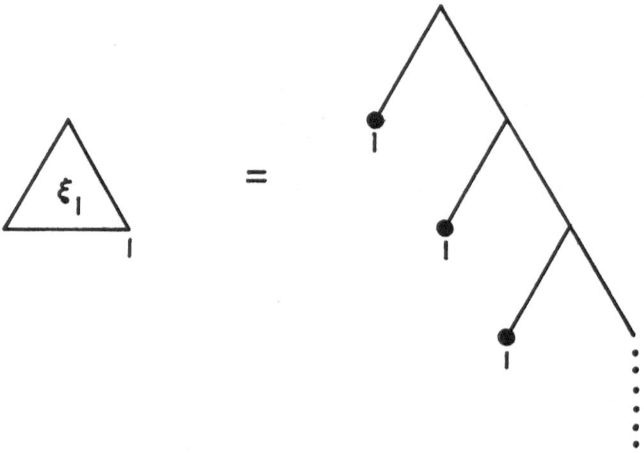

Figure 2.3.3

Then Eq. (2.3.7) is represented by Fig. 2.3.2, and Fig. 2.3.3 is a solution to (2.3.7). Thus, we have

THEOREM 2.3.2. *If* $f = ([BC]; a)$ *is in* SUR *and* C *is positive, then Eq.* (2.3.7) *has a unique solution in* SUR.

Permitting ourselves the extravagance of a different name, *Tr*, for the algebraic theory isomorphic to SUR whose elements ("morphisms") are normal trees, we have

COROLLARY 2.3.3. *If* $f: n \to p + n$ *is in Tr and* $\mathbf{i}_n \cdot f$ *is not a root tree for each* $i \in [n]$, *then Eq.* (2.3.7) *has a unique solution in Tr.*

By virtue of Corollary 2.3.3, *Tr* is an "iterative algebraic theory" in the sense of [3] (cf. Remark 2.3.4).

We may describe the unique solution to Eq. (2.3.7) in more detail using the notion of "profile." If $T: n \to p$ is a normal tree and (i, v) $i \in [n]$, $v \in [\omega]^*$, is a vertex of T, let the *length* of (i, v) be the length of the word v. The *profile of* T *at length* d, $P_d(T)$, is the sequence of non-negative integers $w_1\rho$, $w_2\rho, \ldots, w_n\rho$, where $w_1, \ldots, w_n$, $n \geqslant 0$ is the sequence (from left to right) of vertices of T of length d.

Now if $\xi = f \cdot (1_p, \xi)$, where f satisfies the hypotheses of Corollary 2.3.3, it follows that for any $m \geqslant 0$,

$$\xi = f \cdot (1_p \oplus 0_n, f)^m \cdot (1_p, \xi)$$

where $1_p \oplus 0_n : p \to p + n$ is the p-rooted normal tree such that the ith root is a terminus labelled i, $i \in [p]$. Thus, as an unlabelled tree, ξ may be described by the fact that for every m the profile of ξ at length $d \leqslant m$ equals the profile of $f \cdot (1_p \oplus 0_n, f)^m$ at length d.

Remark 2.3.4. As was noted in [5], either Eq. (2.3.7) or the equation

$$\xi = f \cdot (\xi, 1_p) \tag{2.3.13}$$

where $f : n \to n + p$ is ideal, may be used to characterize iterative algebraic theories. Equations of the form (2.3.7) have unique solutions iff those of the form (2.5.13) do. The unique solution to (2.3.7) is denoted $f^\dagger$ and was called [5] the right iterate of f. The unique solution to (2.3.13) was called the left iterate of f.

In order to use some results of [3] without translation, we will rely on a temporary expedient. Namely we will adopt the (unsatisfactory) convention that if the source of a morphism f is $[n]$ and the target of f is *written* $[n + p]$ then $f^\dagger$ indicates the left iterate while if the target of f is written $[p + n]$ then $f^\dagger$ is the right iterate of f. This convention is clearly unsatisfactory as a permanent measure, e.g. ambiguity results when $n = p$. Despite this we believe that using this convention *here* will not cause any confusion.

For the sequel we require the following.

DEFINITION 2.3.5. If $f : n \to p$ is in Tr and $\mathbf{i}_n \cdot f$ is a root tree for each $i \in [n]$, then f is called *base*.

2.4. Γ *Tr*

By a *genus* Γ we mean a family of pairwise disjoint sets Γ_i, $i \in N$. Clearly there is a "canonical bijection" between genera and ranked sets, the choice between the two being mainly a matter of notational convenience. Thus, a genus Γ gives rise to a ranked set (A, ρ) where $A = \Gamma_0 \cup \Gamma_1 \cup \Gamma_2 \cup \cdots$ and $\rho: A \to N$ has the value i on the elements of Γ_i. Conversely, a ranked set (A, ρ) gives rise to the genus Γ, where $\Gamma_i = \rho^{-1}(i)$. Moreover, the compositions

$$\text{Genera} - \text{Ranked sets} \to \text{Genera}$$

and

$$\text{Ranked sets} - \text{Genera} \to \text{Ranked sets}$$

are respectively the identity Genera $\to$ Genera and the identity Ranked sets $\to$ Ranked sets.

Let Γ be a genus. By a Γ-tree $T: n \to p$ we mean a tree $T': n \to p$ in the sense of Section 2.2 together with a ("labelling") function λ: $V^{-} \to \Gamma_0 \cup \Gamma_1 \cup \Gamma_2 \mathscr{I} \ldots$, where V^{-} is the set of non-termini satisfying

$$\lambda(v) \in \Gamma_{\rho(v)} . \tag{2.4.1}$$

Thus specification of a Γ-tree $n \to p$ involves specifying a ranked set (V, ρ), roots $\epsilon_1, \ldots, \epsilon_n$, a successor function σ, terminus function τ and the function λ. Even if the tree is normal, so that the roots as well as ρ and σ need not be specified, one still has (V, τ, λ). The information, we shall see, can be compectly secured in an appropriate kind of augmented matrix which, as before, will serve as a convenient surrogate for "normal tree."

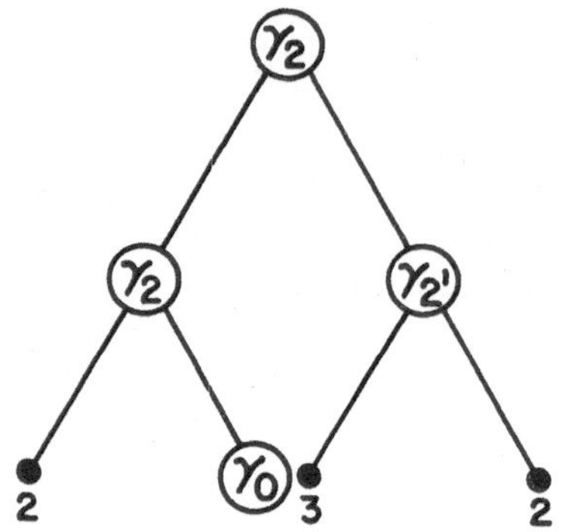

Figure 2.4.1

An example will facilitate the exposition. Consider Figure 2.4.1, where $\gamma_2, \gamma_2' \in \Gamma_2$ and $\gamma_0 \in \Gamma_0$, which is intended to represent a tree $1 \to 3$. The normal tree represented by Figure 2.4.1 is specified by the first four columns of the following table.

V	ρ	τ	λ	New names for vertices	Abbreviations for new names
Λ	2		γ_2	Λ	Λ
1	2		γ_2	$(\gamma_2, 1)$	$\gamma_2 1$
2	2		γ_2'	$(\gamma_2, 2)$	$\gamma_2 2$
11	0	2		$(\gamma_2, 1)(\gamma_2, 1)$	$\gamma_1 1 \gamma_2 1$
12	0		γ_0	$(\gamma_2, 1)(\gamma_2, 2)\gamma_0$	$\gamma_2 1 \gamma_2 2 \gamma_0$
21	0	3		$(\gamma_2, 2)(\gamma_2', 1)$	$\gamma_2 2 \gamma_2' 1$
22	0	2		$(\gamma_2, 2)(\gamma_2', 2)$	$\gamma_2 2 \gamma_2' 2$

If T: $1 \to 3$ is the Γ-tree specified by the above table, i.e. the Γ-tree depicted by Figure 2.4.1, we define $\bar{T}$ to be the 1×3 augmented matrix $(A; a)$ where $A_{11} = \varnothing$, $A_{12} = \{\gamma_2 1 \gamma_2 1, \gamma_2 2 \gamma_2' 2\}$, $A_{13} = \{\gamma_2 2 \gamma_2' 1\}$, $a_1 = \{\gamma_2 1 \gamma_2 2 \gamma_0, \gamma_2 1, \gamma_2 2, \Lambda\}$. In general, given any

Γ-tree $n \to p$, we define the augmented matrix $\bar{T}$ as in Section 2.2 but interpret the phrase "labels of paths" in the manner indicated by the above example (see the discussion preceding Proposition 2.2.1).

In the case of finite Γ-trees $n \to^{T} p$ a more efficient representation is possible (cf. Section 2.2) while preserving the injectiveness of the map $T \mapsto \bar{T}$. The more efficient representation applied to Fig. 2.4.1 yields $a_1 = \{\gamma_2 1 \gamma_2 2 \gamma_0\}$ while A_{1j} is unchanged, $j \in [3]$.

By a Γ-surmatrix we mean an augmented matrix of the form $\bar{T}$ where T is a Γ-tree.

PROPOSITION 2.4.1. *Proposition* 2.2.1 *holds for Γ-trees T, U and Γ-surmatrices are closed under multiplication.*

PROPOSITION 2.4.2. *Proposition* 2.2.1 *holds in the domain of finite Γ-trees even if $\bar{T}$ is interpreted as "the more efficient" representation of T.*

The discussion of Section 2.3 carries over to Γ-trees and their representations leading to

THEOREM 2.4.3. Γ SUR *and* Γ Tr *are (isomorphic) iterative algebraic theories.*

In the special case that Γ_i is a one element set for each $i \in N$ each tree $T: n \to p$ in the sense of Section 2.2 may be made into a Γ-tree in eactly one way, i.e. there is exactly one function $\lambda: V^- \to \bigcup_{i=0}^{\infty} \Gamma_i$ satisfying (2.4.1). Thus the notion "Γ-tree $n \to p$" may be regarded as a generalization of the notion "tree $n \to p$". Thus

PROPOSITION 2.4.4. *In the case that Γ is a family of singletons* Γ SUR $\approx$ SUR *and* Γ *Tr* $\approx$ *Tr*.

In the case that Γ is a family of singletons Fig. 2.4.1 may be "abbreviated" by the tree $1 \to 3$ represented by Fig. 2.4.2, in particular $\gamma_2 = \gamma_2'$.

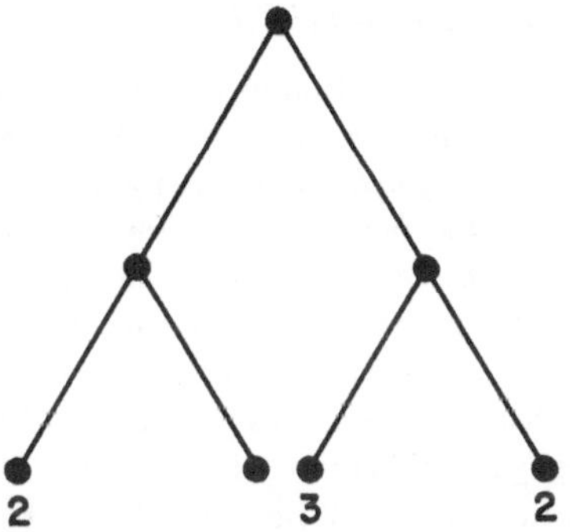

Figure 2.4.2

For use in Section 3, we point out a fundamental property of the theory Γ *Tr*. Call a Γ-tree T: $1 \to n$ *atomic* if T has $n + 1$ vertices: a root ϵ of rank n, and n immediate successors, all of which are termini; the ith successor is labelled i, $i \in [n]$; the value $\lambda(\epsilon) = \gamma$

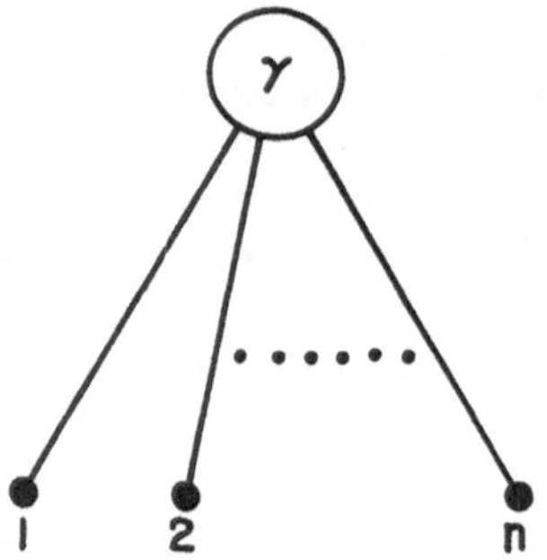

Figure 2.4.3

belongs to Γ_n. Such a tree may be represented by Fig. 2.4.3 (Clearly for each n there is a bijection between the set of normal atomic Γ-trees $1 \to n$ and the set Γ_n.)

The property of $\Gamma\ Tr$ we want to call attention to is the following:

PROPOSITION 2.4.5. *Let* T_1, T_2: $1 \to n$ *be* atomic (*normal*) *Γ-trees, and let* U_1, U_2: $n \to p$ *be arbitrary* (*normal*) *Γ-trees. If* $T_1 \cdot U_1 = T_2 \cdot U_2$, *then* $T_1 = T_2$ *and* $U_1 = U_2$.

We express this fact briefly by saying $\Gamma\ Tr$ has the *unique factorization property*.

We also require the notion "primitive tree." A tree T: $n \to p$ in $\Gamma\ Tr$ is *primitive* if for each $i \in [n]$,

$$\mathbf{i}_n \cdot T = \gamma_i \cdot f_i$$

for some atomic γ_i and base f_i.

2.5. *Trees of Finite Index and Iterates of Primitive Trees*

Call a Γ-tree T: $1 \to p$ *trivial* if T is isomorphic to the root tree $\mathbf{j}_p$, for some $j \in [p]$. Thus a tree T: $1 \to p$ with only one vertex ϵ (so $\epsilon\rho = 0$) is non-trivial iff ϵ is labelled with an element in Γ_0. By the *descendency index* of T (briefly, the *index* of T) we mean the number of distinct normal trees isomorphic to *non-trivial* descendency trees of T; i.e. the index of T is the cardinality of the set of non-trivial normal Γ-trees T': $1 \to p$ such that there is a vertex v with T' isomorphic to vD_T (see Proposition 2.1.4). Since the Γ-trees we are dealing with are locally finite, the index of T is either finite or denumerably infinite.

For example, the indices of the trees in Figs. 2.1.2–2.1.9 are respectively 1, 2, 2, 2, 3, 1, 1, 0, when regarded as trees $1 \to 1$ (where Γ is a family of singletons). The tree in Figure 2.1.11 has infinite index. When regarded as trees $1 \to 0$, the index of the trees in Figures 2.1.2–2.1.6, 2.1.8 and 2.1.9 are increased by one. The index of the tree in Fig. 2.1.7 remains one.

Now suppose T: $1 \to p$ has finite index $n > 0$ and let $T_1, T_2, \ldots, T_n$ be an enumeration without repetition of the non-trivial normal Γ-trees isomorphic to descendency trees of T, and suppose T_1 is isomorphic to T. We shall construct a primitive Γ-tree τ: $n \to p + n$

(i.e. a primitive morphism in $\Gamma\,\mathrm{Tr}$) such that $\mathbf{1}_n \cdot \tau^{\dagger} = T_1$. (Recall $\mathbf{1}_n$: $1 \to n$ is the root tree.)

For $i \in [n]$, let v_i be a vertex of T such that $v_i D_T$ is isomorphic to T_i. Suppose the label of v_i is $\gamma_i \in \Gamma_{v_i\rho}$. We define τ by the requirement that

$$\mathbf{i}_n \cdot \tau = \gamma_i \cdot f_i, \tag{2.5.1}$$

where γ_i is the atomic Γ-tree $1 \to v_i\rho$, and f_i (defined below) is base; i.e.

$$\mathbf{i}_n \cdot \tau\colon [1] \xrightarrow{\gamma_i} [v_i\rho] \xrightarrow{f_i} [p+n]. \tag{2.5.2}$$

Thus, in the case that $v_i\rho = 0$, $f_i = 0_{p+n}$: $[0] \to [p+n]$. Otherwise, let $k \in [v_i\rho]$. The kth successor $\sigma(v_i, k) = v'$ of v_i in T is either a terminus or not. If, in the former case, v' is labelled $j \in [p]$, we define $kf_i = j \in [p+n]$. Otherwise, if $v'D_T$ is isomorphic to T_l, $l \in [n]$, we define $kf_i = p + l \in [p+n]$.

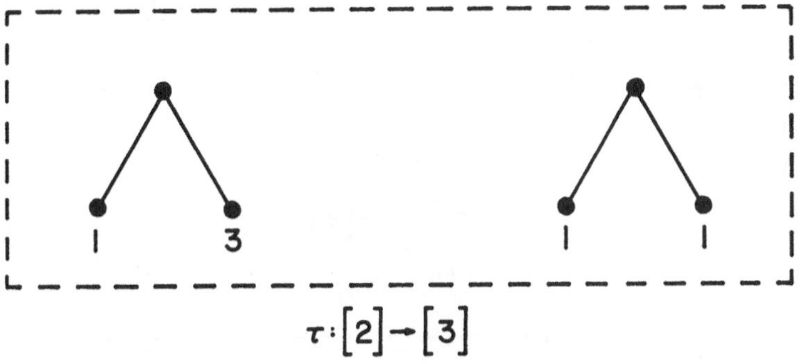

Figure 2.5.1

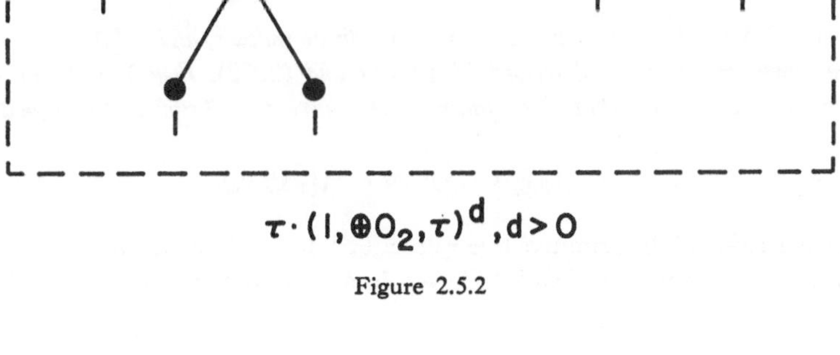

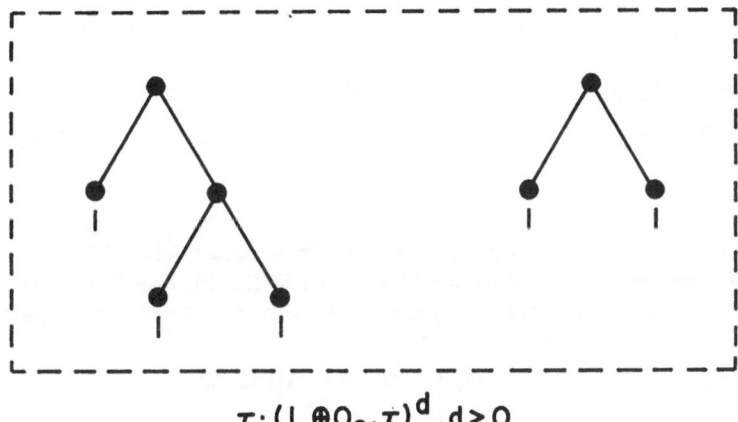

Figure 2.5.2

For example, suppose that the tree T indicated in Fig. 2.1.4 is treated as a tree $1 \to 1$, and that all the vertices of rank 2 are labelled $\Delta \in \Gamma_2$. Then τ: $[2] \to [3]$ is indicated in Fig. 2.5.1, where again, the vertices of rank 2 are labelled Δ. Here $\mathbf{i}_2 \cdot \tau = \Delta \cdot f_i$, $i \in [2]$, where f_i: $[2] \to [3]$ and $1f_1 = 1$, $2f_1 = 3$; $1f_2 = 1$, $2f_2 = 1$.

$\tau \cdot (1_1 \oplus 0_2, \tau)^d$, for $d > 0$ is indicated in Fig. 2.5.2, where, as before, the vertices of rank 2 are labelled Δ.

As another example, if we treat the tree indicated in Fig. 2.1.8 as a tree T: $1 \to 0$, where each vertex of rank 2 is labelled $\Delta \in \Gamma_2$ and each vertex of rank 0 is labelled $\perp \in \Gamma_0$, then the index of T is 2, and τ: $[2] \to [2]$ is given by Fig. 2.5.3. τ^2, τ^3, τ^4 are indicated in Fig. 2.5.4. These examples illustrate the following theorem.

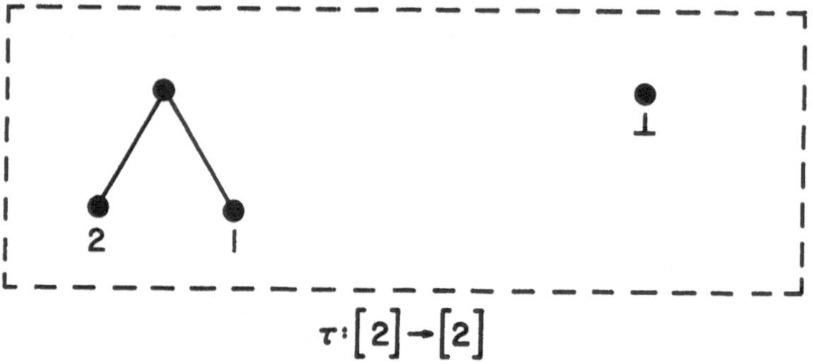

Figure 2.5.3

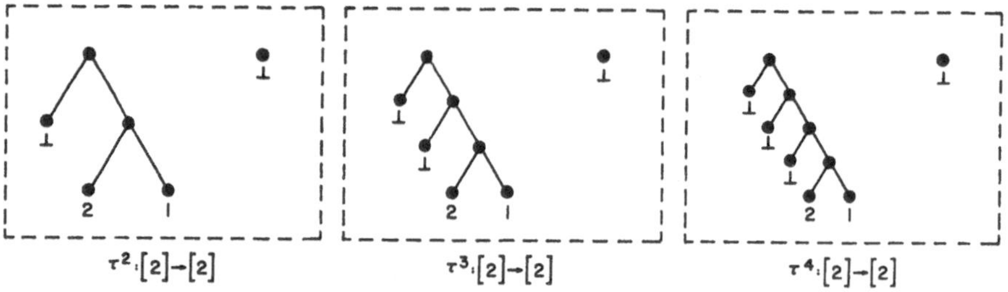

Figure 2.5.4

THEOREM 2.5.1. *If T: $1 \to p$ is a Γ-tree with finite index s, and τ: $[s] \to [p + s]$ is the primitive morphism in ΓTr described by* (2.5.1) *and* (2.5.2), *then $\mathbf{1}_s \cdot \tau^\dagger$: $[1] \to [p]$ is isomorphic to T. In the case that T is a finite tree, we have, for all sufficiently large d,*

$$\tau \cdot (1_p \oplus 0_s, \tau)^d = \tau^\dagger \cdot (1_p \oplus 0_s).$$

The construction of the primitive tree τ described in the discussion preceding Theorem 2.5.1 "works" even when the Γ-tree T: $1 \to p$ does not have finite index. Indeed, suppose the index of T is ω. Define the primitive (infinitely rooted) normal tree $\tau: [\omega] \to [\omega]$ exactly as before. $\tau^\dagger: [\omega] \to [p]$ is to be taken as a "limit" of the trees

$$\tau \cdot (1_p \oplus 0_\omega, \tau)^d,$$

as $d \to \infty$. Then, as before $\mathbf{1}_\omega \cdot \tau^\dagger$ is isomorphic to T. Thus, we have

THEOREM 2.5.2. *Theorem* 2.5.1 *holds even when the descendency index of T is* ω.

We call attention to the fact that the iterate of $\tau: [\omega] \to [\omega]$ is ambiguous until one specifies "the p", $0 \leqslant p < \omega$, which is to be the target of $\tau^\dagger$.

Remark 2.5.3. *It is important to note that "infinite vector iteration" is used to obtain Theorem* 2.5.2 *while only "finite vector iteration" is used in Theorem* 2.5.1.

Remark 2.5.4. The collection of trees of finite index is closed under composition, (finite) source-tupling and iteration. Indeed, if T has index n and U has index p, then $T \cdot U$ and (T, U) have (when defined) index at most $n + p$ and $T^\dagger$ has (when defined) index at most n.

3. INJECTIVITY

In this section, it is assumed that the reader is familiar with [3]. It would be helpful to the reader to have read [1] as well, but this is not essential. In [1] it was shown that for any genus Γ, there is an iterative theory $\Gamma\mathscr{I}$, freely generated by Γ; i.e. for any iterative theory J and any family h of functions mapping Γ_n into ideal morphisms $[1] \to [n]$ in J there is a unique ideal theory morphism $\Gamma\mathscr{I} \to J$ extending h. Furthermore, it was shown that $\Gamma\mathscr{I}$ contained $\Gamma\mathscr{T}$, the algebraic theory freely generated by Γ. The argument given in [1] showed that $\Gamma\mathscr{I}$ may be constructed as certain equivalence classes of "normal descriptions" (see below). A more concrete description of $\Gamma\mathscr{I}$ is obtained in this section (Corollary 3.2). Another description of $\Gamma\mathscr{I}$ is obtained in Section 4 (Corollary 4.1.2).

The objective of this section is to prove the following.

THEOREM 3.1. *The ideal theory morphism from* $\Gamma\mathscr{I}$ *into* $\Gamma\, Tr$ *induced by the map which takes the generator* $\gamma \in \Gamma_n$ *into the tree* $\gamma: 1 \to n$ *for each* γ *in* Γ_n *(and each* $n \in N$*) is injective.*

COROLLARY 3.2. *The iterative subtheory of* $\Gamma\, Tr$ *generated by* Γ *denoted* $\Gamma\, tr$ *is (a description of) the iterative theory freely generated by* Γ.

In [1] it was shown (cf. Theorem 4.1 end last paragraph of Section 6) that $\Gamma\mathscr{I}$ may be described as $\mathrm{ND}(\Gamma\mathscr{I})/\sim$ where, by definition, $D \sim D'$ iff for all ideal theory-morphisms $\phi \mapsto \bar{\phi}$ from $\Gamma\mathscr{T}$ into a arbitrary iterative theory J, we have $|\bar{D}|_J = |\bar{D}'|_J$. [If $D = (\beta; \tau)$ then $|\bar{D}|_J = \beta \cdot (\bar{\tau}^\dagger, 1_p)$.]

A morphism $n \xrightarrow{\phi} p$ in $\Gamma\mathscr{T}$ will be called *primitive* if for all $i \in [n]$, $i \cdot \phi$ has degree 1, i.e., $i \cdot \phi = \gamma_i g_i$, where γ_i is in Γ and g_i is a base morphism.[3] A normal description $D = (\beta; \tau)$ in ND($\Gamma\mathscr{T}$) will be called *primitive* if τ is primitive. By $\Gamma \cdot \mathscr{N}$ we mean the collection of all primitive morphisms and by ND($\Gamma \cdot \mathscr{N}$) we mean the collection of all primitive normal descriptions; $\Gamma \cdot \mathscr{N}$ is a "sort" in the sense of [3] and ND($\Gamma \cdot \mathscr{N}$) is the collection of all normal descriptions of sort $\Gamma \cdot \mathscr{N}$. Since the collection of all normal descriptions of a given sort is closed under composition, source tupling and iteration and contains for each n, p and for each function $[n] \to^{b} [p]$ the base normal description $(b; 0_p)$, we have ND($\Gamma \cdot \mathscr{N}$)/$\sim$ is a sub-iterative theory of ND($\Gamma\mathscr{T}$)/$\sim$ containing "a copy of Γ" and so ND($\Gamma\mathscr{T}$)/$\sim$ = ND($\Gamma \cdot \mathscr{N}$)/$\sim$. Thus *ND($\Gamma \cdot \mathscr{N}$)/$\sim$ is a description of the free iterative theory $\Gamma\mathscr{I}$.*

To prove Theorem 3.1 (and with it Corollary 3.2), we wish to show for primitive normal descriptions $[n] \to^{D_i}_{s_i} [p]$, $i \in [2]$: if $| D_1 |_{\Gamma Tr} = | D_2 |_{\Gamma Tr}$ then $D_1 \sim D_2$, i.e., for any iterative theory J and for any ideal theory-morphism $\phi \mapsto \bar{\phi}$ of $\Gamma\mathscr{T}$ into J, $| \bar{D}_1 |_J = | \bar{D}_2 |_J$. In fact, it is enough to prove this for $n = 1$. It is then sufficient to prove:

$$\text{if} \quad [2] \xrightarrow[s]{D} [p] \quad \text{and} \quad 1 \cdot | D |_{\Gamma Tr} = 2 \cdot | D |_{\Gamma Tr} \quad \text{then} \quad 1 \cdot | \bar{D} |_J = 2 \cdot | \bar{D} |_J$$

for this reduces to the former assertion by taking $D = (D_1, D_2)$: $[2] \to [p]$. Now if $D = (\beta; \tau)$ and $1 \mapsto^{\beta} i$, $2 \mapsto^{\beta} j$, then the latter assertion reduces to:

$$i \cdot \tau^{\dagger} = j \cdot \tau^{\dagger} \text{ in } \Gamma\, Tr \Rightarrow i \cdot \bar{\tau}^{\dagger} = j \cdot \bar{\tau}^{\dagger} \text{ in } J, \text{ which is Proposition 3.4 (3.8).}$$

We pause to make the following observation.

PROPOSITION 3.3. (a) *Let τ: $[s] \to [s + p]$ be an ideal morphism in an iterative theory. Define the base morphism $[s] \to^{\alpha} [s]$ by the requirement $i \cdot \alpha = \inf\{k \in [s] \mid k \cdot \tau^{\dagger} = i \cdot \tau^{\dagger}\}$. Then*

$$\alpha \cdot \tau^{\dagger} = \tau^{\dagger} \tag{3.1}$$

$$i \cdot \tau^{\dagger} = j \cdot \tau^{\dagger} \quad \text{iff} \quad i \cdot \alpha = j \cdot \alpha. \tag{3.2}$$

(b) *The conjunction of* (3.1) *and* (3.2) *is equivalent to* α: $[s] \to^{\nu} [s]/\equiv \to^{c} [s]$, (*i.e.* $\alpha = \nu \cdot c$) *for some c where $i \equiv j \Leftrightarrow i \cdot \tau^{\dagger} = j \cdot \tau^{\dagger}$, $[s]/\equiv$ is the partition induced by the equivalence relation $\equiv$ on $[s]$, ν takes $i \in [s]$ into its equivalence class $i/\equiv$ and c is a choice function, i.e. $c(E) \in E$ where E is an $\equiv$-equivalence class.*

THEOREM 3.4. (a) *In an iterative theory let $[s] \to^{\tau} [s + p]$ be an ideal morphism and $[s] \to^{\alpha} [s]$ the base morphism of Proposition* 3.3(a). *Define ψ by*

$$\psi\colon [s] \xrightarrow{\tau} [s + p] \xrightarrow{\alpha \oplus 1_p} [s + p], \qquad \text{i.e.} \quad \psi = \tau \cdot [\alpha \oplus 1_p].$$

[3] In this section we write $i \cdot \phi$ in place of $\mathbf{i}_n \cdot \phi$ since the source of the morphism ϕ will be clear from context.

Then

$$\psi^\dagger = \tau^\dagger. \tag{3.3}$$

(b) *Suppose the iterative theory is $\Gamma\ Tr$ and the ideal morphism τ is primitive.*

Then ψ is standard, i.e.

$$i \cdot \psi^\dagger = j \cdot \psi^\dagger \Rightarrow i \cdot \psi = j \cdot \psi: \tag{3.4}$$

Furthermore

$$\alpha \cdot \psi = \psi, \tag{3.5}$$

$$\psi \cdot (\psi, 0_s \oplus 1_p) = \tau \cdot (\psi, 0_s \oplus 1_p), \tag{3.6}$$

$$\bar{\psi}^\dagger = \bar{\tau}^\dagger \text{ in } J, \tag{3.7}$$

$$i \cdot \tau^\dagger = j \cdot \tau^\dagger \text{ in } \Gamma\ Tr \Rightarrow i \cdot \bar{\tau}^\dagger = j \cdot \bar{\tau}^\dagger \text{ in } J. \tag{3.8}$$

Proof. (a)
$$\begin{aligned} \psi \cdot (\tau^\dagger, 1_p) &= \tau \cdot [\alpha \oplus 1_p] \cdot (\tau^\dagger, 1_p) && \text{by definition} \\ &= \tau \cdot (\alpha \cdot \tau^\dagger, 1_p) && \text{by [3, (2.5.16)]} \\ &= \tau \cdot (\tau^\dagger, 1_p) && \text{by (3.1)} \\ &= \tau^\dagger. \end{aligned}$$

Thus (3.3) follows by the unique solution property i.e. by [3, (4.1.3)].

(b) Assume $i \cdot \psi^\dagger = j \cdot \psi^\dagger$ and suppose $i \cdot \tau = \gamma_i \cdot f_i$ for each $i \in [s]$ where f_i is base. By the assumption, (3.3) and definition of ψ we have

$$\gamma_i \cdot f_i \cdot [\alpha \oplus 1_p] \cdot (\tau^\dagger, 1_p) = \gamma_j \cdot f_j \cdot [\alpha \oplus 1_p] \cdot (\tau^\dagger, 1_p).$$

From the unique factorization property in $\Gamma\ Tr$, we have $\gamma_i = \gamma_j$ and $f_i \cdot (\alpha \cdot \tau^\dagger, 1_p) = f_i \cdot [\alpha \oplus 1_p] \cdot (\tau^\dagger, 1_p) = f_j \cdot [\alpha \oplus 1_p] \cdot (\tau^\dagger, 1_p) = f_j \cdot (\alpha \cdot \tau^\dagger, 1_p)$. Using (3.1), we obtain

$$f_i \cdot (\tau^\dagger, 1_p) = f_j \cdot (\tau^\dagger, 1_p). \tag{3.9}$$

Let $k \in [s]$ and suppose $k \overset{f_i}{\mapsto} l$, $k \overset{f_j}{\mapsto} l'$. If $l \in [s]$, then so is l' and $l \cdot \tau^\dagger = l' \cdot \tau^\dagger$ from the last equality. Using (3.2), we conclude that $l \cdot \alpha = l' \cdot \alpha$. If $l \in s + [p]$, we obtain $l = l'$ from (3.9). It follows then for $l \in [s+p]$, $l \cdot [\alpha \oplus 1_p] = l' \cdot [\alpha \oplus 1_p]$ so that $f_i \cdot [\alpha \oplus 1_p] = f_j \cdot [\alpha \oplus 1_p]$, $\gamma_i \cdot f_i \cdot [\alpha \oplus 1_p] = \gamma_j \cdot f_j \cdot [\alpha \oplus 1_p]$, $i \cdot \tau \cdot [\alpha \oplus 1_p] = j \cdot \tau \cdot [\alpha \oplus 1_p]$ and $i \cdot \psi = j \cdot \psi$. Thus (3.4) is proved.

Now (3.5) readily follows from (3.1), (3.3) and (3.4).

To obtain (3.6) we calculate:

$$\begin{aligned} \psi \cdot (\psi, 0_s \oplus 1_p) &= \tau \cdot [\alpha \oplus 1_p] \cdot (\psi, 0_s \oplus 1_p) \\ &= \tau \cdot (\alpha \cdot \psi, 0_s \oplus 1_p) \\ &= \tau \cdot (\psi, 0_s \oplus 1_p) \qquad \text{by (3.5).} \end{aligned}$$

It remains to prove (3.7) and (3.8). Note first that if τ and ψ are any ideal morphisms satisfying (3.6) (in *any* iterative theory) then

$$\begin{aligned}\psi^\dagger &= \psi \cdot (\psi, 0_s \oplus 1_p) \cdot (\psi^\dagger, 1_p)\\ &= \tau \cdot (\psi, 0_s \oplus 1_p) \cdot (\psi^\dagger, 1_p) = \tau \cdot (\psi^\dagger, 1_p)\end{aligned}$$

so that $\psi^\dagger = \tau^\dagger$. From (3.6), $\bar{\psi} \cdot (\bar{\psi}, 0_s \oplus 1_p) = \bar{\tau} \cdot (\bar{\psi}, 0_s \oplus 1\theta)$ and so $\bar{\psi}^\dagger = \bar{\tau}^\dagger$ by the argument immediately above, which proves (3.7). For the final assertion (3.8) assume $i \cdot \tau^\dagger = j \cdot \tau^\dagger$. By (3.3) and (3.4), $i \cdot \psi = j \cdot \psi$ so that $i \cdot \bar{\psi} = j \cdot \bar{\psi}$ and $i \cdot \bar{\psi}^\dagger = j \cdot \bar{\psi}^\dagger$ from the defining equation for $\bar{\psi}^\dagger$. Since $\bar{\psi}^\dagger = \bar{\tau}^\dagger$, we have $i \cdot \bar{\tau}^\dagger = j \cdot \bar{\tau}^\dagger$. ∎

4

4.1. *Characterizing* $\Gamma\, tr$

In Section 3, it was proved that the morphism $\Gamma\mathcal{I} \to \Gamma\, Tr$ from the iterative theory $\Gamma\mathcal{I}$, freely generated by the genus Γ into $\Gamma\, Tr$, which takes $\gamma \in \Gamma_p$ to the primitive tree $\gamma: 1 \to p$, is injective. This yielded one "concrete" description of Γ, namely that given in Corollary 3.2. In this section, we note that the elementary observation contained in Theorem 2.5.1 gives an even simpler description of $\Gamma\mathcal{I}$. Theorem 4.1.1 together with its Corollary constitute the main result of this paper.

Recall that $\Gamma\, tr$ is the iterative subtheory of $\Gamma\, Tr$ generated by the atomic Γ-trees.

THEOREM 4.1.1. *The morphisms in* $\Gamma\, tr$ *consist precisely of those normal* Γ*-trees of finite index.*

Proof. Each primitive Γ-tree $1 \to p$ has finite index, and this property is preserved by the operations of composition, source-tupling and iteration (cf. 2.5.4). Thus every morphism in $\Gamma\, tr$ has finite index.

Conversely, suppose $T: 1 \to p$ is a normal Γ-tree with finite index s. If $s = 0$, T is j_p ,for some $j \in [p]$. Otherwise, by Theorem 2.5.1, T is $1_s \cdot \tau^\dagger$, where $\tau: s \to p + s$ is a primitive Γ-tree. Thus, in either case, T is a morphism in $\Gamma\, tr$, completing the proof.

COROLLARY 4.1.2. *The iterative theory,* $\Gamma\mathcal{I}$ *is isomorphic to* $\Gamma\, tr$.

Proof. By Theorem 4.1.1 and Corollary 3.2.

In section 2.5 it was noted that every Γ-tree of finite index is a component of a finite vector iterate of a primitive Γ-tree while every Γ-tree of infinite index is a component of an infinite vector iterate of a primitive Γ-tree. In order to provide an algebraic theory setting for the discussion of infinite vector iteration, we "replace" (in the next section) the base category consisting of the skeletal category of finite sets by the category of all sets (of all cardinalities). We do not employ a skeletal category here in order to avoid getting involved with cardinal or ordinal arithmetic. Indeed in Section 2.5, in the case that the index of the tree T was ω, we were tempted to describe the primitive tree τ by

$\tau: [\omega] \to [p + \omega]$, even though $p + \omega = \omega$. Nevertheless this notation may be useful. The reader may recall from ordinal arithmetic that $\omega + p \neq \omega$. Thus ordinal arithmetic here suggests a spurious distinction between "right" and left" infinite vector iteration.

4.2. *Algebraic Theories with Base $\mathscr{S}$; Completely Iterative Theories*

The algebraic theories used in Sections 2, 3, and 4.1 (see also [1, 3, 9]) might be more precisely described as "algebraic theories with base $\mathscr{N}$," where $\mathscr{N}$ is the category whose morphisms are functions $[n] \to [p]$. Indeed, in [3], the functions $[n] \to [p]$ were called the "base morphisms" in any algebraic theory. In this section we define the notion of an algebraic theory with base $\mathscr{S}$, where $\mathscr{S}$ is the category of sets. The category $\mathscr{N}$ may be described as the full subcategory of $\mathscr{S}$ determined by restricting the objects of $\mathscr{S}$ to be $[n]$, for $n = 0, 1, 2,\ldots$. We will also indicate how the definitions and most of the results of [3] extend to theories with base $\mathscr{S}$. Theorem 3.1 and its corollary have an interesting generalization in this setting.

DEFINITION 4.2.1. An algebraic theory T with base $\mathscr{S}$ (briefly "$\mathscr{S}$-theory") is a category having the class of all sets as its class of objects. Furthermore, for each set A and each $a \in A$, there is a *distinguished* morphism

$$\mathbf{a}: [1] \to A$$

satisfying

for any family $\phi_a: [1] \to B$ of morphisms indexed by $a \in A$, there is a *unique* morphism (4.2.1)

$$\phi: A \to B$$

such that for each $a \in A$

$$\phi_a: [1] \xrightarrow{\mathbf{a}} A \xrightarrow{\phi} B.$$

ϕ is called the source-tupling of the family $(\phi_a: a \in A)$.

The $\mathscr{S}$-theory T is *nondegenerate* if for $a \neq a'$ in $A = [2]$, the distinguished morphisms $\mathbf{a}, \mathbf{a}': [1] \to A$ are distinct. It follows that if T is nondegenerate and a, a' are distinct members of any set A, then the distinguished morphisms $\mathbf{a}, \mathbf{a}': [1] \to A$ are also distinct.

If T is nondegenerate, a function $f: A \to B$ may be identified with the source-tupling of the morphisms $(\phi_a: [1] \to B \mid a \in A)$, where for each $a \in A$, if $af = b$ then ϕ_a is the distinguished morphism $\mathbf{b}: [1] \to B$. In this way, $\mathscr{S}$ is (isomorphic to) a subtheory of any nondegenerate $\mathscr{S}$-theory.

Henceforth, all $\mathscr{S}$-theories are assumed nondegenerate.

The (isomorphic images of) functions $f: A \to B$ in $\mathscr{S}$ are called the *base morphisms* in T. The distinguished $\mathbf{a}: [1] \to A$ is of course base, being identified with the function $1 \mapsto a$.

A morphism $\phi: [1] \to A$ in an $\mathscr{S}$-theory T is *ideal* if for any morphism $\psi: A \to B$, the composition $\phi \cdot \psi: [1] \to B$ is not base. A morphism $\phi: A \to B$ is *ideal* if for each $a \in A$, $\mathbf{a} \cdot \phi$ is ideal. The algebraic theory T itself is ideal if every nondistinguished morphism $[1] \to A$ is ideal, for every set A.

Assume a *choice* for forming the "disjoint union" $C_1 + C_2$ of the sets C_1, C_2 has been made, along with the corresponding base injections $\iota_i: C_i \to C_1 + C_2$, $i = 1, 2$. It may be shown that the following "universal property" holds in any $\mathscr{S}$-theory.

For any morphisms $\phi_i: C_i \to D$, $i = 1, 2$ (with a common target) there is a unique morphism (4.2.2)

$$\theta: C_1 + C_2 \to D$$

such that

$$\phi_i: C_i \xrightarrow{\iota_i} C_1 + C_2 \xrightarrow{\theta} D, \qquad i = 1, 2, \qquad \text{i.e. } \phi_i = \iota_i \cdot \theta.$$

The morphism θ is called the *source pairing* of ϕ_1, ϕ_2 and is denoted (ϕ_1, ϕ_2).

DEFINITION 4.2.2. An ideal $\mathscr{S}$-theory T is *completely iterative* if for each morphism $\phi: A \to B + A$ there is a unique morphism $\phi^\dagger: A \to B$ satisfying

$$\phi^\dagger = \phi \cdot (1_B, \phi^\dagger). \tag{4.2.3}$$

The morphism $\phi^\dagger$ is the *infinite vector iterate* of ϕ if the cardinality of A is infinite.

Note that Eq. (4.2.3) is the analogue of Eq. (2.3.7) (see Remark 2.3.4). The morphism $(1_B, \phi^\dagger)$ is the source pairing of the identity morphism (function) $1_B: B \to B$ and $\phi^\dagger: A \to B$. It may be shown that the property of being "completely iterative" does not depend on the "choice" made above (preceding (4.2.2)).

We now indicate briefly how the main results of [1, 3] extend to ideal and completely iterative $\mathscr{S}$-theories.

Let T be an ideal $\mathscr{S}$-theory. An *$\mathscr{S}$-normal description $D = (\beta; \tau): A \to_S B$ over T of weight S* consists of a morphism

$$\beta: A \to B + S$$

and an ideal morphism

$$\tau: S \to B + S$$

(where A, B, S are sets; i.e., objects of T). If T is completely iterative, the *behavior* of D, denoted $| D |$, is the morphism

$$| D |: A \xrightarrow{\beta} B + S \xrightarrow{(1_B, \tau^\dagger)} B$$

A *sort* Σ in an ideal $\mathscr{S}$-theory T is a collection $\{\Sigma_A: A \text{ an object in } T\}$, where for each set A, Σ_A is a set of ideal morphisms $[1] \to A$, such that if $[1] \to^\sigma A$ is in Σ_A and $A \to B$

is base, then $[1] \xrightarrow{\sigma} A \xrightarrow{f} B$ is in Σ_B. If Σ is a sort, we let Σ^0 be the collection of all morphisms $\sigma: A \to B$ such that, for each distinguished morphism $\mathbf{a}: [1] \to A$,

$$\mathbf{a} \cdot \sigma \in \Sigma_B .$$

THEOREM 4.2.3. (An analogue for completely iterative theories of part of the Main Theorem in [3]). *Let Σ be a sort in the completely iterative theory T. The least completely iterative subtheory of T containing Σ consists precisely of the behaviors of all normal descriptions $D = (\beta; \tau)$ over T such that $\tau \in \Sigma^0$.*

The proof of Theorem 4.2.3 may be obtained by essentially notational changes in the proof given in [3].

The part of the Main Theorem of [3] that does not generalize to completely iterative theories concerns the relation between "scalar" and "vector" iteration because the generalization admits "infinite vector" iteration. (See Remark 4.2.7). For iterative theories (with base $\mathscr{N}$), scalar iteration is as powerful as vector iteration (see [2]). For completely iterative theories, scalar iteration is weaker than vector iteration, as we will explain below.

By making use of the constructions involved in the proof indicated above of Theorem 4.2.3, and with only minor changes in the argument in [1, Sections 5, 6], one can prove

For any genus $\Gamma = (\Gamma_n: n \in N)$, there is a completely iterative theory freely generated by Γ. (4.2.4)

In fact, by generalizing the notion of genus (or, equivalently, ranked set) a stronger theorem may be proved with no additional labor. An *$\mathscr{S}$-ranked set* consists of a set Γ and a function

$$\rho: \Gamma \to \mathscr{S}$$

where $\mathscr{S}$, here, is merely the class of all sets. Thus an $\mathscr{S}$-ranked set is equivalent to an "$\mathscr{S}$-genus": a collection $\{\Gamma_i: i \in \mathscr{S}\}$ of pairwise disjoint set indexed by $\mathscr{S}$.

THEOREM 4.2.4. (Analogue of [1]). *For any $\mathscr{S}$-genus $\{\Gamma_i: i \in \mathscr{S}\}$, there is a completely iterative theory $\Gamma\mathscr{C}(\mathscr{S})$, freely generated by Γ; i.e., for any completely iterative theory J and any function F taking $\gamma \in \Gamma_i$, $i \in \mathscr{S}$ to an ideal morphism $\gamma F: [1] \to i$ in J, there is a unique $\mathscr{S}$-ideal theory morphism $\bar{F}: \Gamma\mathscr{C}(\mathscr{S}) \to J$ extending F.*

Arguing as in [1], one first shows there is an $\mathscr{S}$-theory, $\Gamma\mathscr{T}(\mathscr{S})$, freely generated by Γ. Then the elements of $\Gamma\mathscr{C}(\mathscr{S})$ can be described as certain equivalence classes of those (primitive) normal descriptions $D = (\beta; \tau): A \to_S B$, over $\Gamma\mathscr{T}(\mathscr{S})$, where for each distinguished $\mathbf{s}: [1] \to S$, $s \in S$, $\mathbf{s} \cdot \tau: [1] \to B + S$ factors uniquely as

$$\mathbf{s} \cdot \tau: [1] \xrightarrow{\gamma} i \xrightarrow{f} B + S$$

for some $\gamma \in \Gamma_i$, $i \in \mathscr{S}$, and some base morphism f.

Again using trees, a "concrete" description of $\Gamma\mathscr{T}(\mathscr{S})$ and $\Gamma\mathscr{C}(\mathscr{S})$ may be given.

DEFINITION 4.2.5. For any set A, an *A-rooted $\mathscr{S}$-tree* consists of:

an $\mathscr{S}$-ranked set $\rho: V \to \mathscr{S}$; (4.2.4)

a ("root") function $r: A \to V$ (4.2.5)

a ("successor") function $\sigma: E \to V$, where $E = \{(v, i) \mid v \in V, i \in v\rho\}$ satisfying the following requirements: (4.2.6)

σ and r are injective functions; (4.2.7)

no element ar, $a \in A$, is in the image of σ; (4.2.8)

any subset V' of V, containing each element ar, $a \in A$ and closed under σ, coincides with V. (4.2.9)

Clearly Definition 4.2.5 is a generalization of "n-rooted tree." Isomorphism of A-rooted $\mathscr{S}$-trees is defined analogously to the numeric case so that if two A-rooted $\mathscr{S}$-trees are isomorphic, they are uniquely isomorphic. In the obvious way now, we may define the notion of an $\mathscr{S}$-tree $T: A \to B$, and a *normal* $\mathscr{S}$-tree $T: A \to B$. (The vertices of a normal $\mathscr{S}$-tree are elements of I^*, the set of finite sequences of elements of I, where $I = \bigcup_{v \in V} \rho(v)$). Note that in general, $\mathscr{S}$-trees are neither locally finite nor locally ordered but they are locally indexed and this indexing, to a great extent, serves as a substitute for the order. If $\rho_\Gamma: \Gamma \to \mathscr{S}$ is an $\mathscr{S}$-ranked set, then a (normal) *$\Gamma\mathscr{S}$-tree* $T: A \to B$ consists of a (normal) $\mathscr{S}$-tree $T': A \to B$ together with a labelling function $\lambda: V^- \to \Gamma$ (where V^- is the set of non-termini) such that the following diagram commutes.

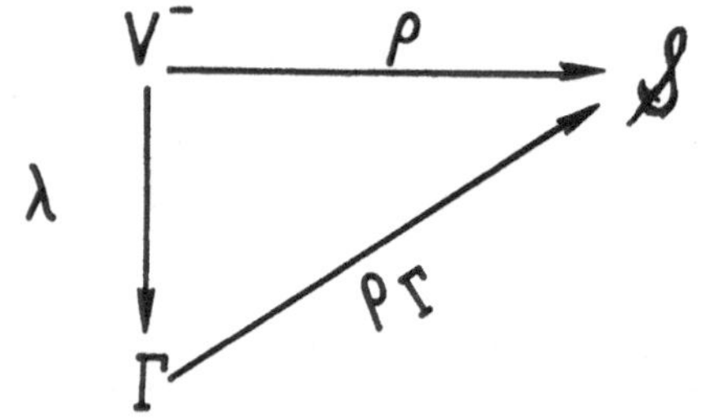

The atomic $\Gamma\mathscr{S}$-tree $\gamma: [1] \to i$ corresponding to $\gamma \in \Gamma_i$, $i \in \mathscr{S}$ is indicated in Fig. 4.2.1.[4] The set of vertices of the normal $\Gamma\mathscr{S}$-tree indicated in the figure consists of the empty sequence Λ, and all words a in i^* of length one such that $a \in i$; the termini function takes $a \in i$ to a.

[4] In accordance with the discussion preceding Theorem 4.2.4, Γ_i is the set of elements of Γ whose rank is i; i.e. $x \in \Gamma_i$ if $x\rho_\Gamma = i$.

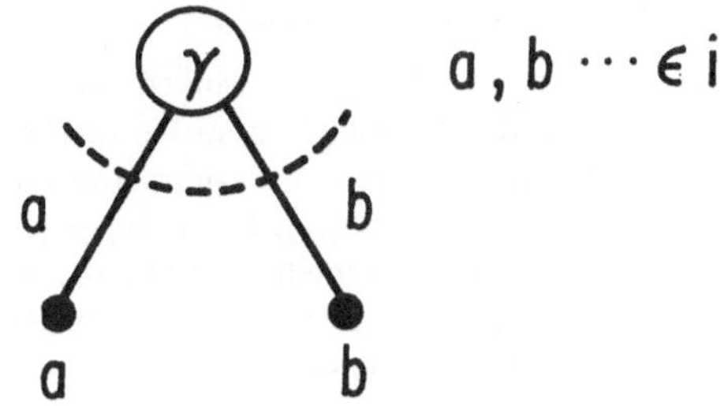

Figure 4.2.1

With the by now familiar definitions of composition and the distinguished ("root") trees $\mathbf{a}$: $[1] \to A$, it may be shown in a straightforward manner that the collection of normal $\Gamma\mathscr{S}$-trees T: $A \to B$ forms a completely iterative theory; we denote this theory $\Gamma\, Tr(\mathscr{S})$.

Clearly, the least subtheory (note: *not* completely iterative subtheory) of $\Gamma\, Tr(\mathscr{S})$ containing the normal primitive $\Gamma\mathscr{S}$-trees γ: $[1] \to i$, for $\gamma \in \Gamma_i$, is (isomorphic to) the $\mathscr{S}$-theory $\Gamma\mathscr{T}(\mathscr{S})$, freely generated by Γ. The morphisms in this copy of $\Gamma\mathscr{T}(\mathscr{S})$ consist of those normal $\Gamma\mathscr{S}$-trees having no infinite paths.

The argument of Section 3 carries over to prove

THEOREM 4.2.6. *The unique ideal $\mathscr{S}$-theory morphism F: $\Gamma\mathscr{C}(\mathscr{S}) \to \Gamma\, Tr(\mathscr{S})$, taking $\gamma \in \Gamma_i$, to the primitive normal $\Gamma\mathscr{S}$-tree γ: $[1] \to i$, for all $i \in \mathscr{S}$, in $\Gamma\, Tr(\mathscr{S})$, is an injection.*

The morphism F takes the equivalence class of the primitive normal description $D = (\beta; \tau)$: $A \to_S B$ to the morphism $\beta \cdot (1_B, \tau^\dagger)$ in $\Gamma\, Tr(\mathscr{S})$.

The idea used to prove Theorem 4.1.1 (and Theorem 2.5.1) can be used to show that F is not only injective, but surjective as well.

Indeed, let ϕ: $[1] \to A$ be any normal $\Gamma\mathscr{S}$-tree T. Let S be the set of all non-trivial normal trees isomorphic to descendancy trees of T. Define the primitive morphism τ: $S \to A + S$ by the requirement that for each $s \in S$

$$\mathbf{s} \cdot \tau\colon [1] \xrightarrow{\gamma_s} i \xrightarrow{f_s} A + S$$

where $\gamma_s \in \Gamma_i$ is the label of the vertex v, where $vD_T \approx s$, and f_s is the base function $i \to A + S$ defined by:

For $b \in i$ if edge (v, b) points to v' in T then

(a) if v' is a non-terminus and $v'D_T \approx s'$ define $bf_s = s'$

(b) if v' is a terminus of T labelled $a \in A$ then $bf_s = a$.

Let β: $[1] \to A + S$ be the base function taking 1 to s_0 where s_0 is the normal tree isomorphic to T.

It may be verified that $\beta \cdot (1_A, \tau^\dagger) = \phi$ in $\Gamma\, Tr(\mathscr{S})$. Thus for $\mathscr{S}$-theories we have

THEOREM 4.2.7. *$\Gamma\mathscr{C}(\mathscr{S})$ is isomorphic to Γ Tr($\mathscr{S}$).*

Remark 4.2.8. Even when $\rho: \Gamma \to \mathscr{S}$ is an $\mathscr{S}$-ranked set, such that for each $\gamma \in \Gamma$, $\gamma\rho = [n]$ for some n, we can use the completely iterative theory Γ *Tr*($\mathscr{S}$) to show that "scalar" and infinite vector iteration are not equivalent. Indeed, suppose $\gamma\rho = [1]$ for all γ in the countably infinite set $\Gamma = \{\gamma_1, \gamma_2, \ldots\}$. Let $\tau: [1] \to \phi$ be the normal $\Gamma\mathscr{S}$-tree indicated by Fig. 4.2.2. τ does not belong to the least subtheory T of Γ *Tr*($\mathscr{S}$) containing the atomic trees γ closed under *scalar iteration* since τ has infinitely many non-isomorphic subtrees. These matters are more fully explained below.

Figure 4.2.2

An ideal $\mathscr{S}$-theory is *finitely iterative* (respectively *scalar iterative*) if for any finite set A (respectively, any singleton set A) and any ideal morphism $\phi: A \to B + A$ there is a unique morphism $\phi^\dagger: A \to B$ such that $\phi^\dagger = \phi \cdot (1_B, \phi^\dagger)$. The morphism $\phi^\dagger$ is called the "finite vector iterate of ϕ" (respectively, the "scalar iterate of ϕ").

Call an ideal morphism ϕ "numerical" if $\phi: [s] \to B + [s]$ for some set B, some number $s \geqslant 0$; it may be easily shown that an ideal $\mathscr{S}$-theory T is finitely iterative iff every numerical ideal morphism has a finite vector iterate; also T is scalar iterative iff every numerical ideal morphism with source [1] has a scalar iterate. Using these facts and the argument of [2], one obtains

PROPOSITION 4.2.9. *Every scalar iterative $\mathscr{S}$-theory is finitely iterative.*

If the rank of each γ in the ranked set Γ is a finite set, Γ is called *finitary.*

THEOREM 4.2.10. *Suppose that Γ is a finitary ranked set. Let T be the least subtheory of Γ Tr($\mathscr{S}$) closed under scalar iteration* (*or equivalently, by* 4.2.9, *closed under finite iteration*). *Then*

(a) *Every function taking* $\gamma\colon [1] \to A$, $\gamma \in \Gamma$ *to* $\bar{\gamma}\colon [1] \to A$ *in a scalar iterative theory* J *extends uniquely to a* $\mathcal{S}$*-theory morphism* $\mathbf{F}\colon T \to J$ *(briefly,* T *is the scalar iterative* $\mathcal{S}$*-theory freely generated by* Γ*).*

(b) *The morphisms* $A \to B$ *in* T*, where* A *is a singleton set, consist precisely of those normal* $\Gamma\mathcal{S}$*-trees in* Γ *Tr with finite descendency index.*

The proof of this theorem makes use of the observation that any Γ tree $T\colon [1] \to A$ of finite index can be written as a composition

$$T\colon [1] \xrightarrow{T'} [n] \xrightarrow{f} A$$

where f is an injective function.

Problem. If the ranked set Γ is not finitary and if T is the scalar iterative subtheory of $\Gamma\, Tr(\mathcal{S})$ generated by Γ, is T freely generated by Γ?

We suspect the answer to the problem is "yes."

APPENDIX I: Comparison of Definitions of Rooted Trees

A popular definition of "tree" (cf. e.g. [10]) has it that a tree is a connected acyclic (undirected) graph. In this appendix, the connection between this popular definition and the definition of "rooted tree" given in Section 2.1 is discussed.

In [10] a graph G consists of a finite set V (of vertices or nodes) and a set E of (edges or lines) i.e., doubletons $\{v, v'\}$, $v, v' \in V$, $v \neq v'$. We immediately delete the requirement that V be finite (since our trees are permitted to be infinite) but otherwise embrace this definition. A *path* from v to v' in $G = (V, E)$ is a word $p = v_0 v_1 \cdots v_n$ in V^+ (the set of all finite sequences of elements of V of positive length) such that $v = v_0$, $v' = v_n$, $\{v_{i-1}, v_i\} \in E$ for all $i \in [n]$ and whenever $i \neq j$, $v_i \neq v_j$. (*Remark*: if p is a path in G from v to v then $n = 0$; thus a path from v to v is unique.) The *edge count* of p is n; its *node count* is $1 + n$. A *cycle* in a graph G with vertex set V is a word of edge count $\geqslant 3$ of the form vwv, where $v \in V$, $w \in V^+$ and both vw and wv are paths in G; a graph G is *acyclic* if there are no cycles in G. The graph G is *connected* if for any $v, v' \in V$, there is a path from v to v'.

We formally record the "popular" definition of "tree."

Definition A. A graph $G = (V, E)$, is a *tree* if G is connected and acyclic.

Proposition B. *For a graph* $G = (V, E)$ *the following conditions are equivalent.*

(B1) *G is connected and acyclic; i.e., G is a tree.*

(B2) *For any vertices $a, b \in V$, there is a unique path from a to b.*

Proof. Since the implication (B2) $\Rightarrow$ (B1) is easily established we prove only that if p_1 and p_2 are paths in G from a to b then $p_1 = p_2$. Suppose the edge counts of p_1

and p_2 are e_1 and e_2. The proof proceeds by induction on $e_1 + e_2$. The case $e_1 + e_2 \leqslant 2$ is easily disposed of using the parenthetical remark above.

Suppose now $e_1 + e_2 \geqslant 3$ so that $a \neq b$. Suppose too, $p_1 = u_1 b$, $p_2 = u_2 b$, where $u_1, u_2 \in V^*$. Then the word $u_1 b u_2^{\smile}$ (where $u_2^{\smile}$ is the word "u_2 in reverse order") which has edge count $e_1 + e_2$ would be a cycle if no vertex, other than a, occurred more than once in that word. But G is acyclic. Hence there is a vertex $v \neq a$ which occurs in $u_1 b u_2^{\smile}$ more than once. Now $v \neq b$ since no vertex occurs more than once in p_i for each $i \in [2]$. Making use of the "distinctness" property of p_i again, we conclude there is an occurrence of v "to the left of b" and one "to the right of b" in $u_1 b u_2^{\smile}$. Thus we have

$$p_1 = aw_1 v w'_1 b$$

$$p_2 = aw_2 v w'_2 b.$$

By inductive assumption, $aw_1 v = aw_2 v$ and $vw'_1 b = vw'_2 b$ so that $p_1 = p_2$. ∎

Let $G = (V, E)$ be a graph and suppose $\epsilon \in V$. We call (G, ϵ) a *rooted* graph; ϵ is the *root* of (G, ϵ). Suppose G is *connected* and *acyclic.* We define the *immediate successor relation* $s \subseteq V \times V$ in G *as follows*: $(v_1, v_2) \in s$ iff the unique path (cf. Proposition B) from ϵ to v_2 is of the form $\cdots v_1 v_2$. Now suppose (G, ϵ) is *locally finite* i.e., for each $v \in V$, the set of $v' \in V$ such that $(v, v') \in s$, is finite. Let $\rho(v)$ be the number of immediate successor of v. (Actually "local finiteness" is independent of ϵ; it may be stated: for each $v \in V$, the number of doubletons $e \in E$ such that $v \in e$, is finite.) Suppose further that for each $v \in V$, the (finite) set of immediate successors of v is *ordered.* We then define $\sigma(v, i) = v'$ if v' is the ith successor of v, $i \in [\rho(v)]$. We take it as generally known that the data $(V, \rho, \sigma, \epsilon)$ satisfies Definition 2.1.2 in the case of singly rooted trees and focus attention on the reverse direction.

Now suppose $T = (V, \rho, \sigma, \epsilon)$ satisfies Definition 2.1.2 in the case $n = 1$. We define the graph $G = (V, E)$ by taking $E = \{\{v, v'\} \mid \sigma(v, i) = v' \text{ for some } i \in [\rho(v)]\}$ and wish to show that G is connected and acyclic.

We first observe that by the principle of tree induction (cf. 2.1.13), there exists at most one function $l: V \to N$ satisfying: $l(\epsilon) = 0$, $l(v) = n \Rightarrow l(\sigma(v, i)) = 1 + n$ for each $i \in [\rho(v)]$. The proof that there exists at least one function satisfying these conditions makes use of (2.1.1), (2.1.2) and (2.1.3). (This function is sometimes called, level or length or depth or $\cdots$; if T is normal $l(v)$ is the length of v.) To show that G is connected, assume $v, v' \in V$ and suppose $l(v) \leqslant l(v')$. We construct the sequence (of *ancestors* of v')

$$(v_0, v_1, v_2, \ldots, v_n)$$

with $v_0 = v'$ and v_i an immediate successor of v_{i+1}, $0 \leqslant i < n$, $v_n = \epsilon$, (so that $n = l(v')$). Let i, $0 \leqslant i \leqslant n$, be the smallest index such that v_i is an ancestor of v (such an i exists since ϵ is a common ancestor of v and v'). Then the sequence

$$(v_0, v_1, v_2, \ldots, v_i = w_i, w_{i+1}, w_{i+2}, \ldots, w_{i+p}), \qquad p \geqslant 0,$$

where w_{i+j} is an immediate successor of w_{i+j-1} for $j \in [p]$ and $w_{i+p} = v$, is a path in G.

It remains to show that G is acyclic. Suppose the contrary so that $(v_1, v_2, ..., v_n)$, $v_1 = v_n$, $n \geqslant 4$, is a cycle. Since σ is injective, either all the (ordered) pairs (v_1, v_2), $(v_2, v_3), ..., (v_n, v_1)$ are in the immediate successor relation s or all the reversed pairs are in s. Suppose the former. Then $l(v_1) < l(v_2) < \cdots < l(v_n) < l(v_1)$. The contradiction compels the conclusion that G is acyclic.

APPENDIX II: Replacing Finite Unsuccessful Paths by Infinite Unsuccessful Paths

In Subsections 2.2, 2.3, 2.4 certain augmented matrices, the "surmatrices," were used to (faithfully) represent rooted trees and rooted Γ-trees. When T: $1 \to 1$ is a tree, T is represented by the surmatrix $(A; a)$ where A is the set of all labels of paths from the root of T to a terminus; a is the set of all labels of paths from the root to a non-terminus. In 2.2 and 2.3, both A and a are subsets of Σ^* where $\Sigma = [\omega]$; in 2.4, A and a are subsets of Σ^* where $\Sigma = \Gamma_0 \cup \bigcup_{n \geqslant 1} (\Gamma_n \times [n])$.

If we call a path in a tree from a root to a leaf *successful* then the elements of A are labels of successful paths. The elements of a may be partitioned into the set a_1 of labels of successful paths (which end with a non-terminus) and the set a_2 of labels of unsuccessful paths (which begin with the root). Thus $a_1 \cup a_2 = a$, $a_1 \cap a_2 = \varnothing$.

The two cases mentioned above may be (essentially) subsumed under a single case by passing to $\mathscr{S}$-ranked sets Γ. The case $\Sigma = [\omega]$ is then replaced by the case $\Gamma = \{\gamma, \gamma_0\}$ with $\rho(\gamma) = [\omega]$, $\rho(\gamma_0) = [0]$. Then i gets replaced by the "letter" (γ, i) so that the "new" $A \cup a_1 \subseteq (\{\gamma\} \times [\omega])^*$ while $u \in a_2$ gets replaced by $u\gamma_0$ so that the "new" $a_2 \subseteq (\{\gamma\} \times [\omega])^* \{\gamma_0\}$.

The main objective of this Appendix is to show that the tree T: $1 \to 1$ may equally well be represented by $(A; a_1 \cup b)$ where b is the set of labels of infinite paths in T which begin with the root.

Let Σ be any set. As usual Σ^* devotes the set of all finite sequences (words) of elements of Σ while Σ^∞ denotes the set of all infinite sequences (functions) f: $[\omega] \to \Sigma$, where $[\omega] = \{1, 2, 3, ...\}$ is the set of positive integers. Thus an element of Σ^∞ may be called an "infinite word on Σ." There is a partial ordering $\leqslant$ on $\Sigma^* \cup \Sigma^\infty$: $u \leqslant v$ iff u is a prefix of v (i.e., u is an initial segment of v). With respect to this ordering all elements of Σ^∞ are maximal and two distinct infinite words are incomparable. We write $u < v$ if u is a *proper prefix* of v, i.e., $u \leqslant v$ and $u \neq v$. If X is a set, we write $X^\wedge$ for the set of all subsets of X.

We define the function

$$pref\colon (\Sigma^* \cup \Sigma^\infty)^\wedge \to \Sigma^{*\wedge}$$

by

$$pref(M) = \{u \in \Sigma^* \mid u \leqslant v, \text{ for some } v \in M\}, \qquad M \subseteq \Sigma^* \cup \Sigma^\infty.$$

Thus, *pref*(M) is the set of all finite prefixes of words in M. If M is a singleton consisting of v alone, we write "*pref*(v)" for *pref*(M). Clearly,

PROPOSITION A. $M_1 \subseteq M_2 \subseteq \Sigma^* \cup \Sigma^\infty \Rightarrow \mathit{pref}\, M_1 \subseteq \mathit{pref}\, M_2 \subseteq \Sigma^*$.

A subset F of Σ^* is *prefix closed* if $F = \mathit{pref}\, F$. Let $\mathit{Pref}(\Sigma^*)$ be the set of all prefix closed subsets of Σ^*.

We define the function

$$\mathit{lim}\colon \Sigma^{*\wedge} \to \Sigma^{\infty\wedge}$$

by the following requirement: for $v \in \Sigma^\infty$, $F \subseteq \Sigma^*$

$$v \in \mathit{lim}\, F \quad \text{iff} \quad \mathit{pref}\, v \subseteq F.$$

The following two propositions are obvious.

PROPOSITION B. $F_1 \subseteq F_2 \subseteq \Sigma^* \Rightarrow \mathit{lim}\, F_1 \subseteq \mathit{lim}\, F_2 \subseteq \Sigma^\infty$.

PROPOSITION C. *For* $I \subseteq \Sigma^\infty$, $I \subseteq \mathit{lim}\,(\mathit{pref}\, I)$.

Before stating the theorem which leads to our main objective, we require two more functions. The function

$$\mathit{max}\colon (\Sigma^* \cup \Sigma^\infty)^\wedge \to \Sigma^{*\wedge}$$

is defined as follows: for $M \subseteq \Sigma^* \cup \Sigma^\infty$

$$\mathit{max}(M) = \{u \in M \cap \Sigma^* \mid u < v \text{ for no } v \in M\}$$

i.e., $\mathit{max}(M)$ is the set of finite words in M which are maximal in M. Clearly,

PROPOSITION D. $F \in \mathit{Pref}\, \Sigma^* \Rightarrow (F - \max(F)) \in \mathit{Pref}\, \Sigma^*$; $\lim F = \mathit{lim}(F - \max(F))$ *for* $F \in \mathrm{Pref}\, \Sigma^*$.

The function

$$\mu\colon \mathit{Pref}\, \Sigma^* \to (\Sigma^* \cup \Sigma^\infty)^\wedge$$

is defined by the following, where $F \subseteq \Sigma^*$, $F = \mathit{pref}\, F$:

$$\mu(F) = \mathit{max}(F) \cup \mathit{lim}(F).$$

THEOREM E. *The function* μ *is injective i.e. for* $F_1, F_2 \in \mathit{Pref}\, \Sigma^*$, $\mu(F_1) = \mu(F_2) \Rightarrow F_1 = F_2$.

With A, a, a_1, a_2, b as in the beginning of this Appendix, we have $A \cup a$ is prefix closed, $\max(A \cup a) = A \cup a_1$, $a_2 = (A \cup a) - (A \cup a_1)$ is prefix closed (by Proposition D) and $b = \mathit{lim}\, a_2$.

COROLLARY F. *With A, a_1, a_2, b as above, the function which takes $(A; a_1 \cup a_2)$ into $(A; a_1 \cup b)$ is injective.*

Proof of Theorem E. Suppose $F_1, F_2 \in Pref\,\Sigma^*$ and $\mu(F_1) = \mu(F_2)$. Then $max\,F_1 = max\,F_2$ and $lim\,F_1 = lim\,F_2$. Suppose $u \in F_1$. Either $u \leqslant v$ for some $v \in \max F_1$ or else there is an infinite chain: $u < u_1 < u_2 < \cdots$, $u_i \in F_1$, so that the unique $v \in F_1^\infty$ satisfying $u_i < v$ for all i, is in $lim\,F_1$. In the former case, $v \in max\,F_2 \subseteq F_2$ and so, by prefix closure, $u \in F_2$. In the latter case, $v \in lim\,F_2$ and $u \in pref(v) \subseteq F_2$. Thus, in either case, $u \in F_1 \Rightarrow u \in F_2$, i.e., $F_1 \subseteq F_2$. By symmetry we obtain $F_2 \subseteq F_1$ which concludes the proof.

We now ask: Which subsets of $\Sigma^* \cup \Sigma^\infty$ are in the image of μ?

THEOREM G. (a) *Suppose $F \cup f \in Pref\,\Sigma^*$ and $F = \max(F \cup f)$. Let $g = lim(F \cup f)$ so that $\mu(F \cup f) = F \cup g$. Then $F \subseteq max(F \cup g)$ and $lim\,pref(F \cup g) \subseteq g$.*

(b) *Suppose $F \subseteq \Sigma^*$, $g \subseteq \Sigma^\infty$, $F \subseteq max(F \cup g)$, $lim\,pref(F \cup g) \subseteq g$. Then $\mu(pref(F \cup g)) = F \cup g$.*

Proof. (a) Let $u \in F$. If $u \notin max(F \cup g)$ then $u < v \in g$ for some v and so $u < w \in pref\,v \subseteq F \cup f$. We conclude $u \notin max(F \cup f)$ which contradicts the supposition $F = max(F \cup f)$. Thus: $F \subseteq max(F \cup g)$. Now:

$$\begin{aligned} pref(F \cup g) &\subseteq pref\,F \cup pref\,g \\ &\subseteq pref\,F \cup pref(F \cup f) \\ &\subseteq F \cup f. \end{aligned}$$

Thus: $lim\,pref(F \cup g) \subseteq lim(F \cup f) = g$.

(b) From the supposition $F \subseteq \max(F \cup g)$, we conclude $F \subseteq max\,pref(F \cup g)$. Now:

$$\begin{aligned} max\,pref(F \cup g) &\subseteq \max(pref\,F \cup pref\,g) \subseteq max\,pref\,F \cup max\,pref\,g \\ &\subseteq F \cup \varnothing \subseteq F. \end{aligned}$$

Thus: $F = max\,pref(F \cup g)$.

Now, $g \subseteq lim\,pref\,g \subseteq lim\,pref(F \cup g)$ while by supposition the opposite inclusion holds. Thus $g = lim\,pref(F \cup g)$ which concludes the proof.

Observation. As a point of independent interest suggested by the condition $lim\,pref(F \cup g) \subseteq g$, $F \subseteq \Sigma^*$, $g \subseteq \Sigma^\infty$, we observe that the condition $lim\,pref\,g \subseteq g$, (which is implied by the previous condition), i.e., $g = lim\,pref\,g$, is equivalent to the condition that g is topologically closed if Σ is given the discrete topology and Σ^∞ the induced product topology. Explicitly, $lim\,pref\,g \subseteq g$ iff g is the complement of an arbitrary union of sets of the form

$$A_1 \times A_2 \times \cdots \times A_n \times \Sigma^\infty = A_1 \times A_2 \times \cdots \times A_n \times \Sigma \times \Sigma \times \cdots$$

where, for each $i \in [n]$, $A_i \subseteq \Sigma$.

If $F, G \subseteq \Sigma^*$ and $f, g \subseteq \Sigma^* \cup \Sigma^\infty$, we define $(F; f) \cdot (G; g) = (FG; f \cup Fg)$. It is straightforward but tedious to verify directly that

PROPOSITION H. *The function that takes* $(A; a_1 \cup a_2)$ *into* $(A; a_1 \cup b)$, *where* $b = \mathit{lim}\, a_2 = \lim(A \cup a_1 \cup a_2)$, *preserves composition (thus giving rise to an isomorphism).*

The above considerations extend without difficulty to the case $T: n \to p$ and $n \times p$ augmented matrices.

Certain trees do not require *augmented* matrices for their representation. The fact that finite trees have a "more efficient" representation was mentioned in Section 2.4. There are other trees as well, however, which can be faithfully represented by labels of successful paths only.

DEFINITION. Call a tree $T: n \to p$ *biaccessible* if every vertex of the tree lies on a successful path. (Thus finite trees are biaccessible.) The corresponding notion for 1×1 surmatrices $(A; a_1 \cup a_2)$ where $A \cup a_1 \cup a_2)$ is prefix closed and $A \cup a_1 = \max(A \cup a_1 \cup a_2)$ is given by the following.

PROPOSITION I. *A tree* $T: 1 \to 1$ *is biaccessible iff its surrogate* $(A; a_1 \cup a_2)$ *satisfies* $a_2 \subseteq \mathit{pref}(A \cup a_1)$ *or, equivalently,* $A \cup a_1 \cup a_2 = \mathit{pref}(A \cup a_1)$.

PROPOSITION J. $\mathit{lim}\, a_2 \subseteq \mathit{lim}\, \mathit{pref}(A \cup a_1) \Leftrightarrow a_2 \subseteq \mathit{pref}(A \cup a_1)$.

Proof $\Rightarrow$. $\mathit{lim}\, a_2 \subseteq \mathit{lim}\, \mathit{pref}(A \cup a_1) \subseteq \mathit{lim}\, \mathit{pref}(A \cup a_1 \cup a_2) \subseteq \mathit{lim}(A \cup a_1 \cup a_2) \subseteq \mathit{lim}\, a_2$ so that $\mathit{lim}\, \mathit{pref}(A \cup a_1) = \mathit{lim}\, a_2$. Notice that $A \cup a_1 = \mathit{max}\, \mathit{pref}(A \cup a_1)$ so that $\mu(\mathit{pref}(A \cup a_1)) = A \cup a_1 \cup \lim a_2 = \mu(A \cup a_1 \cup a_2)$ and by the injectiveness (Theorem E) of μ, we have $(A \cup a^0 \cup a_2) = \mathit{pref}(A \cup a^0)$ which proves $\Rightarrow$. The opposite implication is obvious (Proposition B).

ACKNOWLEDGMENT

We are indebted to Martha Mierce for her painstaking attention to the manuscript (including locating misprints) resulting in the production of a document pleasing to the eye.

REFERENCES

1. S. L. BLOOM AND C. C. ELGOT. The Existence and Construction of Free iterative Theories, *J. Comput. System Sci.* **12** (1976), 305–318.
2. S. L. BLOOM, S. GINALI, AND J. D. RUTLEDGE, Scalar and Vector Iteration, *J. Comput. System Sci.* **14** (1977), 251–256.
3. C. C. ELGOT, Monadic computation and iterative algebraic theories, *in* "Logic Colloquium 1973," Vol. 80, "Studies in Logic" (H. E. Rose and J. C. Shepherdson, Eds.), North-Holland, Amsterdam, 1975.
4. C. C. ELGOT, Matricial theories, *J. Algebra* **42** (1976), 391–421.

5. C. C. Elgot, Structured programming with and without GO TO statements, *IEEE Trans. Software Eng. SE-2*, No. 1 (March 1976); Erratum and Corrigendum (September 1976).
6. D. E. Knuth, "The Art of Computer Programming," Vol. I, Addison-Wesley, Reading, Mass. 1969.
7. E. Engeler, Structure and meaning of elementary programs, *in* "Symposium on Semantics of Algorithmic Languages Proceedings," pp. 89–101, Springer-Verlag, New York/Berlin, 1971.
8. E. G. Manes, "Algebraic Theories," Academic Press, New York, 1976.
9. S. Eilenberg and J. B. Wright, Automata in general algebras, *Inform. Contr.* **11** (1967), 452–470.
10. F. Harary, "Graph Theory," Addison-Wesley, Reading, Mass. 1969.
11. J. Goguen, J. Thatcher, E. Wagner, and J. Wright, Initial algebra semantics and continuous algebras, *J. Assoc. Comput. Mach.*, to appear.
12. M. Wand, "Mathematical Foundations of Language Theory," Dissertation. Project MAC, MIT, 1973.
13. S. Ginali, "Iterative Algebraic Theories, Infinite Trees, and Program Schemata," Dissertation, University of Chicago, June 1976.
14. W. Lawvere, Functorial semantics of algebraic theories, *Proc. Nat. Acad. Sci. U.S.A.* **50** (1963), 869–872.

SIAM J. COMPUT.
Vol. 9, No. 1, February 1980

0097-5397/80/0901-0002 $01.00/0

SOLUTIONS OF THE ITERATION EQUATION AND EXTENSIONS OF THE SCALAR ITERATION OPERATION*

STEPHEN L. BLOOM,† CALVIN C. ELGOT‡ AND JESSE B. WRIGHT‡

Abstract. We study the solutions to a (vector) equation somewhat analogous to the traditional equations of linear algebra. Whereas, in introductory linear algebra the domain of discourse is the field of real numbers (or an arbitrary field) our domain of discourse is the algebraic theory of (multi-rooted, leaf-labeled) trees (or, more generally, any iterative theory).

As in linear algebra, we obtain a necessary and sufficient condition for our equations to have unique solutions and we can describe "parametrically" the totality of solutions. However, whereas in linear algebra, there is no way of giving $1 \div 0$ meaning in such a way that all the "old laws" hold, we can give meaning to the "iteration operation" (the analogue of division into 1) in such a way that all the "old laws" still hold. Indeed, we can describe "parametrically" all such ways of extending the (partially defined) scalar iteration operation to all trees (more generally, morphisms).

Key words. iteration, semantics, flowchart schemes, iterative theory

1. Introduction.

1.1. An example. The notion of iterative theory (reviewed in this section) is discussed in [5], [3], [11], [4] and [7] the principal reference for this paper. Its relationship to the theory of computation is indicated in [5]. An important example of an iterative theory is provided by multi-rooted, (locally) ordered trees (not necessarily finite) with *termini* (a distinguished subset of the set of all leaves of a tree) which are "labeled" by positive integers. An n-rooted (vector) tree f (or "forest" of n singly rooted trees) of this kind whose termini labels come from the set $\{1, 2, 3, \cdots, p\}$—abbreviated $[p]$—is indicated as follows

$$f: n \to p.$$

When $p = 0$, there are no termini, i.e., all the leaves (if any) are unlabeled. If $g: p \to q$ is another such tree, by $f \cdot g: n \to q$, we mean the tree which is obtained by "attaching", to each terminus of f labeled j, for each $j \in [p]$, a "copy" of the jth singly rooted tree of g; $f \cdot g$ is the *composition* of f and g. (More strictly speaking, this operation is on isomorphism classes of trees rather than on individual trees.)

The tree $I_p \oplus 0_n: p \to p + n$, $n \geqq 0$, consists of a sequence of p vertices; the jth vertex is both the jth root and a terminus labeled j, for each $j \in [p]$. Notice that $f \cdot I_p = f$ and $I_p \cdot g = g$.

Another basic operation, besides composition, is "source pairing". If $f_i: n_i \to p$, $i \in [2]$, are forests, the forest $(f_1, f_2): n_1 + n_2 \to p$ has f_1 as its first n_1 singly rooted trees and f_2 as its next n_2 singly rooted trees. This operation (again, strictly speaking on isomorphism classes), as well as composition, is associative. If $r > 2$, we write

$$(f_1, f_2, \cdots, f_r)$$

for the extension of the binary operation to an r-ary operation.

The very familiar operation of constructing a new singly rooted (or scalar) tree from n single rooted trees by adjoining a new vertex may be described as

$$\gamma_n \cdot (f_1, f_2, \cdots, f_n)$$

* Received by the editors May 23, 1978.

† Department of Mathematics, Stevens Institute of Technology, Hoboken, New Jersey 07030.

‡ Mathematical Sciences Department, IBM T. J. Watson Research Center, Yorktown Heights, New York 10598.

where f_i is a scalar tree, for each $i \in [n]$ and γ_n: $1 \to n$ is the tree depicted by

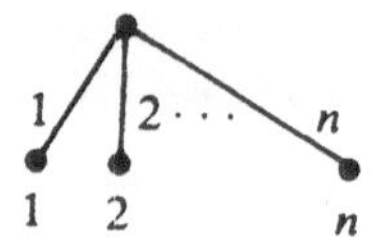

The trees γ_n, $n = 0, 1, 2, 3, \cdots$, are called *atomic*. A *primitive tree* is one such that every successor of the root is a terminus. A vector tree is *primitive* if each of its component scalar trees is primitive.

The collection of (isomorphism classes of) of trees $n \to p$ for variable n, p form a category Tr whose *objects* are $0, 1, 2, \cdots$ and whose *morphisms* $n \to p$ are trees $n \to p$. Each object n in Tr is a coproduct of the object 1 with itself n times and thus forms an *algebraic theory* in the sense of Lawvere (cf. [13]). This algebraic theory is *ideal* (cf. [5]) in that if f: $1 \to n$ is a tree which is *nontrivial* (in the sense that its root is not a terminus) then so is $f \cdot g$ for every g: $n \to p$ in Tr.

The ideal algebraic theory Tr is *closed with respect to conditional iteration*, briefly, is *iterative*, in that the equation in Tr

$$\xi = g \cdot (I_p, \xi),$$

where ξ: $n \to p$ and where g: $n \to p+n$ is *ideal* (i.e. is a forest of n nontrivial singly rooted trees), has a unique solution for the "unknown" ξ. This "vector" equation is, in a sense we will not make clear here, linear and may be "rewritten" as n (linear) equations in n scalar "unknowns" ξ_i: $1 \to p$, $i \in [n]$. This equation is the *iteration equation determined by g*.

In the case $p = 0$, the iteration equation for g takes the simpler form:

$$\xi = g \cdot \xi.$$

The collection of trees of "finite index" (i.e. trees having only a finite number of nonisomorphic subtrees, cf. [7]) forms an iterative subtheory tr of Tr.

One reason for the importance of the iteration equation is this: for each (possibly infinite) tree f: $1 \to p$ in tr there is a primitive g: $n \to p+n$ in *tr* such that f is the first component of the solution to the iteration equation for g. Another reason is that in models of computation, the solution to the iteration equation expresses "looping".

1.2. Solutions. While the solution (in a given iterative theory) to the iteration equation for g: $n \to p+n$ is unique in the case that g is ideal, in general, there are many solutions. For example, in the case that $p = 0$ and $g = I_n$, each morphism f: $n \to n$ satisfies the iteration equation for g. We give, in 2 (cf. 2.15), a kind of parametric description of all solutions to the iteration equation for g, where g: $n \to p+n$ is an arbitrary morphism. We obtain as a corollary (cf. 2.19) a necessary and sufficient condition on g for its iteration equation to have a unique solution. This corollary is equivalent to Theorem B of J. Tiuryn [8].

When applied to Tr (cf. 2.21) the corollary states that g's iteration equation has a unique solution iff the tree

$$g^n = g \cdot (I_p \oplus 0_n, g) \cdot (I_0 \oplus 0_n, g) \cdot (I_p \oplus 0_n, g) \cdots, \quad (n \text{ occurrences of "}g\text{"})$$

has the property that each trivial component of g^n carries a label from the set $[p]$ (out of the set $[p+n]$ of possible labels). In the case $p = 0$,

$$g^n = g \cdot g \cdot g \cdots \quad (n \text{ occurrences of "}g\text{"})$$

$[p]$ is the empty set and the property of g becomes simply:

each component of g^n is nontrivial.

1.3. Extension. In iterative theories we may define $g^\dagger$ (the *iterate of* g) to be the unique solution of g's iteration equation, if g's iteration equation has a unique solution; in the contrary case we may regard "$g^\dagger$" as meaningless. It is natural then to raise the question: it is possible to extend the meaning of $g^\dagger$ to *all* g in such a way that all the old "laws", i.e., "identities" still hold. The answer to the question is "yes". This matter is taken up in § 3 (cf. 3.7 and 3.8) for the case that g is scalar, i.e., g has source 1; the vector case is reserved for a sequel to this paper (in preparation).

The "full answer" to the question is rather neat and perhaps surprising. It is this: if one chooses $\perp : 1 \to 0$ in the iterative theory arbitrarily[1] and defines $I_1^\dagger = \perp$, then the requirement that the old laws still persist, uniquely determines $g^\dagger$ for all g. Thus, in Tr, we may choose $\perp : 1 \to 0$ to be any infinite tree (without termini) and we may define the iterate of the trivial tree $I_1 : 1 \to 1$ to be $\perp$ without violating any laws!

There is another sense in which the result is surprising, viz: while the results mentioned in § 1.2 are reminiscent of and roughly analogous to results for linear equations over a field, and while the question raised also has an analogue in the field domain (replace "$g^\dagger$" by "$1 \div g$"), the answer in the field domain is "no".

1.4. The sequel. The sequel [1] to this paper (to which we've already alluded) centers around a formula from [4] which is an iterative theory identity. This formula expresses the iterate of an $(n+1)$ dimensional vector morphism in terms of the iterates of n-dimensional and 1-dimensional morphisms. The formula may be used to define the vector iteration operation (on all morphisms) in terms of the scalar iteration operation. It turns out that *all* the old laws valid in iterative theories are still valid.

The morphism $g^\dagger$, for the arbitrary vector morphism g in Tr (and other "tree theories"), may be expressed as a metric limit, no matter how $\perp : 1 \to 0$ is selected. (The trees $n \to p$ in Tr form a complete metric space as was noted independently in [9].) This contrasts with the fact that it is not always possible to define a partial ordering (compatible with composition) in Tr such that $g^\dagger$ is the least (or greatest) solution of the iteration equation for g. Whether or not one can define such an ordering depends on the choice of $\perp$. This fact is of interest in connection with the many mathematical treatments of the semantics of flowchart schemes that make fundamental use of partial orderings, (e.g. [10], [12], [2]).

1.5. Elementary properties of algebraic theories. An *algebraic theory* T is a category whose objects are the nonnegative integers, having for each $n > 0$, n "distinguished morphisms" $\mathbf{i}: 1 \to n$, $i \in [n]$[2] (where $[n] = \{1, 2, \cdots, n\}$; $[0] = \varnothing$) such that: for each family of morphisms $f_i : 1 \to p$, $i \in [n]$, $n \geqq 0$, there is a unique morphism $f: n \to p$ such that for each $i \in [n]$, f_i is the composition

$$f_i : 1 \xrightarrow{i} n \xrightarrow{f} p.$$

The morphism f is called the *source tupling* of the morphisms f_i, $i \in [n]$, and is denoted $(f_1, f_2, \cdots, f_n)$. In the case $n = 0$, this condition requires the existence of a unique

[1] Every iterative theory except $\mathcal{N}$ contains at least one morphism $1 \to 0$. The theory $\mathcal{N}$ may be described as the subtheory of Tr which consists of all the trivial trees in Tr; it may also be described as the theory whose morphisms $n \to p$ are the functions $[n] \to [p]$.

[2] We admit this notation is ambiguous. $\mathbf{2}$ may have target 2 or any number larger than 2, but in context the target should be clear.

morphism 0_p: $0 \to p$. All morphisms $n \to p$, n, $p \geqq 0$, formed by source tupling the distinguished morphisms are called *base* morphisms. When the distinguished morphisms are distinct, (which is the case in every algebraic theory but two: the "trivial theories") the base morphisms are in 1–1 correspondence with the collection of functions $[n] \to [p]$. The function $f: [n] \to [p]$ corresponds to the source tupling $(\mathbf{f}(\mathbf{1}), \mathbf{f}(\mathbf{2}), \cdots, \mathbf{f}(\mathbf{n}))$, where $\mathbf{f}(\mathbf{i})$: $1 \to p$ is distinguished.

If "f" denotes a function $[n] \to [p]$, then "$f: n \to p$" will also denote the morphism in T corresponding to f. The base morphism corresponding to the identity function on $[n]$ is denoted I_n.

The *composition* of $f: n \to p$ and $g: p \to q$ is denoted either $f \cdot g$ or $n \xrightarrow{f} p \xrightarrow{g} q$.

It is convenient to define an operation, derived from source tupling, which pairs two morphisms with arbitrary sources. If f_i: $n_i \to p$, $i = 1, 2$, then the *source pairing* (f_1, f_2): $n_1 + n_2 \to p$ is the unique morphism satisfying

$$\mathbf{i} \cdot (f_1, f_2) = \begin{cases} \mathbf{i} \cdot f_1, & \text{if } i \in [n_1], \\ \mathbf{j} \cdot f_2, & \text{if } i = n_1 + j, j \in [n_2]. \end{cases}$$

Let $\kappa: [p_1] \to [p_1 + p_2]$, $\lambda: [p_2] \to [p_1 + p_2]$ be the inclusion and translated inclusion functions (i.e., $i\kappa = i$, $i\lambda = p + i$). If f_i: $n_i \to p_i$, $1 = 1, 2$, then we define the "circle plus" of f_1 and f_2 as follows: $f_1 \oplus f_2$: $n_1 + n_2 \to p_1 + p_2$ is the morphism $(f_1 \cdot \kappa, f_2 \cdot \lambda)$.

Whenever the expressions below are meaningful, i.e., for the appropriate sources and targets, the following assertions hold in any algebraic theory (see [5]).

(1.5.1) $$(f_1 \oplus f_2) \cdot (g_1 \oplus g_2) = f_1 \cdot g_1 \quad \oplus \quad f_2 \cdot g_2,$$

(1.5.2) $$(f_1 \oplus f_2) \cdot (g_1, g_2) = (f_1 \cdot g_1, f_2 \cdot g_2),$$

(1.5.3) $$(0_p, g) = g = (g, 0_p),$$

(1.5.4) $$(f_1, g_2) \cdot g = (f_1 \cdot g, f_2 \cdot g).$$

The algebraic theory T is *ideal* if for each nondistinguished morphism $g: 1 \to n$, $g \cdot f$ is nondistinguished, for any $f: n \to p$. A nondistinguished morphism $g: 1 \to n$ in T is called ideal.

An *iterative theory* is a nontrivial ideal theory such that for each ideal morphism $g: 1 \to p + 1$ there is a unique morphism $g^\dagger: 1 \to p$ such that

(*) $$g^\dagger = g \cdot (I_p, g^\dagger).$$

In an iterative theory, if $g: n \to p + n$, $n > 1$ is ideal, i.e., for each $\mathrm{i} \in [n]$, $\mathbf{i} \cdot g$ is ideal, it can be shown there is a unique morphism $g^\dagger$: $n \to p$ such that (*) holds [4].

If $g: n \to p + n$ is any morphism in an algebraic theory, the "powers" of g are defined as follows:

$$g^0 = 0_p \oplus I_n$$

and

$$g^{r+1} = g^r \cdot (I_p \oplus 0_n, g): n \to p + n.$$

The following facts about g^r are quite useful.

(1.5.5) $$(I_p \oplus 0_n, g)^r = (I_p \oplus 0_n, g^r), \quad \text{all } r \geqq 0;$$

(1.5.6) $$\text{if } \xi = g \cdot (I_p, \xi), \quad \text{then } \xi = g^r \cdot (I_p, \xi), \quad \text{all } r \geqq 0.$$

1.6. Γtr. In §§ 3 and 4, Γ will denote a ranked set; i.e., Γ is the disjoint union $\bigcup(\Gamma_k: k \geqq 0)$.

In [7] it was shown that Γtr, the collection of Γ-trees of "finite index", formed an iterative theory which is freely generated by Γ. This means that for any iterative theory J and any function F that maps $\gamma \in \Gamma_k$ into an *ideal* morphism $\bar{\gamma}: 1 \to k$ in J, there is a unique theory morphism $\mathbf{F}: \Gamma\text{tr} \to J$ extending F. This result is generalized by the Universality Theorem in § 3.

2. All solutions of the iteration equation. The determination of all solutions of the iteration equation depends upon classifying the component positions of a morphism $g: n \to p+n$ into three disjoint categories. A position $i \in [n]$ of g is either "singular", "power successful" or "power ideal" (see § 2.3). Singular positions are responsible for the existence of more than one solution of the iteration equation. On the other hand, if all the component positions are either power successful or power ideal, there is a unique solution of the iteration equation for g.

2.1. Let $g: n \to p+n$ be an arbitrary morphism in an iterative theory I. The *iteration equation for* g is the equation in the "variable" $\xi: n \to p$

$$\xi = g \cdot (I_p, \xi). \tag{2.1.1}$$

By definition of "iterative theory" (more fully, "ideal theory closed under conditional iteration") whenever g is an ideal morphism, the equation (2.1.1) has a unique solution, i.e., there is a unique morphism $\xi: n \to p$ which satisfies (2.1.1). This solution is denoted $g^{\dagger}$. If g is not ideal, the iteration equation for g may have many solutions. In this section, solutions to (2.1.1) are determined.

In the case $n > 1$ it is useful to rewrite (2.1.1) as a system of simultaneous equations. For $i \in [n]$, let $\xi_i = \mathbf{i} \cdot \xi = 1 \xrightarrow{\mathbf{i}} n \xrightarrow{\xi} p$ and let $g_i = \mathbf{i} \cdot g: 1 \xrightarrow{\mathbf{i}} n \xrightarrow{g} p+n$. Then (2.1.1) is equivalent to the system

$$\begin{aligned}
\xi_1 &= g_1 \cdot (I_p, \xi_1, \xi_2, \cdots, \xi_n),\\
\xi_2 &= g_2 \cdot (I_p, \xi_1, \xi_2, \cdots, \xi_n),\\
&\vdots\\
\xi_n &= g_n \cdot (I_p, \xi_1, \xi_2, \cdots, \xi_n).
\end{aligned} \tag{2.1.2}$$

When g is not ideal, there is at least one $i \in [n]$ for which g_i is a base morphism $1 \to p+n$. If g_i is $0_p \oplus \mathbf{i}'$ for some $i' \in [n]$, then the ith equation (2.1.2) is a "pure variable equation":

$$\xi_i = \xi_{i'}. \tag{2.1.3}$$

Similarly, if g_i is $\mathbf{j} \oplus 0_n$, for some $j \in [p]$, then the ith equation in (2.1.2) determines the value of ξ_i:

$$\xi_i = \mathbf{j}: 1 \to p. \tag{2.1.4}$$

To illustrate these possibilities, we will discuss an example in some detail.

2.2. Example. Throughout § 2 we will discuss the solution of the iteration equation for a morphism G in the iterative theory of Γ-trees (see [7]). We recall that a tree $f: n \to p$ is an n-tuple $(f_1, \cdots, f_n)$ of rooted (locally-ordered, locally-finite) trees, some of whose leaves are labeled with elements of $[p]$. To compose $f: n \to p$ with $g: p \to q$ one attaches to each leaf of f labeled $i \in [p]$ a copy of the ith component of g. In our example, we assume all vertices of G of outdegree one have the same (unindicated) label in Γ_1.

Let $G: 9 \to 10$ be the tree

(2.2.1) $G =$

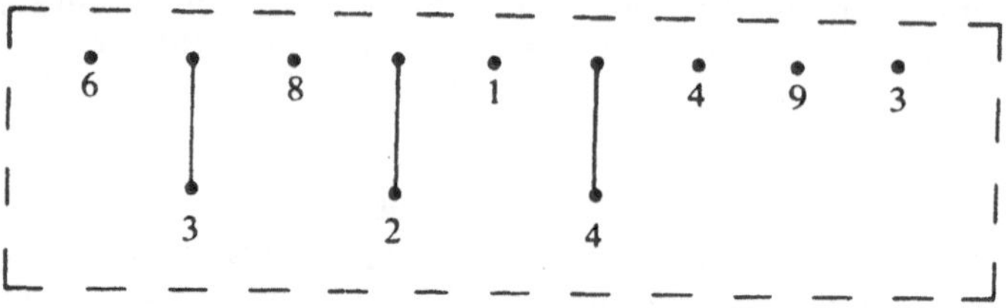

Then the system of equations corresponding to the iteration equation for G (the solution is depicted by (2.6.2)) is

$$\xi_1 = \xi_5 \quad (1)$$
$$\xi_2 = G_2' \cdot \xi_2 \quad (2)$$
$$\xi_3 = \xi_7 \quad (3)$$
$$\xi_4 = G_4' \cdot \xi_1 \quad (4)$$
(2.2.2)
$$\xi_5 = \mathbf{1}: 1 \to 1 \quad (5)$$
$$\xi_6 = G_6' \cdot \xi_3 \quad (6)$$
$$\xi_7 = \xi_3 \quad (7)$$
$$\xi_8 = \xi_8 \quad (8)$$
$$\xi_9 = \xi_2. \quad (9)$$

Here, for $i = 2, 4, 6$

$$G_i': 1 \xrightarrow{G_i} 10 \xrightarrow{\text{base morphism}} 1,$$

i.e. $G_2' (= G_4' = G_6')$ is the tree

so that $G_i = G_i' \cdot \mathbf{j}$, for some unique $j \in [10]$.

Although these equations are sufficiently clear to allow one to determine all solutions of the iteration for G, before doing so we note that the equations fall into three distinct groups. Equations (1) and (5) of (2.2.2) may be grouped together, since they determine the value of ξ_1 and ξ_5 as the base morphism $\mathbf{1}: 1 \to 1$. The "pure variable" equations (3), (7), and (8) may be grouped together, since they determine only that $\xi_3 = \xi_7$; $\xi_3 (= \xi_7)$ and ξ_8 may be any trees $1 \to 1$. The last group is the remaining equations (2), (4), (6), and (9). In § 2.4 we will rearrange the equations so that these groups occur together. The classification of these three groups of equations is discussed in general in the next section.

2.3. The initial and final classification. Let $g: n \to p + n$ be a morphism in an iterative theory. The *initial classification* of the component positions $i \in [n]$ of g is the following.

2.3.1. The position i is a *successful* position (for g) if for some (unique) $j \in [p]$, $\mathbf{i} \cdot g = \mathbf{j}: 1 \to p + n$. We define the function $g^e: S_g \to [p]$, where S_g is the set of successful positions of g, by $ig^{\varepsilon} = j$. Thus if i is a successful position for g, the ith equation in (2.1.2) has the form (2.1.4). In the example § 2.2, only 5 is a successful position of G so that $S_g = \{5\}$ and from the fifth equation of the example, $5G^{\varepsilon} = 1$.

2.3.2. The position i is *ideal* if the morphism $\mathbf{i} \cdot g$ is ideal. Let I_g be the set of ideal positions of g. In the example of § 2.2, positions 2, 4, and 6 are ideal for G so that $I_G = \{2, 4, 6\}$.

2.3.3 The position i is *potentially singular* if, for some (unique) $i' \in [n]$, $i \cdot g = 0_p \oplus \mathbf{i}'$. We define the function $g^{\nu}: P_g \to [n]$, where P_g is the set of potentially singular positions of g, by $ig^{\nu} = i'$. Thus if i is potentially singular, the ith equation in (2.1.2) has the form (2.1.3). In example of § 2.2, the positions 1, 3, 7, 8, and 9 are potentially singular for G so that $P_G = \{1, 3, 7, 8, 9\}$ and from the first, third, seventh, eighth, and ninth equations of the example: $1G^{\nu} = 5$, $3G^{\nu} = 7$, $7G^{\nu} = 3$, $8G^{\nu} = 8$, $9G^{\nu} = 2$,

For the *final classification* of the component positions of g, we note that exactly one of three possibilities can occur. Either for every r, i is a potentially singular position for g^r or not. If not, either $\mathbf{i} \cdot g^r$ is $\mathbf{j} \oplus 0_n$ for some $j \in [p]$, or $\mathbf{i} \cdot g^r$ is ideal. Thus the possibilities are:

(2.3.4) For every $r \geqq 1$ there is an $i_r \in [n]$ such that $\mathbf{i} \cdot g^r = 0_p \oplus \mathbf{i}_r$.

(Recall the definition of g^r, given in 1.5.)

(2.3.5) There is some $r \geqq 1$ and some $j \in [p]$ such that $\mathbf{i} \cdot g^r = \mathbf{j} \oplus 0_n$.

(2.3.6) There is some $r \geqq 1$ such that $\mathbf{i} \cdot g^r$ is ideal.

2.3.7. We say i is a *singular* position of g if (2.3.4) occurs. In the example of § 2.2, positions 3, 7 and 8 are singular for G. (The singular positions are responsible for the existence of more than one solution of the iteration equation, as will be seen below.)

2.3.8. We say i is a *power successful* position of g if (2.3.5) holds. Note that if $\mathbf{i} \cdot g^r = \mathbf{j} \oplus 0_n$, for some $j \in [p]$, then $\mathbf{i} \cdot g^s = \mathbf{j} \oplus 0_n$ for all $s > r$. In the example of § 2.2, the power successful positions are 1 and 5. (Since 5 is a successful position for G, it is clearly power successful: (2.3.5) holds with $r = 1$. Position 1 is power successful, since $\mathbf{1} \cdot G^2 = G_5 = \mathbf{1}: 1 \to 1$.)

2.3.9. Lastly, we say i is a *power ideal* position if (2.3.6) holds. Note that if $\mathbf{i} \cdot g^r$ is ideal, so is $\mathbf{i} \cdot g^s$, all $s > r$. In the example of § 2.2, positions 2, 4, 6 and 9 are power ideal for G. (Indeed 2, 4 and 6 are ideal, and $\mathbf{9} \cdot G^2 = G_2$, an ideal tree.)

It is convenient to define a function $g^{\sigma}: [n] \to [n]^+ \cup [n]^{\infty}$ where $[n]^+$($[n]^{\infty}$) is the set of nonempty finite (infinite) sequences of elements of $[n]$.

2.3.10. *Definition of* $g^{\sigma}: [n] \to [n]^+ \cup [n]^{\infty}$. Suppose i is a singular position of g. Then ig^{σ} is the infinite sequence in $[n]^{\infty}$

$$ig^{\sigma} = u_1 u_2 u_3 \cdots$$

where $u_1 = i$ and for each $r \geqq 1$, $u_r g^{\nu} = u_{r+1}$ or equivalently,

$$\mathbf{i} \cdot g^r = \quad \mathbf{u}_r \cdot g \quad = 0_p \oplus \mathbf{u}_{r+1}. \tag{2.3.11}$$

A formal proof of (2.3.11) is by induction on r. The truth of the equation is, on reflection, obvious.

Suppose i is a power successful position of g. Then ig^{σ} is the finite sequence $u_1 u_2 \cdots u_t$ where $t \geqq 1$, $u_1 = i$, $u_t \in S_g$ and where, for $r < t$, (2.3.11) holds.

Lastly, suppose i is a power ideal position of g. We define ig^σ to be the finite sequence $u_1u_2\cdots u_t$, where $u_1 = i$, $u_t \in I_g$ and for $r < t$, (2.3.11) holds. Thus $\mathbf{i}\cdot g^t = \mathbf{u}_t \cdot g$ is ideal.

We note that for any potentially singular position i of g if $ig^\nu = i$, then i is a singular position and $ig^\sigma = iii\cdots$.

2.3.12. We let M_g, K_g, and L_g denote the set of power successful, singular and power ideal positions of g, respectively. We will, however, usually drop the subscripts.

For the sake of illustration, we will compute several values of the function $G^\sigma: [9]\to[9]^+\cup[9]^\infty$, where G is the tree of example in § 2.2. The position 3 is singular for G, and

$$3G^\sigma = 3\ 7\ 3\ 7\ 3\ 7\cdots.$$

The position 1 is power successful for G, and

$$1\cdot G^\sigma = 1\ 5.$$

The positions 9 and 2 are power ideal for G, and

$$9\cdot G^\sigma = 9\ 2;$$

$$2\cdot G^\sigma = 2.$$

2.3.13. Note that if $ig^\sigma = u_1u_2u_3\cdots = i\,u_2u_3\cdots$ and $u_1 = i$ is in M (resp., K, L), so is u_r, all $r \geqq 1$; similarly, if u_r, $r > 1$ is in M (resp., K, L), so is u_s, for all s, $1 \leqq s < r$. Note also that $u_2g^\sigma = u_2u_3u_4\cdots$, so that $ig^\sigma = iv$, where $v = u_2g^\sigma$.

If $i\in M$ or L and $ig^\sigma = u_1u_2\cdots u_{t+1}$, $t \geqq 0$, then the elements $u_1,\cdots,u_t$ are *distinct* potentially singular positions of g. Thus $t < n$ and (*using the finiteness of n*) it follows that

(2.3.14) i is a power successful position of g iff i is a successful position of g^n; in this case $\mathbf{i}\cdot g^n = \mathbf{j}$ for some $j\in[p]$;

(2.3.15) i is a power ideal position of g iff i is an ideal position of g^n; in this case $\mathbf{i}\cdot g^n$ is ideal;

(2.3.16) i is a singular position of g iff i is a potentially singular position of g^n; in this case $\mathbf{i}\cdot g^n = 0_p \oplus \mathbf{i}'$ for some $i'\in[n]$.

In particular, no position of g is singular iff for each $i\in[n]$, $\mathbf{i}\cdot g^n$ is ideal or successful.

2.3.17. Because the description of a morphism g in an iterative theory is often given in such a way that one can immediately "read off" the set P_g (as, for example, in (2.2.1) or (2.2.2)), we note that in (2.3.14), (2.3.15) and (2.3.16), the superscript "n" may be replaced by $\|P_g\|+1$, where $\|P_g\|$ is the cardinality of P_g.

In the example, $\|P_G\| = 5$. Thus, in this case, we have $n = 9$ and $\|P_G\|+1 = 6$.

2.4. We return to the example of § 2.2. As mentioned, it would be convenient to rearrange the system of equations (2.2.2) so that the power successful, singular and power ideal positions occur together. This amounts to "rearranging" G. Recall from 2.3.7–2.3.9 that the set M of power successful positions of G is $\{1, 5\}$; K, the set of singular positions, is $\{3, 7, 8\}$, and the set L of power ideal positions of G is $\{2, 4, 6, 9\}$. Let $\pi: [9]\to[9]$ be a permutation that maps M bijectively onto $\{1, 2\}$, K bijectively onto $\{3, 4, 5\}$ and L bijectively onto $\{6, 7, 8, 9\}$. For definiteness, suppose that π is the unique such permutation which is order preserving so that π is given by reading top-down the following table

i	1	2	3	4	5	6	7	8	9
$i\pi$	1	6	3	7	2	8	4	5	9

and π^{-1} is given by reading the table bottom-up. Then the tree $\bar{G} = \pi^{-1} \cdot G \cdot (I_1 \oplus \pi)$ is

(2.4.1) $\bar{G} =$

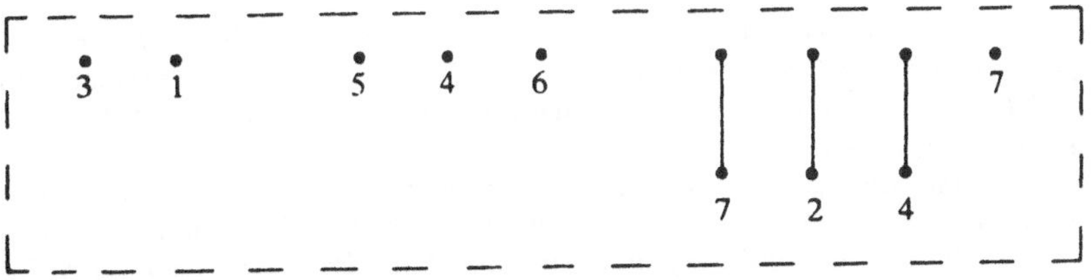

and the system of equations corresponding to the iteration equation for $\bar{G}$ (the solution is depicted by (2.6.1)) is

$$\xi_1 = \xi_2, \quad \xi_2 = \mathbf{1}: 1 \to 1,$$

(2.4.2) $$\xi_3 = \xi_4, \quad \xi_4 = \xi_3, \quad \xi_5 = \xi_5,$$

$$\xi_6 = \bar{G}'_6 \cdot \xi_6, \quad \xi_7 = \bar{G}'_7 \cdot \xi_1, \quad \xi_8 = \bar{G}'_8 \cdot \xi_3, \quad \xi_9 = \xi_6.$$

See (2.2.2) for the meaning of the prime.

The function $\bar{G}^\nu$ is given by Table 1 and $\bar{G}^\sigma$ is given by Table 2.

TABLE 1

i	1	3	4	5	9
$i\bar{G}^\nu$	2	4	3	5	6

TABLE 2

	$\bar{G}^\sigma$
1	1 2
2	2
3	3 4 3 4 ···
4	4 3 4 3 ···
5	5 5 5 ···
6	6
7	7
8	8
9	9 6

The $\bar{G}^\sigma$-table verifies that the power successful positions for $\bar{G}$ are 1 and 2; the singular positions are 3, 4 and 5 and the remaining positions are power ideal. It is much easier to compute the solutions $\xi = (\xi_1, \xi_2, \cdots, \xi_9): 9 \to 1$ of the iteration equation for $\bar{G}$ from (2.4.2) than to find the solutions to (2.2.2). The relation between G, $\bar{G}$ and the solutions of their iteration equations is discussed in general below.

2.5. Conjugation. Let $g, \bar{g}: n \to p + n$ be morphisms in the iterative theory I, and let $\pi\ [n] \to [n]$ be a permutation.

2.5.1. DEFINITION. (a) $\bar{g}$ is the *π-conjugate* of g if $\bar{g} = \pi^{-1} \cdot g \cdot (I_p \oplus \pi)$.

(b) $\bar{g}$ is *conjugate* to g if for some permutation π of $[n]$, $\bar{g}$ is the π-conjugate of g.

Note that in the case $p = 0$, if $\bar{g}$ is the π-conjugate of g, $\bar{g} = \pi^{-1} \cdot g \cdot \pi$; (thus the use of group theoretic terminology).

2.5.2. *Remark.* We observe that the relation "$\bar{g}$ is conjugate to g" is an equivalence relation. Indeed, if $\bar{g}$ is the π-conjugate of g, g is the π^{-1}-conjugate of $\bar{g}$; if g_2 is the π_1-conjugate of g_1, and g_3 is the π_2-conjugate of g_2, then g_3 is the $\pi_1\pi_2$-conjugate of g_1.

2.5.3. PROPOSITION. *Let $\bar{g}$ be the π-conjugate of $g: n \to p+n$. Then*: (a) *for each* $r \geqq 0$, $\bar{g}^r = \pi^{-1} \cdot g^r \cdot (I_p \oplus \pi)$;

(*b*) *for each $i \in [n]$, i is a power successful* (*respectively*: *singular, power ideal*) *position for g iff $i\pi$ is a power successful* (*respectively*: *singular, power ideal*) *position of $\bar{g}$*;

(c) *if $\xi: n \to p$ is a solution of the iteration equation for $\bar{g}$, then $\pi \cdot \xi$ is a solution of the iteration equation for g*;

(d) *the map $\xi \mapsto \pi \cdot \xi$ is a bijection between the set of solutions $\xi: n \to p$ of the iteration solution for $\bar{g}$ and the set of solutions of the iteration equation for $\bar{g}$.*

Proof. (a) The proof is by induction on r. The case $r = 0$ is trivial. Now

$$\bar{g}^{1+r} = \bar{g}^r \cdot (I_p \oplus 0_n, \bar{g})$$
$$= \pi^{-1} \cdot g^r \cdot (I_p \oplus \pi) \cdot (I_p \oplus 0_n, \bar{g})$$

by the induction assumption

$$= \pi^{-1} \cdot g^r \cdot (I_p \oplus \pi) \cdot (I_p \oplus 0_n, \pi^{-1} \cdot g \cdot (I_p \oplus \pi))$$
$$= \pi^{-1} \cdot g^r \cdot (I_p \oplus 0_n, g \cdot (I_p \oplus \pi)) \quad \text{by (1.5.2)}$$
$$= \pi^{-1} \cdot g^r \cdot (I_p \oplus 0_n, g)(I_p \oplus \pi) \quad \text{by (1.5.4) and (1.5.1)}$$
$$= \pi^{-1} \cdot g^{r+1} \cdot (I_p \oplus \pi),$$

which completes the induction.

(b) By part (a), for every $r \geqq 1$

$$\pi \cdot \bar{g}^r = g^r \cdot (I_p \oplus \pi).$$

Thus if i is a power successful position for g, for some $r \geqq 1$, $j \in [p]$, $\mathbf{i} \cdot g^r = \mathbf{j} \oplus 0_n$, and $\mathbf{i} \cdot g^r \cdot (I_p \oplus \pi) = \mathbf{j} \oplus 0_n = \mathbf{i} \cdot \pi \cdot \bar{g}^r = \mathbf{i\pi} \cdot \bar{g}^r$. Hence $i\pi$ is a power successful position for $\bar{g}$. By 2.5.2 and what was just shown, if $i\pi$ is a power successful position of $\bar{g}$, then $(i\pi)\pi^{-1} = i$ is a power successful position of g.

The remaining statements of (b) follow from part (a) in the same way.

(c) Suppose $\xi = \bar{g} \cdot (I_p, \xi)$. Then $\xi = \pi^{-1} \cdot g \cdot (I_p \oplus \pi) \cdot (I_p, \xi) = \pi^{-1} \cdot g \cdot (I_p, \pi \cdot \xi)$, by (1.5.2). Thus $\pi \cdot \xi = g \cdot (I_p, \pi \cdot \xi)$, so that $\pi \cdot \xi$ is a solution of the iteration equation for g.

(d) The map $\xi \mapsto \pi \cdot \xi$ is a surjection, by 2.5.2 and part (c). If $\pi \cdot \xi = \pi \cdot \xi'$, clearly $\xi = \xi'$, so the map is also injective. This completes the proof.

2.6. Example continued. From the system of equations (2.4.2) it follows that every solution to the iteration equation of the tree $\bar{G}$, given in (2.4.1) is a nine-tuple of trees

$$\xi = (\xi_1, \xi_2, \xi_3, \xi_4, \xi_5, \xi_6, \xi_7, \xi_8, \xi_9)$$

such that $\xi_1 = \xi_2 = \mathbf{1}$; $\xi_3 = \xi_4$, ξ_5 is arbitrary, ξ_6 is the infinite tree

$\xi_6 =$ 

$$\xi_7 = \bar{G}'_7 \cdot \mathbf{1}; \qquad \xi_8 = \bar{G}'_8 \cdot \xi_3 \quad \text{and} \quad \xi_9 = \xi_6.$$

Conversely, every nine-tuple of trees satisfying the above conditions will be a solution of the iteration equation for $\bar{G}$.

The "general" solution of the iteration equation for $\bar{G}$ is depicted by Fig. 2.6.1; cf. (2.4.1).

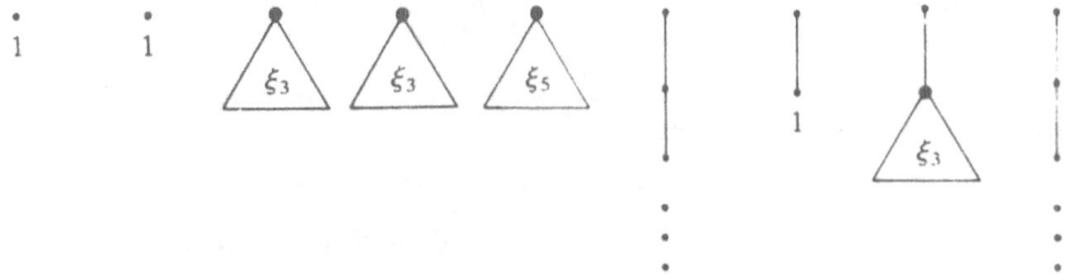

FIG. 2.6.1. *Solution to the iteration equation* $\bar{G}$ (*cf.* (2.4.1)).

Thus, by 2.5.3(d) all of the solutions of the iteration equation for G are the trees $\pi \cdot \xi$, where π is the permutation of [9] given in § 2.4 and ξ is a solution of the iteration equation for $\bar{G}$ (see Fig. 2.6.2).

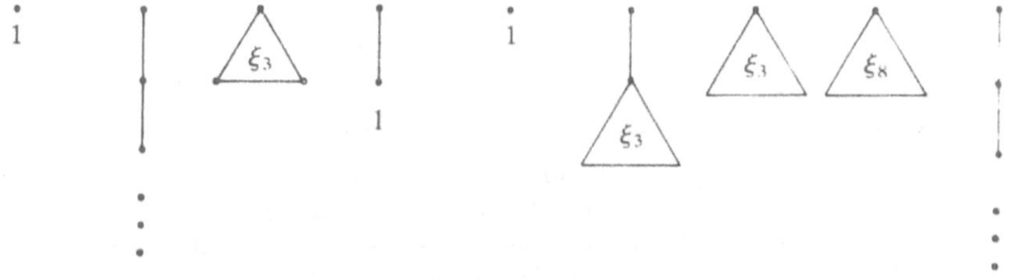

FIG. 2.6.2. *Solution to the iteration equation for G.*

The generalization of this method for solving the iteration equation is contained in Theorem of §2.8.

2.7. Let $g: n \to p+n$ be a morphism in an iterative theory. Let $M, K, L \subseteq [n]$ be respectively the set of power successful, singular and power ideal positions of g. Let the cardinalities of these sets be m, k and l respectively, so that $m+k+l=n$. Let π be a permutation of $[n]$ which maps M bijectively onto $[m]$, K onto $m+[k]$, and L onto $m+k+[l]$. Let $\bar{g}$ be the π-conjugate of g. Then, by 2.5.3(b),

(a) the set of power successful positions of $\bar{g}$ is [m];

(b) the set of singular positions of $\bar{g}$ is $m+[k]$;

(c) the set of power ideal positions of $\bar{g}$ is $m+k+[l]$.

We call a morphism $\bar{g}$ with the properties (a), (b), (c) a *mkl-morphism* ("*mkl*" is pronounced to rhyme with "nickle").

2.8. THEOREM. *Let* $\bar{g}: n \to p+n$ *be a mkl-morphism in an iterative theory. Then*

(a) $\bar{g}$ *is uniquely expressible in the form*

(2.8.1) $$\bar{g} = (a \oplus b \oplus 0_l, h),$$

where

(2.8.2) $a: m \to p+m$ *is base and* a^m "*factors through p*"; *i.e., for some (unique) base morphism* $\bar{a}: m \to p$, $a^m: m \xrightarrow{a} p \xrightarrow{I_p \oplus 0_m} p+m$ (*thus every position of a is power successful); note that* $m = 0$ *if* $p = 0$;

(2.8.3) $$b: k \to k \text{ is base};$$

(2.8.4) $$h: l \to p + m + k + l \text{ is power ideal};$$

(2.8.5) *moreover, for any morphisms* $\xi_1: m \to p$, $\xi_2: k \to p$, *the morphism* $f = h \cdot [(I_p, \xi_1, \xi_2) \oplus I_l]: l \to p + l$ *is power ideal.*

(b) *The solutions of the iteration equation for* $\bar{g}$ *are expressible in the form* (ξ_1, ξ_2, ξ_3) *where* $\xi_1: m \to p$, $\xi_2: k \to p$, $\xi_3: l \to p$ *satisfy*

(2.8.6) $\xi_1 = a \cdot (I_p, \xi_1)$; *i.e.,* ξ_1 *is a solution of the iteration equation for* a;

(2.8.7) $\xi_2 = b \cdot \xi_2$; *when* $p = 0$, ξ_2 *is a solution of the iteration equation for* b;

(2.8.8) $\xi_3 = f \cdot (I_p, \xi_3)$; *with* f *as in* (2.8.5), ξ_3 *is a solution of the iteration equation for* f.

(c) *The solution of* (2.8.6) *is unique (and is denoted* $a^\dagger$; *in fact,* $a^\dagger = \bar{a}$). *For each choice of* ξ_1 *and* ξ_2, *the solution of* (2.8.8) *is unique (and is denoted* $f^\dagger$). *Further,* $f^\dagger = h^\dagger \cdot (I_p, \xi_1, \xi_2)$.

(d) *The solutions of the iteration equation for* $\bar{g}$ *are all expressible in the form*

$$(a^\dagger, \xi_2, \quad h^\dagger \cdot (I_p, a^\dagger, \xi_2))$$

where ξ_2 *satisfies* (2.8.7).

The proof of the Theorem 2.8 occupies §§2.9–2.14.

2.9. On (2.8.2). Let $i \in [m]$, so that i is a power successful position of $\bar{g}$. If i is not a successful position of $\bar{g}$, $\mathbf{i} \cdot \bar{g} = 0_p \oplus \mathbf{i}': 1 \to p + n$, for some $i' \in [m + k + l] = [n]$. By 2.3.12, i' must belong to $[m]$. The morphism $a: m \to p + m$ is thus the base morphism corresponding to the function $a: [m] \to [p + m]$ determined by the requirement that

(2.9.1) $ia = p + i'$ if i is not a successful position of $\bar{g}$, $\mathbf{i} \cdot \bar{g} = 0_p \oplus \mathbf{i}'$;

otherwise,

(2.9.2) $$ia = j \quad \text{if } \mathbf{i} \cdot \bar{g} = \mathbf{j} \oplus 0_n, \quad j \in [p].$$

Note that for $i \in [m]$,

(2.9.3) $i\bar{g}^\sigma = ia^\sigma = u_1 u_2 \cdots u_t$, where $u_1 = i$, and u_t is a successful position of $\bar{g}$, and for $1 \leqq r \leqq t$, $\mathbf{i} \cdot g^r = \mathbf{i} \cdot a^r = 0_p \oplus \mathbf{u}_{r+1}$.

Lastly, since all of the entries $u_1, \cdots, u_t$ are necessarily distinct elements of $[m]$, we infer $t \leqq m$; thus

(2.9.4) $$\mathbf{i} \cdot a^t = \mathbf{i} \cdot a^m = \mathbf{j} \oplus 0_m, \quad \text{for some } j \in [p].$$

If we define $\bar{a}: [m] \to [p]$ by

$$i\bar{a} = j \qquad \text{if (2.9.4) holds,}$$

we have $a^m = \bar{a} \cdot (I_p \oplus 0_m)$, so that a^m factors through p, as claimed.

2.10. On (2.8.3). Let $i \in m + [k]$, so that i is a singular position of $\bar{g}$. Thus if

(2.10.1) $$i \cdot \bar{g}^\sigma = u_1 u_2 \cdots$$

each item u_r is a singular position of $\bar{g}$, and hence in particular, $u_2 \in m + [k]$. We define the function $b: [k] \to [k]$ by the requirement that for $i \in m + [k]$

$$(i-m)b = i' - m$$

where $i' = u_2$ in (2.10.1). Then the morphism α, where

$$\alpha = a \oplus b \oplus 0_l : m + k \to p + m + k + l = p + n$$

is the morphism such that $\mathbf{i} \cdot \alpha = \mathbf{i} \cdot \bar{g}$, for all $i \in [m+k]$. Thus the first $m+k$ components of $\bar{g}$ are writable as $(a \cdot (I_{p+m} \oplus 0_{k+l}), b \cdot (0_{p+m} \oplus I_k \oplus 0_l))$ which equals $a \oplus b \oplus 0_l$.

2.11. On (2.8.4) and (2.8.5). The morphism $h: l \to p + m + k + l$ is defined by the requirement that its l-components are the last l-components of $\bar{g}$.

Thus, by 2.3.13, if $i \in [l]$, and if $(m+k+i)\bar{g}^\sigma = u_1 u_2 \cdots u_t$, then

$$ih^\sigma = v_1 v_2 \cdots v_t$$

where $m + k + v_r = u_r$, for $r \in [t]$. Also $\mathbf{u}_t \cdot \bar{g} = \mathbf{m+k+i} \cdot \bar{g}'$ is ideal; hence $\mathbf{v}_t \cdot h = \mathbf{i} \cdot h'$ is ideal. Since $t \leqq l$, $\mathbf{i} \cdot h'$ is ideal, for all $i \in [l]$. Thus h is power ideal, proving (2.8.4).

The assertion (2.8.5) follows immediately from the following more general fact (and from (3.1.2) of [5]).

2.11.1. Proposition. *Let $h: l \to q + l$ and $\beta: q \to s$ be morphisms in an algebraic theory. If $f = h \cdot (\beta \oplus I_l)$, then for all $r \geqq 1$, $f^r = h^r \cdot (\beta \oplus I_l)$.*

Proof. The proof is by induction on r. When $r = 1$, there is nothing to prove. Assume $f^r = h^r \cdot (\beta \oplus I_l)$. Then

$$\begin{aligned} f^{r+1} &= f^r \cdot (I_s \oplus 0_l, f) = f^r \cdot (I_s \oplus 0_l, h \cdot (\beta \oplus I_l)) \\ &= h^r \cdot (\beta \oplus I_l) \cdot (I_s \oplus 0_l, h \cdot (\beta \oplus I_l)). \\ &= h^r \cdot (\beta \oplus 0_l, h \cdot (\beta \oplus I_l)) \qquad \text{by (1.5.2), since } \beta \cdot (I_s \oplus 0_l) = \beta \oplus 0_l; \\ &= h^r \cdot (I_1 \oplus 0_l, h) \cdot (\beta \oplus I_l) \qquad \text{by (1.5.4)} \\ &= h^{r+1} \cdot (\beta \oplus I_l). \end{aligned}$$

The proof of (a) of the Theorem 2.8 is now complete.

2.12. Proof of (b). The iteration equation for $\bar{g}: n \to p + n$ is

$$\xi = \bar{g} \cdot (I_p, \xi), \quad \text{where } \xi: n \to p. \tag{2.12.1}$$

Using (2.8.1) and writing $\xi = (\xi_1, \xi_2, \xi_3)$ (as in the Theorem 2.8(b)) the equation (2.12.1) becomes

$$\begin{aligned} (\xi_1, \xi_2, \xi_3) &= (a \oplus b \oplus 0_l, h) \cdot (I_p, \xi_1, \xi_2, \xi_3) \\ &= (a \cdot (I_p, \xi), \quad b \cdot \xi_2, \quad h \cdot (I_p, \xi_1, \xi_2, \xi_3)) \end{aligned} \tag{2.12.2}$$

by (1.5.4), (1.5.2) and (1.5.3).

Thus the equation (2.12.2) is equivalent to the three equations (2.8.6), (2.8.7) and

$$\xi_3 = h \cdot (I_p, \xi_1, \xi_2, \xi_3). \tag{2.12.3}$$

This latter equation is the same as (2.8.8), since

$$h \cdot (I_p, \xi_1, \xi_2, \xi_3) = h \cdot [(I_p, \xi_1, \xi_2) \oplus I_l] \cdot (I_p, \xi_3) \tag{2.12.4}$$

by (1.5.2).

2.13. Proof of (c). Every solution of (2.8.6) is also a solution to

$$\xi_1 = a^m \cdot (I_p, \xi_1). \tag{2.13.1}$$

By (2.8.2), there is a base morphism $\bar{a}: m \to p$ such that $a^m = \bar{a} \cdot (I_p \oplus 0_m)$. Thus, if ξ_1 is a solution of (2.13.1),

$$\xi_1 = \bar{a} \cdot (I_p \oplus 0_m) \cdot (I_p, \xi_1) = \bar{a}.$$

Thus (2.13.1) has at most one solution, and (2.8.6) has at most one solution, viz., $\bar{a}$. But, by the definition of $\bar{a}$ in § 2.9,

$$a \cdot (I_p, \bar{a}) = \bar{a},$$

which proves (2.8.6) has at least one solution.

Since f is power ideal, by (2.8.5), there is a unique solution $f^{\dagger}$ to (2.8.8). The fact that $f^{\dagger} = h^{\dagger} \cdot (I_p, \xi_1, \xi_2)$ follows immediately from the following easily proved fact.

2.13.2. Proposition. *Let $h: l \to q + l$ be a power ideal morphism in an iterative theory I, and let $\beta: q \to s$ be an arbitrary morphism in I. Then $[h \cdot (\beta \oplus I_l)]^{\dagger} = h^{\dagger} \cdot \beta$.*

Proof of 2.13.2. By 2.11.1, $h \cdot (\beta \oplus I_l): l \to s + l$ is power ideal since h is, and thus there is a unique solution $[h \cdot (\beta + I_l)]^{\dagger}$ of its iteration equation. But

$$\begin{aligned} h \cdot (\beta \oplus I_l) \cdot (I_s, \quad h^{\dagger} \cdot \beta) &= h \cdot (\beta, h^{\dagger} \cdot \beta) && \text{by (1.5.2)} \\ &= h(I_q, h^{\dagger}) \cdot \beta && \text{by (1.5.4)} \\ &= h^{\dagger} \cdot \beta. \end{aligned}$$

Thus $h^{\dagger} \cdot \beta$ is one solution of the iteration equation for $h \cdot (\beta \oplus I_l)$.

2.14. On (d). The proof of (d) is immediate from parts (b) and (c).

This completes the proof of the Theorem 2.8.

A number of facts follow easily from the theorem and § 2.5.

2.15. Corollary. *Let $g: n \to p + n$ be an arbitrary morphism in an iterative theory. Let $\pi: [n] \to [n]$ be a permutation such that the π conjugate $\bar{g}$ of g is a mkl-morphism $(a \oplus n \oplus 0_l, h)$ as in § 2.8. Then every solution of the iteration equation for g is expressible as $\pi \cdot (a^{\dagger}, \xi_2, h^{\dagger} \cdot (I_p, a^{\dagger}, \xi_2))$ where $\xi_2: K \to p$ satisfies $\xi_2 = b \cdot \xi_2$.*

2.16. Corollary. *Let $\bar{g} = (a \oplus b \oplus 0_l, h)$ be a mkl-morphism in an iterative theory. The set of solutions of the iteration equation for $\bar{g}$ is in bijective correspondence with the set of solutions $\xi_2: k \to p$ of the equation*

(2.16.1) $$\xi_2 = b \cdot \xi_2.$$

The proof of the Corollary 2.16 is immediate from the Theorem 2.8(d). Note that in the case $k = 0$ (i.e., when there are no singular positions for $\bar{g}$) there is a unique morphism $0 \to p$; thus the iteration equation for $\bar{g}$ (and hence for all morphisms conjugate to $\bar{g}$) will have a unique solution, i.e., will have exactly one solution. A morphism $g: n \to p + n$ having no singular positions is called *nonsingular.* In this terminology, we have proved the next corollary.

2.17. Corollary. *If $g: n \to p + n$ is a nonsingular morphism in an iterative theory, the iteration equation for g has a unique solution (denoted $g^{\dagger}$).*

The converse of this corollary is almost always true. The statement of the corollary really tells the whole story. There are, however, some "exceptional" combinations of iterative theory I and target p of ξ_2 such that the equation (2.16.1) has a unique solution for any value of k. First we tabulate the cases which yield a multiplicity of solutions—or none—independent of k so long as k is positive. To this end let $\mathcal{N}$ be the only iterative theory which has no morphisms $1 \to 0$, i.e., the iterative theory all of whose morphisms are base morphisms.

2.18. PROPOSITION. *The equation* $\xi = b \cdot \xi$ *(where* $b: k \to k$ *is a base morphism in the iterative theory* I, *where* $\xi: k \to p$ *and where* $k > 0$*) has two or more solutions in* I *provided that*

(a) $\perp: 1 \to 0$, $\perp': 1 \to 0$ *are in* I *and* $\perp \neq \perp'$, *or*

(b) $I \neq \mathcal{N}$ *and* $p \geqq 1$, *or*

(c) $I = \mathcal{N}$ *and* $p \geqq 2$.

The equation has no solution when

(d) $I = \mathcal{N}$ *and* $p = 0$.

Proof. Note that the equation $\xi = b \cdot \xi$ requires only that certain components of ξ be equal to certain other components of ξ. Hence, if each of the k components of ξ are equal, the equation in question is satisfied. The solutions we will identify will have all their components equal so we need specify only what the components are.

Case (a). $\perp \cdot 0_p$, $\perp' \cdot 0_p$,

Case (b). $\perp \cdot 0_p$, $I_1 \oplus 0_{p-1}$,

Case (c). $I_1 \oplus 0_{p-1}$, $0_1 \oplus I_1 \oplus 0_{p-2}$.

The final case is obvious since there are no morphisms with target 0.

The following is essentially Theorem B of [8].

2.19. COROLLARY. *Let* $g: n \to p+n$ *be a morphism in an iterative theory* I. *A necessary and sufficient condition that the iteration equation for* g *have a unique (i.e., one and only one) solution is that*

(2.19.1) $$g \text{ be nonsingular, i.e., } K_g = \varnothing,$$

or

(2.19.2) $$I \text{ has exactly one morphism } 1 \to 0 \text{ and } p = 0,$$

or

(2.19.3) $$I = \mathcal{N} \quad \text{and} \quad p = 1.$$

Proof. See §§ 2.18, 2.16, and 2.5.

2.20. COROLLARY. *Let* I *be an iterative theory having at least two distinct morphisms* $1 \to 0$. *Then the iteration equation for a morphism* $g: n \to n$ *has a unique solution iff* g *is power ideal.*

Proof. If g is power ideal, g is nonsingular, so that by the corollary 2.17 the iteration equation for g has a unique solution. Conversely, by the corollary 2.19, if the iteration equation for g has a unique solution, g is nonsingular. But a nonsingular morphism $g: n \to n$ is clearly power ideal. This completes the proof.

The following is a special case of the Corollary 2.19.

2.21. In Tr, a necessary and sufficient condition that the iteration equation for g have a unique solution is that g be nonsingular.

3. The category Pit of $I_1^\dagger$-iterative theories. The goal of this section is to show that given any morphism $\perp: 1 \to 0$ in an iterative theory, by requiring only that $(0_p \oplus I_1)^\dagger = 0_p \oplus \perp$ (for each $p \geqq 0$) the scalar iteration operation may be consistently extended to all scalar morphisms. This fact follows from 3.7, Universality Theorem.

3.1. PROPOSITION. *Let* $F: J_1 \to J_2$ *be an ideal theory-morphism between ideal theories* J_1, J_2 *(i.e., if* f *is an ideal morphism in* J_1 *then* fF *is an ideal morphism in* J_2*). If* $g: n \to p+n$ *is a nonsingular morphism in* J_1 *then* gF *is a nonsingular morphism in* J_2.

Proof. The proof is immediate from the fact (cf. 2.3.13) that g is nonsingular iff for each $i \in [n]$, $i \cdot g^n$ is ideal or successful. □

In particular, if each J_i is iterative and g is nonsingular then the iteration equations for g and gF have unique solutions, $g^\dagger$ and $(gF)^\dagger$, respectively. Moreover, $g^\dagger F = (gF)^\dagger$.

This nice behavior of F contrasts sharply with the case that the theory-morphism F is not ideal. This case is taken up in the Lemma 3.2. For that discussion we recall that each morphism $f: 1 \to p$ (in any algebraic theory) is a solution to the iteration equation for the base morphism $0_p \oplus I_1: 1 \to p+1$ in that theory. This base morphism corresponds to the function $[1] \to [p+1]$ whose value is $p+1$.

3.2. LEMMA. *If $F: J_1 \to J_2$ is a theory-morphism between ideal theories which is* not *ideal, then there exists a biscalar ideal morphism h in J_1 such that $hF = I_1$.* ("Biscalar" means "$1 \to 1$".)

Proof. By hypothesis there exists a scalar ideal morphism $f: 1 \to p$ in J_1 such that fF is base (so that $p > 0$). Let $h = f \cdot \mu$, where $\mu: p \to 1$ is base. Then h is ideal and $hF = (f \cdot \mu)F = fF \cdot \mu F = fF \cdot \mu = I_1$.

PROPOSITION. *Let $F: J_1 \to J_2$ be a theory-morphism between iterative theories J_1, J_2 which is not ideal. The requirement* R1 *on extending the iteration operation in J_2 implies the requirement* R2.

R1. *For all scalar ideal f in J_1, if $fF = 0_p \oplus I_1$, then $(fF)^\dagger$ is defined and $(fF)^\dagger = f^\dagger F$.*

R2. *For all p, $(0_p \oplus I_1)^\dagger$ is defined in J_2 and $(0_p \oplus I_1)^\dagger = 0_p \oplus I_1^\dagger$.*

Proof. According to the lemma, there exists an ideal $h: 1 \to 1$ in J_1 such that $hF = I_1$. According to R1, we must define $I_1^\dagger = h^\dagger F$ (where, of course, $h^\dagger$ is the unique solution to the iteration equation for h). For each $p > 0$, let $g_p = h \cdot (0_p \oplus I_1) = 0_p \oplus h$. Then, $g_p: 1 \to p+1$ is ideal and $g_p F = 0_p \oplus I_1$. According to R1, we must define $(0_p \oplus I_1)^\dagger = g_p^\dagger F$. Now, as may readily be shown, $g_p^\dagger = 0_p \oplus h^\dagger$. Hence

$$(0_p \oplus I_1)^\dagger = g_p^\dagger F = (0_p \oplus h^\dagger)F = 0_p \oplus h^\dagger F = 0_p \oplus I_1^\dagger.$$

Note that when $p = 0$, R2 asserts only that $I_1^\dagger$ is defined.

3.3. DEFINITION. By an *$I_1^\dagger$-iterative theory,* we mean a *pointed iterative theory* (i.e., a pair $(J, \perp)$ where $\perp: 1 \to 0$ is a morphism in the iterative theory J) *in which we complete the scalar iteration operation by defining for all p, $(0_p \oplus I_1)^\dagger = 0_p \oplus \perp$.* (We note that in an iterative theory the only scalar morphism $1 \to p+1$ whose iteration equation does not have a unique solution is $0_p \oplus I_1$.) *Thus any pointed iterative theory may be regarded as an $I_1^\dagger$-iterative theory.* If it is necessary or desirable to be very explicit, the $I_1^\dagger$-iterative theory $(J, \perp)$ may be described by: $(J, I_1^\dagger = \perp)$.

Before introducing the category of $I_1^\dagger$-iterative theories we note the following.

3.4. PROPOSITION. *Let $F: (J_1, \perp) \to (J_2, \perp)$ be a theory-morphism $F: J_1 \to J_2$ between iterative theories such that $\perp F = \perp$. Then the conditions* (3.4.1) *and* (3.4.2) *on F are equivalent. In particular, if F preserves idealism, then* (3.4.1) *is satisfied and so* (3.4.2) *is too.*

(3.4.1) *For all scalar ideal morphisms $f: 1 \to p+1$, $fF = 0_p \oplus I_1 \Rightarrow f^\dagger F = 0_p \oplus \perp$.*

(3.4.2) *Regarding $(J_i, \perp)$ as an $I_1^\dagger$-iterative theory, $i \in [2]$, F preserves scalar iteration.*

Proof. The proof is obvious.

Notice that (3.4.1) does not require defining $(0_p \oplus I_1)^\dagger$ while (3.4.2) does.

3.5. Definition. The category Pit[3] of $I_1^\dagger$-iterative theories has $I_1^\dagger$-iterative theories as objects. A morphism F: $(J_1, \perp) \to (J_2, \perp)$ in this category is a theory morphism such that $\perp F = \perp$ and such that (3.4.1) holds (or, equivalently, (3.4.2) holds).

Remark. The proof that morphisms between $I_1^\dagger$-iterative theories are closed under composition is a bit more obvious using (3.4.2) rather than (3.4.1).

3.6. We employ the following conventions. Γ_Δ is obtained from Γ (see §1.6) by adjoining a "new" element Δ to Γ_1; Δ: $1 \to 1$ is an atomic tree in $\Gamma_\Delta$tr.

The following is fundamental. Recall from [7] that Γtr is the iterative theory freely generated by Γ.

Δ-Lemma. *For any tree* $\perp$: $1 \to 0$ *in* Γtr *there is a unique theory morphism* Φ: $\Gamma_\Delta$tr $\to \Gamma$tr *satisfying*

(a) $\Delta\Phi = I_1$,

(b) $\Delta^\dagger\Phi = \perp$,

(c) $\gamma\Phi = \gamma$, *all* γ *in* Γ,

(d) *if* $g\Phi$ *is base, say* $g\Phi = \mathbf{j}$: $1 \to p+1$, *then* $g = \Delta^r \cdot \mathbf{j}$ *for some* $r \geqq 0$.

The proof of the Δ-lemma is given in §4.

Remark. If we distinguish $\Delta^\dagger$: $1 \to 0$ in Γ_Δ tr and distinguish $\perp$ in Γ tr and treat the two theories as $I_1^\dagger$-iterative theories, then, as may readily be seen from (d), Φ: (Γ_Δ tr, $I_1^\dagger = \Delta^\dagger$) $\to$ (Γ tr, $I_1^\dagger = \perp$) becomes a morphism in the category Pit.

3.7. Universality Theorem. *Let* $\square \in \Gamma_0$ *and let* J *be an iterative theory. Given an arbitrary function* $\mathbf{F}$: $\Gamma \to J$ *taking* $\gamma \in \Gamma_n$ *into* $\gamma\mathbf{F}$: $1 \to n$ *such that* $\perp = \square F$, *there is a unique morphism*

$$(\Gamma\ \mathrm{tr}, I_1^\dagger = \square) \xrightarrow{\mathbf{F}'} (J, I_1^\dagger = \perp) \quad \textit{in}\ \mathrm{Pit}$$

which makes the following diagram commute:

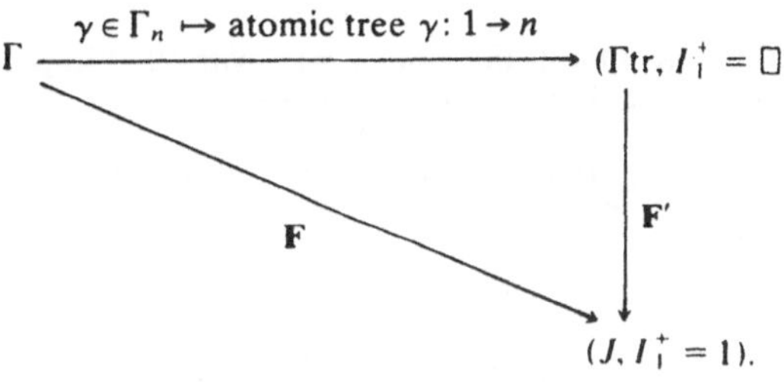

Proof. Let $\Gamma' = \{\gamma \text{ in } \Gamma | \gamma\mathbf{F} \text{ is ideal}\}$ so that $\square \in \Gamma'_0$. Let $\mathbf{U}$: $\Gamma \to \Gamma'_\Delta$ tr be defined as follows:

$$\gamma\mathbf{U} = \Delta \cdot \mathbf{i}, \quad \text{if } \gamma\mathbf{F} = \mathbf{i}\text{: } 1 \to n \text{ is base;}$$

$$\square\mathbf{U} = \Delta^\dagger;$$

$$\gamma\mathbf{U} = \gamma \quad \text{otherwise.}$$

Notice that, for all γ, $\gamma\mathbf{U}$ is an ideal morphism. Then there is a unique theory morphism

[3] "Pit" is suggested by "pointed iterative theory". As noted in definition 3.3 any pointed iterative theory "may be regarded" as an $I_1^\dagger$-iterative theory. "Pit", as used here, may be regarded as an abbreviation for "Pit($I_1^\dagger$)".

(since Γ tr is, as an iterative theory, freely generated by Γ; cf. [7])

$$\mathbf{U}' \colon \Gamma\,\mathrm{tr} \to \Gamma'_\Delta\,\mathrm{tr}$$

which extends **U**—or more exactly—makes the following diagram commute:

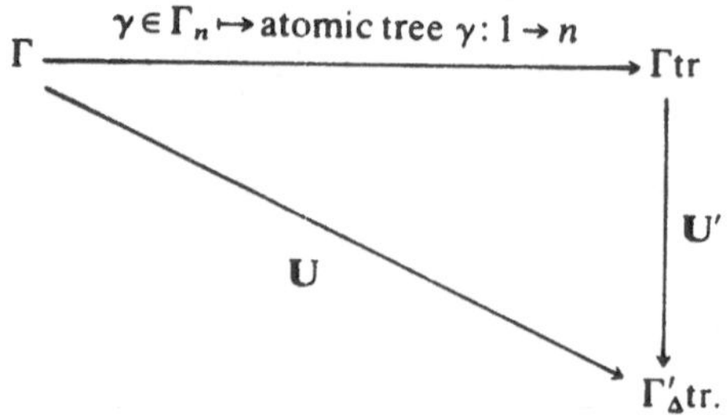

Notice that, by proposition 3.4,

$$\text{(3.7.1)} \qquad \mathbf{U}' \colon (\Gamma\,\mathrm{tr},, I_1^\dagger = \Box) \to (\Gamma'_\Delta\,\mathrm{tr}, I_1^\dagger = \Delta^\dagger)$$

is a Pit morphism since $\Box\mathbf{U}' = \Delta^\dagger$ and U' preserves idealism.

Similarly, if we define $\mathbf{E}\colon \Gamma' \to J$ by:

$$\gamma\mathbf{E} = \gamma\mathbf{F} \text{ for } \gamma \text{ in } \Gamma'; \text{ in particular, } \Box\mathbf{E} = \bot;$$

we obtain a theory morphism $\mathbf{E}'\colon \Gamma'\,\mathrm{tr} \to J$ which "extends" **E**. Again, we notice

$$\text{(3.7.2)} \qquad \mathbf{E}' \colon (\Gamma'\,\mathrm{tr}, I_1^\dagger = \Box) \to (J, I_1^\dagger = \bot)$$

is a Pit morphism.

Applying the Δ-lemma in the case $\bot = \Box$, with "Γ'" in place of "Γ", we readily conclude (as already noted in the Remark of 3.6

$$\text{(3.7.3)} \qquad \Phi \colon (\Gamma'_\Delta\,\mathrm{tr}, I_1^\dagger = \Delta^\dagger) \to (\Gamma'\,\mathrm{tr}, I_1^\dagger = \Box)$$

is a Pit morphism.

We now define

$$\text{(3.7.4)} \qquad \mathbf{F}' \colon \Gamma\,\mathrm{tr} \xrightarrow{\mathbf{U}'} \Gamma'_\Delta\,\mathrm{t} \xrightarrow{\Phi} \Gamma'\,\mathrm{tr} \xrightarrow{\mathbf{E}'} J.$$

One readily checks that $\mathbf{F}'$ "extends" **F** and notes that $\mathbf{F}'$ is a Pit morphism since $\mathbf{U}', \Phi, \mathbf{E}'$ all are, i.e.,

$$\mathbf{F}' \colon (\Gamma\,\mathrm{tr}, I_1^\dagger = \Box) \to (J, I_1^\dagger = \bot)$$

is a Pit morphism.

The uniqueness of $\mathbf{F}'$ follows from the fact that Γ tr is the smallest subtheory of Γ Tr which contains the atomic trees Γ and is closed under scalar iteration of ideal (scalar) morphisms.

3.8. Significance of the Universality Theorem. The Universality Theorem implies that all "laws" of iterative theories remain true when the meaning of $g^\dagger$ is extended to all scalar g. Indeed, if "E_1" and "E_2" are iterative theory expressions involving only the scalar† and if the assertion

(3.8.1) for all scalar ideal $f, g, \cdots (E_1 = E_2)$

is valid in all iterative theories then the assertion

(3.8.2) for all scalar $f, g, \cdots (E_1 = E_2)$

is valid in all $I_1^\dagger$-iterative theories.

(By a (scalar) *iterative theory expression*, we mean an expression constructed from letters $f, g, \cdots$ to denote arbitrary scalar ideal morphisms, symbols to denote base scalar morphisms, symbols for the algebraic theory operations, and a symbol $(^\dagger)$ for scalar iteration of ideal morphisms. For convenience one may include "0_p", "I_p", etc.).

Thus, while the extended scalar iteration operation was uniquely determined by one iterative theory identity (viz., $[0_p \oplus g]^\dagger = 0_p \oplus g^\dagger$) once $I_1^\dagger$ was assigned a meaning, it turns out that the extended operation satisfies *all* scalar iterative theory identities (see [1]).

4. Proof of the "Δ-lemma". We will use the somewhat more suggestive notation Δ^∞ for the tree $\Delta^\dagger$: $1 \to 0$ in Γ_Δ tr, and we will say a tree g in Γ_Δ tr "has no Δ^∞-subtrees" if there is no vertex v of g such that the tree of descendants of v (see [7, p. 9]) is isomorphic to Δ^∞.

Before defining the function Φ we define a function $T \to \Gamma$ tr whose domain T is the set (actually, subtheory) of trees in Γ_Δ tr having no Δ^∞-subtrees.

4.1. Definition. If $g: n \to p$ is a tree in T, then $g_0: n \to p$ is the tree in Γ tr obtained from g by replacing every occurrence of $\Delta: 1 \to 1$ in g by $I_1: 1 \to 1$.

For example, $(\Delta^2 \cdot b)_0 = b$, for any base $b: 1 \to p$.

We list the following obvious properties of the function $g \mapsto g_0$.

4.2. Proposition. (a) *Suppose both $g: n \to p$ and $h: p \to q$ are in T. Then so is $g \cdot h$, and $(g \cdot h)_0 = g_0 \cdot h_0$.*

(b) *Suppose both $g: n \to p$ and $h: m \to p$ are in T. Then so is $(g, h): n + m \to p$ and $(g, h)_0 = (g_0, h_0)$; similarly if $g_i: n_i \to p_i$, $i = 1, 2$, are in T, then so is $g_1 \oplus g_2$ and $(g_1 \oplus g_2)_0 = (g_1)_0 \oplus (g_2)_0$.*

(c) *For any base morphism $b: n \to p$, $b_0 = b$; in particular, $(0_p)_0 = 0_p$, where 0_p is the unique tree $0 \to p$.*

Remark. As a consequence of this proposition, T is a subtheory of Γ_Δ tr and the function

$$T \xrightarrow{g \mapsto g_0} \Gamma \text{ tr}$$

is a theory morphism.

Proofs of all of these statements are straightforward if one first represents a Γ_Δ-tree by a "surmatrix" (as in [7]) and uses the fact that the substitution of I_1 for Δ is a monoid homomorphism on the sets of words which occur as entries in the surmatrices.

Let $g: n \to p$ be any Γ_Δ-tree. We may specify those vertices v such that the tree of descendants of v is isomorphic to Δ^∞ in the following way.

4.3. Proposition. *Let $g: n \to p$ be any Γ_Δ-tree. There is a tree $\hat{g}: n \to p + 1$ such that g is the composition*

$$g: n \xrightarrow{\hat{g}} p + 1 \xrightarrow{I_p \oplus \Delta^\infty} p$$

and such that $\hat{g}$ has no Δ^∞-subtrees. Further, if g is ideal, $\hat{g}$ may be chosen to be ideal.

4.4. *Remark.* Note that there may be several ways of choosing the tree $\hat{g}$. However if

$$\hat{g}_1 \cdot (I_p \oplus \Delta^\infty) = \hat{g}_2 \cdot (I_p \oplus \Delta^\infty)$$

and if $\hat{g}_1$ and $\hat{g}_2$ have no Δ^∞-subtrees, then $(\hat{g}_1)_0 = (\hat{g}_2)_0$.

We now are in a position to define the map Φ.

4.5. *Definition of* Φ: Γ_Δ tr $\to \Gamma$ tr. Let $g: n \to p$ be an arbitrary tree in Γ_Δ tr. Write g as $\hat{g} \cdot (I_p \oplus \Delta^\infty)$, using Proposition 4.3. We define

$$g\Phi = (\hat{g})_0 \cdot (I_p \oplus \perp).$$

Note that $g\Phi$ is well-defined by remark 4.4. Note also that if g has no Δ^∞-subtrees, $g\Phi = g_0$, so that in particular

$$\Delta\Phi = I_1;$$

$$\gamma\Phi = \gamma, \qquad \gamma \in \Gamma_n, \quad n \geqq 0;$$

also $\Delta^\infty\Phi = \perp$. Thus to prove the Δ-lemma we must show both that Φ is a theory morphism and that (d) (**3.6**) holds.

Before beginning this task, we observe that $g\Phi$ is described in geometric terms as the tree obtained from g by first replacing all Δ^∞-subtrees by $\perp$ and then replacing all remaining Δ's by I_1.

4.6. Proposition. *Φ preserves composition.*

Proof. Suppose $g: n \to p$ and $h: p \to q$ are trees in Γ_Δ tr. Writing g and h as in 4.3 as

(4.6.1)
$$g: n \xrightarrow{\hat{g}} p+1 \xrightarrow{I_p \oplus \Delta^\infty} p, \qquad h: p \xrightarrow{\hat{h}} q+1 \xrightarrow{I_q \oplus \Delta^\infty} q$$

we have

$$\begin{aligned} g \cdot h &= \hat{g} \cdot (I_p \oplus \Delta^\infty) \cdot \hat{h} \cdot (I_q \oplus \Delta^\infty) && \text{by (4.7.1)} \\ &= \hat{g} \cdot (\hat{h} \cdot (I_q \oplus \Delta^\infty) \oplus \Delta^\infty) && \text{by (1.5.1) when } g_2: 0 \to 0 \\ &= \hat{g} \cdot (\hat{h} \oplus I_1) \cdot (I_q \oplus (I_1, I_1)) \cdot (I_q \oplus \Delta^\infty) && \text{by two applications of (1.5.1).} \end{aligned}$$

Using the fact that $\hat{g} \cdot (\hat{h} \oplus I_1)(I_q \oplus (I_1, I_1))$ has no Δ^∞-subtrees, by 4.2.,

$$\begin{aligned} (g \cdot h)\Phi &= \hat{g}_0 \cdot (\hat{h}_0 \oplus I_1) \cdot (I_q \oplus (I_1, I_1)) \cdot (I_q \oplus \perp) \\ &= \hat{g}_0 \cdot (I_p \oplus \perp) \cdot \hat{h}_0 \cdot (I_q \oplus \perp) \\ &= g\Phi \cdot h\Phi. \end{aligned}$$

□

4.7 Proposition. *Φ preserves source tupling.*

Proof. Let $g: n \to p$ and $h: m \to p$ be trees in Γ_Δ tr. Using 4.3, we write

$$g: n \xrightarrow{\hat{g}} p+1 \xrightarrow{I_p \oplus \Delta^\infty} p \quad \text{and} \quad h: m \xrightarrow{\hat{h}} p+1 \xrightarrow{I_p \oplus \Delta^\infty} p.$$

Hence

$$(g, h) = (\hat{g} \cdot (I_p \oplus \Delta^\infty), \hat{h} \cdot (I_p \oplus \Delta^\infty)) = (\hat{g}, \hat{h}) \cdot (I_p \oplus \Delta^\infty)$$

and $(\hat{g}, \hat{h})$ has no Δ^∞-subtrees. Hence, by 4.2,

$$
\begin{aligned}
(g, h)\Phi &= (\hat{g}_0, \hat{h}_0) \cdot (I_p \oplus \perp) \\
&= (\hat{g}_0 \cdot (I_p \oplus \perp), \hat{h}_0 \cdot (I_p \oplus \perp)) \\
&= (g\Phi, h\Phi).
\end{aligned}
$$

4.8. PROPOSITION. *If* $b: n \to p$ *is base,* $b\Phi = b$.

Proof. Indeed, since b has no Δ^∞-subtrees, $b\Phi = b_0 = b$, by 4.2.

4.9. COROLLARY. $\Phi: \Gamma_\Delta \operatorname{tr} \to \Gamma \operatorname{tr}$ *is a theory morphism.*

Proof. Use propositions 4.6, 4.7 and 4.8.

To prove the Δ-lemma, it only remains to show that (d) holds. Suppose: $g: 1 \to p+1$ is ideal while $g\Phi$ is base and $g = \gamma \cdot g'$ where γ is atomic. Since $(\gamma \cdot g')\Phi = \gamma\Phi \cdot g'\Phi$ is base we must have $\gamma = \Delta$. Similarly, $g' = \Delta \cdot g''$, $g'' = \Delta \cdot g'''$, etc. This process must terminate, for otherwise $g\Phi = \perp \cdot 0_{p+1}$ which is ideal.

REFERENCES

[1] S. L. BLOOM, C. C. ELGOT AND J. B. WRIGHT, *Vector iteration in pointed iterative theories*, IBM Res. Rep. RC7322, IBM T. J. Watson Research Center, Yorktown Heights, NY, Sept. 1978.

[2] J. B. WRIGHT, J. W. THATCHER, E. G. WAGNER AND J. A. GOGUEN, *Rational algebraic theories and fixed point solutions*, IEEE Symp. Foundations of Comp. Sci., Houston, TX, October 1976.

[3] S. L. BLOOM AND C. C. ELGOT, The existence and construction of free iterative theories, J. Comput. System Sci., 12 (1976), no. 3.

[4] S. L. BLOOM, S. GINALI AND J. RUTLEDGE, *Scalar and vector iteration*, Ibid., 14 (1977), pp. 251–256.

[5] C. C. ELGOT, *Monadic computation and iterative algebraic theories*, Logic Colloquium '73, 80, Studies in Logic, North-Holland, Amsterdam, 1975.

[6] ———, *Structured programming with and without GO-TO statements*, IEEE Trans. Software Engrg., SE-2 (1976), pp. 41–54; Erratum and Corrigendum, Ibid., SE-2 (1976), p. 232.

[7] C. C. ELGOT, S. L. BLOOM AND R. TINDELL, *The algebraic structure of rooted trees*, IBM Res. Rep. RC-6230, IBM T. J. Watson Research Center, Yorktown Heights, NY, 1976; extended abstract in Proc. Johns Hopkins 1977 Conf. Inf. Sci. and Systems; J. Comput. System. Sci., 16 (1978) No. 3, pp. 362–399.

[8] J. TIURYN, *On the domain of iteration in iterative algebraic theories*, Proc. Math. Found. in Comput. Sci. 1976, Gdansk, Lecture Notes in Comput. Sci. No. 45, Springer-Verlag, New York, 1976.

[9] J. MYCIELSKI AND W. TAYLOR, *A compactification of the algebra of terms*, Algebra Universalis, 6 (1976), pp. 159–163.

[10] D. SCOTT, *The lattice of flow diagrams*, Semantics of Algorithmic Languages, Lecture Notes in Math. 182, Springer-Verlag, New York, 1971, pp. 311–366.

[11] S. GINALI, *Iterative algebraic theories, infinite trees, and program schemata*, Dissertation, University of Chicago, June 1976.

[12] M. WAND, *A concrete approach to abstract recursive definitions*, Automata Languages and Programming, M. Nivat, ed., 1972, North-Holland pp. 331–341.

[13] W. LAWVERE, *Functional semantics of algebraic theories*, Proc. Nat. Acad. Sci. U.S.A., 50 (1963), pp. 869–872.

[14] S. EILENBERG AND J. B. WRIGHT, *Automata in general algebras*, Information and Control, 11 (1967), pp. 52–70.

SIAM J. COMPUT.
Vol. 9, No. 3, August 1980

0097-5397/80/0903-0006 $01.00/0

VECTOR ITERATION IN POINTED ITERATIVE THEORIES*

STEPHEN L. BLOOM†, CALVIN C. ELGOT‡ AND JESSE B. WRIGHT‡

Abstract. This paper is a sequel to a previous paper (S. L. Bloom, C. C. Elgot and J. B. Wright, Solutions of the iteration equation and extensions of the scalar iteration operations, SIAM J. Comput., 9 (1980), pp. 25–45. In that paper it was proved that for each morphism $\perp : 1 \to 0$ in an iterative theory J there is exactly one extension of the scalar iteration operation in J to all scalar morphisms such that $I_1^\dagger = \perp$ and all scalar iterative identities remain valid. In this paper the scalar iteration operation in the pointed iterative theory $(J, \perp)$ is extended to vector morphisms while preserving all the old identities.

The main result shows that the vector iterate $g^\dagger$ in $(J, \perp)$ satisfies the equation $g^\dagger = (g_\perp)^\dagger$, where $g_\perp$ is a nonsingular morphism simply related to g (so that $(g_\perp)^\dagger$ is the unique solution of the iteration equation for $g_\perp$).

In the case that $J = \Gamma\mathrm{Tr}$, the iterative theory of Γ-trees, it is shown that the vector iterate $g^\dagger$ in $(J, \perp)$ is a metric limit of "modified powers" of g.

Key words. Algebraic iterative theory, computation semantics iteration, flowcharts

Introduction. This paper is a sequel to [3]. In that paper it was shown that each morphism $\perp : 1 \to 0$ in an iterative theory J determines uniquely an extension of the scalar iterative operation in J to all scalar morphisms in such a way that $I_1^\dagger = \perp$ and all scalar iterative identities remain valid. In the current paper the scalar iteration operation is extended to all vector morphisms while preserving all old identities.

In § 2, this extension is defined in an inductive way. It is shown (Theorem 2.4) that four identities characterize this extension. Our main result (Theorem 2.7) provides an explicit description of extended iteration in any pointed iterative theory $(J, \perp)$. The special case of the iterative theory of labeled trees $(\Gamma\mathrm{Tr}, \perp)$ [6], [1] is considered in § 5. The Γ-trees $n \to p$ form a complete metric space in a natural way and we show that the vector iterate $g^\dagger$ of the Γ-tree $g : n \to p + n$ is a metric limit of the sequence $g \circ (I_p \oplus \perp_n), g^2 \circ (I_p \oplus \perp_n), g^3 \circ (I_p \oplus \perp_n) \cdots$.

Section 3 is devoted to showing that extended iteration satisfies all iterative theory identities. To this end the category of "iteration theories" is introduced and these theories are characterized in five ways (Theorem 3.3).

In section 4, some examples are given relating extended iteration to ordered algebraic theories [1]. Two examples are given of a choice for $\perp : 1 \to 0$ in the theory of Γ-trees so that in the resulting iteration theory $(\Gamma\mathrm{Tr}, \perp)$ there is some partial ordering of the trees such that the value of extended iteration $g^\dagger$ is the least solution of the iteration equation for g. For other choices of $\perp : 1 \to 0$ in $\Gamma\mathrm{Tr}$, no such ordering need exist (see Theorem 4.4).

1. Preliminaries. While familiarity with [3] would be very helpful, we will not need to depend on it heavily except in the appendix. However the reader should have some knowledge of algebraic and iterative theories as defined in [5]: the elementary properties of these theories are summarized in § 1.6 of [3].

For the reader's convenience we will recall here some facts, definitions and notation from [3]. An *algebraic theory* J is a category whose objects are the nonnegative integers, having for each $n \geqq 0$, n "distinguished morphisms" $i : 1 \to n$, $i \in [n]$ (where $[n] = \{1, 2, \cdots, n\}$; $[0] = \phi$) such that:

* Received by the editors November 28, 1978, and in final revised form June 30, 1979.

† Department of Pure and Applied Mathematics, Stevens Institute of Technology, Hoboken, New Jersey 07030. During the preparation of this paper this author was partially supported by the National Science Foundation under Grant MGS-78-00882.

‡ Mathematical Sciences Department, IBM T. J. Watson Research Center, Yorktown Heights, New York 10598.

for each family $f_i: 1 \to p$, $i \in [n]$ $n \geqq 0$, of morphisms in J there is a unique morphism $f: n \to p$ such that for each $i \in [n]$, f_i is the composition of $i: 1 \to n$ and $f: n \to p$.

In the case $n = 0$, this condition amounts to the requirement that there is a unique morphism $O_p: 0 \to p$, for each $p \geqq 0$. The morphism f is the *source tupling* of the morphisms f_i and is denoted $(f_1, \cdots, f_n)$. All morphisms $n \to p$, $n, p \geqq 0$ formed by source tupling the distinguished morphisms are *base*. When the distinguished morphisms are distinct (which is the case in every algebraic theory but the two "trivial theories") the base morphisms are in bijective correspondence with the functions $[n] \to [p]$: the function $f: [n] \to [p]$ corresponds to the source tupling $(f(1), \cdots, f(n))$, where $f(i): 1 \to p$ is distinguished. The base morphism corresponding to the identity function on $[n]$ is denoted I_n.

The *composition* of $f: n \to p$ and $g: p \to q$ is denoted either $f \circ g: n \to q$ or $n \overset{f}{-} p \overset{g}{\to} q$ (with an arrowhead missing). Source tupling permits defining *source pairing* of two morphisms with a common target. If $f_i: n_i \to p$, $i = 1, 2$, then the *source pairing* $(f_1, f_2): n_1 + n_2 \to p$ satisfies $i \circ (f_1, f_2) = i \circ f_1$, if $i \in [n]$; $i \circ (f_1, f_2) = j \circ f_2$ if $i = n_1 + j$, $j \in [n_2]$. The "circle-sum" $f_1 \oplus f_2: n_1 + n_2 \to p_1 + p_2$ of $f_i: n_i \to p_i$, $i \in [2]$, is the source pairing $(f_1 \circ \kappa, f_2 \circ \lambda)$, where $\kappa: p_1 \to p_1 + p_2$ and $\lambda: p_2 \to p_1 + p_2$ are base morphisms corresponding to the inclusion and translated inclusion functions. (See [3, § 1.6] for a list of some elementary properties of these operations.)

A morphism $g: 1 \to p$ in an algebraic theory J is *ideal* if for any $h: p \to q$, $g \circ h: 1 \to q$ is not distinguished; J itself is *ideal* if every nondistinguished morphism $1 \to p$ in J is ideal.

An *iterative theory* is an ideal theory J such that for any ideal morphism $g: 1 \to p + 1$, the *iteration equation* for g (in the variable $\xi: 1 \to p$)

$$\xi = g \circ (I_p, \xi) \tag{1.1}$$

has a unique solution. In [2] it was shown that in an iterative theory, when $n > 1$ and the morphism $g: n \to p + n$ has the property that $i \circ g$ is ideal for each $i \in [n]$ (we say "g is ideal") then the iteration equation (1.1) for g (now in the variable $\xi: n \to p$) also has a unique solution, denoted $g^\dagger$. Thus

$$g^\dagger = g \circ (I_p, g^\dagger).$$

The *powers* g^k of a morphism $g: n \to p + n$ in an algebraic theory J are defined inductively:

$$g^0 = O_p \oplus I_n: n \to p + n,$$
$$g^{k+1} = g^k \circ (I_p \oplus O_n, g).$$

Two useful facts about g^k are:

$$(I_p \oplus O_n, g)^k = (I_p \oplus O_n, g^k), \text{ all } k \geqq 0;$$
$$\text{if } \quad \xi: n \to p \text{ satisfies } \xi = g \circ (I_p, \xi),$$
$$\text{then } \quad \xi = g^k \circ (I_p, \xi), \text{ all } k \geqq 0.$$

Let $g: n \to p + n$ be a morphism in an algebraic theory. The morphism g is *power ideal* if for some $k \geqq 1$, g^k is ideal; g is *power successful* if g is base and for some $k \geqq 1$, g^k may be written as $n \overset{a}{-} p \xrightarrow{I_p \oplus O_n} p + n$, for some base morphism $a: n \to p$. A number $i \in [n]$ is a *singular position* of g if for each $k \geqq 1$ there is some number u_{k+1} in $[n]$ such that $i \circ g = O_p \oplus u_{k+1}$. The set of all singular positions of g is denoted K_g. If K_g is empty,

g is a *nonsingular* morphism. In [3, **2.17**] it was shown that in an iterative theory, the iteration equation for a nonsingular morphism g has a unique solution.

A morphism $g: n \to p + n$ in an algebraic theory J is a "*mkl*-morphism" [3, **2.7**] if $m + k + l = n$ and there are base morphisms $a: m \to p + m$, $b: k \to k$ and a power ideal morphism $h: l \to p + n$ such that

$$g = (a \oplus b \oplus O_l, h), \quad \text{and} \quad a^m = a^\dagger \oplus O_m, \quad \text{where } a^\dagger: m \to p$$

and in fact $a^\dagger$ is the unique solution of the iteration equation for a. The properties of nonsingular, ideal, power ideal and power successful morphisms are discussed in detail in [3].

In §§ 3–5, Γ denotes a ranked alphabet; i.e. Γ is the union $\bigcup_{n=0}^{x} \Gamma_n$ of the pairwise disjoint sets Γ_n. The iterative theory of all Γ-trees, mentioned in the introduction, was studied in [6]. (Algebraic theories of trees were studied also in [1].) The subtheory Γtr of ΓTr consists of those Γ-trees having (up to isomorphism) a finite number of descendancy trees. In [6] it was shown that Γtr is the iterative theory freely generated by Γ: i.e. for any iterative theory J and any function $F: \Gamma \to J$ taking $\gamma \in \Gamma_n$ to an ideal morphism $\gamma F: 1 \to n$ in J there is a unique "theory morphism" $\mathbf{F}': \Gamma$tr $\to J$ such that the diagram

$$\begin{array}{ccc} \Gamma & \xrightarrow{\iota} & \Gamma\text{tr} \\ & \searrow_{F} & \downarrow_{\mathbf{F}'} \\ & & J \end{array}$$

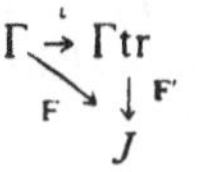

commutes, where if $\gamma \in \Gamma_n$, $\gamma\iota$ is the tree having a root labeled γ and n leaves; the ith leaf of $\gamma\iota$ is labeled i, as indicated by this figure.

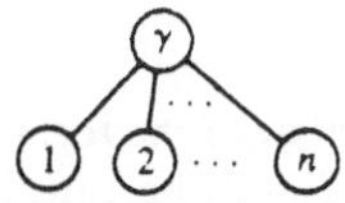

If J and J' are algebraic theories, a *theory morphism* $\mathbf{F}: J \to J'$ is a family of functions taking each morphism $f: n \to p$ in J to a morphism $f\mathbf{F}: n \to p$ in J' such that $(f \circ g)\mathbf{F} = f\mathbf{F} \circ g\mathbf{F}$ and $i\mathbf{F} = i$ for all composible f, g in J and distinguished morphisms $i: 1 \to n$.

2. The uniqueness and explicit description of extended iteration. The first task of this section is to define an extension of the iteration operation in an iterative theory J in such a way that the "identities" or "equations" which hold for the restricted operation (defined only on those morphisms whose iteration equation has a unique solution) will continue to hold. The definition of this operation is inductive. In Def. 2.7 an explicit description of this operation is given.

Two equations valid in all iterative theories are given by (2.1) and (2.2).

(2.1) $$[O_m \oplus g]^\dagger = O_m \oplus g^\dagger$$

where $g: n \to p + n$ is ideal. This identity may be verified by showing that the righthand side is a solution of the iteration equation for $O_m \oplus g$.

The following equation, established in [2], will be referred to as the "pairing identity".

(2.2) $$(f_1, f_2)^\dagger = (g^\dagger, f_2^\dagger \circ (I_p, g^\dagger))$$

where $f_1: n \to p + n + 1$ and $f_2: 1 \to p + n + 1$ are ideal, and where g is defined to be the composition

$$(2.3) \qquad g: n \xrightarrow{f_1} p+n+1 \xrightarrow{(I_{p+n}, f_2^\dagger)} p+n.$$

We will repeat the argument of [2] below (in 2.5) which shows that the pairing identity holds in all iterative theories.

A special case of the equation (2.1) occurs when $n = 1$; i.e. $g: 1 \to p+1$. In this case, if g is base and $g = j: 1 \to p+1$, where $j < p+1$, then $g^\dagger$ is meaningful since the unique solution of the iteration equation for g is $j: 1 \to p$ (note the change of target). Of course if g is ideal, $g^\dagger$ is meaningful. But if $g = O_p \oplus I_1: 1 \to p+1$, every morphism $1 \to p$ is a solution of the iteration equation for g. The requirement that (2.1) remain valid for the extended operation (also denoted †) means that extended † should satisfy

$$(2.1') \qquad [O_m \oplus I_1]^\dagger = O_m \oplus I_1^\dagger.$$

These considerations yield the uniqueness part of the following theorem.

THEOREM 2.4. *Let $\perp: 1 \to 0$ be a morphism in the iterative theory J. There is one and only one operation $g \mapsto g^\dagger$, defined for all $g: n \to p+n$ (all $n, p \geqq 0$, yielding a morphism $g^\dagger: n \to p$) which satisfies* (2.1'), (2.2) *as well as*

$$(2.4.1) \qquad I_1^\dagger = \perp,$$

$$(2.4.2) \qquad g^\dagger = g \circ (I_p, g^\dagger),$$

for all scalar $g: 1 \to p+1$, all $p \geqq 0$.

Proof. For scalar g which is ideal or of the form $j \oplus O_1$, where $j: 1 \to p$ is base, we define $g^\dagger$ by the requirement that (2.4.2) be satisfied. (Thus, on scalar ideal g the extended and unextended iteration operations agree.) For scalar g of the form $O_p \oplus I_1$, we define $g^\dagger$ by (2.1') and (2.4.1).

Vector iteration in $(J, I_1^\dagger = \perp)$ is defined inductively. Assume that for all p and all $h: k \to p+k$, $k \leqq n$, $h^\dagger$ is defined. If (as in (2.2)) $f = (f_1, f_2): n+1 \to p+n+1$, we *define* $f^\dagger$ by (2.2).

It is then obvious that (2.2) *holds for all* $n, p \geqq 0$. Note that when $n = 0$, (2.2) becomes:

$$(O_{p+1}, f_2)^\dagger = (O_p, f_2^\dagger \circ (I_p, O_p))$$

which is trivially true.

From the above theorem, we obtain

COROLLARY 2.5. *In $(J, I_1^\dagger = \perp)$, extended vector iteration satisfies (and is determined by) the identities* (2.1'), (2.2), (2.4.1) *and* (2.4.2).

We now repeat the argument of [2] to show that the following holds.

THEOREM. *In $(J, I_1^\dagger = \perp)$ the condition* (2.4.2) *is valid for all vectors g.*

Proof. The proof is by induction. Assume that (2.4.2) holds for all $h: k \to p+k$, all $k \leqq n$. Let $f = (f_1, f_2): n+1 \to p+n+1$, where $f_1: n \to p+n+1$, $f_2: 1 \to p+n+1$. Define $g: n \to p+n$ by (2.3). We then substitute

$$\xi = (g^\dagger, f_2^\dagger \circ (I_p, g^\dagger))$$

in the right-hand side of the iteration equation for f yielding:

$$f \circ (I_p, g^+, f_2^+ \circ (I_p, g^+)) = (f_1, f_2) \circ (I_{p-n}, f_2^+) \circ (I_p, g^+) \quad \text{by (1.6.4) [3]},$$
$$= (f_1 \circ (I_{p-n}, f_2^+) \circ (I_p, g^+), f_2^+ \circ (I_p, g^+)) \quad \text{by [3, (1.6.4)] and } f_2^+ = f_2 \circ (I_{p-n}, f_2^+),$$
$$= (g^+, f_2^+ \circ (I_p, g^+)) \quad \text{by (2.3) and } g^+ = g \circ (I_p, g^+).$$

We record here without proof the following miscellaneous fact.

PROPOSITION 2.6. *With $f = (f_1, f_2)$, g as in (2.2) and (2.3): f is nonsingular iff both g and f_2 are nonsingular.*

We turn now to the second task of this section and one of the main points of the paper—the explicit description of the extended iteration operation. Recall from the introduction the definition of the set K_g of the singular positions of the morphism $g: n \to p+n$.

DEFINITION 2.7. Let $g: n \to p+n$ be a morphism in $(J, I_1^+ = \perp)$. The morphism $g_\perp: n \to p+n$ is defined by the following requirements:

(2.7.1) $$i \circ g_\perp = \perp \circ O_{p-n}, \quad \text{if } i \in K_g;$$

(2.7.2) $$i \circ g_\perp = i \circ g, \quad \text{if } i \notin K_g.$$

LEMMA. *If i is a power ideal (respectively, power successful) position of g, then i is a power ideal (respectively, power successful) position of $g_\perp$.*

Proof. Suppose that i is a power ideal position of g, so that for some k, $i \circ g^k$ is ideal. Let r be the least integer such that $i \circ g^{r-1}$ is ideal. If $r = 0$, then $i \circ g_\perp = i \circ g$ is ideal. If $r > 0$, then $i \circ g^r = p + i'$, for some $i' \in [n]$, and $i' \circ g = i \circ g^{r-1}$ is ideal. Then for all $k \leq r$, $i \circ (g_\perp)^k = i \circ g^k$. In particular, $i \circ (g_\perp)^r = p + i'$ so that $i \circ (g_\perp)^{r-1} = i' \circ g_\perp = i' \circ g$ is ideal. Thus i is a power ideal position of $g_\perp$.

The proof in the case that i is a power successful position of g is similar.

We may now give an explicit description of g^+ in $(J, I_1^+ = \perp)$.

THEOREM. *In $(J, I_1^+ = \perp)$, for any $g: n \to p+n$, $g_\perp$ is nonsingular and*

(2.7.3) $$g^+ = (g_\perp)^+.$$

Proof. By (2.7.1) and the Lemma, $g_\perp$ is nonsingular. Thus, by [3, **2.17**], the iteration equation for $g_\perp$ has a unique solution, and we need verify only that g^+ is one such solution.

By the Theorem in § 2.5,

(2.7.5) $$g^+ = g \circ (I_p, g^+).$$

Thus from (2.7.2) if $i \notin K_g$,

(2.7.6) $$i \circ g^+ = i \circ g_\perp \circ (I_p, g^+),$$

while if $i \in K_g$, by (2.7.1) we have

(2.7.7) $$i \circ g_\perp \circ (I_p, g^+) = \perp \circ O_p.$$

We now need the following fact, proved in the Appendix.

LEMMA. *If $i \in K_g$, $i \circ g^+ = \perp \circ O_p$.*

Applying the Lemma, we have

$$i \circ g^+ = i \circ g_\perp \circ (I_p, g^+), \quad \text{for all } i \in [n].$$

Thus $g^+ = g_\perp \circ (I_p, g^+)$, and, since $g_\perp$ is nonsingular, $g^+ = (g_\perp)^+$, by [3, **2.19**].

COROLLARY 2.8. *If $\bar{g}$ is a "mkl morphism", $\bar{g}=(a\oplus b\oplus O_l, h)$, (defined in § 1) then*

(2.8.1) $$\bar{g}=(a^{\dagger}\oplus\perp_k, h\circ(I_p, a^{\dagger}\oplus\perp_k))$$

where $\perp_k=(\overbrace{\perp,\cdots,\perp}^{k}): k\to 0$.

Proof. The proof follows easily from the following fact, used in [3, **2.12**]. If $g: n\to p+n$ is writable in the form

(2.8.2) $$(a\oplus c\oplus O_l, h)$$

where $m+k+l=n$, $a: m\to p+m$, $c: k\to k$ and $h: l\to p+n$ and $\xi=(\xi_1,\xi_2,\xi_3)$ is any solution of the iteration equation for g, where $\xi_1: m\to p$, $\xi_2: k\to p$, $\xi_3: l\to p$, then

(2.8.3) $$\xi_1=a\circ(I_p,\xi_1)$$

(2.8.4) $$\xi_2=c\circ\xi_2$$

(2.8.5) $$\xi_3=h\circ(I_p,\xi_1,\xi_2,\xi_3)=h\circ[(I_p,\xi_1,\xi_2)\oplus I_l]\circ(I_p,\xi_3).$$

Now in the case $\bar{g}$ is a *mkl*-morphism, $\bar{g}_{\perp}$ is writable in the form (2.8.2) where $c: k\xrightarrow{\text{base}}1\xrightarrow{\perp}0\xrightarrow{O_k}k$, and a and h are as in g. Then, by (2.8.3) and the hypothesis on a, $\xi_1=a^{\dagger}$; by (2.8.4), $\xi_2=\perp_k\circ O_p$. Lastly, (2.8.5) forces ξ_3 to be the unique solution to the iteration equation of $h\circ(I_p, a^{\dagger}\oplus\perp_k)\oplus I_l$, which is $h^{\dagger}\circ(I_p, a^{\dagger}\oplus\perp_k)$. The corollary now follows from Theorem 2.7.

3. Iteration theories. By a "preiteration theory" we mean an algebraic theory P (not necessarily an ideal theory) equipped with an operation † ("iteration") which for each $n, p\geqq 0$ takes a morphism $g: n\to p+n$ in P to a morphism $g^{\dagger}: n\to p$ in P (subject to no conditions at all). Of course, strictly speaking † is a family of operations indexed by pairs n, p of nonnegative integers. In this section we will provide five equivalent conditions on a preiteration theory P which ensure that P will satisfy all identities of iterative theories. In this section the "pairing identity" (2.2) plays a major role.

PROPOSITION 3.1. *Let P and Q be preiteration theories which satisfy the pairing identity (i.e. for all f, (2.2) holds). If $\mathbf{G}: P\to Q$ is a theory morphism which preserves scalar iteration (i.e. $f^{\dagger}\mathbf{G}=(f\mathbf{G})^{\dagger}$, all $f: 1\to p+1$ in P) then $\mathbf{G}$ preserves vector iteration.*

The easy proof is by induction.

Remark 3.2. If J is an iterative theory and $\perp: 1\to 0$ is in J, then the preiteration theory $(J, I_1^{\dagger}=\perp)$ satisfies the pairing identity for all $f: n+1\to p+n+1$, (all $n, p\geqq 0$) in J (see the proof of Theorem 2.4).

Before stating the main theorem of this section we introduce some notation. Γ will always denote a ranked set, and $\Gamma_{\square}$ is the ranked set obtained from Γ by adjoining a "new" element $\square$ to Γ_0. Thus

$$(\Gamma_{\square})_0=\Gamma_0\cup\{\square\};\qquad (\Gamma_{\square})_n=\Gamma_n,\qquad n>0.$$

Recall that Γtr is the iterative theory (of trees) freely generated by Γ (see [6]).

THEOREM 3.3. *The following five properties of a preiteration theory P are equivalent.*

(3.3.1) *P satisfies the pairing identity* and *for any Γ, and any function $\mathbf{F}: \Gamma\to P$[1] there is an extension of $\mathbf{F}$ to a theory morphism*

$$\mathbf{F}': \Gamma\text{tr}\to P$$

[1] A function from Γ to P will be assumed to be "rank preserving"; i.e. $\gamma\in\Gamma_n\mapsto\gamma\mathbf{F}: 1\to n$ in P.

which preserves scalar iteration applied to ideal *morphisms* (*i.e. if* $g: 1 \to p+1$ *is ideal*, $g^{\dagger}\mathbf{F}' = (g\mathbf{F}')^{\dagger}$).

(3.3.2) *For any* Γ *and any function* $\mathbf{F}: \Gamma \to P$ *there is an extension of* $\mathbf{F}$ *to a theory morphism*

$$\mathbf{F}': (\Gamma_{\square}\mathrm{tr}, I_1^{\dagger} = \square) \to P$$

which preserves extended vector iteration (*i.e. for all* $g: n \to p+n$, $g^{\dagger}\mathbf{F}' = (g\mathbf{F}')^{\dagger}$).

(3.3.3) *For any* Γ *and any function* $\mathbf{F}: \Gamma \to P$ *there is an extension of* $\mathbf{F}$ *to a theory morphism*

$$\mathbf{F}': \Gamma\mathrm{tr} \to P$$

which preserves vector iteration applied to ideal morphisms.

(3.3.4) *There is an iterative theory* J *and a theory morphism* $\mathbf{H}: J \to P$ *such that every morphism in* P *is the image of an ideal morphism in* J *and such that* $g^{\dagger}\mathbf{H} = (g\mathbf{H})^{\dagger}$ *for all* ideal *morphisms* $g: n \to p+n$.

(3.3.4′) *The same as* (3.3.4) *except that each morphism in* P *is the image of some* (*not necessarily ideal*) *morphism in* J.

Proof. (3.3.1) $\Rightarrow$ (3.3.2). Let P satisfy (3.3.1). We first show that P will satisfy some identities in addition to the pairing identity. Let Γ be a ranked set so large that there is a surjective function $\mathbf{F}: \Gamma \to P$. For any morphisms $h: 1 \to q+1$, $\beta: q \to s$ in P, let $\bar{h}$ and $\bar{\beta}$ be ideal morphisms in Γtr such that $\bar{h}\mathbf{F}' = h$, $\bar{\beta}\mathbf{F}' = \beta$, where $\mathbf{F}': \Gamma\mathrm{tr} \to P$ is the extension of $\mathbf{F}$ guaranteed by (3.3.1). Since it is easy to show that

$$[\bar{h} \circ (\bar{\beta} \oplus I_1)]^{\dagger} = \bar{h}^{\dagger} \circ \bar{\beta}$$

in Γtr, and since $\mathbf{F}'$ is a theory morphism preserving scalar iteration of ideal morphisms,

$$[h \circ (\beta \oplus I_1)]^{\dagger} = h^{\dagger} \circ \beta \tag{3.3.5}$$

in P.

Note that in the case $q = 0$ and $h = I_1$, (3.3.5) specializes to

$$[O_s \oplus I_1]^{\dagger} = I_1^{\dagger} \circ O_s = O_s \oplus I_1^{\dagger}. \tag{3.3.6}$$

In the same way one may verify that for $j \in [p]$ the identity

$$[j \oplus O_1]^{\dagger} = j \tag{3.3.7}$$

is valid in P, where $j: 1 \to p$ is base.

We may now prove (3.3.2). Let Γ and $\mathbf{F}: \Gamma \to p$ be fixed. We must show there is an extension of $\mathbf{F}$ to a theory morphism

$$\mathbf{F}': (\Gamma_{\square}\mathrm{tr}, I_1^{\dagger} = \square) \to P$$

which preserves extended vector iteration.

First we extend $\mathbf{F}$ to $\Gamma_{\square}$ by defining $\square\mathbf{F} = I_1^{\dagger}$ in P. Then by assumption, $\mathbf{F}: \Gamma_{\square} \to P$ extends to a theory morphism $\mathbf{F}': \Gamma_{\square}\mathrm{tr} \to P$ which preserves scalar iteration applied to *ideal* morphisms. We now show that $\mathbf{F}'$ also preserves scalar iteration on *base* morphisms. If $f: 1 \to p+1$ is base, either $f = j \oplus O_1$, some $j: 1 \to p$ or $f = O_p \oplus I_1$. In the first case $f^{\dagger} = j$ in $\Gamma_{\square}$tr and thus $f^{\dagger}\mathbf{F}' = j\mathbf{F}' = j = (f\mathbf{F}')^{\dagger}$, by (3.3.7) since $\mathbf{F}'$ is a theory morphism. Similarly, if $f = O_p \oplus I_1$, $f^{\dagger} = O_p \oplus \square$ in $(\Gamma_{\square}\mathrm{tr}, I_1^{\dagger} = \square)$, so $f^{\dagger}\mathbf{F}' = O_p \oplus \square\mathbf{F}' = O_p \oplus I_1^{\dagger} = (f\mathbf{F}')^{\dagger}$, by (3.3.6). Thus $\mathbf{F}'$ preserves scalar iteration. Thus by Proposition 3.1 $\mathbf{F}'$ preserves extended vector iteration, completing the proof.

Proof (3.3.2) ⇒ (3.3.3). The proof is obvious.

Proof (3.3.3) ⇒ (3.3.1). The fact that the pairing identity is valid in P is proved in the same way it was shown that (3.3.1) implies that the identity (3.3.5) holds in P. The rest of the statement follows trivially.

Proof (3.3.3) ⇒ (3.3.4). The proof is obvious.

Proof (3.3.4) ⇒ (3.3.3). Let $\mathbf{H}: J \to P$ be as in the statement of (3.3.4) and let $\mathbf{F}: \Gamma \to P$ be an arbitrary function. There is a function $\mathbf{G}: \Gamma \to J$ such that for each $\gamma \in \Gamma_n$, $\gamma\mathbf{G}: 1 \to n$ is an ideal morphism in J and $\gamma\mathbf{GH} = \gamma\mathbf{F}$. By [6, 4.1.2] $\mathbf{G}$ has an extension $\mathbf{G}': \Gamma\mathrm{tr} \to J$ to an ideal theory morphism, which necessarily preserves vector iteration applied to ideal morphisms. We may *define* $\mathbf{F}'$ to be the composition

$$\mathbf{F}': \Gamma\mathrm{tr} \xrightarrow{\mathbf{G}'} J \xrightarrow{\mathbf{H}} P.$$

Since both $\mathbf{G}'$ and $\mathbf{H}$ preserve vector iteration applied to ideal morphisms, so does $\mathbf{F}'$.

Proof (3.3.4) ⇒ (3.3.4′). The proof is obvious.

Proof (3.3.4′) ⇒ (3.3.4). Assume that $H: J \to P$ is a surjective theory morphism which preserves vector iteration applied to ideal morphisms. Without loss of generality we may assume that J is $\Gamma\mathrm{tr}$, for some ranked set Γ. Let Γ_Δ be the ranked set obtained from Γ by adding a new element Δ to Γ_1, so that $(\Gamma_\Delta)_1 = \Gamma_1 \cup \{\Delta\}$; $(\Gamma_\Delta)_n = \Gamma_n$, $n \neq 1$. Let $\perp: 1 \to 0$ be a morphism such that $\perp H = I_1^\dagger$ in P. We now apply the

Δ-LEMMA [3, **3.6**]. *There is a unique theory morphism* $\Phi: \Gamma_\Delta\mathrm{tr} \to \Gamma\mathrm{tr}$ *such that* (a) $\Delta\Phi = I_1$; (b) $\Delta^\dagger\Phi = \perp$; (c) $\gamma\Phi = \gamma$, *all* $\gamma \in \Gamma$; (d) *if* $f\Phi$ *is base, say* $j: 1 \to p+1$, *then* $f = \Delta^r \circ j$, *for some* $r \geqq 0$.

It is easy to show, using Prop. 3.1 and parts (b) and (d) of the Δ-Lemma that $g^\dagger\Phi = (g\Phi)^\dagger$ for all ideal $g: n \to p+n$ in $\Gamma_\Delta\mathrm{tr}$ (where $I_1^\dagger = \perp$ in $\Gamma\mathrm{tr}$). Let H_Δ be the composition

$$H_\Delta: \Gamma_\Delta\mathrm{tr} \xrightarrow{\Phi} \Gamma\mathrm{tr} \xrightarrow{H} P.$$

Then H_Δ is a theory morphism which preserves vector iteration on ideal morphisms. Lastly, every morphism in P is the H_Δ-image of an ideal morphism in $\Gamma_\Delta\mathrm{tr}$. Indeed, if $f: 1 \to p$ in P is not base, $f = gH$ for some ideal $g: 1 \to p$ in $\Gamma\mathrm{tr}$; hence $f = gH_\Delta$ (by part (c) of the Δ-Lemma). However if $f: 1 \to p$ is base, then $f = I_1 \circ f = (\Delta \circ f)H_\Delta$. This completes the proof of the theorem.

Remark. In (3.3.1), (3.3.2) and (3.3.3) the extension $\mathbf{F}'$ of $\mathbf{F}$ is necessarily unique, by [6, 2.5.1].

DEFINITION 3.4. An *iteration theory* is a preiteration theory which satisfies any (and hence all) of the properties of Theorem 3.3.

PROPOSITION 3.5. *Let* $\perp: 1 \to 0$ *be a morphism in the iterative theory* J. *Then the preiteration theory* $(J, I_1^\dagger = \perp)$ *defined in* 2.5 *is an iteration theory.*

Proof. The Universality Theorem (**3.7** in [3]) implies that for any function $\mathbf{F}: \Gamma \to J$ there is an extension of $\mathbf{F}$ to a theory morphism

$$\mathbf{F}': (\Gamma_\square\mathrm{tr}, I_1^\dagger = \square) \to (J, I_1^\dagger = \perp)$$

such that $\square\mathbf{F}' = \perp$ and such that $\mathbf{F}'$ preserves (extended) scalar iteration. Both preiteration theories $(\Gamma_\square\mathrm{tr}, I_1^\dagger = \square)$ and $(J, I_1^\dagger = \perp)$ satisfy the pairing identity, so that by 3.1, $\mathbf{F}'$ preserves extended vector iteration. This shows that $(J, I_1^\dagger = \perp)$ has property (3.3.2), and is thus an iteration theory.

COROLLARY. *The iteration theory* $(\Gamma_\square\mathrm{tr}, I_1^\dagger = \square)$ *is freely generated by* Γ *in the class of iteration theories.*

Proof. This is a restatement of property (3.3.2). (Recall the Remark preceding Def. 3.4.)

3.6. As has already been indicated, the intuitive meaning of "iteration theory" is a preiteration theory P which satisfies (according to (3.3.2)) all identities, e.g.

$$g^{\dagger} = g \circ (I_p, g^{\dagger})$$

which are (meaningful and) valid in $(\Gamma_{\square}\text{tr}, I_1^{\dagger} = \square)$. By Theorem 3.3 (3.3.3), a similar statement may be made with "Γtr" in place of "$(\Gamma_{\square}\text{tr}, I_1^{\dagger} = \square)$", but the class of meaningful identities is reduced and "valid" must be properly understood. According to Theorem 3.3 (3.3.1), it is enough to know P satisfies the pairing identity and all scalar iteration identities of Γtr to conclude that P is an iteration theory.

We state, for emphasis, an immediate corollary of this discussion, Theorem 3.3 and Proposition 3.5:

COROLLARY 3.6.1. *If* $\perp : 1 \to 0$ *is any morphism in the iterative theory* J, *every valid iterative theory identity is true in* $(J, I_1^{\dagger} = \perp)$.

3.7. Although the class of iterative theories is not closed under products (since the product of two ideal theories is not ideal), the class of iteration theories is, i.e.

if P *and* Q *are iteration theories, so is* $P \times Q$.

For example, the pairing identity is valid in $P \times Q$, since otherwise it would fail to hold in either P or Q.

Example 3.8. An important example of an iteration theory which is not even ideal is the theory $[X]$ where X is a nonempty set. (This theory is denoted $[X, Q]$ in [5, p. 189], where Q is a singleton set.) A morphism $f: n \to p$ in $[X]$ is a partial function $f: X \times [n] \to X \times [p]$. In $[X]$, if $f: n \to p + n$, the partial function $f^{\dagger}$ defined below is the least (in the sense of set inclusion of their graphs) solution ξ of the iteration equation for f (viz. $\xi = f \circ (I_p, \xi)$).

3.8.1. $f^{\dagger}: X \times [n] \to X \times [p]$ is the partial function defined by: $xif = x'i'$ if there is a sequence $x_0 i_0 x_1 i_1 \cdots x_m i_m$, $m \geqq 0$ such that $x_0 = x$, $i_0 = i$, and for each $j < m$.

$$x_j i_j f = x_{j+1}(p + i_{j+1}); \quad \text{also} \quad x_m i_m f = x' i'.$$

The preiteration theory $[X]$ is an iteration theory since the "forgetful functor" $[[X \circ N, \square]] \to [X]$ is a theory morphism with property (3.3.4). The proof of this statement is in [5]. (The iterative theory $[[X \circ N, \square]]$ of "timed terminal functions" is defined in [5, p. 200].) Hence any equation valid in $[[X \circ N, \square]]$ will hold in $[X]$.

Problem 3.9. Is there any other way of defining $f^{\dagger}$ on $[X]$ so that the resultant preiteration theory is an iteration theory?

Example 3.10. One of the referees asked whether any ideal *iteration* theory is necessarily an *iterative* theory. We answer the question negatively with the following example. Let S be the set consisting of the nonnegative integers N and two "new" points $a \neq b$. Let $g: S \to S$ be the function such that $ng = n + 1$, for $n \in N$ and $ag = a$, $bg = b$. Let P be the least subtheory of $S^{\cdot}$ containing $g: 1 \to 1$, $a: 1 \to 0$ and $b: 1 \to 0$. (A morphism $n \to p$ in $S^{\cdot}$ is a function $S^p \to S^n$; composition is function composition and the distinguished morphism $i: 1 \to n$ is the ith projection function $S^n \to S$.) P is easily seen to be an ideal theory which is not iterative, since the iteration equation for g has the two solutions a and b. Now define $g^{\dagger} = a = I_1^{\dagger}$ in P, and extend † to all other morphisms using the equations (2.1), (2.4.1), (2.4.2) and the pairing identity. We will use (3.3.4′) to show that P is an iteration theory.

Let $\Gamma_0 = \{\bar{a}, \bar{b}\}$, $\Gamma_1 = \{\bar{g}\}$, $\Gamma_n = \phi$, $n \geq 1$. Note that there is only one infinite tree $1 \to 0$ in Γtr, namely $\bar{g}^{\dagger}$. We define $H: \Gamma\text{tr} \to P$ on the finite trees by the requirements that

$\bar{a}H = a$, $\bar{b}H = b$, $\bar{g}H = g$ and that H be a theory morphism. If we now define $\bar{g}^{\dagger}H = a$, H preserves scalar iteration on ideal morphisms and remains a theory morphism. Thus by 3.1, H preserves vector iteration on ideal morphisms. Clearly H is surjective, so that by (3.3.4'), P is an iteration theory.

4. Examples involving partial ordering. Let T be an algebraic theory. A *compatible partial ordering* on T is a family of partial orderings $\sqsubseteq$ on the sets $T_{n,p}$ of morphisms $n \to p$ in T which are "compatible" with the theory operations in that

$$\text{(4.1)} \qquad \begin{array}{ll} \text{if} & f_1 \sqsubseteq f_2 \text{ in } T_{n,p} \text{ and } g_1 \sqsubseteq g_2 \text{ in } T_{p,q} \\ \text{then} & f_1 \circ g_1 \sqsubseteq f_2 \circ g_2 \text{ in } T_{n,q}; \end{array}$$

also

$$\text{(4.2)} \qquad \begin{array}{ll} \text{if } f_i \sqsubseteq g_i \text{ in } T_{1,p}, \text{ each } i \in [n], \\ \text{then} \quad (f_1, \cdots, f_n) \sqsubseteq (g_1, \cdots, g_n) \text{ in } T_{n,p}. \end{array}$$

In § 2 it was shown how to extend the iteration operation to all morphisms in an iterative theory J, depending on an arbitrary morphism $\perp : 1 \to 0$ in J, in such a way that all iterative theory identities are preserved. In particular the value $g^{\dagger}$ of the extended iteration operation on g is always a solution of the iteration equation for g; i.e.

$$g^{\dagger} = g \circ (I_p, g^{\dagger}).$$

In many treatments of the semantics of programming languages solutions of the iterative equation for g are found as "least" fixed points of the operation

$$\xi \mapsto g \circ (I_p, \xi)$$

(see [1] for one example). We give without proof two examples where the extended iteration operation yields the least fixed point of the iteration equation with respect to some compatible partial ordering.

Example 1. Let J be the iterative theory $\Gamma\mathrm{Tr}$ of all Γ-trees, and suppose $\perp \in \Gamma_0$; i.e. is an "atomic" tree $1 \to 0$. For any $f: n \to p$, let B_f be the set of vertices (leaves in this case) of f labeled $\perp$. *Define* $f \sqsubseteq_{\perp} g$ if g can be obtained from f by attaching a Γ-tree $h_v : 1 \to p$ to each v in B_f. We claim the following

4.3. *$\sqsubseteq_{\perp}$ is a compatible partial ordering on $\Gamma\mathrm{Tr}$ and that in* ($\Gamma\mathrm{Tr}$, $I_1^{\dagger} = \perp$), *$g^{\dagger}$ is the $\sqsubseteq_{\perp}$-least solution of the iteration equation for g.*

(The fact that $\sqsubseteq_{\perp}$ is compatible, for this choice of $\perp$, is proved in [1].)

Example 2. Let $J = \Gamma\mathrm{Tr}$ again and this time let "$\perp$" denote the infinite tree $\Delta^{\dagger}$, where $\Delta \in \Gamma_1$. For $f: n \to p$ in J, let B_f be the set of all vertices v of f such that

(a) the tree of descendants of v in f is isomorphic to $\perp$; and

(b) no predecessor of v has property (a).

If we *define* $f \sqsubseteq_{\perp} g$ as in Example 1 (with this new definition of B_f), $\sqsubseteq_{\perp}$ is also a compatible partial ordering on $\Gamma\mathrm{Tr}$ and 4.3 holds on this case also.

Example 3. By way of contrast, if $J = \Gamma\mathrm{Tr}$ and $\perp$ is either the finite tree $\Delta \circ \gamma_0$, where $\Delta \in \Gamma_1$ and $\gamma_0 \in \Gamma_0$, or the infinite tree $\pi^{\dagger} \circ \gamma_0$, (see Fig. 1) where $\pi \in \Gamma_2$ and $\gamma_0 \in \Gamma_0$, there is no partial ordering $\sqsubseteq_{\perp}$ on $\Gamma\mathrm{Tr}$ such that 4.3 holds.

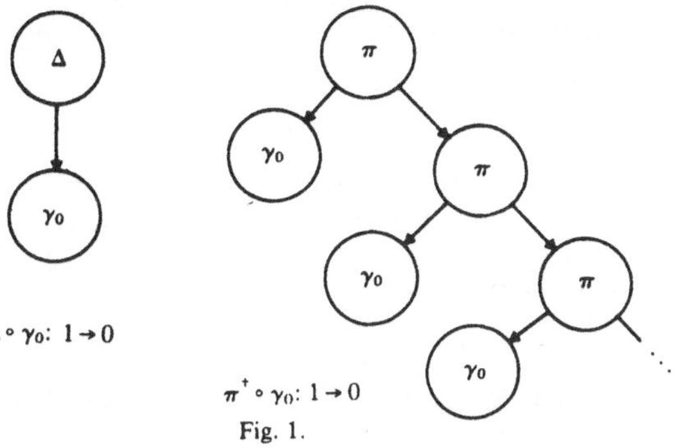

$\Delta \circ \gamma_0 \colon 1 \to 0$

$\pi^{\dagger} \circ \gamma_0 \colon 1 \to 0$

Fig. 1.

The following theorem, which explains these examples, was obtained in collaboration with Professor Ralph Tindell.

THEOREM 4.4. *Let* $\perp \colon 1 \to 0$ *be a tree in* $\Gamma\mathrm{Tr}$. *There is a compatible partial ordering* $\sqsubseteq$ *on* $\Gamma\mathrm{Tr}$ *such that in* $(\Gamma\mathrm{Tr}, I_1^{\dagger} = \perp)$ $g^{\dagger}$ *is the* $\sqsubseteq$*-least solution of the iteration equation for* g *iff* $\perp$ *is homogeneous*; *i.e. for each vertex* v *of* $\perp$, *the tree of descendants of* v *in* $\perp$ *is isomorphic to* $\perp$.

We prove the necessity of the condition. Thus suppose $\sqsubseteq$ is a compatible partial ordering on $\Gamma\mathrm{Tr}$ such that for each $g \colon n \to p + n$ in $(\Gamma\mathrm{Tr}, I_1^{\dagger} = \perp)$ $g^{\dagger}$ is the $\sqsubseteq$-least solution of the iteration equation for g. In the case $g = I_1$, it follows that $\perp$ is the $\sqsubseteq$-least morphism $1 \to 0$ in $\Gamma\mathrm{Tr}$. Let v_0 be any vertex of $\perp$ and let $\tau \colon 1 \to 0$ be the tree of descendants of v_0. Let $\sigma \colon 1 \to 1$ be the tree obtained from $\perp$ by deleting all the successors of v_0 (if any) and relabeling v_0 by "1". Then σ has precisely one leaf labeled "1", and clearly

$$\perp = \sigma \circ \tau. \tag{4.4.1}$$

Since $\perp$ is the $\sqsubseteq$-least morphism $1 \to 0$,

$$\perp \sqsubseteq \tau \tag{4.4.2}$$

and

$$\perp \sqsubseteq \sigma \circ \perp. \tag{4.4.3}$$

By (4.4.2), since $\sqsubseteq$ is compatible,

$$\sigma \circ \perp \sqsubseteq \sigma \circ \tau = \perp. \tag{4.4.4}$$

Thus

$$\sigma \circ \perp = \sigma \circ \tau$$

which shows that $\perp$ is isomorphic to τ (since both are the tree of descendants of v_0 in $\sigma \circ \perp = \sigma \circ \tau = \perp$).

The proof of the sufficiency of the condition will appear in [4].

5. ΓTr as a metric space. The free iterative theory $\Gamma\mathrm{tr}$ is a subtheory of the iterative theory $\Gamma\mathrm{Tr}$ of *all* labeled Γ-trees (studied in [6]). In this section it is observed that $\Gamma\mathrm{Tr}$ is a complete metric space and that for any Γ-tree $\perp \colon 1 \to 0$ and any morphism $f \colon n \to p + n$ in the corresponding iteration theory $(\Gamma\mathrm{Tr}, I_1^{\dagger} = \perp)$, the value of the extended iteration operation $f^{\dagger}$ is a metric limit of the trees $f^k \circ (I_p \oplus \perp_n)$.

In [6] the *profile* of an unlabeled tree was defined. The definition is extended to all labeled Γ-trees in the expected way.

DEFINITION 5.1. Let $g: n \to p$ be a tree in ΓTr. For any natural number $d \geqq 0$, the *profile* of g at length (or "depth") d, in symbols $P_d(g)$, is the sequence of elements in $\Gamma \cup [p]$.

$$l(w_1), l(w_2), \cdots, l(w_k)$$

where $w_1, w_2, \cdots, w_k$, $k \geqq 0$, is the sequence (from left to right) of the vertices of g of distance d from a root and where $l(w_i) \in \Gamma \cup [p]$ is the label of w_i, $i \in [k]$.

Two trees have the same profile at every depth iff they are isomorphic. Hereafter we identify isomorphic trees, so that

$$g = g' \quad \text{iff} \quad P_d(g) = P_d(g') \text{ for all } d \geqq 0.$$

More generally, g and g' are identical up to vertices of length $\leqq l$ iff $P_d(g) = P_d(g')$ for $d \leqq l$.

Using the concept of the profile of a tree, we will make the Γ-trees into a metric space.

5.2. Let $(r_n: n = 0, 1, 2, \cdots)$ be a sequence of positive real numbers. Let $g, g': 1 \to p$ be two scalar Γ-trees.

DEFINITION (the function d). If $g \neq g'$, we define

$$d(g, g') = r_{n_0}$$

where n_0 is the least integer k such that $P_k(g) \neq P_k(g')$; if $g = g'$, we let $d(g, g') = 0$. For vector trees $g, g': n \to p$, $n > 1$, we *define*

$$d(g, g') = \max \{d(i \circ g, i \circ g'): i \in [n]\}.$$

We note immediately that

(5.2.1) $$d(g, g') = 0 \quad \Leftrightarrow \quad g = g' \qquad (\text{since } r_k > 0, \text{ all } k)$$

(5.2.2) $$d(g, g') = d(g', g).$$

PROPOSITION 5.2.3. *For each $n, p \geqq 0$ the function d is a metric on the set of Γ-trees $n \to p$ in ΓTr.*

Proof. It follows from well-known facts that we need prove the assertion only for the case $n = 1$. (See e.g. Munkres, Theorem 1.3, p. 266, *Topology*, Prentice-Hall, Englewood Cliffs, NJ 1975.) In view of (5.2.1) and (5.2.2), we need verify only the triangle inequality (5.2.4) to show d is a metric. If any two of the three trees g, g' and g'': $1 \to p$ are equal, it is clear that

(5.2.4) $$d(g, g'') \leqq d(g, g') + d(g', g'').$$

Now suppose that all three trees are pairwise distinct. Let $d(g, g') = r_m$ and $d(g', g'') = r_{m'}$. In the case $m \leqq m'$ we have by Def. 5.1, $d(g, g'') = r_m$. From this fact (5.2.4) follows.

PROPOSITION 5.2.5. *If the sequence $(r_n: n \geqq 0)$ converges monotonically down to zero, the metric d is complete; i.e. every Cauchy sequence converges.*

Proof. Let $(g_k: k \geqq 0)$ be a Cauchy sequence of trees $1 \to p$ in ΓTr. Thus for every real number $\varepsilon > 0$ there is a natural number $n(\varepsilon)$ such that $d(g_k, g_{k'}) < \varepsilon$ whenever $k, k' > n(\varepsilon)$. If $m(\varepsilon)$ is the least integer such that $r_k < \varepsilon$ for all $k > m(\varepsilon)$, then we have

$$P_d(g_k) = P_d(g_{k'}), \quad \text{all } d \leqq m(\varepsilon)$$

for all $k, k' > n(\varepsilon)$. Thus we may define a tree $g: 1 \to p$ by requiring that for all $i \geqq 1$:

$$P_d(g) = P_d(g_{n(\varepsilon_i)}), \; d \leqq m(\varepsilon_i)$$

where $\varepsilon_i = 1/i$. Then clearly

$$\lim_{k\to\infty} g_k = g.$$

Thus any Cauchy sequence converges.

From now on we will assume the sequence $(r_n: n \geqq 0)$ *converges monotonically to zero.* It is easily shown that any two such sequences yield equivalent metrics, in the sense that the two induced topologies are the same. In the paper [7] the Γ-trees $1\to 0$ were observed to be a complete metric space with the metric of § 5.2 corresponding to the sequence $r_n = 1/n+1$, $n \geqq 0$.

5.3. The metric behaves nicely with respect to the theory operations.

LEMMA 5.3.1. *Let* $g_i: n\to p$, $h_i: p\to q$ *be* Γ-*trees*, $i = 1, 2$. *Then*

$$d(g_1\circ h_1, g_2\circ h_1) \leqq d(g_1, g_2), \tag{5.3.2}$$

and

$$d(g_1\circ h_1, g_1\circ h_2) \leqq d(h_1, h_2). \tag{5.3.3}$$

$$d(g^s, h^s) \leqq d(g, h), \quad \text{any } s \geqq 0, g, h: n\to p+n. \tag{5.3.4}$$

Proof. We prove only (5.3.2) since the proofs of (5.3.3) and (5.3.4) are similar. If $g_1 = g_2$ there is nothing to prove. Otherwise, let $d(g_1, g_2) = r_k$. Since the trees $g_1\circ h_1$ and $g_2\circ h_1$ are obtained by "attaching" h_1 to the termini of g_1 and g_2, $P_d(g_1\circ h_1) = P_d(g_2\circ h_1)$ all $d \leqq k$. Thus $d(g_1\circ h_1, g_2\circ h_1) \leqq d(g_1, g_2)$.

Note that it is possible to have $g_1 \neq g_2$ but $g_1\circ h_1 = g_2\circ h_1$; thus in (5.3.2) we cannot in general assert equality. Indeed if $g: 1\to 1$ is any finite ideal Γ-tree, we can let $g_1 = g$, $g_2 = g\circ g$ and $h_1 = g^{\dagger}$.

COROLLARY 5.3.5. *The theory operations of composition and source-pairing are continuous.*

Proof. Suppose (g_k) is a sequence of Γ-trees $n\to p$ with limit g, and (h_k) is a sequence of Γ-trees $p\to q$ with limit h. We show that $\lim_{k\to\infty} g_k\circ h_k = g\circ h$. Indeed, from the triangle inequality,

$$d(g_k\circ h_k, g\circ h) \leqq d(g_k\circ h_k, g_k\circ h) + d(g_k\circ h, g\circ h).$$

But by Lemma 5.3.1, the righthand side is less than $d(h_k, h) + d(g_k, g)$, which goes to zero as $k\to\infty$.

The proof that source pairing is continuous is simpler and is omitted.

It will be shown later that iteration and extended iteration are continuous as well.

5.4. The main result of this section is an explicit description of extended iteration in the iteration theory $(\Gamma\mathrm{Tr}, I_1^{\dagger} = \perp)$, for any Γ-tree $\perp: 1\to 0$. We will show that $g^{\dagger}$ is a metric limit of the trees $g^k\circ(I_p\oplus \perp_n)$, for any Γ-tree $g: n\to p+n$. ($\perp_n$ was defined in Cor. 2.8.)

Call a morphism $g: n\to p+n$ in any iterative theory "quasi-ideal" if for each $i\in[n]$, $i\circ g$ is either ideal or $j\oplus O_n$, for some base $j\in[p]$. In other words, every component position of a quasi-ideal morphism is successful or ideal.

LEMMA. *Suppose* $f: n\to p+n$ *is a quasi-ideal morphism in* $\Gamma\mathrm{Tr}$. *Then for any* $\alpha: n\to p$ *in* $\Gamma\mathrm{Tr}$,

$$\lim_{k\to\infty} f^k\circ(I_p, \alpha) = f^{\dagger}.$$

Proof. It is easily seen that for any $k \geqq 0$ and $d \leqq k$,

$$P_d(f^{k+1}) = P_d(f^{k+1} \circ (I_p, \alpha)).$$

But since $f^\dagger = f^{k+1} \circ (I_p, f^\dagger)$, we have for $d \leqq k$

$$P_d(f^\dagger) = P_d(f^{k+1} \circ (I_p, \alpha)),$$

from which the result follows.

PROPOSITION. *If $h: n \to p+n$ is a nonsingular morphism in ΓTr, then for any $\alpha: n \to p$, the sequence*

$$h \circ (I_p, \alpha), h^2 \circ (I_p, \alpha), \cdots \tag{5.4.1}$$

is Cauchy and converges to $h^\dagger$.

Proof. Let $h = g_\perp$ (defined in 2.7), so that $h^\dagger = g^\dagger$, by Theorem 2.7. By the proposition of § 5.4 and the fact that $(I_p, \perp_n \circ O_p) = I_p \oplus \perp_n$, we have

$$P_k(h^s \circ (I_p, \alpha)) = P_k(h^s) = P_k(h^t) = P_k(h^t \circ (I_p, \alpha)).$$

Thus the sequence (5.4.1) is Cauchy and therefore converges. Let $f = h^n$. Then $f^\dagger = h^\dagger$, and $\lim_{k\to\infty} f^k \circ (I_p, \alpha) = f^\dagger$, by the Lemma. But since $f \circ (I_p, \alpha), f^2 \circ (I_p, \alpha), \cdots$ is a subsequence of (5.4.1), it follows that the limit of (5.4.1) is $h^\dagger = f^\dagger$.

We may now prove the following:

THEOREM 5.5. *Let $g: n \to p+n$ be an arbitrary Γ-tree in (ΓTr, $I_1^\dagger = \perp$). Then*

$$\lim_{k\to\infty} g^k \circ (I_p \oplus \perp_n) = g^\dagger.$$

Proof. Let $h = g_\perp$ (defined in 2.7), so that $h^\dagger = g^\dagger$, by Theorem 2.7. By the proposition of § 5.4 and the fact that $(I_p, \perp_n \circ O_p) = I_p \oplus \perp_n$, we have

$$\lim_{k\to\infty} h^k \circ (I_p \oplus \perp_n) = h^\dagger = g^\dagger.$$

In order to prove the theorem, we prove that

$$h^k \circ (I_p \oplus \perp_n) = g^k \circ (I_p \oplus \perp_n), \quad \text{all } k \geqq 1. \tag{5.5.1}$$

We prove (5.5.1) by induction on k.

When $k = 1$, we have to show

$$g_\perp \circ (I_p \oplus \perp_n) = g \circ (I_p \oplus \perp_n). \tag{5.5.2}$$

If $i \notin K_g$, $i \circ g_\perp = i \circ g$, so $i \circ g_\perp \circ (I_p \oplus \perp_n) = i \circ g \circ (I_p \oplus \perp_n)$. If $i \in K_g$, $i \circ g_\perp = \perp \circ O_{n+p}$, so that $i \circ g_\perp \circ (I_p \oplus \perp_n) = \perp_n \circ O_p$. Also $i \circ g = O_p \oplus i'$, some $i' \in [n]$, since $i \in K_g$, and thus $i \circ g \circ (I_p \oplus \perp_n) = \perp_n \circ O_p$. This completes the proof of (5.5.2).

Now assume (5.5.1) holds for k. Then

$$\begin{aligned} h^{k+1} \circ (I_p \oplus \perp_n) &= h \circ (I_p \oplus O_n, h^k) \circ (I_p \oplus \perp_n) \\ &= h \circ (I_p, h^k \circ (I_p \oplus \perp_n)) \\ &= h \circ (I_p, g^k \circ (I_p \oplus \perp_n)), \end{aligned}$$

by the induction hypothesis. We now show

$$g_\perp \circ (I_p, g^k \circ (I_p \oplus \perp_n)) = g \circ (I_p, g^k \circ (I_p \oplus \perp_n)). \tag{5.5.3}$$

Let L and R be the morphisms on the left and right sides of the equation (5.5.3). We show $i \circ L = i \circ R$ for all $i \in [n]$. We clearly need only consider the case $i \in K_g$. In this case $i \circ g_{\perp} = \perp \circ O_{n+p}$, so that $i \circ L = \perp \circ O_p$. But since $i \in K_g$, $i \circ g^{k+1} = O_p \oplus i'$, for some $i' \in [n]$. Hence $i \circ R = i \circ g^{k+1} \circ (I_p \oplus \perp_n) = \perp \circ O_p$ completing the proof.

5.6. The continuity of extended iteration will be shown to follow from Theorem 5.5 and the Lemma 5.3.1.

PROPOSITION. *Let $g_k: n \to p+n$, $k \geq 0$, be a sequence of Γ-trees converging to the tree $g: n \to p+n$ in $(\Gamma\mathrm{Tr}, I_1^{\dagger} = \perp)$. Then the sequence*

$$g_0^{\dagger}, g_1^{\dagger}, \cdots,$$

converges to $g^{\dagger}$.

Proof. By the triangle inequality, for any $k, s > 0$

$$(5.6.1)\qquad \begin{aligned} d(g_k^{\dagger}, g^{\dagger}) \leq d(g_k^{\dagger}, (g_k)^s \circ (I_p \oplus \perp_n)) + d((g_k)^s \circ (I_p \oplus \perp_n), g^s \circ (I_p \oplus \perp_n)) \\ + d(g^s \circ (I_p \oplus \perp_n), g^{\dagger}). \end{aligned}$$

By (5.3.2) and (5.3.4), the middle term is not greater than $d(g_k, g)$, for any s. Thus, given any real $\varepsilon > 0$, first choose k such that $d(g_k, g) < \varepsilon/3$ and for that k, choose s so large that both the first and third summand in (5.6.1) are less than $\varepsilon/3$, using Theorem 5.5. This completes the proof.

Appendix. In this section we will prove the fact stated in the proof of Theorem 2.7.

LEMMA. *If $f: n \to p+n$ is a morphism in $(J, I_1^{\dagger} = \perp)$ and $i \in K_f$, then $i \circ f^{\dagger} = \perp \circ O_p$.*

We will use the function f^{σ} (defined in [3, 2.3.10]) and we prove first the following fact:

PROPOSITION A. *In $(J, I_1^{\dagger} = \perp)$ if $i \in K_f$, and if $f^{\sigma} = u_1 u_2 u_3 \cdots$ then $i \circ f^{\dagger} = u_r \circ f^{\dagger}$, for any $r \geq 1$.*

Proof. When $r = 1$, $u_r = i$ and there is nothing to prove. Assume $i \circ f^{\dagger} = u_r \circ f^{\dagger}$. From [3, (2.3.11)] and the fact $f^{\dagger} = f^r \circ (I_p, f^{\dagger})$, we obtain

$$\begin{aligned} i \circ f^{\dagger} &= i \circ f^r \circ (I_p, f^{\dagger}) = [O_p \oplus u_{r+1}] \circ (I_p, f^{\dagger}) \\ &= u_{r+1} \circ f^{\dagger}, \quad \text{by [3, (1.6.2) and (1.6.3)]}, \end{aligned}$$

completing the induction.

We now prove the Lemma by induction on $n \geq 1$. If $n = 1$ and $i = 1 \in K_f$, then $f = O_p \oplus I_1$, so that $i \circ f^{\dagger} = \perp \circ O_p$ by (2.1') and (2.4.1). Now assume $f: n+1 \to p+n+1$. Write $f = (f_1, f_2)$ where f_1 has source n and f_2 has source 1, and define g by (2.3). Then (2.2) holds. Assume now $i \in K_f$ and assume inductively: $i \in K_g \Rightarrow i \circ g^{\dagger} = \perp \circ O_p$. We distinguish three cases.

Case 1. *if* $f^{\sigma} = u_1 u_2 \cdots u_r (n+1)(n+1) \cdots$, $r \geq 0$, where $u_i \in [n]$ for each $i \in [r]$.

Case 2. *if* $f^{\sigma} = u_1 u_2 \cdots u_r (n+1) u_{r+2} \cdots$, $r \geq 0$, where $u_i \in [r]$ and $u_{r+2} \in [n]$.

Case 3. *if* $f^{\sigma} = u_1 u_2 \cdots u_i \cdots$, where $u_i \in [n]$ for all i.

On Case 1. $(n+1) \circ f = p+n+1 = O_{p+n} \oplus I_1 = f_2$. Hence by (2.1') and (2.4.1): $f_2^{\dagger} = O_{p+n} \oplus \perp = \perp \circ O_{p+n}$ so that by (2.2) we infer $(n+1) \circ f^{\dagger} = \perp \circ O_p$. Since $u_{r+1} = (n+1)$, it follows (cf. (2.3.11) of [3]) that

$$(\text{A.1})\qquad i \circ f^{r+1} = (n+1) \circ f.$$

From Proposition A we infer

$$(\text{A.2})\qquad i \circ f^{\dagger} = (n+1) \circ f^{\dagger} = \perp \circ O_p.$$

Thus in this case (without the aid of the inductive assumption) we've shown $i \in K_f \Rightarrow i \circ f^{\dagger} = \perp \circ O_p$.

In preparation for the remaining cases we make the following
Observation: where $i, i' \in [n]$.

(A.3) $i \circ f = p + i' \Rightarrow i \circ g = p + i'$; in the notation of [3, 2.3.3], $if^{\nu} \in [n] \Rightarrow ig^{\nu} = if^{\nu}$;

(A.4) $[i \circ f = p + n + 1 \;\&\; (n+1) \circ f = p + i'] \Rightarrow i \circ g = p + i'$;
$[if^{\nu} = n + 1 \;\&\; (n+1)f^{\nu} \in [n]] \Rightarrow ig^{\nu} = (n+1)f^{\nu}$.

Assertion (A.3) follows from (2.3); (A.4) follows from (2.3) together with the observation that since $f_2 = p + i'$ and $i' \in [n]$, we have $f_2^{\dagger} = p + i'$. It follows that in Case 2 we have

$$ig^{\sigma} = u_1 u_2 \cdots u_r u_{r+2} \cdots$$

while in Case 3, we have $ig^{\sigma} = if^{\sigma}$. Thus in Cases 2 and 3

(A.5) $$i \in K_f \quad \Rightarrow \quad i \in K_g.$$

With the aid of the inductive assumption we conclude $i \circ g^{\dagger} = \perp \circ O_p$, and thus from (2.2), $i \circ f^{\dagger} = \perp \circ O_p$. This shows that for Cases 2 and 3:

(A.6) $$\text{if } i \in K_f \cap [n], \quad \text{then } i \circ f^{\dagger} = \perp \circ O_p.$$

It remains to show for Case 2 where $i = n + 1$ that $i \circ f^{\dagger} = \perp \circ O_p$. But

$$(n+1) \circ f^{\dagger} = (n+1) \circ f \circ (I_p, f^{\dagger}) = (O_p \oplus u_{r+2}) \circ (I_p, f^{\dagger}) = u_{r+2} \circ f^{\dagger}.$$

By (A.6), $u_{r+2} \circ f^{\dagger} = \perp \circ O_p$.

The proof is complete since the three cases considered exhaust all possibilities. Thus if $if^{\sigma} = u_1 u_2 \cdots$ and there is a t such that $u_{t+1} = n + 1$, we choose r to be the smallest such t. Case 1 obtains if $(n+1)f^{\nu} = n + 1$ while Case 2 occurs if $(n+1)f^{\nu} \neq n + 1$. If there is no such t, then Case 3 obtains.

REFERENCES

[1] J. B. WRIGHT, J. W. THATCHER, E. G. WAGNER AND J. A. GOGUEN, *Rational algebraic theories and fixed point solutions*, Proc. IEEE Symp. Foundations of Comp. Sci. (Houston, Texas), Oct. 1976.

[2] S. L. BLOOM, S. GINALI AND J. RUTLEDGE, *Scalar and vector iteration*, J. Comput. System Sci., 14 (1977), pp. 251–256.

[3] S. L. BLOOM, C. C. ELGOT AND J. B. WRIGHT, *Solutions of the iteration equation and extensions of the scalar iteration operations*, this Journal, 9 (1980), pp. 25–45.

[4] S. L. BLOOM AND R. TINDELL, *Compatible orderings on the metric theory of trees*, this Journal, to appear.

[5] C. C. ELGOT, *Monadic computation and iterative algebraic theories*, *Logic Colloq. '73*, 80, Studies in Logic, North Holland, 1975.

[6] C. C. ELGOT, S. L. BLOOM AND R. TINDELL, *The algebraic structure of rooted trees*, J. Comput. System Sci., 16 (1978), no. 3, pp. 362–399; extended abstract in Proc. Johns Hopkins 1977 Conf. Inf. Sci. and Systems.

[7] J. MYCIELSKI AND W. TAYLOR, *A compactification of the algebra of terms*, Algebra Universalis, 6 (1976), pp. 159–163.

Structured Programming With and Without GO TO Statements

CALVIN C. ELGOT

Abstract—While "Dijkstra flow-chart schemes" (built out of assignment statement schemes by means of composition, IF—THEN and WHILEDO) are simple and perspicuous, they lack the descriptive power of flow-chart schemes (provided additional "variables" are not permitted). On the other hand, the analogous multiexit composition binary alternation-conditional iteration (CACI) schemes introduced below, which are virtually as simple and perspicuous as Dijkstra schemes, describe exactly the same computational processes as flow-chart schemes (without the aid of additional variables).

Theorem 9.1 makes contact with "reducible flow-graphs" an active area in its own right.

Index Terms—Algebra of flow-chart schemes, Böhm-Jacoppini theorem, composition, conditional iteration, Dijkstra program, flow-chart scheme, multientry, multiexit, structured programming.

I. INTRODUCTION

MUCH OF the discussion carried out under the banner "structured programming" focuses on eliminating "GO TO statements" on the one hand and programs constructed out of "assignment statements" by means of composition, alternation (i.e., IF—THEN), and WHILEDO on the other. (These "Dijkstra programs"—or rather the corresponding schemata—are defined and discussed in Section 5.) While it is not at all clear (outside the context of a particular programming language) what it means for a given program to contain "GO TO statements," there appears to be tacit agreement that the "Dijkstra programs" contain no such statements even though these "programs" are often expressed, e.g., by means of labeled directed graphs, without reference to any particular programming language.

It has been shown (cf., for example, [3], [8], [4], [1]) that every "flow-chart program" can be "translated" into a "Dijkstra program" *provided* one adjoins additional "variables"—or as I prefer to put it—*provided* one adjoins additional "memory." In any case there is considerable question whether these transformations are "desirable"; in particular, whether or not they promote perspicuity.

If one denies oneself additional memory this result no longer holds (cf. Section 5 and [1]).

Manuscript received November 26, 1975; revised December 1, 1975. The main body of this paper is a modification of notes for some of the lectures delivered at the NSF-CBMS Regional Research Conference on An Algebraic Analysis of Flowchart Algorithms, Stevens Institute of Technology, Hoboken, NJ, July 28–August 1, 1975. An earlier version of this paper appeared as IBM Research Report RC 5626 in 1975. There has been one change in terminology: "exit collapsing" has become "exit merging." Section 4 has been significantly augmented.

The author is with the Department of Mathematical Sciences, IBM Thomas J. Watson Research Center, Yorktown Heights, NY 10598.

The context of our discussion is "flow-chart algorithm schemes" defined in the next section. These schemes have n begin-vertices and p exits. They are sometimes referred to as *vector schemes*.

In the case that $n = 1$, we refer to the schemes as *scalar* (following terminology introduced in [6], for an analogous situation). In the case that $p = 1$ as well, we refer to the schemes as *biscalar* (following a suggestion of J. B. Wright).

Section 4 contains the main result. We define a subset of the set of scalar schemes called the CACI schemes. A CACI scheme is a near scalar analog of the Dijkstra scheme (which is biscalar). These schemes are built up by composition, binary alternation, and "conditional iteration." We indicate an algorithm which operates on a scalar scheme to produce a CACI scheme with the "same memory" which is "strongly equivalent" to it. A similar result holds for the closely related $\mathcal{G}$-schemes also defined here.

In Section 6 we define a set of scalar schemes called the CASCI schemes by replacing "conditional iteration" by a generalized WHILEDO operation and give an example of a scalar scheme which is not "weakly equivalent" to any CASCI scheme. Our point here is to argue that "conditional iteration" is "stronger" than generalized WHILEDO.

In Section 9 we argue briefly that one can deal with the totality of flow-chart schemes in a "structured way" and indicate one clear cut benefit to be derived from the use of the totality of scalar schemes as opposed to the subset of CACI schemes or $\mathcal{G}$-schemes. We also announce here a result obtained jointly with J. C. Shepherdson which characterizes a class $\mathcal{R}$ of scalar schemes properly larger than the class $\mathcal{G}$ (of $\mathcal{G}$-schemes). The class $\mathcal{R}$ enjoys pleasant structural properties making it rather *flexible*.

In Section 3 we define some basic operations on flow-chart schemes. All other operations on schemes are defined by using these basic operations.

The significance of "structured programming" in an industrial environment has been discussed by H. D. Mills and F. T. Baker, but is outside the scope of this paper.

The message we wish to convey, as succinctly as possible, is this. While "Dijkstra schemes" are simple and perspicuous, they lack the descriptive power of all flow-chart schemes (provided additional "variables" are not permitted). On the other hand the analogous CACI schemes (respectively, $\mathcal{G}$-schemes), which are virtually as simple and perspicuous as Dijkstra Schemes, describe exactly the same computational processes as flow-chart schemes (respectively, accessible flow-chart schemes) without the aid of "additional variables."

2. Definition of Monadic Flow-Chart Algorithm Scheme

The notion "flow-chart algorithm scheme" is obtained from the relationship between a flow-chart algorithm description and its interpretation by abstracting the form of the description. This involves, in particular, replacing in the (flow-chart algorithm) description operation letters, denoting specific operations on "data," by uninterpreted operation letters (similarly, for predicate letters). An interpretation of the algorithm scheme is uniquely determined by assigning to each operation letter an operation and assigning to each predicate letter a predicate.

Each operation letter and each predicate letter has a certain arity—the number of arguments it takes. Our considerations, however, are independent of this degree of detail. Thus we treat the operations and predicates as if they depend on one "vector variable" which varies over "state vectors." Indeed a flow-chart algorithm so abstracted would involve several occurrences of exactly one vector variable. Thus it is simpler to omit the variable altogether. This abstraction explains the term "monadic." It is analogous to the use of "monadic" in *monadic algebra* as treated by Halmos.

We abbreviate "monadic flow-chart algorithm scheme" by "flow-chart scheme" or sometimes, simply "scheme." Assume then a set Π given, whose elements we call *predicate letters*, and a set Ω given, whose elements we call *operation letters*. We shall define $\mathcal{F}(\Pi, \Omega)$ as the set of all flow-chart schemes relative to Π, Ω. Each flow-chart scheme F in $\mathcal{F}(\Pi, \Omega)$ will have a number of beginning vertices (frequently one) and a number of exits (frequently one). The notation $n \xrightarrow[s]{F} p$ will indicate a flow-chart scheme F with n begin-vertices, p exits, and s occurrences of "instructions" where n, p, s are nonnegative integers.

In what follows we shall use the word "graph" in the sense of "directed graph" and in a sense in which Fig. 1 qualifies as a graph. This graph consists of vertices v_1, v_2 and edges e_1, e_2, e_3 with edges e_1, e_2 pointing from v_1 to v_2 and edge e_3 pointing from v_1 to v_1. Vertex v_1 has *in-degree* 1 and *out-degree* 3 while vertex v_2 has in-degree 2 and out-degree 0.

A *flow-chart scheme* $n \xrightarrow[s]{F} p$ in $\mathcal{F}(\Pi, \Omega)$ consists of the following:

1) a *(finite) graph* whose vertices have out-degree at most two, the number (called the *weight* of F) of vertices of positive out-degree being s and the number of vertices of zero out-degree being p;

2) *a labeling* as follows: each vertex of out-degree 2 is labeled with an element from the set Π, one edge emanating from such a vertex is labeled τ (for true) and the other ϕ (for false); each vertex of out-degree 1 is labeled with an element of Ω;

3) *an enumeration* $\mathbf{e}_1, \mathbf{e}_2, \cdots, \mathbf{e}_p$ of the vertices of out-degree 0 (called *exits*); and

4) *a begin-function* b with domain $\{1, 2, \cdots, n\}$ whose values are vertices of the graph.

Hereafter, where n is a nonnegative integer, we use $[n]$ as an abbreviation for $\{1, 2, \cdots, n\}$.

Fig. 1.

In the case that $n = 1$, i.e., in the case of scalar schemes, specifying the begin-function amounts to specifying a single vertex, viz., $b(1)$, as the *begin-vertex*. In this case the data provided by 1) and 4) may be called a *pointed graph*. In the case that $p = 1$, there is only one enumeration 3) and so it need not be specified. Thus in the case $n = 1 = p$, i.e., in the case of biscalar schemes, a flow-chart scheme is given by a labeled pointed graph with exactly one vertex of out-degree 0.

In drawings representing flow-chart schemes we use a rhombus (diamond) to represent a vertex of out-degree 2, a circle to represent a vertex of out-degree 1 and a rectangle to represent a vertex of out-degree 0.

Let V be the set of vertices of positive out-degree and let E be the set of exits of the flow-chart scheme F. Then the labeled graph of F is fully specified by a function (called the *table of instructions*) of F which associates with $v \in V$ either a pair (ω, a) where $\omega \in \Omega$, $a \in V \cup E$ or a triple $(\pi; v_\tau, v_\phi)$ where $\pi \in \Pi$ and $v_\tau, v_\phi \in V \cup E$. If, for example, $(\pi; v_\tau, v_\phi)$ is associated with v, then v has out-degree 2 and its τ-edge is directed to vertex v_τ. The intended meaning (in this context) of the triple (π, v_τ, v_ϕ) may be suggested by the following. "Execute π. If outcome is τ go to v_τ; if outcome is ϕ go to v_ϕ." Similarly, the intended meaning of the pair (ω, a) is suggested by the following. "Execute ω. Go to a."

3. Operations on Flow-Chart Schemes

Our discussion will be facilitated by indicating a variety of operations on flow-chart schemes in $\mathcal{F}(\Pi, \Omega)$. We take it for granted that there is an obvious notion of isomorphism between schemes and we shall deal with schemes "up to isomorphism." An alternative approach (as we used in [6]) normalizes the notion of scheme.

With each $\omega \in \Omega$ there is a unique scheme $1 \xrightarrow[1]{\omega} 1$ which we may indicate by Fig. 2. This scheme will usually simply be described by: ω. The consequent ambiguity in the meaning of "ω" causes no difficulty since it may be resolved by context.

With each $\pi \in \Pi$, there is a unique scheme $1 \xrightarrow[1]{\pi} 2$ which we may indicate by Fig. 3. Notice the association of $1 \xrightarrow[1]{\pi} 2$ with $\pi \in \Pi$ involves, in effect, the convention of associating τ with 1 and ϕ with 2. We call these schemes *atomic*. Besides these we deal with some trivial schemes.

With each positive integer p and with each $j \in [p]$ there is a *trivial scheme* $1 \xrightarrow[0]{j_p} p$ consisting of exits $\mathbf{e}_1, \mathbf{e}_2, \cdots, \mathbf{e}_p$ with begin vertex $\mathbf{e}_j$. In the case $p = 3, j = 2$, the trivial scheme j_p may be indicated by Fig. 4.

Given flow-chart schemes $n \xrightarrow[s]{F} p$ and $p \xrightarrow[t]{G} q$, the operation of *composition* produces a scheme $n \xrightarrow[s+t]{F \cdot G} q$. The scheme $F \cdot G$ is obtained from F and G by identifying exit j of F with

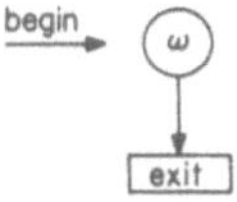

Fig. 2.

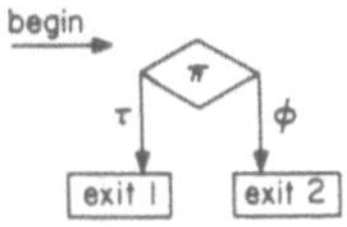

Fig. 3.

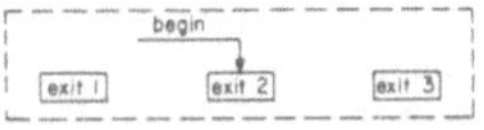

Fig. 4.

begin j of G for each $j \in [p]$. Pictorially, $F \cdot G$ (sometimes written FG) may be suggested by Fig. 5 where the heavy line marked "p" represents a "cable of p lines" (see [5] or [6] for an exact definition).

Given schemes $n_i \xrightarrow[s_i]{F_i} p_i$, $i \in [2]$, the operation of *separated source-pairing*, also called *circle sum*, produces a scheme $n_1 + n_2 \xrightarrow[s_1+s_2]{F_1 \oplus F_2} p_1 + p_2$ which may be suggested as shown by Fig. 6.

Given schemes $n_i \xrightarrow[s_i]{G_i} p$, $i \in [2]$, the operation of *source-pairing* produces a scheme $n_1 + n_2 \xrightarrow[s_1+s_2]{(G_1, G_2)} p$ which may be suggested as shown by Fig. 7 (see [2] for an exact definition). The scheme (G_1, G_2) may be obtained from the scheme $G_1 \oplus G_2$ by identifying exits j and $p + j$ of $G_1 \oplus G_2$, for each $j \in [p]$.

We remark incidentally that all three operations are associative but that none are commutative.

For example, the scheme $2 \xrightarrow[0]{(2_2, 1_2)} 2$ may be represented by Fig. 8., the scheme $\pi \cdot (2_2, 1_2)$ by Fig. 9., the scheme $\pi \cdot [1_1 \oplus \omega]$ by Fig. 10, and the scheme $\pi \cdot (2_2, 1_2) \cdot [1_1 \oplus \omega]$ by Fig. 11. Figs. 9-11 incorporate a *convention*: the topmost vertex is the begin.

We observe that $(1_2; 2_2) = [1_1 \oplus 1_1] = I_2$ is the identity $2 \longrightarrow 2$ (similarly for any $n > 0$).

It is outside the scope of this paper to systematically discuss how these operations are related to each other. We do, however, note that the separated source-pairing operation is expressible in terms of source-pairing, composition, and the trivial schemes. For example, if $n_1 \xrightarrow[s_1]{F_1} 2$, $n_2 \xrightarrow[s_2]{F_2} 3$; then $F_1 \oplus F_2 = (F_1 \cdot (1_5, 2_5), F_2 \cdot (3_5, 4_5, 5_5))$. We also observe that *for each n, p there is a one-to-one correspondence which preserves composition between functions* $[n] \xrightarrow{f} [p]$ *and vector schemes of weight zero*. In particular, the identity function I_n corresponds to the identity scheme $n \rightarrow n$. For example, if $n = 3$, $p = 5$, and $f(1) = 2$, $f(2) = 1$, $f(3) = 2$; then the corresponding scheme is $3 \xrightarrow[0]{(2_5, 1_5, 2_5)} 5$. Identifying these functions with their corresponding schemes enables us to express the verbal description given above concerning how

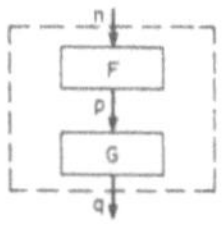

Fig. 5.

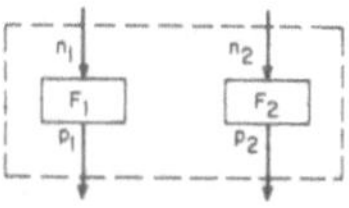

Fig. 6.

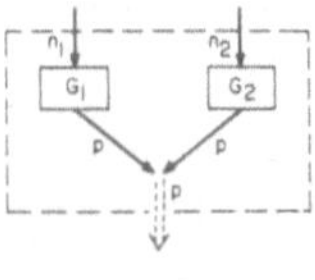

Fig. 7.

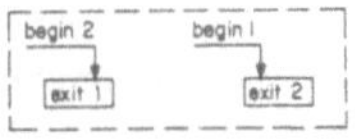

Fig. 8.

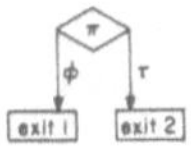

Fig. 9.

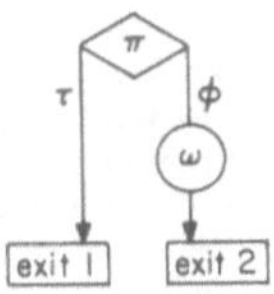

Fig. 10.

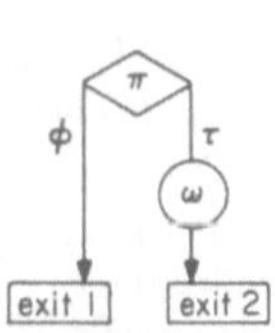

Fig. 11.

(G_1, G_2) is obtained from $G_1 \oplus G_2$ by the following equation:

$$(G_1, G_2) = [G_1 \oplus G_2] \cdot h$$

where $h: [2p] \longrightarrow [p]$ is the function which takes j and $p + j$ into j and $j \in [p]$.

The operation on schemes mentioned above, when applied to schemes which contain no closed nontrivial path, produce schemes which contain no closed nontrivial path. The following operations fail to satisfy that property.

Let $1 \xrightarrow[s]{F} p + 1$ be a flow-chart scheme and suppose the begin vertex is not an exit. For each j, $j \in [p + 1]$, by $1 \xrightarrow[s]{F^{(j)}} p$ we mean the scheme obtained from F by identifying the jth exit with the begin-vertex (or delete the jth exit and redirect the edges pointing to the jth exit, to the begin-vertex) and renumbering the exits to maintain the same order. We call these operations on schemes *(scalar) conditional iterations.*

Each of the p operations, $F^{(j)}$, $j \in [p]$, is expressible in terms of $F^{(p+1)}$ with the help of the other operations. Indeed, $F^{(j)} = (F \cdot h)^{(p+1)}$ where h is a function from $[p + 1]$ into itself which keeps $1, 2, \cdots, j - 1$ fixed, takes j into $p + 1$, $j + 1$ into j, $j + 2$ into $j + 1, \cdots, p + 1$ into p. Following [5], we shall write $F^{\dagger}$ in place of $F^{(p+1)}$. Similarly, $F^{(j)}$, $2 \leqslant j \leqslant p + 1$ is expressible in terms of $F^{(1)}$. (In [6] we have used $F^{\dagger}$ to mean $F^{(1)}$.)

Pictorially, the dagger operation applied to the scheme shown by Fig. 12 yields the scheme shown by Fig. 13. The intended meaning of $F^{\dagger}$ may be suggested by the following. "Repeat F until an exit other than $p + 1$ is reached."

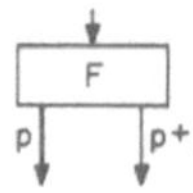

Fig. 12.

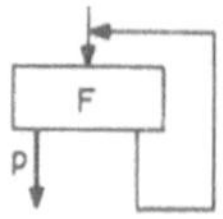

Fig. 13.

Let $G: s \longrightarrow s + p$ be a flow-chart scheme. Let ${}^{\dagger}G: s \longrightarrow p$ be the flow-chart scheme obtained from G by identifying, for each $i \in [s]$, the ith begin vertex with the ith exit. This is one form of *vector iteration*. Another form of vector iteration, of which the scalar iteration discussed immediately above is a special case, operates on the above $G: s \longrightarrow p + s$ to produce $G^{\dagger}: s \longrightarrow p$ by identifying, for each $i \in [s]$, the ith begin vertex with the $(p + i)$th exit. Each of these two operations is definable in terms of the other in the presence of composition (of vector schemes) and trivial schemes.

Theorem 3.1: Every flow-chart scheme can be built up from j_p, Π and Ω by means of composition, source-pairing, and vector iteration.

Proof Indication: Let $n \xrightarrow[s]{F} p$ be a flow-chart scheme. Let the set of vertices of F be $[s + p]$ with $s + 1, s + 2, \cdots, s + p$ the exits in order. Let $s + p \xrightarrow[s]{H} p$ be the flow-chart scheme whose begin function is the identity $I_{s+p}: [s + p] \longrightarrow [s + p]$, but is otherwise identical to F. Then $F = b \cdot H$ where $b: [n] \longrightarrow [s + p]$ is the begin function of F and $H = ({}^{\dagger}G, I_p)$, where $s \xrightarrow[s]{G} s + p$ so that $s \xrightarrow[s]{{}^{\dagger}G} p$ and $G = (G_1, G_2, \cdots, G_s)$ where, for each $i \in [s]$, $1 \xrightarrow[1]{G_i} s + p$ is defined by the following.

$\omega \cdot j_{s+p}$ if vertex i (in F or in H) has label ω and its edge goes to vertex j; and $\pi \cdot (j_{s+p}, k_{s+p})$ if vertex i (in F or in H) has label π, its τ-edge goes to j and its ϕ-edge goes to k.

Thus $F = b \cdot ({}^{\dagger}G, I_p)$. (Compare this formula with the definition of $|D|$ in "Monadic Computation . . . ," Section 5.)

The description of $G = (G_1, G_2, \cdots, G_s)$ above bears a strong resemblance to the "table of instructions" described in Section 2. Indeed, G_i corresponds to the ith "instruction" in the table of s instructions. The "intended meaning" of an instruction then reveals—or at least suggests—that the trivial schemes j_p, $p > 1$, play the role of "unconditional GO TO statements," while the pairings (j_p, k_p) of two such schemes play the role of "conditional GO TO statements."

This raises the following question. Is it possible to obtain a result similar to Theorem 4.1 or Corollary 4.2 without the use of the trivial schemes $j_p, p > 1$? This question is answered in the affirmative in Theorem 4.3. First, however, we observe that j_p is not an accessible scheme if $p > 1$, e.g., exit 2 is not accessible from the begin (vertex) in the scheme $1 \xrightarrow[0]{1_2} 2$. Indeed the scheme $n \xrightarrow[0]{f} p$ is accessible (i.e., each vertex is accessible from some begin) if the function $[n] \xrightarrow{f} [p]$ is surjective (i.e., ONTO). We further observe that the property of accessibility is preserved by composition, circle sum, source-pairing, and iteration (vector as well as scalar); moreover, the atomic schemes are accessible.

4. The Adequacy of CACI Schemes

Given $\pi \in \Pi$, $1 \xrightarrow[s_i]{F_i} p$, $i = 1, 2$, the operation of *binary alternation* produces the scheme

$$1 \xrightarrow[1 + s_1 + s_2]{\pi \cdot (F_1, F_2)} p$$

which may be suggested by the (simplified) picture (cf., Fig. 7) shown by Fig. 14. ("Alternation," as used for example in [19], is this same operation but restricted to certain biscalar schemes.)

A *CACI scheme* in $\mathfrak{F}(\Pi, \Omega)$ is a scalar scheme which can be built up from the trivial schemes j_p, $p > 0$, $j \in [p]$ and the atomic schemes ω, $\omega \in \Omega$, by a finite number of applications of composition (C), binary alternation (A), and conditional iteration (CI). In particular, if F and G are CACI schemes $F \cdot G$ makes sense iff F is a biscalar scheme. The set of CACI schemes may be described as the smallest set of flow-chart schemes which contains the trivial schemes j_p, $j \in [p]$, the

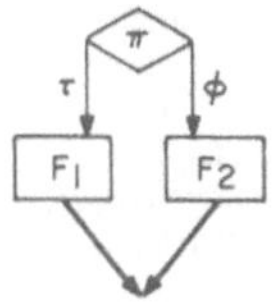

Fig. 14.

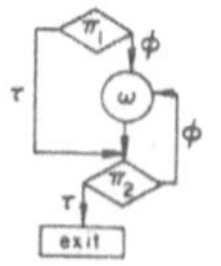

Fig. 15.

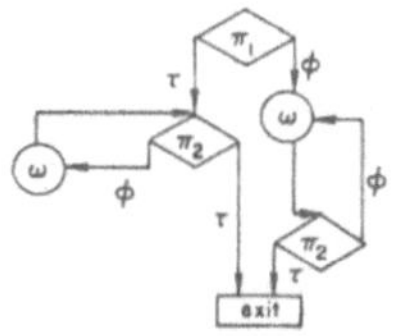

Fig. 16.

atomic schemes ω and is closed with respect to composition, binary alternation, and conditional iteration. ("*Closed* with respect to binary alternation" means: if F_1 and F_2 are in the set, then so is $\pi \cdot (F_1, F_2)$ for each $\pi \in \Pi$.)

It should first be observed that not every scalar flow-chart scheme is a CACI scheme. Indeed Fig. 15, suggested by E. G. Wagner, gives an example of a biscalar scheme which is not a CACI scheme. (This follows from Theorems 4.6 and 9.1.)

The scheme of Fig. 15 is "strongly equivalent" (in a sense which we shall straightaway make precise) to the scheme $\pi_1 \cdot ([\pi_2 \cdot [1_1 \oplus \omega]^\dagger,\ [\omega \cdot \pi_2]^\dagger)$ which may be depicted by Fig. 16.

Let $n \xrightarrow[s]{F} p$ be a flow-chart scheme in $\mathcal{F}(\Pi, \Omega)$. We wish to associate with F an $n \times p$ matrix of sets of words together with an n-tuple of sets of infinite "words" on a certain alphabet A. The alphabet $A = \Omega \cup \Pi_\tau \cup \Pi_\phi$ where $\Pi_\tau = \{\pi_\tau | \pi \in \Pi\}$ and $\Pi_\phi = \{\pi_\phi | \pi \in \Pi\}$. Thus, e.g., if Ω has r elements and Π has s then A has $r + 2s$ elements. Given any path in F, finite or infinite, we associate with the path in an obvious way a finite or infinite word (the *label of the path*) in the alphabet A. By the *finite strong behavior* of F we mean the $n \times p$ matrix whose (i, j)th entry is the set of labels of paths from the ith begin vertex to the jth exit; by the *infinite strong behavior* of F we mean the n-tuple of sets whose ith term is the set of labels of infinite paths beginning with begin-vertex i; by the *strong behavior* $\|F\|$ of F we mean the pair consisting of the finite and infinite strong behaviors of F. If $n \xrightarrow[s_i]{F_i} p$, $i = 1, 2$, is in $\mathcal{F}(\Pi, \Omega)$ as well, we say F_1 and F_2 are *strongly* equivalent if $\|F_1\| = \|F_2\|$.

The schemes given by Figs. 15 and 16 are strongly equivalent.

Theorem 4.1: Given any flow-chart scheme $1 \longrightarrow p$, there is a strongly equivalent CACI scheme $1 \longrightarrow p$. Indeed there is an algorithm which operates on the given scheme to produce the strongly equivalent CACI scheme.

The proof of this theorem is very similar to the proof of Theorem 4.3 in [5] and is not given here. We do, however, illustrate in the Appendix application of the algorithm to a flow-chart scheme given in [1, fig. 1]. This illustration, we expect, is sufficient to convince the reader of the truth of the theorem.

Corollary 4.2: Given any accessible flow-chart scheme $1 \longrightarrow p$, there is a strongly equivalent accessible CACI scheme $1 \longrightarrow p$. Indeed the algorithm alluded to above produces the CACI scheme.

In building up such an equivalent accessible CACI scheme we may pass through "intermediate" inaccessible CACI schemes. The class $\mathcal{G}$ defined consists only of accessible schemes and the analogue of Corollary 4.2 (given by Theorem 4.3) holds.

Given $\pi \in \Pi$ and scalar schemes $1 \xrightarrow[s_i]{F_i} p_i, i \in [2]$, *separated binary alternation* produces the scheme $1 \xrightarrow[s_1 + s_2]{\pi \cdot [F_1 \oplus F_2]} p_1 + p_2$.

Given schemes $n \xrightarrow[s]{F} p$ and $p \xrightarrow[0]{f} q$, where $[p] \xrightarrow{f} [q]$ is surjective, *exit-merging*, or briefly, *merging*, produces the scheme $F \cdot f$, e.g., if $p = 2$ and $q = 1$, $F \cdot f$ is obtained from F by identifying the two exits of F. Let $\mathcal{G}$ be the smallest set of scalar schemes which contains 1_1, Ω and is closed under composition, separated binary alternation, merging and scalar iteration. Call a scheme in $\mathcal{G}$ a $\mathcal{G}$*-scheme.*

Theorem 4.3: Every accessible scalar flow-chart scheme is strongly equivalent to a $\mathcal{G}$-scheme.

The proof is similar to that of Theorem 4.1. With reference to the accessible scheme S_1 of the Appendix we obtain

$$S_1 = a \cdot S_2, S_2 = S_3^\dagger,$$
$$S_3 = p \cdot [e \oplus S_4] \cdot \alpha, \text{where } [3] \xrightarrow{\alpha} [2], \alpha(1) = 2, \alpha(2) = \alpha(3) = 1,$$
$$S_4 = q \cdot [S_7 \oplus g] \cdot \beta, \text{where } [3] \xrightarrow{\beta} [2], \beta(1) = \beta(3) = 1, \beta(2) = 2,$$
$$S_7 = b \cdot S_8, S_8 = S_9^\dagger,$$
$$S_9 = r \cdot [d \oplus S_{10}] \cdot \gamma, [3] \xrightarrow{\gamma} [3], \gamma(1) = 3, \gamma(2) = 1, \gamma(3) = 2,$$
$$S_{10} = s \cdot [c \oplus f] \cdot \delta, [2] \xrightarrow{\delta} [2], \delta(1) = 2, \delta(2) = 1.$$

We make the following observation.

Proposition 4.4: Any set of schemes (respectively, scalar schemes) closed under $\oplus$ (respectively, separated binary alternation) and closed under merging is also closed under source pairing (respectively, binary alternation).

Corollary 4.5: The set $\mathcal{G}$ is closed under binary alternation.

We do not have a thorough comparison of $\mathcal{G}$-schemes and CACI schemes. However, we do have Theorem 4.6.

Theorem 4.6: Every accessible CACI scheme is a $\mathcal{G}$-scheme.

The theorem may be proved by proving the stronger result (which we merely suggest): the "accessible part" of a CACI scheme is a $\mathcal{G}$-scheme.

The reader may have noted that the above results admit easy generalization to a broader context where the underlying graphs are not restricted to having an out-degree of at most 2. To be explicit, let $\Gamma = (\Gamma_0, \Gamma_1, \Gamma_2, \Gamma_3, \cdots)$ be a sequence of pairwise-disjoint sets. This more general situation will reduce to the one already considered by taking $\Gamma_1 = \Omega, \Gamma_2 = \Pi$ and taking all other sets to be empty. The definition (given in a different style than the earlier one for the special case) of the set $\mathcal{F}(\Gamma)$ of finite flow-chart schemes relative to Γ follows. A *flow-chart scheme* $n \xrightarrow[s]{F} p$ in $\mathcal{F}(\Gamma)$ consists of the following:

1) a finite set V_F (whose elements are called *vertices*),
2) a function $\theta: V_F \longrightarrow N = \{0, 1, 2, \cdots\}$, called the *out-degree function*,
3) an injective (i.e., one to one) function $e: [p] \longrightarrow \theta^{-1}(0) \subseteq V_F$ called the *exit function*, $e(j)$ is the jth exit,
4) a function $\tau: E_F \longrightarrow V_F$, where $E_F = \{(v, i) | v \in V_F, i \in [\theta(v)]\}$ is called the set of *edges* of F and τ is called the *target function*,
5) a function (called the *labeling function*), $\lambda: V_F^- \longrightarrow \Gamma_0 \cup \Gamma_1 \cup \Gamma_2 \cup \cdots$, satisfying $\lambda(v) \in \Gamma_{\theta(v)}$, where V_F^- is the set of nonexit vertices of F, and
6) a function $b: [n] \longrightarrow V_F$ called the *begin function*.

The atomic scheme $1 \xrightarrow[1]{\gamma} n$ corresponding to $\gamma \in \Gamma_n$, $n \geqslant 0$ consists of one nonexit vertex, say v, n exits, and n edges (v, i), $i \in [n]$; $\tau(v, i)$ is the ith exit, $\lambda(v) = \gamma$ and $b(1) = v$.

Given $\gamma \in \Gamma_n$, $1 \xrightarrow[s_i]{F_i} p$, $i \in [n]$, the operation of *atomic substitution* (i.e., substitution in atoms) produces the scheme

$$1 \xrightarrow[1 + s_1 + s_2 + \cdots + s_n]{\gamma \cdot (F_1, F_2, \cdots, F_n)} p.$$

In the case that $n = 0$, we understand $(F_1, F_2, \cdots, F_n)$ to denote the unique scheme $0_p : 0 \xrightarrow[0]{} p$.

The set of *CACI schemes* in $\mathcal{F}(\Gamma)$ is the smallest set of scalar schemes which contains the trivial schemes j_p, $j \in [p]$, and is closed with respect to composition, atomic substitution, and scalar iteration. In particular, for each n and for each $\gamma \in \Gamma_n$ the atomic scheme γ is a CACI scheme.

Given $\gamma \in \Gamma_n$, $1 \xrightarrow[s_i]{F_i} p_i$, $i \in [n]$, the operation of *separated atomic substitution* produces the scheme

$$1 \xrightarrow[1 + s_1 + s_2 + \cdots + s_n]{\gamma \cdot [F_1 \oplus F_2 \oplus \cdots \oplus F_n]} p_1 + p_2 + \cdots + p_n.$$

In the case that $n = 0$, we understand $[F_1 \oplus F_2 \oplus \cdots \oplus F_n]$ to denote the unique scheme $0_0 : 0 \xrightarrow[0]{} 0$ (which is an identity with respect to composition). Thus, if a set of schemes is closed with respect to separated atomic substitution it contains Γ_n for every $n \geqslant 0$.

The set of $\mathcal{G}$-schemes in $\mathcal{F}(\Gamma)$ is the smallest set of scalar schemes which contains 1_1 and is closed under composition, separated atomic substitution, merging, and scalar iteration.

It remains to extend the definition of strong equivalence to schemes in $\mathcal{F}(\Gamma)$. To this end we redefine $A = \bigcup_{n>0} (\Gamma_n \times [n])$. The *strong behavior* $\|F\|$ of $n \xrightarrow{F} p$ in $\mathcal{F}(\Gamma)$ is a pair consisting of an $n \times p$ matrix $\|F\| \P$ and an n-tuple $\|F\| \cdot 0$ (notation taken from [7]); the (i, j)th entry of $\|F\| \P$, $i \in [n]$, $j \in [p]$, is a subset of A^*, viz., the set of words $u \in A^*$ such that there exists a path from begin i to exit j whose label is u; the ith entry of $\|F\| \cdot 0$ is a subset of $A^\infty \cup A^* \cdot \Gamma_0$ (where A^∞ is the set of infinite sequences $a_1 a_2 a_3, \cdots, a_k \in A$), viz., the set of words $v \in A^\infty$ such that there exists an infinite path from begin i whose label is v, *union* the set of words $w \in A^* \cdot \Gamma_0$ such that there exists a (finite) path whose label is w from begin i to a nonexit vertex of out-degree 0. Again, F is *strongly equivalent* to $n \xrightarrow{G} p$ if $\|F\| = \|G\|$. We now review the results of this section.

Theorem 4.1, Corollary 4.2, Theorem 4.3, and Theorem 4.6 all hold unchanged in $\mathcal{F}(\Gamma)$ while Proposition 4.4 and Corollary 4.5 hold in $\mathcal{F}(\Gamma)$ with "binary alternation" replaced by "atomic substitution" and "separated binary alternation" replaced by "separated atomic substitution."

We now indicate by Theorem 4.7 the *semantical significance* of strong equivalence of flow-chart schemes in $\mathcal{F}(\Gamma)$. By an (unrestricted, functional) interpretation I of Γ we mean a function which assigns to $\gamma \in \Gamma_n$, $n > 0$, a pair of functions

$$X \xrightarrow{I_o(\gamma)} [n], \text{ the } \textit{outcome function}, \text{ and}$$

$$X \xrightarrow{I_r(\gamma)} X, \text{ the } \textit{result function} \text{ and}$$

assigns to $\gamma \in \Gamma_0$ a function $X \xrightarrow{I(\gamma)} Z$, where X and Z are sets. The set X may be regarded as the range of values of the interpreted "vector variable" mentioned in Section 2. We leave open the question: "How is the set Z to be regarded?" Of course, we are not confronted with this question in the case that $\Gamma_0 = \phi$, which is the case in the ungeneralized context.

In the case $n > 1$, $I_o(\gamma)$ may be regarded as a "test" with many outcomes. In the case $n = 1$ there is a unique function $X \longrightarrow [1]$ so that in this case the outcome function is redundant. In Section 6 we are interested in *restricted* (functional) *interpretations*. These are interpretations in which for $n > 1$, $I_r(\gamma)$ is the identity function.

Let $F: n \longrightarrow p$ be a flow-chart scheme in $\mathcal{F}(\Gamma)$ and let I be an unrestricted interpretation of Γ. In preparation for defining the "strong behavior" of (F, I) we define "computation" of (F, I). By a (maximal) computation of (F, I) from begin i we mean an infinite sequence or a finite sequence of $l \geqslant 0$ "steps."

$$i_0 v_0 x_0 i_1 v_1 x_1 \cdots i_l v_l x_l \cdots \tag{4.1}$$

satisfying where γ_i is the label of v_i in F

1) $i_0 = i, v_0 = b_F(i), \quad x_0 \in X$
2) for $0 \leqslant j < l$, (4.2)

$$\tau_F(v_j, i_{j+1}) = v_{j+1} \in V_F$$

$$x_j \xmapsto{I_o(\gamma)} i_{j+1} \in [\theta(\gamma_j)]$$

$$x_j \xmapsto{I_r(\gamma_j)} x_{j+1} \in X$$

3) and in the case that the sequence is finite vertex v_l has out-degree 0; if v_l is the jth exit then (4.1) is a computa-

tion to exit j; if v_l is not an exit then (4.1) is $\cdots i_l v_l x_l z$ where $x_l \overset{I(\gamma)}{\longmapsto} z$.

By the *track* (with outcomes) of the computation (4.1) we mean the infinite sequence

$$x_0 i_1 x_1 i_2 x_2 \cdots i_l x_l \cdots \tag{4.3}$$

if (4.1) is infinite, we mean the finite sequence

$$x_0 i_1 x_1 i_2 x_2 \cdots i_l x_l \tag{4.4}$$

if (4.1) is finite and v_l is an exit; and we mean the finite sequence

$$x_0 i_1 x_1 i_2 x_2 \cdots i_l x_l z \tag{4.5}$$

if (4.1) is finite and v_l is a nonexit. By the *strong behavior* $\|F, I\|$ of the interpreted scheme (F, I) we mean the pair $(U, u) = (\|F, I\|\P, \|F, I\| \cdot 0)$ where U is an $n \times p$ matrix of sets and u is an n-vector of sets defined by, where $i \in [n]$, $j \in [p]$:

1) the sequence (4.4) is in U_{ij} if and only if (4.4) is the track of a computation of (F, I) from begin i to exit j; and
2) the sequence (4.3) [respectively, the sequence (4.5)] is in u_i if and only if the sequence (4.3) [respectively, the sequence (4.5)] is the track of a computation from begin i.

Let $n \xrightarrow[s_i]{F_i} p, i \in [2]$ be flow-chart schemes in $\mathcal{F}(\Gamma)$.

Theorem 4.7: $\|F_1\| = \|F_2\| \iff \|F_1, I\| = \|F_2, I\|$ for all I, where I is any unrestricted functional interpretation.

We hint at a proof that $\|F_1\| \neq \|F_2\| \Longrightarrow \|F_1 \cdot I\| \neq \|F_2, I\|$ for some I as follows: choose $X = \overline{\Gamma}^*$ where $\overline{\Gamma} = \Gamma_1 \cup \Gamma_2 \cup \Gamma_3 \cup \cdots$ and choose $Z = \overline{\Gamma}^* \cdot \Gamma_0$; for $\gamma \in \Gamma_n$, $n > 0$ choose $I_r(\gamma)$ so that $w \overset{I_r(\gamma)}{\longmapsto} w \cdot \gamma$ (the choice of $I_0(\gamma)$ may depend on how $\|F_1\|$ and $\|F_2\|$ differ) and choose $I(\gamma)$, for $\gamma \in \Gamma_0$, so that $w \overset{I(\gamma)}{\longmapsto} w \cdot \gamma$, where $w \in \overline{\Gamma}^*$.

5. The WHILEDO Operations and CAWD (Alias D) Scheme

Given $\pi \in \Pi$ and $1 \xrightarrow[s]{F} 1$, we form new schemes

$$1 \xrightarrow[1+s]{[\pi \cdot [1_1 \oplus F]]^\dagger} 1, \quad 1 \xrightarrow[1+s]{[\pi \cdot (2_2, 1_2) \cdot [1_1 \oplus F]]^\dagger} 1$$

whose intended meanings may respectively be suggested by "WHILE NOT π DO F", "WHILE π DO F" and may, respectively, be depicted by Figs. 17 and 18. We call these two operations on schemes the WHILEDO *operations*.

By a *CAWD scheme* (relative to Π, Ω) we mean a biscalar scheme in $\mathcal{F}(\Pi, \Omega)$ built up from the trivial scheme 1_1 and the atomic schemes ω, $\omega \in \Omega$, by a finite number of applications of composition, alternation, and WHILEDO's. These schemes (with the exception of 1_1) are called *D*-schemes (*D* for Dijkstra) in [19].

Knuth and Floyd in [17] have given an example of a biscalar flow-chart scheme such that no CAWD scheme is strongly equivalent to it. The following simpler example of such a scheme is given by Kosaraju in [19]. (Actually, as noted by

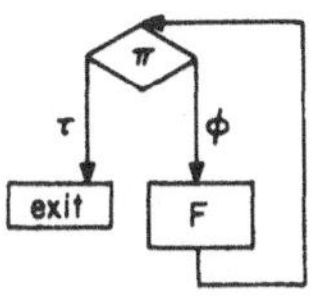

Fig. 17.

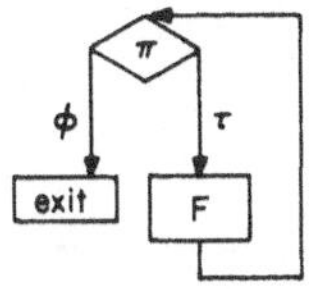

Fig. 18.

Kosaraju, the result is stronger: no CAWD scheme is "weakly equivalent" to Fig. 19.) It remains to define "weak equivalence."

In order to introduce "weak equivalence" in a manner analogous to "strong equivalence" we turn to a description of "weak equivalence contracted flow-chart scheme" or briefly "contracted scheme" and the "contraction" F' of a flow-chart scheme F.

6. Weak Equivalence Contracted Flow-Chart Schemes

For the schemes in question we shall use as "labels" either elements of Ω or elements in the Boolean algebra $\mathcal{B}(\Pi)$ freely generated by Π. The Boolean algebra $\mathcal{B}(\Pi)$ is uniquely determined up to isomorphism by Π. In the case that Π has r elements $\mathcal{B}(\Pi)$ has 2^{2^r} elements and may be described by the set of all subsets of any set with 2^r elements under union, intersection, and complementation. To be more definite, we may take as the 2^r elements the set $\{\tau, \phi\}^\Pi$ of all (truth) functions from Π into $\{\tau, \phi\}$. If the elements of Π are ordered $\pi_1, \pi_2, \cdots, \pi_r$, we may take as the 2^r elements the set of all sequences $t_1 t_2 \cdots t_r$ where t_i is τ or ϕ. We call $t_1 t_2 \cdots t_r$ an *evaluation sequence* for Π.

In order to avoid undue complication *we shall only deal with the case that Π is finite.*

A *weak equivalence contracted flow-chart scheme* (or briefly a *contracted scheme*) $n \longrightarrow p$ relative to Π, Ω consists of an underlying graph with exactly n vertices of in-degree 0 which are ordered (the *entrances* of the scheme), p vertices of out-degree 0 which are ordered (the *exits* of the scheme) and a labeling as follows. Each vertex, not an entrance or an exit, is labeled with an element from Ω. Each edge is labeled with a nonzero element from $\mathcal{B}(\Pi)$, i.e., a nonempty set of evaluation sequences, distinct edges emanating from the same vertex carrying disjoint labels. If, for each vertex, the union of the labels emanating from that vertex is the 1 of $\mathcal{B}(\Pi)$, i.e., the set of all evaluation sequences, then the scheme is said to be *total*.

Let F be a flow-chart scheme $n \longrightarrow p$. By the (*weak equivalence*) *contraction* F' of F we mean the contracted scheme $n \longrightarrow p$ whose set of vertices is the set of vertices of F of out-degree $\leqslant 1$ (i.e., the operation vertices and the exits of F) aug-

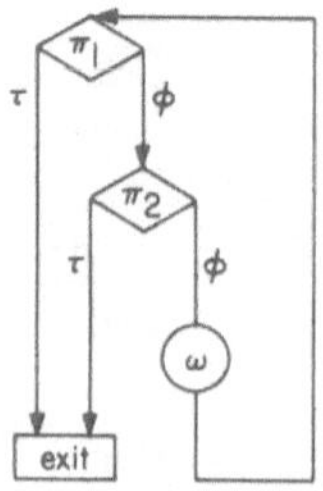

Fig. 19.

mented by n additional vertices (the *entrances* of F') labeled as described below. The operation vertices of F are labeled the same way in F' as in F and the exits of F are ordered the same way in F' as in F. The description of the edges and of the labels on edges of F' is, however, more complicated.

Let v be an operation or predicate vertex of F and let $t_1 t_2 \cdots t_r$ be an evaluation sequence. The evaluation sequence determines a path in F which begins with v, passes through zero or more predicate vertices of F and either

1) terminates in an operation vertex of F, or
2) terminates in an exit of F, or
3) does not terminate

as follows. If v is an operation vertex, let e be its outedge. If v is a predicate vertex with label π_i, $i \in [r]$, let e be the t_i outedge of v. Suppose e points to v' where v' is an operation vertex or an exit of F. Then the path determined by $t_1 t_2 \cdots t_r$ beginning with v is simply $v\, e\, v'$. Suppose on the other hand that v' is a predicate vertex of F and its label is π_j, $j \in [r]$; let e' be the t_j edge of v' and suppose e' points to v''. If v'' is an operation vertex or an exit of F, then $v\, e\, v'\, e'\, v''$ is the path determined by $t_1 t_2 \cdots t_r$ beginning with v. If v'' is a predicate vertex, the process continues.

Let v be an operation vertex of F and let v' be an operation vertex or exit of F. Let β be the set of evaluation sequences which determine a path in F from v to v'. If $\beta \neq \phi$, there is to be exactly one edge from v to v' in F' and the edge is to be labeled β; if $\beta = \phi$, there is to be no edge from v to v' in F'.

Suppose now that v is the ith, $i \in [n]$, begin-vertex of F and it is a predicate vertex while v' and β are as above. If $\beta \neq \phi$ there is to be exactly one edge in F' from the ith entrance of F' to v' and it is to be labeled β; if $\beta = \phi$, there is to be no edge in F' from its ith entrance to v'. In the case the ith begin-vertex $b(i)$ of F is an operation vertex, there is to be exactly one edge from the ith entrance of F' to $b(i)$ and this edge is to be labeled with the 1 of $\mathfrak{B}(\Pi)$. *This completes the description of F'.*

If F is the flow-chart scheme of Fig. 19; then F' is the scheme shown by Fig. 20 where $\overline{\phi\phi} = \{\tau\tau, \tau\phi, \phi, \tau\}$ is the complement of $\phi\phi$ in $\mathfrak{B}(\Pi)$.

To obtain somewhat more illuminating contracted schemes we give in Figs. 21 and 22 contractions of schemes S_8 and S_7 (cf. Appendix); we assume $r, s \in \Pi$ and order r, s alphabetically (ignoring whatever other elements Π may have).

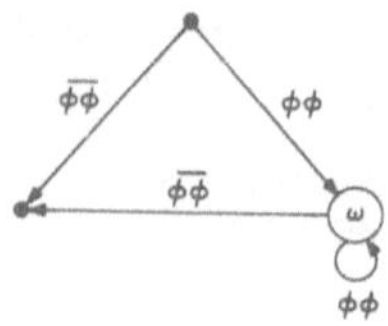

Fig. 20.

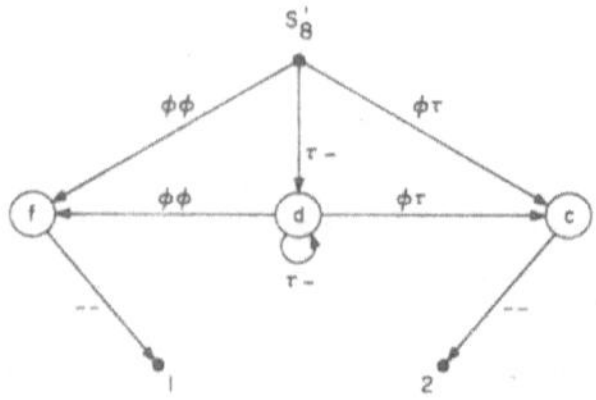

Fig. 21.

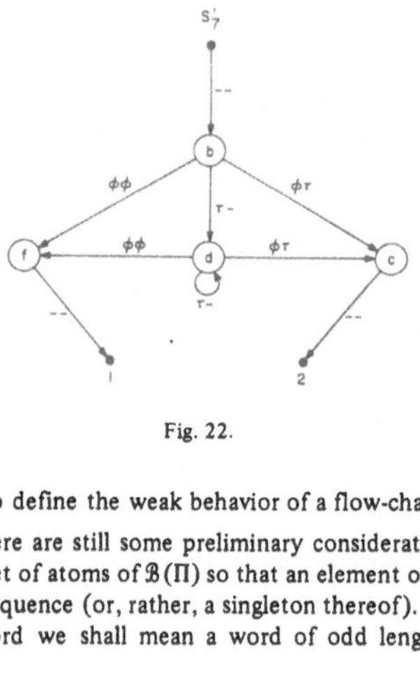

Fig. 22.

In order to define the weak behavior of a flow-chart scheme $n \xrightarrow{F} p$, there are still some preliminary considerations. Let AT be the set of atoms of $\mathfrak{B}(\Pi)$ so that an element of AT is an evaluation sequence (or, rather, a singleton thereof). By an alternating word we shall mean a word of odd length of the form

$$a_1 \omega_1 a_2 \omega_2 \cdots a_l \omega_l a_{l+1}, \tag{6.1}$$

where $l \geq 0$, $a_i \in AT$, $\omega_j \in \Omega$. If $A_i \subset AT$ for each i, then the sequence

$$A_1 \omega_1 A_2 \omega_2 \cdots A_l \omega_l A_{l+1} \tag{6.2}$$

is to be understood as denoting the set of alternating words (6.1) such that $a_i \in A_i$ for each i. If

$$v_0 e_1 v_1 e_2 v_2 \cdots e_l v_l e_{l+1} v_{l+1} \tag{6.3}$$

is a path in F' from an entrance v_0 to an exit v_{l+1}, by the *label of the path* we mean the set

$$e_1' v_1' e_2' v_2' \cdots e_l' v_l' e_{l+1}' \tag{6.4}$$

of alternating words, where $e_i' \in \mathfrak{B}(\Pi)$ is the label of edge e_i in F' and $v_j' \in \Omega$ is the label of vertex v_j in F'.

By the *weak behavior* $|F|$ of F we mean the $n \times p$ matrix whose (i,j)th entry $i \in [n], j \in [p]$, is the union of the labels of paths from the ith entrance to the jth exit of F'.

If F_1, F_2 are flow-chart schemes $n \longrightarrow p$ we say they are *weakly equivalent* if their weak behaviors are the same, i.e., if $|F_1| = |F_2|$.

If $n \xrightarrow{F} p$ is a flow-chart scheme and I is a (restricted) *interpretation*, i.e., a function which assigns to $\pi \in \Pi$ a predicate $I(\pi)$ on a set X and I assigns to $\omega \in \Omega$ an operation $I(\omega)$ on X, then the pair (F, I) has associated with it, in a way which we assume is familiar, a partial function $|F, I| : X \times [n] \longrightarrow X \times [p]$. We call $|F, I|$ the *weak behavior* of (F, I). It is known, essentially from the work of J. D. Rutledge [16] on the one hand and Luckham, Park, and Paterson [18] on the other, that $|F_1| = |F_2|$ iff for all restricted interpretations $I, |F_1, I| = |F_2, I|$.

Call a scheme F *well-formed* iff there is no nontrivial path in F (i.e., no path containing at least one edge) from a predicate vertex to itself which passes only through predicate vertices. We note incidentally that *if F is well formed; then F' is total.*

There is a variant to the notion "contracted scheme," usable in the case that Π is finite, which should be mentioned. The variant is obtained from a contracted scheme by replacing an edge from vertex v to vertex v' labeled $\beta \in \mathfrak{B}(\Pi)$, where $\beta = \{a_1, a_2, \cdots, a_r\}$ and the a_i's are distinct atoms, by r (distinct) edges from v to v' and these edges are respectively labeled by $a_1, a_2, \cdots, a_r$. If there are m atoms in $\mathfrak{B}(\Pi)$ and the given contracted scheme is total; then the variant of this contracted scheme has for each nonexit vertex v and each atom $a \in \mathfrak{B}(\Pi)$ exactly one outedge labeled a.

We have not chosen the variant notion as the notion "contracted scheme" both because the variant notion is not applicable in case Π is infinite and because the labeling of a contracted scheme is more efficient (resulting for example in a graph with fewer edges) and probably more perspicuous. On the other hand, the variant notion is closer to the notion "finite automaton" and "sequential machine" and these similarities may be exploited. In any case the notion and its variant are so simply related that one may be regarded as an "abbreviation" for the other in the case that Π is finite.

7. The Inadequacy of Composition, Alternation, and Generalized WHILEDO

In Section 4 we indicated that for every flow-chart scheme $1 \longrightarrow p$ there is a *strongly* equivalent CACI scheme $1 \longrightarrow p$. In Section 5 it was noted that no CAWD scheme is weakly equivalent to the Kosaraju scheme Fig. 19. The two results taken together might suggest that "conditional iteration" is more powerful than WHILEDO's. The comparison is, however, marred by the fact that the milieu of the CACI result is the totality of scalar schemes while the milieu of the CAWD result is the totality of biscalar schemes. One might suspect, therefore, that the striking difference in the direction of these results is due to the difference in milieu rather than in the relative power of "conditional iteration" and WHILEDO's. Fortunately, it is possible to extend the milieu of CAWD schemes in a natural way to scalar schemes CASCI extending the notion

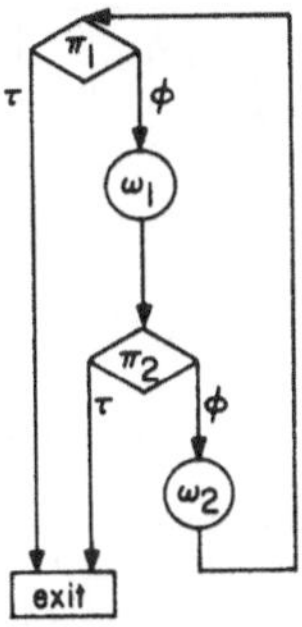

Fig. 23.

of WHILEDO. The resulting weak behaviors of scalar schemes will easily be seen to properly include the weak behaviors of CAWD schemes. Nevertheless, as we shall indicate in this section, there is no CASCI scheme weakly equivalent to the modification of the Kosaraju example given by Fig. 23.

Some insight into why "conditional iteration" is stronger than WHILEDO is given by introducing (in the next section) the "diamond" operation. In the context of this operation "separated iteration," our generalization of WHILEDO becomes as strong as "conditional iteration." But the diamond operation is different in kind from the other introduced operations on schemes since it does not produce a flow-chart scheme but rather a new kind of entity. Now to the business of this section.

By a *test scheme* or briefly a *test* relative to Π we mean an accessible scalar flow-chart scheme which fails to have vertices of out-degree 1. (A scalar scheme is *accessible* if every vertex—even the exits—is accessible from its begin-vertex.) Thus all the labels in a test scheme come from the set Π.

Separated conditional iteration or briefly, *separated iteration* operates on a test $1 \xrightarrow[s]{T} p + 1, p \geqslant 0$, and a flow-chart scheme $1 \xrightarrow[t]{F} 1$ to produce the flow-chart scheme

$$1 \xrightarrow[s+t]{[T \cdot [I_p \oplus F]]^{\dagger}} p,$$

where I_p is the identity scheme $p \longrightarrow p$ (so that $I_1 = 1_1$).

Pictorially the result of separated iteration may be indicated by Fig. 24. We now generalize the binary alternation operation to "alternation."

Alternation operates on a test $T : 1 \longrightarrow n, n > 0$, and n schemes $F_i : 1 \longrightarrow p, i \in [n]$, to produce the scheme $T \cdot (F_1, F_2, \cdots, F_n) : 1 \longrightarrow p$. We defer a pictorial representation of the operation of alternation to the reader.

By a *CASCI scheme* in $\mathfrak{F}(\Pi, \Omega)$, we mean a scalar scheme which can be built up from the atomic schemes $\omega \in \Omega$ and the trivial schemes $j_p, j \in [p]$, by a finite number of applications of composition, alternation and separated iteration. Clearly, every CAWD scheme is a CASCI scheme and every CASCI scheme is strongly equivalent to a CACI scheme.

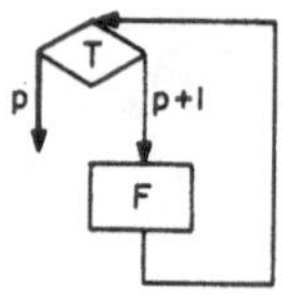

Fig. 24.

We give an example of a biscalar CASCI scheme which is not a CAWD scheme. Let $T : 1 \longrightarrow 2$ be the scheme shown by Fig. 25 whose contraction $T' : 1 \longrightarrow 2$ is given by Fig. 26. Then the scheme $[T \cdot [1_1 \oplus \omega]]^\dagger = [T \cdot (1_2, \ \omega \cdot 2_2)]^\dagger : 1 \longrightarrow 1$ is the CASCI scheme depicted by Fig. 19. It has already been observed that this scheme is not weakly equivalent to any CAWD scheme.

The argument that no CASCI scheme is weakly equivalent to Fig. 23 depends upon noting a certain property that all CASCI schemes have. The statement of this property, however, requires some additional concepts which we now provide.

By a *circle C* in a graph we shall mean a subgraph which looks like Fig. 27, i.e., a subgraph consisting of n vertices $v_1, v_2, \cdots, v_n$ and n edges $e_1, e_2, \cdots, e_n, n \geqslant 1$, such that e_i points from v_i to v_{i+1}, $i \in [n-1]$ and e_n points from v_n to v_1. *Thus a flow-chart scheme F is well-formed iff every circle in F contains at least one operation vertex.* By a *linearization* of the circle C, we mean one of the n sequences

$$v_1 e_1 v_2 e_2 \cdots v_n e_n v_1$$
$$v_2 e_2 v_3 e_3 \cdots v_1 e_1 v_2$$
$$\vdots$$
$$v_n e_n v_1 e_1 \cdots v_{n-1} e_{n-1} v_n,$$

i.e., a shortest path in C from v_i to v_i, $i \in [n]$. Thus a linearization of a circle is determined by selecting a single vertex of the circle (as beginning vertex).

Let v_1, v_2, v_3 be vertices of a circle C. We say *v_2 is between v_1 and v_3 in the circle C* if the shortest path from v_1 to v_3 in C passes through v_2; in particular, $v_2 \neq v_1$ and $v_2 \neq v_3$. *Notice that if $v_1 \neq v_3$ and v_2 is between v_1 and v_3 in C then v_2 is not between v_3 and v_1 in C.*

Let F be a flow-chart scheme. By a *successful path* in F we mean a path in the underlying graph of F from a begin-vertex to an exit.

Theorem 7.1: Let F be a CASCI scheme. Then for every circle C in (the underlying graph of) F, if L_1 and L_2 are linearizations of C, if e_1, e_2 are edges in F not in C and if

$$\cdots L_1 e_1 \cdots \text{ and } \cdots L_2 e_2 \cdots$$

are successful paths in F; then $L_1 = L_2$. If F is a CAWD scheme; then $e_1 = e_2$ as well.

Thus, according to the theorem, if $F : 1 \longrightarrow p+1$ is a CASCI scheme and C is a circle in F, there is a unique vertex v_C in C, (the exit vertex of C in F), such that any successful path in F which "passes through" C must "leave" C via v_C; v_C determines the linearization of C which is part of successful paths in F. Notice this theorem involves only the underlying pointed graph of F.

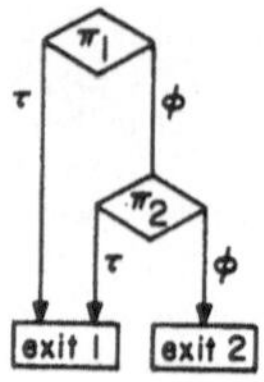

Fig. 25.

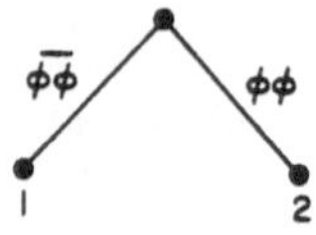

Fig. 26. T'.

Fig. 27.

The form of the proof of the theorem is this. Let $\mathcal{S}$ be the set of flow-chart schemes which satisfy the property "for every circle $C \ldots$" stated in the theorem. One shows the set $\mathcal{S}$ contains the trivial schemes and the atomic schemes and is closed with respect to composition, alternation, and separated iteration. Since the set of CASCI schemes is the *smallest* set having those properties, $\mathcal{S}$ contains the set of CASCI schemes.

We do not give the proof, which is straightforward, here. A typical pair of required observations, however, is this: a circle in $F \cdot G$ is a circle in F or a circle in G; a successful path in $F \cdot G$ uniquely partitions into a pair of paths, viz., a successful path in F and one in G, the "composite" of the pair being the given path.

Theorem 7.2: There is no CASCI scheme weakly equivalent to the one depicted by Fig. 23.

Proof: Let a be the atom $\phi\phi$ and let b be the atom $\tau\tau$ in $\mathcal{F}(\Pi)$ where $\Pi = \{\pi_1, \pi_2\}$. Then the weak behavior of the scheme of Fig. 23 contains, for each $m \geqslant 0$, the alternating word

1) $(a\omega_1 a\omega_2)^m b$ and

2) $(a\omega_1 a\omega_2)^m a\, \omega_1\, b$.

Suppose F is a CASCI scheme weakly equivalent to Fig. 23 and suppose F has n operation vertices. Then

3) $(a\omega_1 a\omega_2)^{n+1} b$

is in the weak behavior of F. Let

4) $v_1 e_1 v_2 e_2 \cdots v_r \cdots v_s \cdots v_t \cdots$ exit

be the successful computation of F whose label has 3) as an element, where $v_r = v_t$ and v_s are operation vertices and further

a) there are no operation vertices to the right of v_t in 4) so that v_t has label ω_2,

b) the sequence in 4) beginning with v_r and ending with v_t is a linearization of a circle C, and

c) v_s is the first operation vertex to the right of v_r in 4) so that v_s has label ω_1.

Using Theorem 7.1 we locate v_C as indicated in 5).

5) $$v_1 e_1 v_2 e_2 \cdots \overbrace{\overset{\downarrow}{\underset{\omega_2}{v_r}} \cdots \overset{v_C}{\downarrow} \cdots \underset{\omega_1}{v_s} \cdots \overset{\downarrow}{\underset{\omega_2}{v_t}}}^{\text{linearization of } C} \overset{v_C}{\downarrow} \cdots \text{exit.}$$

Thus v_C occurs in C between $v_r = v_t$ and v_s. Now

6) $(a\omega_1 a\omega_2)^{n+1} a\, \omega_1\, b$

is in the weak behavior of F. Let its successful computation be

7) $v_1 e_1 v_2 e_2 \cdots v_r \cdots v_s \cdots v_t \cdots v_u \cdots$ exit,

where $v_u = v_s$, there are no operation vertices to the right of v_u in 7) and no operation vertices between v_t and v_u in 7). Thus the part of the sequence 7) beginning with v_s and ending with v_u is a linearization of the same circle C. Again using the theorem we locate v_C between v_s and v_t in C. Thus we have in C: v_C is between v_s and v_t and v_C is between v_t and v_s. Since $v_s \neq v_t$, this is a contradiction and the argument is complete.

Using the part of Theorem 7.1 dealing with CAWD schemes one can prove, in a similar manner, the Kosaraju assertion that no CAWD scheme is weakly equivalent to the scheme of Fig. 19.

8. The Diamond Operation

If $F : 1 \longrightarrow p$ is a flow-chart scheme in $\mathcal{F}(\Pi, \Omega)$ and I is an interpretation with underlying set X, we recall that $|F. I|$ is a partial function $X \longrightarrow X \times [p]$; if $x \in X$ and $|F, I|(x)$ is defined then $|F, I|(x) = (x', j)$ where $x' \in X, j \in [p]$. In the context of weak equivalence, one may identify a flow-chart scheme $F : 1 \longrightarrow p$ with a function (or, some say, functional) which, when applied to an interpretation I as argument, yields as value the partial function $|F, I|$. Thus F becomes the function $I \longmapsto |F, I|$, where I ranges through all interpretations.

The diamond operation takes $F : 1 \longrightarrow p$ into $F^{\diamond} : 1 \longrightarrow p$. The meaning of $F^{\diamond}$ is the function $I \longmapsto |F^{\diamond}, I|$, where $|F^{\diamond}. I|(x) = (x, j)$ iff for some $x' \in X$, $|F, I|(x) = (x', j)$.

Let

$$F^{\circ} : 1 \longrightarrow 1 \quad \text{be} \quad F \cdot (\overbrace{1_1, 1_1, \cdots, 1_1}^{p}),$$

i.e., F° is obtained from F by identifying all the exits of F. We then have F weakly equivalent to

$$G \overset{df}{=} F^{\diamond} \cdot \overbrace{[F^{\circ} \oplus F^{\circ} \oplus \cdots \oplus F^{\circ}]}^{p}$$

and, where p is positive,

$$G^{\dagger} = [F^{\diamond} \cdot \overbrace{[1_1 \oplus 1_1 \oplus \cdots \oplus 1_1}^{p-1} \oplus F^{\circ}]]^{\dagger}$$

$$\cdot \overbrace{[F^{\circ} \oplus F^{\circ} \oplus \cdots \oplus F^{\circ}]}^{p-1}.$$

The $\dagger$ which occurs on the right is an example of separated iteration. Thus, with the help of the diamond operation, we have expressed (in the context of weak equivalence) conditional iteration in terms of separated conditional iteration.

9. Brief Mention

Our discussion up to now has dealt mainly with scalar schemes. Biscalar schemes have been discussed in order to make contact with relevant literature. This literature, for the most part, does not attempt to deal with the totality of flow-chart schemes in a structured way. Indeed, it is sometimes suggested that the "presence of GO TO statements" makes it impossible to deal with flow-charts in a structured way. We believe [5] and [6] provide counterexamples to that suggestion. Briefly put, composition, source-tupling, and iteration are preserved by maps into strong behavior, weak behavior, or interpreted behavior. Moreover, there is a unique map whose source is $\mathcal{F}(\Pi, \Omega)$ and target any of the above, which preserves composition, source-tupling, vector iteration (the uniqueness part of the assertion follows from Theorem 3.1) and prescribes the value of the trivial schemes j_p, $j \in [p]$. (For example, in the domain of interpreted behavior $|j_p, I|$ is the function which takes $x \in X$ into (x, j) and in the domain of weak behavior $|j_p|$ is the $1 \times p$ matrix which has the set A of atoms in the jth position and the empty set elsewhere.

To matrix multiply two weak behaviors: say an $n \times p$ matrix and a $p \times q$ matrix of sets of alternating words, one must know how to multiply two sets of alternating words, say U, V.

Definition: $U \cdot V = \{uav \mid ua \in U, av \in V\}$; here ua is an alternating word terminating with $a \in A$ and av is an alternating word beginning with a. We have (elsewhere) called this operation *fusion*.

A strong behavior of a scheme is of the form (U, u) where U is an $n \times p$ matrix of sets of finite words and u an n tuple of sets of infinite words on the alphabet $\Pi_\tau \cup \Pi_\phi \cup \Omega$. If (V, v) is a $p \times q$ strong behavior, then the appropriate multiplication here is

$$(U. u) \cdot (V. v) = (UV, u + Uv).$$

Here matrix multiplication is involved as well; the underlying set multiplication in this case is that induced by concatenation of words in the uninterpreted case and fusion in the interpreted case. We have

$$\|F \cdot G\| = \|F\| \cdot \|G\|$$

$$\|F \cdot G, I\| = \|F, I\| \cdot \|G, I\|$$

$$|F \cdot G, I| = |F, I| \cdot |G, I|.$$

While we have noted in this section that the totality of flow-chart schemes possesses neat structural properties, there is little doubt that the subclass of CACI schemes constitute a simpler totality. In view of the fact that every scalar flow-chart scheme is strongly equivalent to a CACI scheme, one may well ask what clear-cut benefit, if any, is to be derived from the use of general flow-chart schemes as opposed to CACI schemes. In this connection we call attention to Fig. 15 which has 3 "instructions," i.e., has weight 3. It is not difficult to convince oneself that any CACI scheme weakly equivalent to Fig. 15 (which is not a CACI scheme) has at least 4 instructions (cf. Fig. 16). Indeed if $1 \xrightarrow[3]{F} 1$ is weakly equivalent to Fig. 15; then F is (isomorphic to) Fig. 15. Thus general (i.e., arbitrary) flow-chart schemes at least permit one to save space.

The main results of our discussion up to now have dealt with strong or weak equivalence. The following result, obtained jointly with J. C. Shepherdson, deals with equality. In order to state the result it is necessary to introduce another operation on scalar schemes. Given schemes $1 \xrightarrow{F} n$, $n \geqslant 0$, and $1 \xrightarrow{G_i} p_i$, $i \in [n]$, the operation of ***separated substitution*** produces the scheme

$$1 \xrightarrow{F \cdot [G_1 \oplus G_2 \oplus \cdots \oplus G_n]} p_1 + p_2 + \cdots + p_n.$$

(In order to round out the conceptual distinctions being made, we define another operation even though it is not involved directly in our present discussion.) Given schemes $1 \xrightarrow{F} n$, $n \geqslant 0$ and $1 \xrightarrow{G_i} p$, $i \in [n]$, the operation of ***substitution*** produces the scheme

$$1 \xrightarrow{F \cdot (G_1, G_2, \cdots, G_n)} p.$$

Notice that alternation is the restriction of substitution to the case that F is a *test* scheme and $n > 0$.

Let $\mathcal{R}$ be the smallest set of scalar schemes containing 1_1, Ω, Π and closed under the following operations:

separated substitution
exit merging
scalar iteration.

Clearly every ***$\mathcal{R}$-scheme is accessible*** (cf. Section 7 for definition). Some terminology: Let F be a scalar scheme and C a circle in F. A ***path in F from b_F*** (the begin of F) ***to C*** is a path in F from b_F to a vertex in C. A ***minimal path in F from b_F to C*** is a path in F from b_F to C such that no proper initial segment (i.e., proper prefix) of the path is a path in F from b_F to C.

Theorem 9.1 (Joint with J. C. Shepherdson): An accessible scalar scheme F is an $\mathcal{R}$-scheme iff for every circle C, if P_1, P_2 are minimal paths in F from b_F to C, then P_1 and P_2 terminate with the same vertex (in C).

The theorem admits easy generalization to the broader context $\mathcal{F}(\Gamma)$.

Clearly every $\mathcal{G}$-scheme is an $\mathcal{R}$-scheme. The conclusion is proper since $\pi_1 \cdot [\pi_2 \oplus 1_1] \cdot f \cdot (1_1, \omega)$, where $[3] \xrightarrow{f} [2]$ is given by $f(1) = 1, f(2) = f(3) = 2$, is a scheme in $\mathcal{R}$ which is not in $\mathcal{G}$ (although $1 \xrightarrow{\pi_1 \cdot [\pi_2 \oplus 1_1] \cdot f} 2$ is in $\mathcal{G}$).

If the property stated in the theorem fails to hold for an accessible scalar scheme F; then there exists a circle C in F and there exist minimal paths π_1, π_2 in F from b_F to C such that π_1 and π_2 do not terminate with the same vertex.

From the characterization given in [14, p. 196, theorem 5], it follows that the set of underlying "flow graphs" of $\mathcal{R}$ schemes (which have at most one edge between any pair of vertices) is exactly the set of *reducible* flow graphs (see also [10], [15]).

It may be interesting to note that we singled out the class $\mathcal{R}$ before knowing about its connection with "flow-graph reducibility" thus, perhaps, strengthening the view held by some that $\mathcal{R}$ is a "natural" class.

APPENDIX

This example, drawn from Ashcroft and Manna, uses much of their notation.

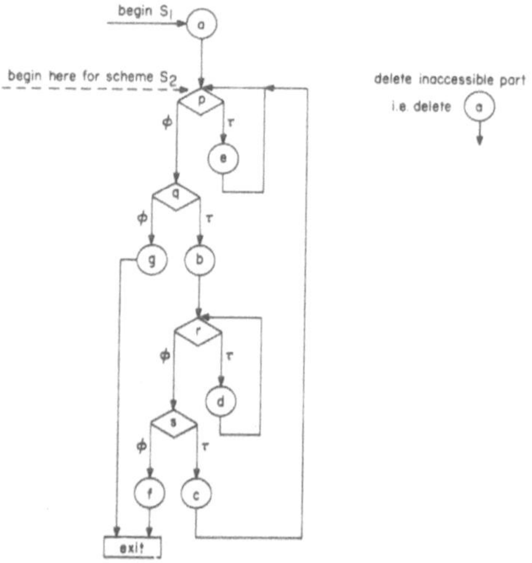

Schemes $S_1, S_2 : 1 \longrightarrow 1$

$S_1 = a \cdot S_2 : 1 \longrightarrow 1$

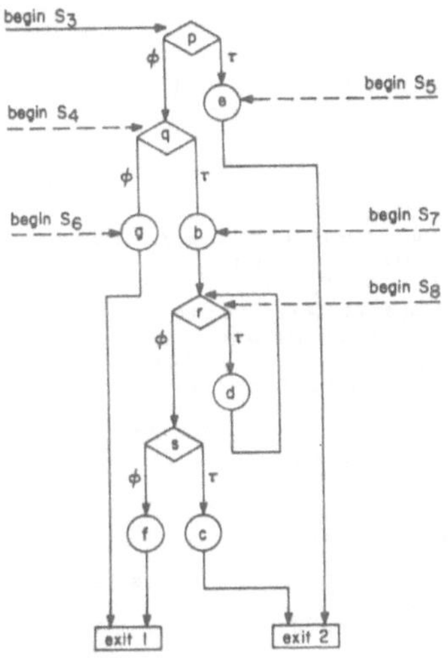

Schemes $S_3, S_4, S_5, S_6, S_7, S_8 : 1 \longrightarrow 2$,

$$S_2 = S_3{}^\dagger, \qquad S_3 = p \cdot (S_5, S_4),$$
$$S_4 = q \cdot (S_7, S_6), \qquad S_5 = e \cdot 2_2,$$
$$S_6 = g \cdot 1_2, \qquad S_7 = b \cdot S_8.$$

In each case delete inaccessible parts (other than exits). Thus S_5 has one vertex of out-degree 1 and two vertices of out-degree 0.

In each case delete inaccessible parts (other than exits).

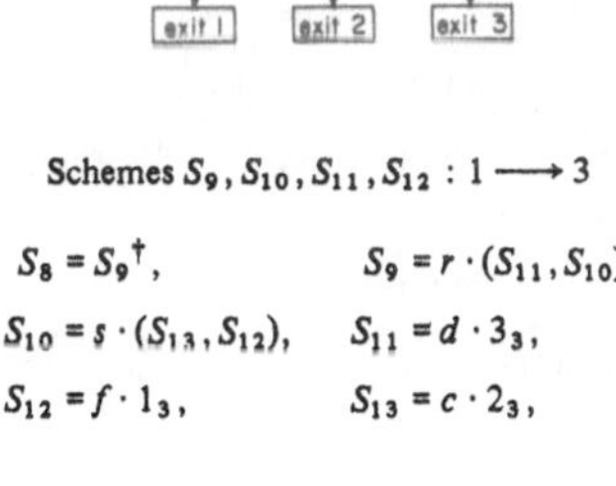

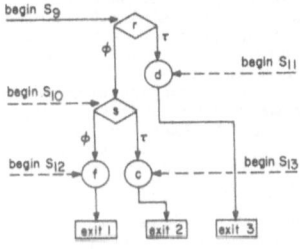

Schemes $S_9, S_{10}, S_{11}, S_{12} : 1 \longrightarrow 3$

$$S_8 = S_9{}^\dagger, \qquad S_9 = r \cdot (S_{11}, S_{10}),$$
$$S_{10} = s \cdot (S_{13}, S_{12}), \qquad S_{11} = d \cdot 3_3,$$
$$S_{12} = f \cdot 1_3, \qquad S_{13} = c \cdot 2_3,$$

$$S_8 = (r \cdot (d \cdot 3_3, s \cdot (c \cdot 2_3, f \cdot 1_3)))^\dagger$$
$$S_1 = a \cdot (p \cdot (e \cdot 2_2, q \cdot (b \cdot S_8, g \cdot 1_2)))^\dagger$$

Acknowledgment

It is a pleasure to acknowledge the many helpful comments of J. C. Shepherdson and E. G. Wagner on the manuscript and a discussion with J. D. Rutledge which ultimately led to Section VI. The author thanks R. Sethi for calling attention to [14, theorem 5].

It is a pleasure to also thank S. L. Bloom (organizer of the NSF-CBMS Regional Research Conference on An Algebraic Analysis of Flowchart Algorithms), Stevens Institute of Technology (the host institution), the National Science Foundation, and the Conference Board of Mathematical Sciences for making the conference possible and the participants for making the conference stimulating.

References

[1] A. Ashcroft and A. Manna, "The translation of 'go to' programs to 'while' programs," in *Proc. Int. Fed. Inform. Processing Congr.*, 1971, vol. 1. Amsterdam, The Netherlands: North-Holland, 1972.

[2] S. L. Bloom and C. C. Elgot, "The existence and construction of free iterative theories," IBM Corp., Yorktown Heights, NY, Res. Rep. RC-4937, July 1974; also to appear in *J. Comput. Syst. Sci.*

[3] A. Bohm and A. Jacopini, "Flow diagrams, turing machines and languages with only two formation rules," *Commun. Ass. Comput. Mach.*, May 1966.

[4] A. Bruno and A. Stieglitz, "The expression of algorithms by charts," *J. Ass. Comput. Mach.*, July 1972.

[5] C. C. Elgot, "The common algebraic structure of exit automata and machines," *Comput.*, Jan. 1971.

[6] —, "Monadic computation and iterative algebraic theories," in *Logic Colloquium '73*, vol. 80, *Studies in Logic*, H. E. Rose and J. C. Shepherdson, Ed. Amsterdam, The Netherlands: North-Holland, 1975.

[7] —, "Matrical theories," IBM Corp., Yorktown Heights, NY, Res. Rep. RC-4833, May 1974; also to appear in *J. Algebra*.

[8] D. C. Copper, "Bohm and Jacopini's reduction of flow charts," *Commun. Ass. Comput. Mach.* (Lett. Editor), Aug. 1967.

[9] D. E. Knuth, "Structured programming with go to statements," *Comput. Surveys*, Dec. 1974.

[10] F. E. Allen, "Control flow analysis," *SIGPLAN Notices*, vol. 5, pp. 1-19, 1970.

[11] F. T. Baker, "Chief programmer team management of production programming," *IBM Syst. J.*, vol. 11, no. 1, 1972.

[12] H. D. Mills, "Mathematical foundations for structured programming," IBM Corp., Gaithersburg, MD, Rep. FSC 72-6012, FSD, 1972.

[13] —, "Top down programming in large systems," in *Debugging Techniques in Large Systems, Courant Computer Science Symp.*, vol. 1, R. Rustin, Ed. Englewood Cliffs, NJ: Prentice-Hall, 1971.

[14] M. S. Hecht and J. D. Ullman, "Flow graph reducibility," *SIAM J. Comput.*, vol. 1, pp. 188-202, June 1972.

[15] J. Cocke, "Global common subexpression elimination," *SIGPLAN Notices*, vol. 5, pp. 20-24, 1970.

[16] J. D. Rutledge, "On Ianov's program schemata," *J. Ass. Comput. Mach.*, Jan. 1964.

[17] A. Knuth and A. Floyd, "Notes on avoiding 'go to' statements," *Inform. Processing Lett.*, vol. 1, 1971; errata, *Inform. Processing Lett.*, vol. 1, 1972.

[18] A. Luckham, A. Park, and A. Paterson, "On formalized computer programs," *J. Comput. Syst. Sci.*, June 1970.

[19] S. R. Kosaraju, "Analysis of structured programs," in *Proc. Fifth Annu. Ass. Comput. Mach. Symp. Theory of Computing*, May 1973; also in *J. Comput. Syst. Sci.*, Dec. 1974.

Erratum and Corrigendum for "Structured Programming With and Without GO TO Statements"

CALVIN C. ELGOT

ERRATUM

The author wishes to thank J. C. Shepherdson for calling to his attention that, in the above paper,[1] Theorem 7.1 is false with Fig. 19 witness to its falsity. Since the proof of Theorem 7.2 depends on Theorem 7.1, it is rendered invalid. Fortunately, Theorem 7.2, the point of the section, remains valid as the proof given below shows. An independent proof, based upon the same idea but different in detail and logical organization, was found by Shepherdson.

CORRIGENDUM

The proof we give of Theorem 7.2 actually proves much more. Before stating the stronger result embodied in the Lemma and Theorem below, we make some preliminary observations.

If $F \in \mathcal{F}(\Pi, \Omega)$, where Π is finite, is a flowchart scheme, v is a vertex of F and a is an atom (i.e., an evaluation sequence) in $\mathcal{B}(\Pi)$, let $\mathrm{p}(F, v, a)$ be the path (cf., Section 6: "The evaluation sequence determines a path in $F \cdots$"[1]) in F starting with v which is compatible with a (i.e., "determined by a"). Let $\mathrm{t}(F, v, a)$ be the terminal vertex (either an operation vertex or an exit of F) $\mathrm{p}(F, v, a)$ if it is finite, otherwise $\mathrm{t}(F, v, a) = \infty$.

Observation 1: If v' is a vertex which occurs, nonterminally, in the path $\mathrm{p}(F, v, a)$, then $\mathrm{p}(F, v', a)$ is a suffix of $\mathrm{p}(F, v, a)$ and $\mathrm{t}(F, v, a) = \mathrm{t}(F, v', a)$.

Observation 2: If $\mathfrak{M}$ is a set of scalar schemes in $\mathcal{F}(\Pi, \Omega)$ *closed under the CASCI operations* (composition, alternation, separated conditional iteration) and $\mathrm{BI}(\mathfrak{M})$ is the set of all biscalar schemes in $\mathfrak{M}$, then $\mathrm{BI}(\mathfrak{M})$ is closed under the CASCI operations. In particular, the set of biscalar CASCI schemes is the smallest set of biscalar schemes containing 1_1, Ω and closed under the CASCI operations.

Now let $\Pi = \{\pi_1, \pi_2\}$, $\pi_1 \neq \pi_2$, $\omega_1 \in \Omega$, $\omega_2 \in \Omega$, $\omega_1 \neq \omega_2$. Let a_1 be the atom $\phi\tau$ and let a_2 be the atom $\tau\phi$. An $a_1\omega_1 a_2 \omega_2$-circle in $F \in \mathcal{F}(\Pi, \Omega)$ is a circle C in F with s operation vertices labeled ω_1, s operation vertices labeled ω_2, where $s > 0$, such that for all $i, j \in [2]$, $i \neq j$ and for all ω_j-vertices v in C:

1) $\mathrm{t}(F, v, a_i)$ is an ω_i-vertex in C; and
2) $\mathrm{t}(F, v, a_j)$ is the exit of F.

Lemma: If $F \in \mathcal{F}(\Pi, \Omega)$ is weakly equivalent to the scheme of Fig. 23, then there is an $a_1\omega_1 a_2\omega_2$-circle in F.

Proof: We note that for each $m \geq 0$, $(a_1\omega_1 a_2\omega_2)^m a_2$ and $(a_1\omega_1 a_2 \omega_2)^m a_1 \omega_1$ are in the weak behavior of Fig. 23 and hence in the weak behavior $|F|$ of F. Since $a_1\omega_1 a_1$, $a_1\omega_1 a_2\omega_2 a_2$, $a_1\omega_1 a_2\omega_2 a_1\omega_1 a_1, \cdots$ are all in $|F|$, if we define $v_1 = \mathrm{t}(F, b_F, a_1)$, where b_F is the begin of F, $v_2 = \mathrm{t}(F, v_1, a_2)$, $v_3 = \mathrm{t}(F, v_2, a_1)$ etc., we ultimately obtain (since F is finite) for some $r > 0, s > 0$ an operation vertex $v_r = v_{r+2s}$. Moreover, v_1 is an ω_1-vertex, v_2 is an ω_2-vertex, v_3 is an ω_1-vertex and $\mathrm{t}(F, v_1, a_1) = \mathrm{t}(F, v_2, a_2) = \mathrm{t}(F, v_3, a_1) = \text{exit } F$, etc. This information is indicated by * below.

$$\begin{array}{lccccccccccc}
 & & \omega_1 & & \omega_2 & & \omega_1 & & \omega_j & & \omega_i & & \omega_j \\
*: & \cdots & v_1 & \cdots & v_2 & \cdots & v_3 & \cdots & v_r & \cdots & v_{r+1} & \cdots & v_{r+2s} = v_r. \\
 & & a_1 & & a_2 & & a_1 & & a_j & & a_i & & a_j \\
 & & \text{to exit} & & \text{to exit} & & \text{to exit} & & \text{to exit} & & \text{to exit} & & \text{to exit}
\end{array}$$

Thus, $v_r \cdots v_{r+1} \cdots v_{r+2s} = v_r$ is a linearization of an $a_1\omega_1 a_2\omega_2$-circle in F. □

Theorem: The set $\mathcal{S}$ of biscalar schemes F in $\mathcal{F}(\Pi, \Omega)$ satisfying

**: there is no $a_1\omega_1 a_2\omega_2$-circle in F

is closed under the CASCI operations.

Proof: Suppose C is an $a_1\omega_1 a_2\omega_2$-circle in F.

Case 1: $F = G \cdot H$, where $G, H \in \mathcal{S}$. Then C is an $a_1\omega_1 a_2\omega_2$-circle in G or in H. Contradiction. Hence, $F \in \mathcal{S}$.

Case 2: $F = T \cdot (G_1, G_2, \cdots, G_n)$, where $G_i \in \mathcal{S}$ for $i \in [n]$. The circle C is not in T since C contains operation vertices. Hence, C is an $a_1\omega_1 a_2\omega_2$-circle in G_i for some $i \in [n]$. Contradiction. Hence, $F \in \mathcal{S}$.

Case 3: $F = [T \cdot [1_1 \oplus G]]^\dagger$, where $G \in \mathcal{S}$ and $T: 1 \to 2$ is a test. The circle C is not in T since C contains operation vertices. Hence, either C lies wholly in G or $b_F = b_T$ is in C. In the former eventuality, we have for each ω_j-vertex v in C, $\mathrm{t}(F, v, a_j) = \text{exit } F$ and $\mathrm{p}(F, v, a_j)$ contains b_F since v is in G. Thus, $\mathrm{t}(G, v, a_j) = \text{exit } G$ and it follows C is an $a_1\omega_1 a_2\omega_2$-circle in G—which contradicts $G \in \mathcal{S}$.

The latter eventuality leads to the crux of our considerations. Suppose then that b_F is a vertex of C. Then for some operation vertex v in C, labeled ω_j, $j \in [2]$, we have assuming $i \neq j$, $\mathrm{t}(F, v, a_i)$ is an ω_i-vertex in C and $\mathrm{p}(F, v, a_i)$ passes through b_F. Thus, by Observation 1, $\mathrm{t}(F, b_F, a_i) = \mathrm{t}(F, v, a_i)$ is an ω_i-vertex. On the other hand, where w is an ω_i-vertex in C, we have $\mathrm{t}(F, w, a_i) = \text{exit } F$ so that $\mathrm{p}(F, w, a_i)$ passes through b_F and, again by Observation 1, $\mathrm{t}(F, b_F, a_i) = \text{exit } F$. But we concluded above that $\mathrm{t}(F, b_F, a_i)$ is an ω_i-vertex. This contradiction implies $F \in \mathcal{S}$ and concludes the argument. □

Corollary: No CASCI scheme is weakly equivalent to the scheme of Fig. 23.

Proof: Suppose F is a biscalar scheme weakly equivalent to Fig. 23. By the lemma there is an $a_1\omega_1 a_2\omega_2$-circle in F. By Observation 2 and the fact that the atomic schemes $\omega \in \Omega$ and trivial scheme 1_1 are in $\mathcal{S}$, it follows from the Theorem that every biscalar CASCI scheme is in $\mathcal{S}$. Thus, F is not a CASCI scheme. □

Manuscript received May 14, 1976.

The author is with the Department of Mathematical Sciences, IBM Thomas J. Watson Research Center, Yorktown Heights, NY 10598.

[1] C. C. Elgot, *IEEE Trans. Software Eng.*, vol. SE-2, pp. 41–54, Mar. 1976.

ACKNOWLEDGMENT

The author wishes to thank J. D. Rutledge and J. B. Wright for their helpful comments in the preparation of the above material.

Annals No. 609SE011

Theoretical Computer Science 8 (1979) 325–357

A SEMANTICALLY MEANINGFUL CHARACTERIZATION OF REDUCIBLE FLOWCHART SCHEMES

Calvin C. ELGOT

Mathematical Sciences Department, IBM Thomas J. Watson Research Center, Yorktown Heights, NY 10598, USA

John C. SHEPHERDSON

School of Mathematics, University of Bristol, Bristol, England

Communicated by E. Engeler
Received September 1977
Revised March 1978

Abstract. A "scalar" flowchart scheme, i.e. one with a single begin "instruction" is *reducible* iff its underlying flowgraph is reducible in the sense of Cocke and Allen or Hecht and Ullman. We characterize the class of reducible scalar flowchart schemes as the smallest class containing certain members and closed under certain operations (on and to flowchart schemes). These operations are "semantically meaningful" in the sense that operations of the same form are meaningful for "the" functions (or partial functions) computed by interpreted flowchart schemes; moreover, the schemes and the functions "are related by a homomorphism." By appropriately generalizing "flowgraph" to (possibly) several begins (i.e. entries) we obtain a class of reducible "vector" flowchart schemes which can be characterized in a manner analogous to the scalar case but involving simpler more basic operations (which are also semantically meaningful). A significant side effect of this semantic viewpoint is the treatment of multi-exit flowchart schemes on an equal footing with single exit ones.

1. Introduction

Roughly speaking, we describe a subclass, $\mathcal{R}_{\text{scal}}$, of the class ΓFl of all flowchart schemes as the smallest subclass of ΓFl which contains certain very simple members of ΓFl and is closed under certain operations of ΓFl and characterize $\mathcal{R}_{\text{scal}}$ by means of a simple property of the underlying pointed-digraph (digraph = directed-graph) of its members. Here $\Gamma = (\Gamma_0, \Gamma_1, \ldots)$ is a sequence of disjoint sets (whose elements are interpreted as operations and multi-exit tests) and the elements of Γ_r are said to have *arity r*. This class recommends itself because of its robustness [9, 10], its power [6], its flexibility, the simplicity of its defining operations, the simple semantics of its defining operations when Γ is interpreted [5, 4] and its relationship to algebraic theories (in the sense of categorical algebra, cf. [3, 7, 14].

The most striking characteristic of our description of $\mathcal{R}_{\text{scal}}$ (and the reason, we believe, for its simplicity) is the way we treat multi-exit flowchart schemes and results from our willingness to treat multi-exit flowchart schemes on an equal footing with single exit ones.

The members of $\mathcal{R}_{\text{scal}}$ are "scalar" in the sense that they have exactly one "begin" instruction location. *A subclass $\mathcal{R}$ of "vector" members of ΓFl is defined in a manner analogous to $\mathcal{R}_{\text{scal}}$ (but involving simpler more basic operations) and characterized by a property of the underlying multi-pointed directed graph.* This property is obtained from the former one by appropriately generalizing from the singly to the multi-pointed case.

We call a (singly) pointed-digraph a "flowgraph" if each vertex of the digraph is "accessible," i.e. can be reached by a path from the distinguished vertex (the "point" of the pointed-digraph). (Our "digraphs," it should be noted, do not rule out more than one edge between a pair of points.) The characterizing property of $\mathcal{R}_{\text{scal}}$ is that its underlying pointed-digraph is a "reducible flowgraph." Our usage of "flowgraph" differs, in a minor way, from some other authors who say a *digraph is a flowgraph* if *there exists* a vertex from which all others are accessible. Because, a digraph G is a flowgraph iff it possesses a vertex v such that the pointed-digraph (G, v) is a flowgraph.

Our usage of "reducible flowgraph" agrees with that of Cocke and Allen (who introduced the concept [11, 8, 1]) provided one disregards the slight difference in use of "flowgraph" and provided one specializes to the case where "edge of a digraph" means an (ordered) pair of vertices of the digraph. To see that this is the case one need only observe that the characterization [9, p. 196, Theorem 5] given by Hecht and Ullman agrees with our description via "ACC" and "CD".

The main point of the paper is to prove our main theorem as well as to make a little more precise some of the concepts in [6]. If desired, the reader may follow a reading of the Introduction by Section 5 or some later section, possibly even Section 11, referring back to earlier sections when necessary. In particular, a reader familiar with "directed graph" may wish to omit Section 2 on a first reading. The point of the formulation "path category" is mainly to avoid an ad hoc choice of the notion "path".

2. Directed graphs and path categories

We distinguish here two kinds of directed graph. A *relational-digraph* consists of a set V (of "vertices") and a subset $E \subseteq V \times V$ (of "edges"). A *two-sorted-digraph* consists of sets V, E together with a pair of functions $\partial_i : E \to V$, $i \in [2]$, the first for the source of an "edge" $e \in E$ and the second for the target of an edge. The latter notion "reduces to" the former in the case that: if e is an edge from v to v' then $e = (v, v')$.

Often interest in digraphs focuses on the paths to which they give rise. The notion "path," however, varies considerably. In most, if not all, choices of the notion, for each pair $v, v' \in V$ of vertices of G, there is a set $[v \to v']$ of *paths* of *G from v to v'*, or alternatively, which *begin with (or in) v and end with (or in) v'* and there is a function

$$[v_0 \to v_1] \times [v_1 \to v_2] \to [v_0 \to v_2] \quad \text{for each triple } v_0, v_1, v_2 \in V, \tag{2.1}$$

which we may call *composition* (of paths). The composition of $\pi_1 \in [v_0 \to v_1]$ and $\pi_2 \in [v_1 \to v_2]$ is written: $\pi_1\pi_2$. The path $\pi_1\pi_2$ begins with v_0 and ends with v_2. We note the following properties of composition of paths.

Composition is associative in the following sense: (2.2)

for $\pi_i \in [v_{i-1} \to v_i]$, $i \in [3]$: $(\pi_1\pi_2)\pi_3 = \pi_1(\pi_2\pi_3)$.

For each $v \in V$, there is a path I_v,(necessarily unique), such that $I_v\pi = \pi$, $\pi' I_v = \pi'$ for all $\pi \in [v \to v']$, $\pi' \in [v' \to v]$. (2.3)

By virtue of satisfying (2.2), (2.3), the data consisting of V, the doubly indexed family $\{[v \to v'] \| v, v' \in V\}$ of sets of paths and the triply indexed family of functions (2.1), is a category. In the nomenclature of categorical algebra, the vertices are *objects* and the paths are *morphisms*. Using the notation of categorical algebra "$\pi \in [v \to v']$" becomes "$\pi : v \to v'$".

Many "path" notions are accompanied by a "length of path" notion (notation: $|\pi|$; $|\pi| \in N$) satisfying:

$$|\pi_1\pi_2| = |\pi_1| + |\pi_2|, \text{ where } \pi_i \in [v_{i-1} \to v_i],\ i \in [2]. \tag{2.4}$$

For any path $\pi : v_0 \to v_2$ and non-negative integers i_1, i_2 satisfying $i_1 + i_2 = |\pi|$, there is a unique vertex v_1 and paths $\pi_1 : v_0 \to v_1$, $\pi_2 : v_1 \to v_2$ such that $\pi_1\pi_2 = \pi$ and $|\pi_j| = i_j$ for each $j \in [2]$. (2.5)

$$|\pi| = 0 \Rightarrow \pi = I_v \text{ for some } v \in V. \tag{2.6}$$

In particular, since $I_v = I_v I_v$, we have from (2.4)

$$|I_v| = 0 \text{ for each } v \in V. \tag{2.7}$$

From (2.5) we obtain:

For any path $\pi : v_0 \to v_r$, where $r = |\pi|$, there is a unique sequence v_1, $v_2, \ldots, v_{r-1}$ of vertices and a unique sequence of paths $\pi_i : v_{i-1} \to v_i$, $i \in [r]$, of length 1 such that $\pi = \pi_1\pi_2 \cdots \pi_r$. (2.8)

Notice that *if a category admits a length function satisfying* (2.4)–(2.6) *that length function is uniquely determined.* To show this, it is sufficient by (2.8) and symmetry to show that

$$|\pi|' = 1 \Rightarrow |\pi| = 1.$$

To establish the latter assertion, assume $|\pi|' = 1$ and $|\pi| = r$. By (2.6) and (2.7), we conclude $r > 0$. If $r > 1$, then by (2.5) $\pi = \pi_1\pi_2$, $|\pi_1| = 1$, $|\pi_2| = r - 1 > 0$; by (2.6) and

(2.7), we infer $|\pi_1|'>0$, $|\pi_2|'>0$ so that by (2.4), we conclude $|\pi|'>1$ which contradicts the assumption $|\pi|'=1$. Hence $|\pi|=1$.

We call a category which admits a length function satisfying (2.4)–(2.6) a *path category*. This is an "elementary" description of the notion (*small*), *free category*, [15, pp. 48–50]. We claim each digraph G (in any sense of "digraph") gives rise to a path category $G^{\rightarrow}$ and a doubly indexed family of injections

> $\iota_{u,v}: E_{u,v} \rightarrow [u \rightarrow v]$, u, $v \in V$, the image of $\iota_{u,v}$ being the subset of $[u \rightarrow v]$ consisting of paths of length l, where $E_{u,v}$ is the set of edges of G from u to v. (2.9)

The point of the formulation of "path category" is to enable one to deal with the path category $G^{\rightarrow}$ of a graph G without having to make an ad hoc choice of "path". For example, for the "relational-digraph" notion or the two-sorted notion with the constraint "at most one edge $v \rightarrow v'$," here are some possibilities for the notion "path of length r":

(i) $(v_0, v_1, v_2, \ldots, v_r)$,

(ii) $((v_0, v_1), (v_1, v_2), \ldots, (v_{r-1}, v_r))$,

(iii) $(v_0, e_1, v_1, e_2, v_2, \ldots, e_r, v_r)$,

where $e_i \in E$ is uniquely determined by (perhaps equal to) the pair (v_{i-1}, v_i), $i \in [r]$. Each of these choices has some point. Yet it is largely irrelevant which choice is made. What is relevant is captured in the notion "path category" and the injections (2.9). In case (iii), for example, edge e_1 gets mapped via (2.9), in the case $u = v_0$, $v = v_1$, into (v_0, e_i, v_1).

It should be clear every path category is (isomorphic to) a path category of the form $G^{\rightarrow}$ (assuming the two-sorted notion "digraph"). We note the following freeness property of $G^{\rightarrow}$. *If $\mathscr{C}$ is an arbitrary category and $v \mapsto v'$ a function from V into the set of objects of $\mathscr{C}$ and $\phi_{u,v}$ any function from the image of $\iota_{u,v}$ into the set $[u' \rightarrow v']$ of morphisms of $\mathscr{C}$ from u' to v', then the functions $\phi_{u,v}$ admit a unique extension to a functor $G^{\rightarrow} \rightarrow \mathscr{C}$, (with object function $v \mapsto v'$).*

Alternatively, making use of the set of edges: given the function $v \mapsto v'$ from Obj $G^{\rightarrow}$ to Obj $\mathscr{C}$ and *given any doubly indexed family of functions $\phi_{u,v}: E_{u,v} \rightarrow [u' \rightarrow v']$, there is a unique functor*

$$\phi^{\rightarrow}: G \rightarrow \mathscr{C} \quad \phi^{\rightarrow} = \{\phi^{\rightarrow}_{u,v}\}_{u,v \in V}, \quad \phi^{\rightarrow}_{u,v}: [u \rightarrow v] \rightarrow [u', v'],$$

which (for each u, $v \in V$) makes the following triangle commute

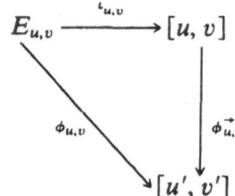

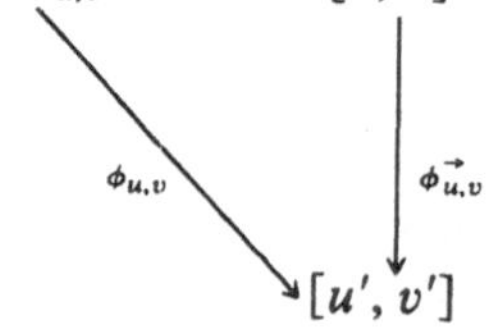

3. Concept of flowscheme

A definition of "flow-chart scheme" (together with some background discussion) was given on p. 42 of [6] and another given "in a different style" and a more general context (i.e. arbitrary finite outdegree, instead of merely outdegrees 1, 2) on p. 46. The definition of "flowscheme" given in this section differs slightly from the p. 46 definition. The difference stems chiefly from two sources. One: the usual (in computer science), informal "up to isomorphism" approach (first paragraph, Section 3) of [6] is being dropped here on the ground that it sometimes leads to serious confusions. Accordingly (in the next section) we define "flowchart scheme" (unhyphenated) and "flowtable scheme" and note, in precise terms, that each gives the "same information" as "flowscheme". Two: the distinction between the two "styles" (in the sense of two different ways of giving the "same information") is being taken more seriously (in keeping with the heavy emphasis on "representation" in computer science).

Let $\Gamma = (\Gamma_0, \Gamma_1, \Gamma_2, \ldots)$ be an infinite sequence of pairwise disjoint sets and let $\bigcup \Gamma$ be an abbreviation for $\Gamma_0 \cup \Gamma_1 \cup \Gamma_2 \cup \cdots$.

Definition 3.1. A *Γ-flowscheme*, briefly *Γ-flow*, F consists of the following data

(i) a set V, (the set of *vertices*);

(ii) a non-negative integer n, (the length of the sequence of *begins*, i.e. *begin-vertices*);

(iii) a subset V^{ex} (the set of *exits* or *exit-vertices*) of V, together with a bijection $V^{ex} \approx [p]$, where p is a non-negative integer, (so that we may speak of exit j or the jth-exit); let $S = V - V^{ex}$ (the set of *internal* or *non-exit vertices*);

a function $\theta : V \to N$, (the *out degree function*) satisfying $v\theta = 0$ for $v \in V^{ex}$; (3.1.1)

a function $\sigma : E \to V$, (the *immediate successor* or *target function*, where $E = \{(s, i) \mid s \in S,\ i \in [s\theta]\}$ = the set of *edges*, (s, i) is the ith *outedge of* s, $e = (s, i)$ begins with s and ends with $e\sigma$); (3.1.2)

a function $b : [n] \to V$, (the *begin function*); (3.1.3)

a function $\lambda : S \to \bigcup \Gamma$, *θ-compatible* in the sense $s\lambda \in \Gamma_{s\theta}$, (the *labeling function*). (3.1.4)

Let $\Gamma\mathrm{Fl}$ be the class of all Γ-flows; $F \in \Gamma\mathrm{Fl}$ is *finite* if S is finite.

We often specify certain constituents of (i), (ii), (iii) associated with F as follows

$$n \xrightarrow[S]{F} p. \tag{3.1.5}$$

The numbers n and p are respectively called the *source* and *target* of F. Often F is specified only with its source and target as follows

$$n \xrightarrow{F} p \quad \text{or} \quad F : n \to p.$$

The Γ-flow

$$1 \xrightarrow[\{5,6,7,8\}]{H} 1$$

detailed below (Tables 1a, b) is depicted by Fig. 5.10, where $V^{\text{ex}} = [p] = [1]$ and the bijection is the identity.

Table 1a

S_H	θ_H	λ_H
5	2	γ_2
6	1	γ_1
7	1	γ_1'
8	1	γ_1

Table 1b

E_H	σ_H
(5, 1)	6
(5, 2)	7
(6, 1)	8
(7, 1)	8
(8, 1)	1

$b_H : [1] \to \{1, 5, 6, 7, 8\}$

$b_H(1) = 5$

Definition 3.2. The Γ-flows

$$n_1 \xrightarrow[S_1]{F_1} p_1, \qquad n_2 \xrightarrow[S_2]{F_2} p_2$$

are *isomorphic* if $n_1 = n_2$, $p_1 = p_2$ and there is a bijection $S_1 \xrightarrow{\beta} S_2$ such that for each $s \in S_1$,

$$s\theta_1 = s\beta\theta_2; \tag{3.2.1}$$

if $(s, i)\sigma = s' \in S_1$, then $(s\beta, i)\sigma_2 = s'\beta \in S_2$,

if $(s, i)\sigma_1 = \text{exit } j$ of F_1, then $(s\beta, i)\sigma_2 = \text{exit } j$ of F_2,

for all $i \in [s\theta_1]$; (3.2.2)

if $ib_1 = s_1 \in S_1$, then $ib_2 = s\beta \in S_2$,

if $ib_1 = \text{exit } j$ of F_1, then $ib_2 = \text{exit } j$ of F_2,

for each $i \in [n_1]$; (3.2.3)

$$s\lambda_1 = s\beta\lambda_2. \tag{3.2.4}$$

The bijection β is an *isomorphism from F_1 onto F_2*.

Each figure below (Section 5) is a two dimensional notation for either a particular Γ-flow or an ambiguous member of an isomorphism class of Γ-flows, i.e. a variable whose range is an isomorphism class of Γ-flows.

Figures 5.9 and 5.11 depict an ambiguous member, $H_{\approx}$, of the isomorphism class $H/\approx$ determined by the Γ-flow H described above and depicted by Fig. 5.10. Notice that the device used by Figs. 5.9 and 5.11 to describe $H_{\approx}$, in particular its target function, involves "pointing" rather than "naming."

4. Charts and tables

The language we have introduced (e.g. "vertices", "edges") to facilitate talking about flows suggests that a flow has an "underlying digraph". Indeed, referring to Definition 3.1, *the underlying digraph of the Γ-flow F* consists of

(D1) $V = V^{ex} \cup S$,

(D2) E as given in (3.1.2),

(D3) the source function: $(s, i) \overset{\partial_1}{\mapsto} s$,

(D4) the target function: $(s, i) \overset{\partial_2}{\mapsto} \sigma(s, i)$, i.e. $\partial_2 = \sigma$.

Notice that the function $\theta: V \to N$ gives the outdegree of the vertices $s \in S$ of the (two-sorted-) digraph $(V, E, \partial_1, \partial_2)$. Notice, too, that the exits have outdegree 0 in this digraph. Thus $\theta: V \to N$ is the outdegree function of $(V, E, \partial_1, \partial_2)$. If $\Gamma_0 = \emptyset$, then for every flow F *only* the exits of F have outdegree 0; if $p = 0$ all the vertices of outdegree 0 are non-exits.

The digraphs $(V, E, \partial_1, \partial_2)$ obtained as underlying digraphs of flows $F: n \to p$, are, however, "special" in certain ways:

They are *locally finite* in the sense that the number of outedges of each vertex is finite; thus the digraph has an outdegree function $\theta: V \to N$. (4.1)

$$E = \{(v, i) \mid i \in [v\theta]\}. \tag{4.2}$$

$$(v, i)\partial_1 = v. \tag{4.3}$$

The underlying digraph of F is enriched by

a distinguished subset V^{ex} of V together with a bijection $V^{ex} \approx [p]$, satisfying (4.4)

$$v\theta = 0 \quad \text{for} \quad v \in v^{ex}. \tag{4.5}$$

We call a locally finite digraph $D = (V, E, \partial_1, \partial_2)$ an *outedge digraph* if it satisfies (4.1), (4.2) and (4.3). If D is a digraph we say (D, p) is a *p-exit-digraph* if D is enriched by (4.4) and (4.5) is satisfied. A digraph D together with a *pointing function*

$b:[n]\to V$ is an *n-pointed-digraph.* A *vector pointed-digraph* of briefly *pointed-digraph* is an n-pointed-digraph for some $n \geq 0$. Let

$$n \xrightarrow[S]{D} p, \text{ where } S = V - \{\text{exit } j \mid j \in [p]\}, \tag{4.6}$$

be an n-pointed-p-exit-outedge digraph. By a *Γ-flowchart scheme* (briefly, *Γ-flowchart* or *Γ-chart*) with *source n* and *target p* we mean (4.6) together with an outdegree-compatible (i.e. $s\lambda \in \Gamma_{s\theta}$) function $\lambda: S \to \bigcup \Gamma$. The following is obvious.

Proposition 4.1. *The function which takes a Γ-flow with p exits into the p-exit-Γ-chart whose underlying digraph is given by* (D1)–(D4), *whose pointing function is the begin function of F, whose exit function is the exit function of F and whose labeling function is the labeling function of F, is bijective.*

Roughly speaking, Proposition 4.1 implies that the "information" given by a Γ-flow and a Γ-chart is the same.

The bijection described in Proposition 4.1 has the effect that any operations (e.g. "composition" defined in the next section) or relations (e.g. the isomorphism of Section 3) defined on Γ-flows have corresponding "induced" operations or relations on Γ-charts (and vice-versa). Notice that if Γ_n is a singleton (i.e. $\Gamma_n = \{\gamma_n\}$) for each $n \in N$ then each n-pointed-p-exit-outedge digraph admits exactly one compatible function λ – so, in this case, λ need not be specified. In the case that $p = 0$ as well (resp. $n = 0$ as well) a chart is, essentially, an n-pointed-outedge digraph (resp. a p-exit-outedge digraph). Finally, *in the case that all three specializations are made* (i.e. $n = 0 = p$, $\Gamma_n = \{\gamma_n\}$ *for each* n) *a chart is, essentially, an outedge digraph.*

Thus, the notion of isomorphism of charts, induces, in particular a notion of isomorphism between outedge digraphs. This isomorphism notion is stricter than the general-algebraic notion of isomorphism for two-sorted-digraphs in that the ith outedge of a digraph is required to go into the ith outedge of the corresponding vertex. Specifically, where D_1, D_2 are outedge digraphs, the notion is this (cf. Definition 3.2): a bijection $\beta: V_1 \to V_2$ (note that $V_i = S_i$, $i \in [2]$ since $p = 0$) which preserves outdegree, i.e. satisfies (3.2.1), and preserves target-function, i.e. $(s, i)\sigma_1\beta = (s\beta, i)\sigma_2$ for all $i \in [s\theta]$, cf. (3.2.2).

The Γ-flow H of Section 3 (defined by Fig. 5.10), for example, may also be described with the aid of the following "pointed Γ-table", (a notion which is defined precisely just below).

Table 2.
The pointed-Γ-table H.

	H	
$b1 \rightarrow$	5	γ_2; (6, 7)
	6	γ_1; (8)
	7	γ_1'; (8)
	8	γ_1; (1)

Here the notation "(6, 7)" denotes the particular function:

$$[2] \rightarrow \{1, 5, 6, 7, 8\}, 1 \mapsto 6, 2 \mapsto 7.$$

(The target of (6, 7) is here indicated by the table context.) Notice that the pointed-table H provides "the same information" as the flow $H: 1 \rightarrow 1$ (cf. Section 3).

In preparation for a definition and a proposition, we introduce the following notation for a flow (3.1.5). Where $s \in S$, σ_s is the finite sequence of elements of V whose length is $s\theta$, whose ith item, $i \in [s\theta]$, is $\sigma(s, i)$ i.e. $\sigma_s : [s\theta] \rightarrow V$ is a function satisfying $\sigma_s(i) = \sigma(s, i)$. We call σ_s a *local successor* (*or target*) *function* of F and $\{\sigma_s\}_{s \in S}$, *the family of local successor functions of F.*

By a *p-exit, Γ-table* we mean a set V, a distinguished subset V^{ex} of V, a bijection $V^{ex} \approx [p]$, together with, where $S = V - V^{ex}$, a function $\lambda : S \rightarrow \bigcup \Gamma$ and a family $\{\sigma_s\}_{s \in S}$ of finite sequences of elements of V, *λ-compatible* in the sense: if $\lambda(s) \in \Gamma_n$ then the length of σ_s is n.

By an *n-pointed, p-exit Γ-table F* or *Γ-table F* : $n \rightarrow p$, we mean a p-exit, Γ-table together with a function $b : [n] \rightarrow V$.

Proposition 4.2. *The function which takes the Γ-flow* (3.1.5) *into the Γ-table* $F : n \rightarrow p$ *where*

$V = V^{ex} \cup S$, *the set of vertices of the Γ-table F*

λ, *the labeling function of the Γ-table F*

$\{\sigma_s\}_{s \in S}$, *the family of local successor functions of the Γ-table F*

is bijective.

The bijections of Propositions 4.1 and 4.2 yield, of course, a bijection between Γ-charts and Γ-tables.

The isomorphism notion between Γ-flows induces an isomorphism notion between Γ-tables. According to this notion, the Γ-table $H: 1 \rightarrow 1$ is (uniquely) isomorphic to the Γ-tables $H', H'': 1 \rightarrow 1$ described by Tables 3a, b.

Table 3a

	H′	
b1→	6	γ_2; (5, 7)
	5	γ_1; (8)
	7	γ_1'; (8)
	8	γ_1; (1)

Table 3b

	H″	
b1→	6	γ_2; (7, 8)
	7	γ_1; (9)
	8	γ_1'; (9)
	9	γ_1; (1)

Exploiting the "natural" well-ordering of the internal set of H, we might make the convention that if $\sigma_2(i)=s+1$ then we will fail to record "$s+1$". This convention corresponds to programming language practice and clearly has some point.

Thus, if we demanded – in the case of a finite table – that its internal set of vertices be totally ordered it would not be unreasonable to require that an isomorphism preserve order of internal vertices. The isomorphism $H \approx H''$ preserves order but the isomorphism $H \approx H'$ does not.

Employing the convention in the description of H and H'' we obtain Tables 4a, b

Table 4a

	H	
$b1$→	5	γ_2; (, 7)
	6	γ_1; (8)
	7	γ_1';
	8	γ_1; (1)

Table 4b

	H''	
$b1$→	6	γ_2; (, 8)
	7	γ_1; (9)
	8	γ_1';
	9	γ_1; (1)

5. Operations on flowschemes

We describe below, what we shall call, the *combinatorial operations* of "separated pairing," "composition" and "scalar conditional iteration". Each of the first two operations takes two Γ-flows, subject to disjointness constraints (designed to avoid undesired coalescence), as argument and produces a Γ-flow as value. There are other operations, derivative from these, important to our discussion. These, too, will be "combinatorial" in the sense[1] that they will be defined only when certain disjointness constraints are met. These combinatorial operations induce, in an obvious way, corresponding *algebraic operations* on isomorphism classes of Γ-flows. [In order to avoid set-theoretic difficulties due to "largeness", all sets may be assumed to be subsets of a suitably chosen infinite set.] The induced operations are "algebraic" in the sense that their definition depends on the fact that our combinatorial operations

[1] The term "set-theoretic" might also be used in this connection. "Combinatorial" reflects, perhaps, an emphasis on finiteness.

"preserve isomorphism of Γ-flows" e.g., if $F_i \approx F'_i$, $i \in [2]$, and "$\circ$" is one of our combinatorial binary operations, then $F_1 \circ F_2 \approx F'_1 \circ F'_2$.

Since a Γ-flow is specified by nine items of data (cf. Definition 3.1), it is, perhaps, not surprising that a formal description of operations on Γ-flows should be rather lengthy. The operations are, however, conceptually simple and may readily be pictured. Moreover, we shall not in fact employ these formal definitions in proofs. Thus, we give informal descriptions of the operations – accompanied by pictures. Actually we employ two "modes" in two of these informal descriptions. Both have formal counterparts which, however, don't produce the *same* result but, rather, *isomorphic* results. Thus either mode yields the *same* algebraic operation. For the properly sceptical reader, though, we do provide formal definitions (following informal mode 2) so that such a reader will have at his disposal (part of) the requisite means to deny or affirm an unproved assertion about these operations.

For the first definition there is only one informal mode.

Definition 5.1. (*Combinatorial*) *separated pairing*. Separated pairing applied to the data

$$n_i \xrightarrow[S_i]{F_i} p_i, \qquad i \in [2], \qquad V_1 \cap V_2 = \emptyset,$$

produces the Γ-flow (cf. Fig. 5.1)

$$n \xrightarrow[S]{F} p \quad \text{where } n = n_1 + n_2,\ p = p_1 + p_2,\ F = F_1 \oplus F_2,\ S = S_1 \cup S_2$$

further described below.

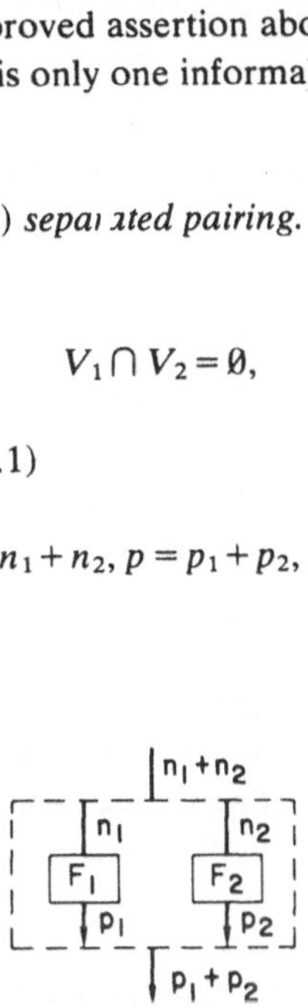

Fig. 5.1. Separated pairing. $F_1 \oplus F_2 : n_1 + n_2 \to p_1 + p_2$.

Informal (Modes 1 and 2) F is obtained by regarding the two flows F_1, F_2 as one and taking the sequence of begins of F to be the sequence of begins of F_1 followed by the sequence of begins of F_2.

Formal

(SP1) $\quad \theta : s \to N$ is the common extension of θ_i, $i \in [2]$; thus $E = E_1 \cup E_2$.

(SP2) $\quad \sigma : E \to V$ is the common extension of σ_1 and σ_2,

(SP3) $b:[n]\to V$
$b(i)=b_1(i)$ for $i\in[n_1]$,
$b(n_1+i)=b_2(i)$ for $i\in[n_2]$.
(SP4) $\lambda:S\to\bigcup\Gamma$ is the common extension of λ_1,λ_2.
(SP5) exit $(j)=\text{exit}_1\,(j)$ for $j\in[p_1]$,
exit $(p_1+j)=\text{exit}_2\,(j)$ for $j\in[p_2]$.

Definition 5.2. (*Combinatorial*) *composition.* Composition applied to the data

$$n_i \xrightarrow[S_i]{F_i} p_i, \qquad i\in[2], \qquad V_1\cap V_2=\emptyset,$$

is defined when and only when $p_1=n_2$ and produces the Γ-flow (cf. Fig. 5.2)

$$n_1 \xrightarrow[S=S_1\cup S_2]{F=F_1\circ F_2} p_2$$

further described below. (Hereafter the small circle is used for composition.)

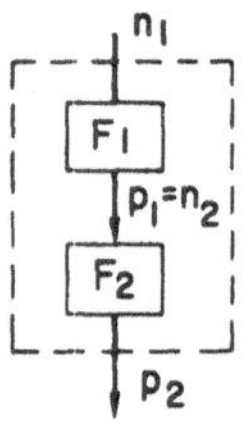

Fig. 5.2. Composition. $F_1\circ F_2:n_1\to p_2$.

Informal (Mode 1). F is obtained by identifying exit j of F_1 with begin j of F_2 for each $j\in[p_1]=[n_2]$; begin i of F is begin i of F_1 for each $i\in[n_1]$.

(Mode 2). F is obtained by deleting the exits of F_1, redirecting the edges of F_1 which end with exit j of F_1 to begin j of F_2 and, if begin i of F_1 is exit j of F_1, taking begin i of F to be begin j of F_2.

Formal

(COMP 1) $\theta:S\to N$ is the common extension of θ_i, $i\in[2]$; thus $E=E_1\cup E_2$.

(COMP 2) $\sigma:E\to V$

$$\sigma(e)=\begin{cases}\sigma_1(e) & \text{if } e\in E_1 \text{ and } \sigma_1(e)\in S_1,\\ b_2(j) & \text{if } e\in E_1 \text{ and } \sigma_1(e)=\text{exit } j \text{ of } F_1,\\ \sigma_2(e) & \text{if } e\in E_2.\end{cases}$$

(COMP 3) $b:[n_1]\to V$

$$b(i)=\begin{cases}b_1(i) & \text{if } b_1(i)\in S_1,\\ b_2(j) & \text{if } b_1(i)=\text{exit } j \text{ of } F_1.\end{cases}$$

(COMP 4) $\lambda : S \to \bigcup \Gamma$ is the common extension of λ_1, λ_2.
(COMP 5) $V^{ex} = V_2^{ex}$ and the bijection $V^{ex} \approx [p_2]$ of F is the bijection $V_2^{ex} \approx [p_2]$ of F_2.

Definition 5.3. (*Combinatorial*) *Scalar conditional iteration.* Conditional iteration applied to the datum

$$1 \xrightarrow[S]{F} p+1$$

is defined when and only when exit $p+1$ is not the begin and produces the Γ-flow cf. Fig. 5.3)

$$1 \xrightarrow[S]{F^{\dagger}} p$$

further described below.

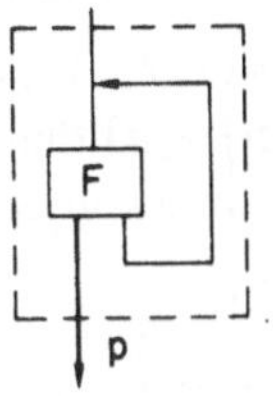

Fig. 5.3. Scalar conditional iteration. $F^{\dagger} : 1 \to p$.

Informal (Mode 1). $F^{\dagger}$ is obtained by identifying exit $p+1$ of F with its begin.
(Mode 2). $F^{\dagger}$ is obtained by deleting exit $p+1$ of F and redirecting the edges of F which end with exit $p+1$ of F to the begin of F.
Formal
(SCI 1) $\theta_{\dagger} = \theta : S \to N$.
(SCI 2) $\sigma_{\dagger} : E \to V_{\dagger} = V - \{\text{exit } p+1\}$.

$$\sigma_{\dagger}(e) = \begin{cases} \sigma(e) & \text{if } \sigma(e) \neq \text{exit } p+1, \\ \text{begin of } F & \text{if } \sigma(e) = \text{exit } p+1. \end{cases}$$

(SCI 3) $b_{\dagger} = b : [1] \to V_{\dagger}$.
(SCI 4) $\lambda_{\dagger} = \lambda : S \to \bigcup \Gamma$.

Two kinds of Γ-flows play a prominent role in our discussion: the "trivial" ones and the "atomic" ones.

Definition 5.4. A Γ-flow (3.1.5) is *trivial* if $S = \emptyset$.

In this case, since $E = \emptyset$ as well, F is determined up to (unique) isomorphism by its begin function $b : [n] \to [p]$. We may then write "$b : n \to p$" to indicate "the" associated Γ-flow (unique up to isomorphism). As a notation for the function or the flow b we may use

$$(b(1), b(2), \ldots, b(n))_p$$

or, in the case that $p = b(i)$ for some $i \in [n]$ (in particular, in the case that b is surjective), simply

$$(b(1), b(2,), \ldots, b(n)).$$

For example, in the case $b:[3]\to[4]$ where $b(1)=b(2)=3$ and $b(3)=1$, $b=(3,3,1)_4$, (cf. Fig. 5.4) and, in the case, $f:[3]\to[2]$ where $f(1)=f(2)=2$ and $f(3)=1$, $f=(2,2,1)$. Thus the flow $(3,3,1)$ is the same as the flow $(3,3,1)_3$. For the *identity*

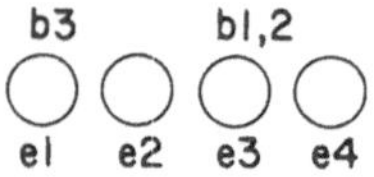

Fig. 5.4. The trivial flow $(3, 3, 1)_4$.

function $I_n:[n]\to[n]$, we have $I_n=(1,2,\ldots,n)$. Note that if $h:[n]\to[q]$ is the composition of $f:[n]\to[p]$ and $g:[p]\to[q]$ then the Γ-flow $f\cdot g: n\to q$ is $h:n\to q$. Some examples of separated pairing and composition applied to trivial flows:

$$(2,2,1)_3\circ(2,2,1)=(2,2,2), \tag{5.1}$$

$$(2,2,1)_3\oplus(2,2,1)=(2,2,1,5,5,4), \tag{5.2}$$

$$(3,3,1)\circ(2,2,1)=(1,1,2), \tag{5.3}$$

$$(3,3,1)\oplus(2,2,1)=(3,3,1,5,5,4), \tag{5.4}$$

$$(2,1)\circ(2,1)=(1,2)=I_2, \tag{5.5}$$

$$(2,1)\oplus(2,1)=(2,1,4,3). \tag{5.6}$$

Notice that the notation used in the above examples enables one to determine the source and target of each of the trivial flows, e.g. $(2,2,1,5,5,4):6\to5$, and that the expression

$$(4,4,4)\circ(1,2,2) \tag{5.7}$$

is "ill formed" since the target of $(4,4,4)$ is 4 and the source of $(1,2,2)$ is 3. Thus, in our context, (5.7) is meaningless.

Definition 5.5. A Γ-flow

$$1 \xrightarrow[S]{F} p \tag{5.8}$$

is *atomic* if $S=\{b_F(1)\}$, the outdegree of $b_F(1)$ is p and the jth edge from $b_F(1)$ goes to exit j, for each $j\in[p]$.

If the begins of two atomic Γ-schemes have the same label, then they also have the same outdegree. Hence two atomic schemes, whose begins have the same label, are isomorphic, indeed uniquely so. Clearly for each $\gamma_p \in \Gamma_p$ there is an atomic schemes (5.8) whose label is γ_p. Thus, for each p, there is a bijection

$$\Gamma_p \to \{F/\approx \mid F: 1 \to p \text{ an atomic scheme}\} \tag{5.9}$$

which takes γ_p into the isomorphism class $F/\approx$, where $F: 1 \to p$ is an atomic scheme whose begin is labelled γ_p; we use the phrase "the Γ-flow γ_p" or "the flow γ_p" to mean the isomorphism class $F/\approx$ or "an ambiguous member $F_\approx$ of" $F/\approx$ (cf. Fig. 5.5, 5.6).

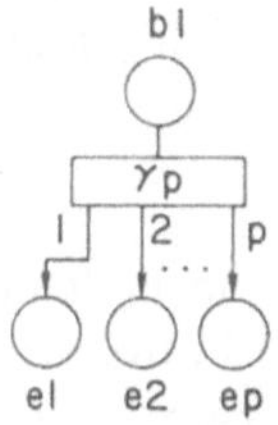

Fig. 5.5. The atomic flow γ_p, $p \geqslant 0$.

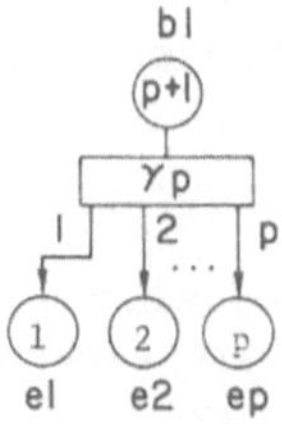

Fig. 5.6. A representative of the flow γ_p.

$$1 \xrightarrow[\{p+1\}]{\gamma_p} p.$$

Proposition 5.1. *When the expressions below are meaningful, the equalities indicated hold.*

$$[F_1 \oplus F_2] \oplus F_3 = F_1 \oplus [F_2 \oplus F_3], \tag{5.10}$$

$$I_0 \oplus F = F = F \oplus I_0, \tag{5.11}$$

$$[F \circ G] \circ H = F \circ [G \circ H], \tag{5.12}$$

$$I \circ F = F = F \circ I, \tag{5.13}$$

$$I_n \oplus I_p = I_{n+p}, \tag{5.14}$$

$$[F_1 \oplus F_2] \circ [G_1 \oplus G_2] = [F_1 \circ G_1 \oplus F_2 \circ G_2], \tag{5.15}$$

$$[F \oplus I] \circ [I \oplus G] = F \oplus G = [I \oplus G] \circ [F \oplus I]. \tag{5.16}$$

Let $\Gamma\mathrm{Fl}/\approx = \{F/\approx \mid F \in \Gamma\mathrm{Fl}\}$. The algebraic operations above on $\Gamma\mathrm{Fl}/\approx$ induced by the combinatorial operations on $\Gamma\mathrm{Fl}$ are always defined so long as the "target-source" constraint for composition is satisfied. thus (5.15) becomes

$$(\phi_1 \oplus \phi_2) \circ (\psi_1 \oplus \psi_2) = \phi_1 \circ \psi_1 \oplus \phi_2 \circ \psi_2 \tag{5.15'}$$

where $\phi_i : n_i \to p_i$, $\psi_i : p_i \to q_i$ are in $\Gamma\mathrm{Fl}/\approx$ for $i \in [2]$.

By virtue of (5.12′) and (5.13′), $\Gamma\mathrm{Fl}/\approx$ is a category (with objects $n \in N$, and morphisms $F/\approx : n \to p$). By virtue of satisfying (5.14′) and (5.15′), $\oplus$ is a bifunctor

$$\Gamma\mathrm{Fl}/\approx \times \Gamma\mathrm{Fl}/\approx \to \Gamma\mathrm{Fl}/\approx.$$

By virtue of (5.10′) and (5.11′), the bifunctor ⊕ is associative and satisfies the "unit law". Thus [15, p. 158], the category ΓFl/≈ together with the bifunctor ⊕ and the unit I_0 is a *strict monoidal category*. Briefly,

Proposition 5.2. *(ΓFl/≈, ⊕, I_0) is a strict monoidal category.*

(The Appendix indicates an extension of this proposition).

There are two additional operations, derivative from composition and separated pairing which are relevant to our discussion.

Definition 5.6. (*Combinatorial, vector*) *separated substitution*, SS. Separated substitution applied to the data $F: n \to p$ and $G_j: 1 \to q_j$, $j \in [p]$, produces the Γ-flow (cf. Fig. 5.7)

$$F \circ [G_1 \oplus G_2 \oplus \cdots \oplus G_p]: n \to q_1 + q_2 + \cdots + q_p;$$

it is defined when and only when the above expression is defined. (Note that $[G_1 \oplus G_2] \oplus G_3$ is defined iff $G_1 \oplus [G_2 \oplus G_3]$ is defined.) In the case that $p = 0$, we understand $[G_1 \oplus G_2 \oplus \cdots \oplus G_p] = I_0$, so that in this case, $F \circ [G_1 \oplus G_2 \oplus \cdots \oplus G_p] = F$.

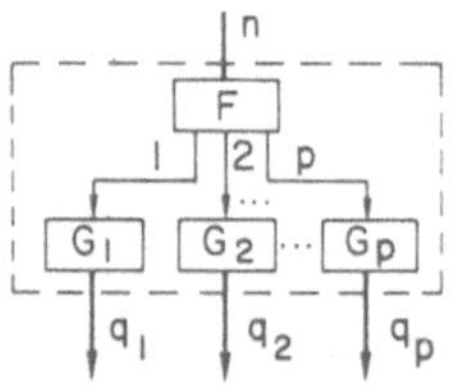

Fig. 5.7. (Vector) Separated substitution $F \circ [G_1 \oplus G_2 \oplus \cdots \oplus G_p]$.

In the case that the above operation is restricted to $n = 1$, the restricted operation is *scalar separated substitution*.

Definition 5.7. (*Combinatorial, vector*) *singular separated substitution*, SSS.

Singular separated substitution applied to the data $F: n \to p + 1$ and $G: 1 \to q$ produces the Γ-flow (cf. Fig. 5.8)

$$F \circ [I_p \oplus G]: n \to p + q;$$

it is defined when and only when the above expression is defined.

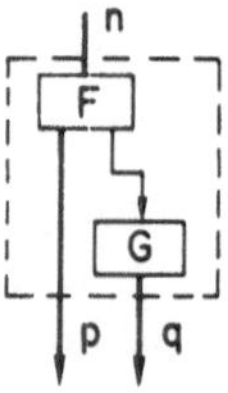

Fig. 5.8. (Vector) Singular separated substitution $F \circ [I_p \oplus G]$.

In the case that the above operation is restricted to $n = 1$, the restricted operation is *scalar singular separated substitution.*

Proposition 5.3. *Any class $\mathscr{C}$ of Γ-flows closed under exit permutations and containing the identity I_1 is closed under separated substitution iff it is closed under singular separated substitution.*

Proof. (*if*). Rather than formulate a careful induction we illustrate the argument for "if $F:n\to p$ and $G_j:1\to q_j$ are in $\mathscr{C}$ then so is $F\circ[G_1\oplus\cdots\oplus G_p]$" for $p=3$. For appropriate permutations (or rather, trivial flows corresponding to permutations, cf. Definition 5.4) $\pi_1, \pi_2, \pi_3, \pi_4$ we have

$$\pi_1\circ[I\oplus G_2\oplus I]=[I\oplus I\oplus G_2]\circ\pi_2$$
$$\pi_3\circ[G_1\oplus I\oplus I]=[I\oplus I\oplus G_1]\circ\pi_4. \tag{5.17}$$

Thus

$$\begin{aligned} F\circ[G_1\oplus G_2\oplus G_3] &= F\circ[I\oplus I\oplus G_3]\circ[I\oplus G_2\oplus I]\circ[G_1\oplus I\oplus I] \\ &= F\circ[I\oplus I\oplus G_3]\circ\pi_1^{-1}\circ[I\oplus I\oplus G_2] \\ &\quad\circ\pi_2\circ\pi_3^{-1}\circ[I\oplus I\oplus G_1]\circ\pi_4 \end{aligned} \tag{5.18}$$

and $F\circ[I\oplus I\oplus G_3]$ is in $\mathscr{C}$ by SSS; $F\circ[I\oplus I\oplus G_3]\,\pi_1^{-1}$ is in $\mathscr{C}$ by exit permutation closure, etc. so that $F\circ[G_1\oplus G_2\oplus G_3]$ is in $\mathscr{C}$.

(*only if*). With I_1 in $\mathscr{C}$, SSS becomes a special case of SS.

Definition 5.8. *Exit-merging* or briefly, *merging*. Exit-merging applied to the Γ-flow $F:n\to p$ and trivial flow $f:p\to q$, where the function $f:[p]\to[q]$ is surjective, produces the Γ-flow $F\circ f$. We call the trivial *flow f surjective* when the function f is surjective.

Corollary 5.4. *Any class of Γ-flows closed under exit merging and containing the identity I_1 is closed under* SS *iff it is closed under* SSS.

The underlying (outedge) exit-digraph of the Γ-flow (depicted by) Fig. 5.10 is (depicted by) Fig. 5.14. Fig. 5.9 (and Fig. 5.11 in alternate style) depicts an ambiguous member of the isomorphism class determined by Fig. 5.10. Fig. 5.13 is the underlying exit-digraph of Fig. 5.9 (and Fig. 5.11); it is an ambiguous member of the isomorphism class determined by the exit-digraph Fig. 5.14. Figs. 5.15 and 5.12 are similarly related. Fig. 5.16 is the underlying pointed-exit-digraph of Fig. 5.9 (and Fig. 5.11). Fig. 5.17 is the underlying exit-digraph of I_1 while Fig. 5.18 is the underlying exit-digraph of the flow γ_p (Fig. 5.5), the case $p=0$ being Fig. 5.19.
These figures incorporate the *convention* $\gamma_i, \gamma_i'\in\Gamma_i$.

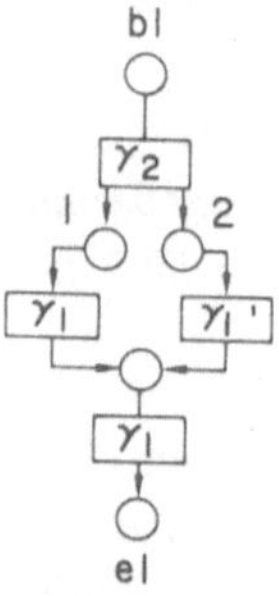

Fig. 5.9. $\gamma_2 \circ [\gamma_1 \oplus \gamma_1'] \circ (1, 1) \circ \gamma_1 : 1 \to 1$.

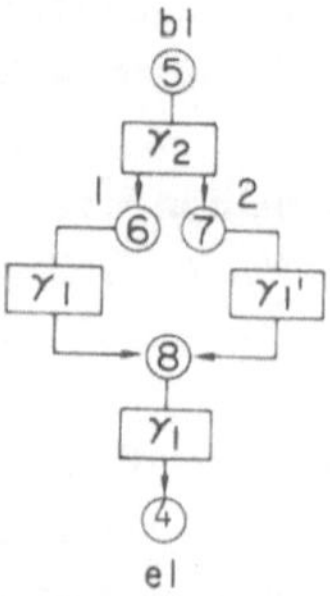

Fig. 5.10. $1 \xrightarrow[\{5,6,7,8\}]{\gamma_2 \circ [\gamma_1 \oplus \gamma_1'] \circ (1, 1) \circ \gamma_1} 1$.

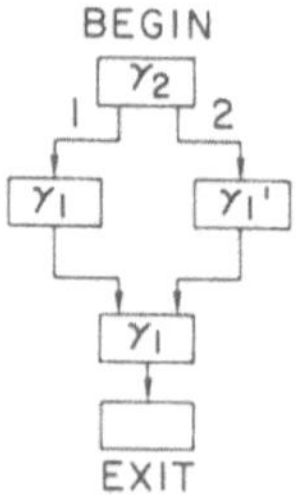

Fig. 5.11. Alternate Style for flow Fig. 5.9.

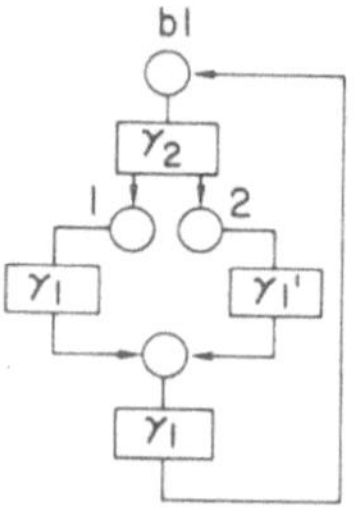

Fig. 5.12. $[\gamma_2 \circ [\gamma_1 \oplus \gamma_1'] \circ (1, 1) \circ \gamma_1]^\dagger$: $1 \to 0$. Anti-iterate of this is Fig. 5.9.

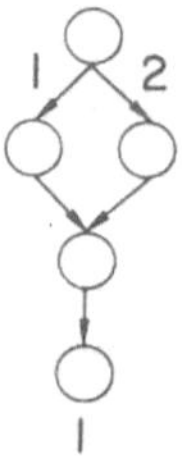

Fig. 5.13.

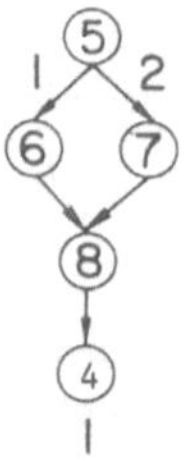

Fig. 5.14.

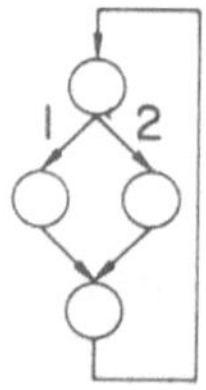

Fig. 5.15.

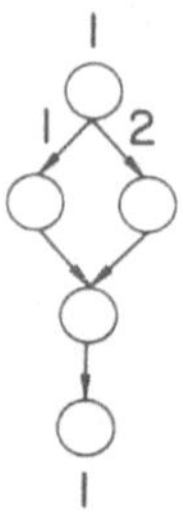

Fig. 5.16.

Fig. 5.17.

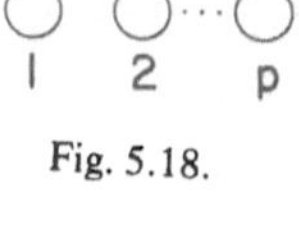

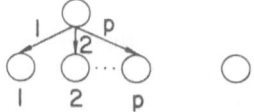

Fig. 5.18.

Fig. 5.19.

6. Some digraph notions: strongly connected, circular, circuit

Let G be a digraph with set V of vertices and let $\pi : v_1 \to v_2$ be a path of G. We define *the set*, $\boldsymbol{v}(\pi)$, *of vertices of* π as follows

$$\boldsymbol{v}(\pi) = \{v \in V \mid \pi : V_1 \xrightarrow{\pi_1} v \xrightarrow{\pi_2} v_2 \text{ for some } \pi_1, \pi_2,$$

$$\text{i.e. } \pi = \pi_1\pi_2 \text{ for some paths } v_1 \xrightarrow{\pi_1} v \xrightarrow{\pi_2} v_2\}$$

so that $v \in \boldsymbol{v}(\pi)$ iff the path π *meets* v; in particular $v_1, v_2 \in \boldsymbol{v}(\pi)$. We say $S \subseteq V$ *is strongly connected in* G if $S \neq \emptyset$ and for all $v_1, v_2 \in S$, there is a path $\pi : v_1 \to v_2$ in G such that $\boldsymbol{v}(\pi) \subseteq S$; S is *trivial* if its cardinality, $\kappa(S)$, is 1.

The path $\pi : v_1 \to v_2$ is a *circuit* if $v_1 = v_2$. The following proposition is obvious.

Proposition 6.1. *In G: a set S of vertices is finite and strongly connected iff there exists a circuit π such that $\boldsymbol{v}(\pi) = S$.*

A path $\pi : v_1 \to v_2$ is *elementary* provided

$$\pi : v_1 \xrightarrow[\pi_1]{} v \xrightarrow{\pi'} v \xrightarrow{\pi_2} v_2, \qquad |\pi'| > 0 \Rightarrow \pi' = \pi.$$

In particular, paths of length zero are elementary.

Definition 6.1. A set $C \subseteq V$ *is circular in* G if there exists an elementary circuit π in G such that $\boldsymbol{v}(\pi) = C$.

Proposition 6.2. *If $M \subseteq V$ is non-trivial, strongly connected and is minimal among all non-trivial, strongly connected sets in G, then M is circular in G.*

Proof. Let $\pi : v \to v$ and suppose $\boldsymbol{v}(\pi) = M$. There exists a factorization

$$\pi : v \xrightarrow{\pi_1} v' \xrightarrow{\pi'} v' \xrightarrow{\pi_2} v, \quad |\pi'| > 0$$

since M is non-trivial and strongly connected. Without loss of generality we may assume π' is elementary. Thus $\boldsymbol{v}'(\pi')$ is non-trivial and circular. Since $\boldsymbol{v}(\pi') \subseteq M$, by the minimality of M, we have $\boldsymbol{v}(\pi') = M$.

There exist digraphs, however, which have non-trivial circular sets which are not minimal in the above sense. For example, (cf. Fig. 6.1), both $\{v_1, v_2, v_3\}$ and $\{v_2, v_3\}$ are circular in the digraph depicted by Fig. 6.1.

However: if C is circular in G and π is an elementary circuit in G such that $\mathbf{v}(\pi) = C$ then C is the only strongly connected set in the subgraph G' of G consisting of the vertices and edges of π.

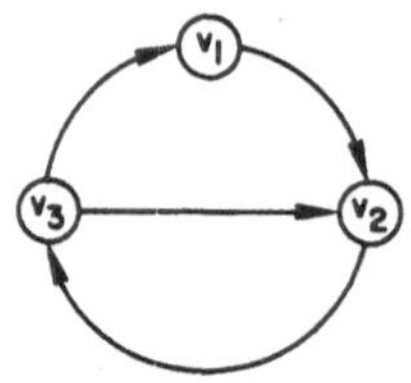

Fig. 6.1.

Corollary 6.3. *If M is strongly connected in G and $\kappa(M) = 2$, then M is circular in G.*

These notions carry over to Γ-flows F via the underlying digraph of F. Thus, for example:

$\pi : v_1 \rightarrow v_2$ is a *path in F* if it is a path in its underlying digraph.

$M \subseteq V$ is *strongly connected in F* if it is strongly connected in its underlying digraph, etc.

For future use, we make the following definitions for a digraph G, where $U \subseteq V$.

Definition 6.2. If π is a path to U(i.e. to some $u \in U$), we say *π enters U at v* if the minimum left prefix (i.e. left factor) π' of π to U, ends in v.

Definition 6.3. We say a path to π (i.e. to $\mathbf{v}(\pi)$) *enters π at v*, if the path enters $\mathbf{v}(\pi)$ at v.

Definition 6.4. We define $E(v, U)$, in words: *v is an entry to U*, by the following condition

$$E(v, U) \Leftrightarrow \text{every edge to } U \text{ from outside } U \text{ ends in } v.$$

Proposition 6.4. *$E(v, U) \Leftrightarrow$ every path to U from a vertex outside U enters U at v.*

Proposition 6.5. *If there exists an edge to U from outside U and $E(v, U)$, then $v \in U$ and for all v', $E(v', U) \Rightarrow v' = v$.*

If the hypothesis of Proposition 6.5 holds and U has an entry then the entry is unique.

7. Some pointed-digraph notions: accessibility, domination

Let P be an n-pointed-digraph whose set of vertices is V. We say a path $\pi : v \to v'$ in P is a *begin path* if v is a begin of P i.e. $v = b(i)$ for some $i \in [n]$; P is *accessible* (abbreviation: ACC) if for every $v \in V$ there is a begin path with target v i.e. a path to v from some begin. An accessible pointed-digraph is called a *flowgraph*. Following much of the "flowgraph literature" (cf. eg. [10]) we say v_1 *dominates* v_2 (notation: v_1 dom v_2) if every begin path to v_2 meets v_1 i.e.

$\pi : b(i) \to v_2$ admits a factorization

$$\pi : b(i) \xrightarrow{\pi_1} v_1 \xrightarrow{\pi_2} v_2.$$

Proposition 7.1. *The relation* dom *(determined by a pointed-digraph) is a pre-order i.e. it is reflexive and transitive.*

Proposition 7.2. *If v_1 or v_2 is accessible (i.e. accessible from some begin), v_1* dom *v_2 and v_2* dom *v_1, then $v_1 = v_2$.*

Proof. Suppose v_2 is accessible. Let π be a shortest begin path to v_2. Since v_1 dom v_2 and v_2 dom v_1, we have

$$\pi : b(i) \xrightarrow{\pi_1} v_2 \xrightarrow{\pi_2} v_1 \xrightarrow{\pi_3} v_2.$$

Since π_1, as well as π, is a begin path to v_2, $|\pi_1| \leq |\pi|$ and π is shortest such, we have $|\pi_1| = |\pi|$ so that $|\pi_2| = 0 = |\pi_3|$ and $v_1 = v_2$.

Corollary 7.3. *If the pointed digraph is accessible, its* dom *is a partial ordering.*

We define, where $U \subseteq V$, $D(v, U) \Leftrightarrow v \in U \;\&\; \forall u \in U\; [v \text{ dom } u]$, more briefly, $v \in U$ & v dom U, and we say *v is a dominator of U.*

Corollary 7.4. *If U is strongly connected and accessible (i.e. some $u \in U$ is accessible), then there is at most one v such that $D(v, U)$. If such a U has a dominator, then U has exactly one dominator (notation:* dom(U)).

Note particularly that dom $U \in U$.

Let π be a circuit. We define $D(d, \pi) \Leftrightarrow D(d, v(\pi))$ and say d is a *dominator of the circuit* π.

Definition 7.1. The pointed-digraph P has the *circuit dominator* property (notation: CD) if every circuit (equivalently, strongly connected set of vertices) in P has a dominator.

Corollary 7.5. *If P is a flowgraph satisfying* CD, *then each circuit π in P has a unique dominator,* dom(π) *and every begin path to π enters π at* dom(π).

The above notions carry over to Γ-flows F via the underlying pointed digraph P of the flow F. For example, "F is accessible" means "P is accessible".

Proposition 7.6. *Two isomorphic, accessible Γ-flows are uniquely isomorphic.*

The isomorphism notion of Γ-flows induces one for pointed-outedge digraphs via the specialization: number of exits $= 0$, $\Gamma_n = \{\gamma_n\}$. Thus,

Corollary 7.7. *Two isomorphic accessible pointed-outedge digraphs are uniquely isomorphic.*

The idea of "dominance" in programming contexts seems to have arisen first as a relation between "blocks of a program". According to [12, p. 15]: "The idea of dominance relations between the blocks of a program was suggested by Prosser" [13] "and refined by Medlock."

8. Circular dominator ⇒ circuit dominator

Let F be an accessible Γ-flow or, alternatively a flowgraph.

Lemma 8.1. *In F: let M_1, M_2, $M_1 \cap M_2 \neq \emptyset$ be strongly connected and have dominators, then $M = M_1 \cup M_2$ is strongly connected and has a dominator. Indeed,* dom $M =$ dom M_1 *or* dom $M =$ dom M_2.

Proof. That M is strongly connected is obvious.

Let $v_i = \text{dom}\, M_i$, $i \in [2]$. Suppose v_1 does not dominate M. Then v_1 does not dominate v_2 so that $v_2 \notin M_1$ and there is a begin path π_2 to v_2 which does not meet v_1. Since M_2 is strongly connected in F, there is a path π from v_2 to some $v \in M_1 \cap M_2$ whose vertices are in M_2 and whose vertices, except for v, are outside M_1. Then $\pi_2\pi$ is a minimal begin path to M_1. Hence $v = v_1 \in M_2$ so v_2 dom v_1 and hence v_2 dom M.

Theorem 8.2. *In F: if each circular set has a dominator, then each finite*[2] *strongly connected set has a dominator.*

Proof. Let M be strongly connected and suppose $\kappa(M) > 2$ (cf. Corollary 6.3). We argue inductively on $\kappa(M)$. Let $\pi : v \to v$ be a circuit of minimum length such that

[2] As noted by S.L. Bloom, if one argues by contradiction, Theorem 8.2 with "finite" omitted may be proved without the aid of Lemma 8.1.

$v(\pi)=M$. To show M has a dominator we may assume π is not elementary for otherwise M is circular and we are through. Thus π admits a factorization

$$\pi : v \xrightarrow{\pi_1} v' \xrightarrow{\pi'} v' \xrightarrow{\pi_2} v, \quad |\pi'|>0,$$

so that $M=v(\pi)=v(\pi_1\pi_2)\cup v(\pi')$ and by the "minimum length property" of π both $M_1=v(\pi_1\pi_2)$ and $M_2=v(\pi')$ are *proper* subsets of M. Moreover they are both strongly connected since $\pi_1\pi_2$ and π' are circuits. Thus by inductive assumption each of M_1 and M_2 has a dominator. Clearly $M_1 \cap M_2 \neq \emptyset$ since $v' \in M_1 \cap M_2$. Lemma 8.1 is applicable and concludes the argument.

One might suspect that Theorem 8.2 may be strengthened by replacing "circular set" by "minimal strongly connected set". That this is not the case is indicated by Fig. 8.1. In the F of Fig. 8.1, $\{v_1, v_2, v_3\}$ is circular but does not have a dominator; $\{v_2, v_3\}$ is minimal strongly connected (among all non-trivial strongly connected sets in F) and is the only such set; v_2 is the dominator of $\{v_2, v_3\}$.

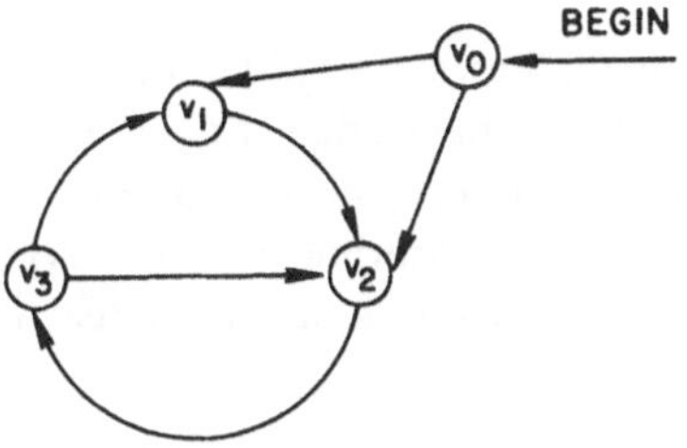

Fig. 8.1.

Remark 8.1. In view of Theorem 8.2, the CD property (Definition 7.1) may be restated as: every circular set has a dominator.

9. Backedges, a Lemma and a Theorem

Definition 9.1. An edge from an accessible vertex u to v in a Γ-flow F is a *back edge* if v dominates u in F.

Lemma 9.1. *If G_1, G_2 are accessible and $F=G_1 \circ G_2$ or $F=G_1 \oplus G_2$ or $F=G_1 \circ f$, where f is a trivial surjective flow, or $F=G_1^\dagger$ then, (if $i=1,2$ in the first two cases, $i=1$ in the last two)*

(i) *F is accessible*;
(ii) *if u, v are nonexit vertices of G_i then u dominates v in G_i iff u dominates v in F*;
(iii) *F satisfies* CD *iff all the G_i do*;

(iv) *an edge e of F is a back edge iff it is a back edge of some G_i or, in case $F = G_1^\dagger$, e is one of the new edges into the begin.*

Proof. (i) Obvious.

(ii) If $F = G_1 \circ G_2$ the result is obvious in the case $G_i = G_1$, and in the case $G_i = G_2$ follows from the fact that a begin path in F to a vertex of G_2 ends in (i.e. has as a right factor) a begin path in G_2 and, because of the accessibility of G_1, every begin path in G_2 can be extended to a begin path in F by prefixing a path in G_1. If $F = G_1 \oplus G_2$ the result is obvious, as it is in the case $F = G_1 \circ f$, since u, v are supposed to be nonexit vertices. If $F = G_1^\dagger$ it follows from the fact that every begin path in G_1 to a nonexit vertex is a begin path in F, and every begin path in F ends in a begin path in G_1.

(iii) Suppose F satisfies CD. Let α be a circuit in G_i. Then α cannot pass through an exit of G_i, and α is a circuit in F. Hence it has a vertex v_α which dominates it in F and, by (ii), also in G_i.

Conversely suppose the G_i satisfy CD and let α be a circuit in F. Then either α is a circuit in G_i, in which case its dominator vertex v_α in G_i will, by (ii), still dominate α in F, or $F = G_1^\dagger$ and α is a 'new' circuit, i.e. one passing through b_F, in which case b_F dominates it in F.

(iv) if e is an edge of some G_i then since neither of the ends of a back edge can be an exit it follows from (ii) that e is a back edge of G_i iff it is a back edge of F. If e is an edge of F but not an edge of some G_i then $F = G_1^\dagger$ and e is one of the new edges which enter b_F in F (instead of entering e_{p+1} in G_1). Since G_1 is accessible b_F dominates all vertices in F so this is a back edge in F. This completes the proof.

Definition 9.2. The class $\Gamma\mathcal{R}$ of *reducible Γ-flows* is the smallest class of Γ-flows containing I_0, I_1, Γ and closed under composition, separated pairing, exit merging, scalar iteration. (We shall usually write "$\mathcal{R}$" in place of "$\Gamma\mathcal{R}$".)

Theorem 9.2. *If F is a reducible flow then every edge of F which is introduced as one of the new edges into the begin in some operation of scalar iteration is a back edge of F. Conversely if e is a back edge of a reducible flow F then in every such way of building up F, e must occur as one of the new edges in some scalar iteration.*

Proof. Immediate consequence of Lemma 1 (iv).

Proposition 9.3. *If e is a back edge in an accessible flow F, then F possesses a non-trivial circuit of which e is a part.*

Proof. Suppose $e : u \to v$. From the accessibility of u and the fact that v dominates u, we conclude there is a path $\pi : v \to u$ in F. Thus (identifying e with the path of length one to which it corresponds) πe is a non-trivial circuit in F.

Corollary 9.4. *If F is an accessible, acyclic (i.e. free of non-trivial circuits) flow, then F possesses no back edges.*

It is easy, however, to produce an example of an accessible flow which possesses no back edges but does possess a non-trivial circuit.

10. Component dominator is an entry

Definition 10.1. We use $\leqslant$ for the usual preorder, i.e., reflexive, transitive relation, associated with the graph of a flow F. Thus $v_1 \leqslant v_2$ means there is a path (of length $\geqslant 0$) from v_1 to v_2. The relation $v_1 \leqslant v_2$ & $v_2 \leqslant v_1$ is an equivalence relation and $\leqslant$ induces a partial ordering on the equivalence classes of this relation. We shall call these equivalence classes *components*. Clearly they are the maximal strongly connected sets of vertices of F. Note that each exit is a component by itself which we shall call an exit component.

We use $v_1 < v_2$ to mean there is a path from v_1 to v_2 but none from v_2 to v_1.

Proposition 10.1. *If F satisfies* ACC *and if C is a component of F which has a dominator v_C, then $E(v_C, C)$, i.e. v_C is an entry to C.*

Proof. Suppose there is a path from a vertex v_1 not in C which enters C at $v \neq v_C$. Then (Fig. 10.1) by ACC there is a begin path $\pi_1 : b \to v_1$, so there is a path $\pi_1\pi_2 : b \to v$ to C. By assumption that C has a dominator, this must go through v_C, i.e. $v_C \in v(\pi_1\pi_2)$. But v_C is not on the path $\pi_2 : v_1 \to v$ so it must be on the path π_1 i.e. $v_C \leqslant v_1$. But $v_1 \leqslant v \leqslant v_C$ so $v_1 \in C$, contrary to hypothesis.

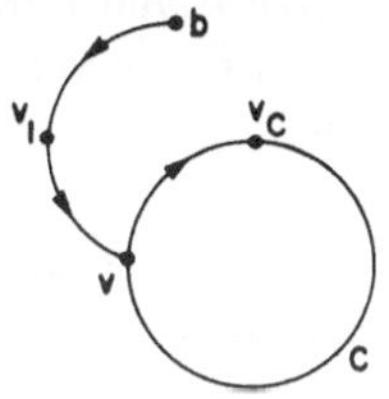

Fig. 10.1.

Corollary 10.2. *If F satisfies* ACC *and* CD, *then for every component C, $E(v_C, C)$, where v_C is the dominator of C.*

Thus F has the stronger property that every path (not merely begin path) to a component enters that component at its dominator.

The following example shows that Proposition 10.1 cannot be strengthened by allowing C to be any strongly connected set of vertices F.

In the flow of Fig. 10.2, let M be the strongly connected set $\{1,2\}$ then the dominator v_M of M is 1 while the entry of M is 2. Thus not $E(v_M, M)$.

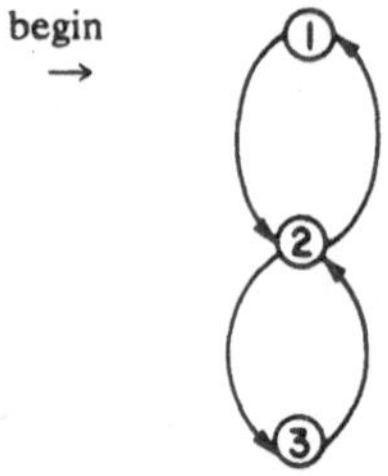

Fig. 10.2.

11. Formulation of the main theorem

The corollary below gives an alternative description – perhaps, a better one – of the class $\mathcal{R}$; however, while Definition 9.2 specializes nicely to the important special case of scalar flows, the description of Corollary 11.2 does not.

Proposition 11.1. *Let $\mathcal{C}$ be a class of (vector) Γ-flows closed under composition and separated pairing. Then $\mathcal{C}$ is closed under exit merging and contains I_0, I_1 iff $\mathcal{C}$ contains the surjective trivial flows.*

Proof. (*if*). I_0 and I_1 are trivial surjective flows. Since $\mathcal{C}$ is closed under composition and contains the surjective trivial flows, it is closed under exit merging.

(*only if*). Let $f: n \to p$ be a surjective trivial flow and suppose $n > 0$. Then, by closure under separated pairing the identity $I_n \in \mathcal{C}$ and by closure under exit merging $f \in \mathcal{C}$. If $n = 0$, then also $p = 0$ and $f = I_0$.

Corollary 11.2. *The class $\mathcal{R}$ is the smallest class of Γ-flows containing the surjective trivial flows, Γ and closed under composition, separated pairing, scalar iteration.*

In order to formulate the main theorem succinctly and with full strength, we define $\mathcal{R}_1$ = smallest class of (vector) Γ-flows containing I_0, I_1, Γ and closed under SSS, separated pairing, exit merging, scalar iteration, $\mathcal{R}_{\text{scal}}$ = smallest class of scalar Γ-flows which contains I_1, Γ and is closed under SSS, exit merging, scalar iteration.

Remark 11.1. It is obvious from the definitions that $\mathcal{R}_1 \subseteq \mathcal{R}$.

(Main) Theorem 11.3. *Let F be a Γ-flow.*

(a) *If $F \in \mathcal{R}$, then F is finite and satisfies* ACC *and* CD.

(b) *If F is finite and satisfies* ACC *and* CD, *then* $F \in \mathscr{R}_1$.
(c) $\mathscr{R} \subseteq \mathscr{R}_1$, $\mathscr{R} = \mathscr{R}_1$.
(d) *A Γ-flow is in $\mathscr{R}$ iff it is finite and satisfies* ACC *and* CD.
(e) *If F is scalar, finite and satisfies* ACC, CD, *then* $F \in \mathscr{R}_{\text{scal}}$.

The proof of the Main Theorem is given in the next section.

Corollary 11.4. $\mathscr{R}_{\text{scal}}$ = *class of scalar Γ-flows in $\mathscr{R}$.*

Proof. If F is scalar and in $\mathscr{R} = \mathscr{R}_1$, then by Theorem 11.3(a) and (e), $F \in \mathscr{R}_{\text{scal}}$. On the other hand, by definition $\mathscr{R}_{\text{scal}} \subseteq \mathscr{R}_1$.

Corollary 11.5. *The class of scalar Γ-flows in $\mathscr{R}$ is the smallest class of scalar Γ-flows which contains I_1, Γ and is closed under* SS, *exit merging, scalar iteration.*

Proof. Corollary 11.4 and Proposition 5.3.

Corollary 11.6. *A scalar Γ-flow is finite and satisfies* ACC *and* CD *iff it is in the smallest class of scalar Γ-flows which contains I_1, Γ and is closed under* SS, *exit merging, scalar iteration.*

Proof. Corollary 11.5 and Main Theorem.
Making use of Remark 8.1, we note this is (essentially) our Theorem 9.1 [6]; cf. also [2].

Corollary 11.7. *As Corollary* 11.6 *except replace* "SS" *by* "SSS".

Corollary 11.8. *A Γ-flow is finite, accessible and acyclic iff it can be built out of I_0, I_1, Γ by means of composition, separated pairing and exit merging.*

Proof. (*if*). Each of the flows I_0, I_1, γ is finite, accessible and acyclic and the operations preserve these properties.

(*only if*). By Corollary 9.4, F has no back edges. Since F is accessible and satisfies CD, it is in $\mathscr{R}$ (Theorem 11.3). But according to Theorem 9.2, † can't be used in the building up process for it introduces back edges.

Corollary 11.9. *A scalar Γ-flow is finite, accessible and acyclic iff it can be built out of I_1, Γ by means of* SSS (*resp.* SS) *and exit merging.*

Proof. By Corollaries 11.4, 11.5 and 11.8. Cf. also [2].
For use in the next section, we informally describe the notion "anti-iterate".

Definition 11.1. The anti-iterate $F\downarrow: 1 \to p+1$ of a Γ-flow $F: 1 \to p$ is obtained from F by adjoining an additional exit as exit $p+1$ and directing to it all edges in F with target the begin of F.

Note that $F\downarrow\dagger = F$.

12. Proof of the main theorem

Proof (a). Clearly I_0, I_1, Γ satisfy ACC and CD. By (i), (iii) of Lemma 9.1 so do all flows in $\mathcal{R}$. Since our operations preserve finiteness and I_0, I_1, Γ are finite, all $\mathcal{R}$-flows are finite.

Proof (b). We wish to prove that if the flow $F: n \to p$ is finite and satisfies ACC and CD then F is an $\mathcal{R}_1$-flow. We note first that we may also assume that b_F is injective. For suppose that F has k $(\leq n)$ distinct begins. Let these be numbered $b_1, \ldots, b_k$ in any way. Let G be the flow with k begins which differs from F only in having $b_G(i) = b_i$. Let $f: [n] \to [k]$ be defined by $f(i) =$ the j such that $b_F(i) = b_j$. Then f is surjective and using the trivial flow f, we have

$$F = f \circ G.$$

Also f is an $\mathcal{R}_1$-flow so F will be an $\mathcal{R}_1$-flow if G is. Clearly G satisfies ACC and CD if F does. And b_G is injective.

We now prove by induction that if F satisfies ACC and CD *and b_F is injective* then F is an $\mathcal{R}_1$-flow. The proof is by induction on $w(F)$ defined by $w(F) = \beta(F) + \varepsilon(F)$ where $\beta(F)$ is the number of begins of F and $\varepsilon(F)$ the number of *internal edges* of F i.e. edges of F which do not end in an exit.

Suppose first that $w(F) = 0$. Then by ACC, $F = I_0$ and so is in $\mathcal{R}_1$.

Suppose next that $w(F) = 1$ i.e. that $\beta(F) = 1$ and $\varepsilon(F) = 0$. Then F must be I_1 or atomic or arise from an atomic flow by exit merging, and all such flows are $\mathcal{R}_1$-flows.

Suppose now that $w(F) > 1$. If there are no internal vertices then since F satisfies ACC and it must be trivial surjective flow and hence in $\mathcal{R}_1$. (See Corollary 11.2 and Theorem 11.3(c)). If there are non-exit vertices let M be a component which is a maximal non-exit component with respect to the partial ordering of components. This implies that all edges from M end either in M or in an exit. And Proposition 10.1 implies that all edges to M from a vertex outside M end in v_M.

Case 1. There is no begin of $F < v_M$. Since F satisfies ACC there must be a begin $\leq v_M$ so in this case such a begin must be in M. By CD every such begin must be identical with v_M.

Subcase 1.1. There are no begins of F other than v_M. Since we are assuming b_F is injective this means that F is a scalar flow with sole begin $b_F(1) = v_M$. Since $w(F) > 1$ we have $\varepsilon(F) > 0$. This means there must be an edge into v_M. For if not then M would consist solely of v_M, all edges from v_M would go to exits and in view of ACC there could be no vertices other than v_M and exits and we would have $\varepsilon(F)=0$. So we may take the anti-iterate F_1 of F and write $F=F_1^{\dagger}$. Now F_1 satisfies ACC for since F satisfies ACC there is a path in F from v_M to every vertex v of F and this factors into a path $\pi_1\pi_2$ where π_1 is a path from begin to begin and π_2 is a begin-path in F_1. The new vertex e_{p+1} of F_1 is also accessible for there is at least one edge into v_M in F so there is at least one edge into e_{p+1} in F_1 and it must be from a vertex of F, all of which we have just seen to be accessible in F_1. Also F_1 satisfies CD by Lemma 9.1 (iii). Finally $\beta(F_1)=\beta(F)=1$ and $\varepsilon(F_1)<\varepsilon(F)$ for there is at least one edge into v_M in F which is an internal edge of F but not of F_1. So by the induction hypothesis F_1, and hence F, is an $\mathcal{R}_1$-flow.

Subcase 1.2. There are begins of F other than v_M. Let G be the flow whose vertices are those in M together with all exits accessible from v_M. Let H be the flow whose vertices are all other vertices of F together with duplicates of all exits accessible both from v_M and another begin. The edges of G (similarly H) are all edges of F joining two vertices of G. The exits of G, H and the begins of H are ordered as in F (since b_F is injective they are all distinct). Since all edges from M to vertices outside go to exits there are no edges from G to H. And there are none from H to G (except to common exits) for since all vertices are accessible from a begin such an edge would imply the existence of a begin of $H<v_M$ contrary to the hypothesis of Case 1. Now, as illustrated in Fig. 12.1 we have $F=f\circ[G\oplus H]\circ g$ for a suitable trivial permutation flow f and trivial surjective flow g. And, clearly, since F satisfies ACC, CD and has an injective begin function so do G, H. Finally $\beta(G)$, $\beta(H)<\beta(F)$ and $\varepsilon(G)$, $\varepsilon(H)\leqslant \varepsilon(F)$ so the inductive step goes through.

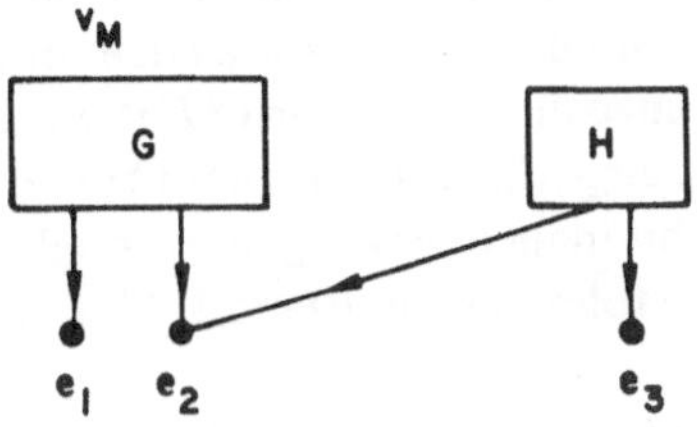

Fig. 12.1.

Case 2. There is a begin of F < v_M. We shall decompose F as indicated in Fig. 12.2 where f is an exit merging flow. The vertices of G are all those $\geqslant v_M$ i.e. all elements of M plus those exits of F to which there are edges from elements of M. These latter, numbered $1,\ldots,p_2$ are the exits of G. The begin of G is v_M and the edges of G are all edges of F which join vertices in G. The vertices of H are all non-exit vertices of F not in M together with those exits which are ends of edges which being in H. Here any exit already in G is replaced by a duplicate; we also take as the last exit of H a duplicate of v_M. The role of f is simply to merge those exits of F which have been duplicated in G, H and to restore their original ordering. The begin function b_H of H

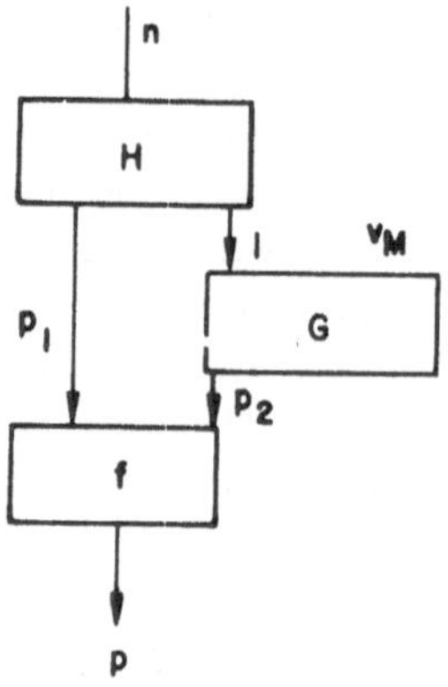

Fig. 12.2.

is the same as that of F (if M contains a begin of F this can only be v_M in which case we use instead the duplicate which is the last exit of H). The edges of H are all edges of F joining two vertices of H.

Now we have

$$F = H \circ [I_{p_1} \oplus G] \circ f.$$

It is easy to see that b_G, b_H are injective and that G, H satisfy ACC, CD. Also $\beta(G)$, $\beta(H) \leq \beta(F)$, $\varepsilon(H) < \varepsilon(F)$ for all edges of H are edges of F and the ones into v_M, of which there is at least one (since there is a begin of $F < v_M$) are internal in F but not in H. Also $\varepsilon(G) < \varepsilon(F)$ for the edges in H into v_M count in $\varepsilon(F)$ but not in $\varepsilon(G)$ for they are not edges in G. Hence this inductive step goes through and the proof of Theorem 11.3(b) is complete. Part (c) follows from Remark 11.1 and parts (a), (b) while part (d) follows from parts (a), (b), (c).

The proof of (e) is obtained from the proof of (b) by deleting the first paragraph of the proof as well as Subcase 1.2. This is so because the closure of $\mathcal{R}_1$ under separated pairing is used only in Subcase 1.2 and the deletion of "separated pairing" and "I_0" in the definition of $\mathcal{R}_1$ yields the class defined as $\mathcal{R}_{\text{scal}}$.

13. Miscellany

The following proposition characterizes a "natural" vector subclass of $\mathcal{R}$. It may be proved by small modifications in the proof of the Main Theorem. (Incorporation of the proposition in the Main Theorem and incorporation of its proof in the proof of the Main Theorem would, however, have resulted in undue complication.)

Proposition 13.1. *Let $\mathscr{R}'$ be the class of vector Γ-flows defined like $\mathscr{R}$ except that*

exit merging is restricted to
(a) *exit merging of ideal flows (i.e., flows none of whose begins are exits)*
(b) *permutations of exits.*

Then a flow F is in $\mathscr{R}'$ iff it satisfies ACC, CD *and* BEG *below.*

BEG. b_F is injective and there is no path between distinct begins.

We now turn briefly to inaccessible flows. Because they appear to have much less relevance to actual programs we shall only sketch the proofs and confine ourselves to the scalar case, though the corresponding theorem for the case of vector flows may well be true.

Definition 13.1. Let $\mathscr{R}''$ be the smallest class of scalar Γ-flows which contains Γ, all the trivial flows j_p and is closed under SSS, exit merging, scalar iteration.

Proposition 13.2. *A scalar flow is in $\mathscr{R}''$ iff by adding additional edges it can be made into a flow satisfying* ACC *and* CD.

Proof (*sketch*). Observe that we are now operating at a purely graph theoretical level where the labelling of the vertices is ignored, for adding edges changes the out-degree of some vertices. We first show that $\mathscr{R}''$ is closed under the operation of removing edges. Then we show, by an argument paralleling the one for Theorem 11.3(a), that every flow in $\mathscr{R}''$ can be extended to a flow satisfying ACC and CD by adding new edges. Some complication is caused here by the fact that adding new edges may remove exits so that the corresponding accessible flows cannot always be built up by exactly the same operations as the original flows. The result then follows from Corollary 11.7.

Appendix

$\mathscr{S}_{\mathrm{Ord}}$ is a strict monoidal category.

With the aid of the Theorem below, Proposition 5.2 (recall that the flows in ΓFl need not be finite) may be extended to the case where the source and target of a Γ-flow is permitted to be an infinite ordinal number.

Let Ord be the class of ordinal numbers. (We do not identify $n \in \mathrm{Ord}$ with the set $\boldsymbol{n} = \{i \in \mathrm{Ord} \mid i < n\}$. Thus we adopt, for notational reasons, the view ([16, p. 269] that *ordinal numbers are order types of well-ordered sets.*) Let N be the set of finite ordinals. Let ω be the smallest infinite ordinal number.

We take the following for granted.

(1) There is a binary relation $<$ on Ord which is a strict well order.

(2) There is a binary operation $+$ on Ord with respect to which Ord is a (non-commutative) monoid.

(3) $n<p \Leftrightarrow \exists x(n+x=p \;\&\; x>0)$. See [16, pp. 274–5, Theorems 1, 2].

(4) $p_1<p_2 \Rightarrow n+p_1<n+p_2$. (For $n \in N$, $n+\omega=\omega$; so right cancellation doesn't hold. From this implication one infers, however, $\alpha+\beta=\alpha+\gamma \Rightarrow \beta=\gamma$.)

Definition. $[n]=\boldsymbol{n}+1=\{i+1 \mid i \in \boldsymbol{n}\}$.

Examples. $[3]=\{1,2,3\}$; $[\omega]=\{1,2,3,\ldots\}$; $[\omega+1]=[\omega]\cup\{\omega+1\}$; note $\omega \notin [\omega+1]$.

Properties. (i) $[n]$ is order isomorphic to $\boldsymbol{n}$.

(ii) The sets $[n_1]$ and $n_1+[n_2]=\{n_1+i \mid i \in [n_2]\}$ are disjoint and their union is $[n_1+n_2]$. In particular, $\omega+[1]=\{\omega+1\}$ so that the statement holds for $n_1=\omega$ and $n_2=1$.

(iii) $\iota_i:[n_i]\to[n_1+n_2]$, $i\in[2]$, where ι_1 is inclusion and $\iota_2(j)=n_1+j$ for $j\in[n_1]$ is a coproduct (*the preferred coproduct*) of $[n_1]$, $[n_2]$ in the category $\mathscr{S}$ of sets.

Definition. The *preferred coproduct* in $\mathscr{S}$ of $[n_j]$, $j\in[3]$, if $\iota_j:[n_j]\to[n]$, where ι_1 is inclusion ι_2 is n_1-translated inclusion $(i\mapsto n_1+i)$, ι_3 is (n_1+n_2)-translated inclusion and $n=n_1+n_2+n_3$.

Definition. In $\mathscr{S}$, where M is any set, $f_i:[n_i]\to M$, $i\in[2]$, define $(f_1,f_2):[n_1+n_2]\to M$ by the requirement: $\iota_i\circ(f_1,f_2)=f_i$ for $i\in[2]$. *Let $\mathscr{S}_{\mathrm{Ord}}$ be the full subcategory of $\mathscr{S}$ (the category of sets) determined by the subclass of objects of the form* $[n]$, $n\in$ Ord.

Theorem. *The category $\mathscr{S}_{\mathrm{Ord}}$ is strict monoidal w.r.t. the preferred coproduct functor $\oplus$: $\mathscr{S}_{\mathrm{Ord}}\times\mathscr{S}_{\mathrm{Ord}}\to\mathscr{S}_{\mathrm{Ord}}$ which takes $[n_1]$, $[n_2]$ into $[n_1+n_2]$ and $f_i:[n_i]\to[p_i]$, $i\in[2]$ into $f_1\oplus f_2:[n_1+n_2]\to[p_1+p_2]$ and unit $I_0:[0]\to[0]$.*

Proof. Follows from observation that the preferred coproduct for $[n_1+n_2]$, $[n_3]$ is the pair of morphisms $(\iota_1,\iota_2):[n_1+n_2]\to n$, $\iota_3:[n_3]\to[n]$, while the preferred coproduct for $[n_1]$, $[n_2+n_3]$ is the pair of morphisms $\iota_1:[n_1]\to[n]$, $(\iota_2,\iota_3):[n_2,n_3]\to[n]$.

Acknowledgement

We gratefully acknowledge the support of the Science Research Council of Great Britain for the research reported on herein.

References

[1] F.E. Allen and J. Cocke, Graph-theoretic constructs for control flow analysis, IBM Research Report RC–3923 (July 1972).

[2] S.L. Bloom and R. Tindell, Algebraic and graph theoretic characterizations of structured flowchart schemes, *Theoret. Comput. Sci.* **9** (1979), to appear.

[3] S.L. Bloom and C.C. Elgot, The existence and construction of free iterative theories, *J. Comput. System Sci.* **12** (3) (June 1976).

[4] C.C. Elgot, The external behavior of machines, Proc. of Third International Conference on System Sciences (1970), also IBM Research Report RC–2740 (December 1969).

[5] C.C. Elgot, Monadic computation and iterative algebraic theories, Logic Colloquium '73, H.E. Rose and J.C. Shepherdson, eds., in: *Studies in Logic* (North Holland, Amsterdam, 1975).

[6] C.C. Elgot, Structured programming with and without GO TO statements, IEEE Trans. on Software Eng. SE-2, No. 1 (March 1976). Erratum and Corrigendum IEEE Trans. on Software Eng. (September 1976).

[7] C.C. Elgot, S.L. Bloom and R. Tindell, On the algebraic structure of rooted trees, IBM Research Report RC–6230 (October 1976). Summary in Proceedings of the 1977 Conference on Information Sciences and Systems, E.E. Dept., Johns Hopkins University.

[8] F.E. Allen, Control flow analysis, Proc. ACM SIGPLAN Symposium on Compiler Optimization, *SIGPLAN Notices* **5** (7) (July 1970) 1–19.

[9] M.S. Hecht and J.D. Ullman, Flowgraph reducibility, *SIAM J. Comput.* **1** (2) (June 1972) 188–202.

[10] M.S. Hecht and J.D. Ullman, Characterizations of reducible flow graphs, *J. Assoc. Comput. Mach.* **21** (3) (July 1974) 367–375.

[11] J. Cocke, Global common subexpression elimination, Proc. ACM SIGPLAN Symposium on Compiler Optimization, *SIGPLAN Notices* **5** (7) (July 1970) 20–24.

[12] E. Lowry and C.W. Medlock, Object code optimization, *Comm. ACM.* (1) (January 1969) 13–22.

[13] R.T. Prosser, Applications of Boolean matrices to the analysis of flow diagrams, Proc. Eastern Joint Computer Conference, December 1959 (Spartan Books, New York) 133–138.

[14] S. Ginali, Iterative algebraic theories, infinite trees and program schemata, Dissertation, University of Chicago (June 1976).

[15] S. Mac Lane *Categories for the working mathematician* (Springer–Verlag, New York, Heidelberg, Berlin, 1971).

[16] W. Sierpinski, *Cardinal and Ordinal Numbers*, (Panstwowe Wydawnictwo Naukowe, Warszawa 1958, Hafner Publishing Co., New York).

RC 8221 (#35758) 4/17/80
Mathematics 48 pages

AN EQUATIONAL AXIOMATIZATION OF THE ALGEBRA OF REDUCIBLE FLOWCHART SCHEMES

Calvin C. Elgot
Mathematical Sciences Department
IBM Thomas J. Watson Research Center
Yorktown Heights, NY 10598

John C. Shepherdson
Department of Mathematics
University of Bristol
Bristol, England BS8 1TW

ABSTRACT The flowchart schemes called "reducible" are of interest because

(1) they schematize programs without "go to" statements;

(2) for every flowchart scheme there is a reducible one which is step-by-step equivalent to it;

(3) they manipulate readily.

The reducible flowchart schemes admit both graph-theoretic and algebraic descriptions. A flowchart scheme F with n begins and p exits is reducible iff both (a) for every vertex v there is a begin-path (i.e. a path from a begin vertex) which ends in v (b) for every closed path C there is a vertex v_C of C such that every begin path to a vertex of C meets v_C. Algebraically: F is reducible iff it can be built up from atomic flowchart schemes (which may be thought of as corresponding to single instructions) and surjective trivial flowchart schemes (which may be thought of as redirecting flow of control) by means of composition ($\circ$), separated pairing ($+$), and scalar iteration ($\dagger$). Dropping $\dagger$ yields the accessible, acyclic flowchart schemes.

The equational axioms referred to in the title relate $\approx$-identified (i.e. isomorphism-identified) flowchart schemes by these three operations. The structure freely generated by the set Γ of atomic flowchart schemes is Γ***Red***, the theory (or multi-sorted algebra) of reducible flowchart schemes. The axioms involving only $+$ and $\circ$ yield Γ***Ac***, the theory of accessible, acyclic flowchart schemes. Semantical domains which may be used to interpret Γ***Red*** also satisfy these axioms. The set of equational statements true for Γ***Red*** (which are also logical consequences of the axioms) determine a congruence on the algebra of expressions built out of Γ, **Sur**, $+$, $\circ$ and $\dagger$ whose quotient is another description of Γ***Red***.

Thus the expressions built up using these three operations provide an alternative notation for reducible flowchart programs; the purpose of the axioms is to determine when two expressions represent the same schematic program.

A similar statement holds for Γ***Ac***.

I INTRODUCTION

I Section 1 Brief description of $\mathcal{F}_\Gamma$, $\circ$, $+$, $\dagger$.

Let Γ be the disjoint union of the family of sets Γ_r indexed by non-negative integers $r \in N$. A graphical Γ-flowchart scheme $F:n \to p$, $n \in N$, $p \in N$ (briefly, a Γ-*flow*) consists of

(i) a locally ordered directed graph (briefly digraph); "locally ordered" means the set of outedges of each vertex is linearly ordered;

(ii) an injection from $[p] = \{j \in N \mid 1 \leq j \leq p\}$ into the set of vertices of F of outdegree zero ((i) and (ii) constitute an exit digraph); the image of the injection is the set of *exit vertices*, or simply, *exits* of F);

(iii) a function $b_F:[n] \to Fv$, the *begin function* of F, where Fv is the set of vertices of (the underlying digraph of) F; ((i) and (iii) constitute a begin digraph);

(iv) a function λ, the *labeling function* of F, which associates with each non-exit vertex v of F, $v\lambda$, (λ applied to v), where $v\lambda \in \Gamma_{v\theta}$ and $v\theta$ is the *outdegree* (the number of *outedges* of v); ((i) and (iv) constitute a labeled digraph).

In the case that $n=1$, the flow F is called *scalar*.

Thus, a flowchart scheme is a begin, exit, labeled digraph. Let $\mathcal{F}_\Gamma$ be the class of all Γ-flows; we drop the index Γ when that doesn't threaten confusion.*

The begin function is *not* assumed injective. The image of b_F is the set of begin vertices of F.

If $F_i:n_i \to p_i$ is in $\mathcal{F}$, for $i \in [2]$ and $F_1v \cap F_2v = \phi$, then $F_1+F_2:n_1+n_2 \to p_1+p_2$ in $\mathcal{F}$ is defined by the requirements:

the underlying labeled digraph of F_1+F_2 is the "union" of F_1 and F_2;

the sequence of begin vertices (resp. exit vertices) of F_1+F_2 is the sequence of begin vertices (resp. exit vertices) of F_1 concatenated with the sequence of begin vertices (resp. exit vertices) of F_2.

*A flow $F:1 \to p$ is *atomic* if it has only one non-exit vertex, which is also its begin, which is labelled with some $\gamma \in \Gamma p$ and has its ith outedge, for $1 \leq i \leq p$, going to the ith exit vertex. (*Added to text, 1982*)

Still, more briefly, if $p_1 = n_2$, then $F_1 \circ F_2$ is obtained by redirecting the edges of F_1 which go into exit $j \in [p_1]$ into the jth begin of F_2 for each j.

Scalar conditional iteration (briefly, scalar iteration) is defined for scalar flows with positive target i.e. for flows $F:1 \to p+1$. If F is non-trivial, $F^{\dagger}:1 \to p$ is obtained from F by redirecting the edges of F which go into exit p+1 into the begin vertex and discarding exit p+1. In [ES], scalar iteration was only defined for non-trivial F. The results of [BEW1], [BEW2] and [BT] suggest that it is desirable to define $j^{\dagger}{}_{p+1} = \perp \circ 0_p = \perp + 0_p : 1 \to p$ in the case that $j = p+1$, where $\perp \in \Gamma_0$ has been distinguished; $j^{\dagger}{}_{p+1} = j_p$ otherwise where j_p is the injection whose value is j.

There is an obvious notion of isomorphism ($\approx$) between two Γ-flows; if $F \approx F'$ then F and F' are uniquely isomorphic. If $F_i':n_i \to p_i$ and $F_i \approx F_i'$ then $F_1+F_2 \approx F_1'+F_2'$ and assuming $p_1 = n_2$, $F_1 \circ F_2 \approx F_1' \circ F_2'$. Similarly, assume $F,F':1 \to p+1$, $F \approx F'$. Then $F^{\dagger} \approx F'^{\dagger}$. Thus, the operations make sense on $\mathcal{F}/\approx$ and, on $\mathcal{F}/\approx$, are everywhere defined i.e., e.g. F_1+F_2 is defined for all F_1,F_2 in $\mathcal{F}/\approx$.

I **Section 2** Brief description of content.

The flowchart schemes called "reducible" are of interest because

(1) they schematize programs without "go to" statements;

(2) for every flowchart scheme there is a reducible one which is step-by-step equivalent to it;

(3) they manipulate readily.

The reducible flowchart schemes admit both graph-theoretic and algebraic descriptions. More precisely: if $\mathscr{R}$ is the class of all finite reducible graphical flowchart schemes then $\mathscr{R}/\approx$ admits a direct algebraic description with respect to suitably chosen operations on $\mathscr{R}/\approx$. A flowchart scheme F with n begins and p exits is reducible iff both

(a) for every vertex v there is a begin-path (i.e. a path from a begin vertex) which ends in v,

(b) for every closed path C there is a vertex v_C of C such that every begin path to a vertex of C meets v_C.

Algebraically: a finite F is reducible iff it can be built up from atomic flowchart schemes (which may be thought of as corresponding to single instructions) and surjective trivial flowchart schemes (which may be thought of as redirecting flow of control) by means of composition ($\circ$), separated pairing ($+$), and scalar iteration ($\dagger$). Dropping $\dagger$ yields the accessible, acyclic flowchart schemes.

The equational axioms referred to in the title relate $\approx$-identified (i.e. isomorphism-identified) graphical flowchart schemes, i.e. algebraic flowchart schemes, by these three operations. The structure freely generated by the set Γ of atomic flowchart schemes is Γ*Red*, the theory (or multi-sorted algebra) of reducible flowchart schemes. The axioms involving only $+$ and $\circ$ yield Γ*Ac*, the theory of accessible, acyclic flowchart schemes. Semantical domains which may be used to interpret Γ*Red* also satisfy these axioms. The set of equational statements true for Γ*Red* (which are also logical consequences of the axioms) determine a congruence on the algebra of expressions built out of Γ, **Sur**, $+$, $\circ$ and $\dagger$ whose quotient is another description of Γ*Red*.

Thus the expressions built up using these three operations provide an alternative notation for reducible flowchart programs; the purpose of the axioms is to determine when two expressions represent the same schematic program.

A similar statement holds for Γ*Ac*, cf III6.

By way of clarification of (1), "go to" statements are understood to correspond to trivial schemes which are properly injective; the trivial schemes may be identified with mappings whose source and target are initial segments of the positive integers.

The notion "flowchart scheme" and the operations $\circ$, $+$, $\dagger$ were briefly described in I1; detailed descriptions are given in [ES], see also [CE-SP].

Once one obtains the algebraic entity Γ*Red*, it displaces the combinatorial i.e. graphical entity $\mathcal{R}_\Gamma$ as the entity of major interest both because it is much simpler and much more intrinsic. This point is emphasized by describing in I3, a substantially different graphical model of the algebraic entity Γ*Red* although, conceptually, it is but a variant of the model of [ES]. The *c*-graphs of [JR] may be used to provide still another graphical model of Γ*Red*.

The term "reducible" comes from "reducible flowgraphs" as used by [AC]. We do not, however, make use of their definition of "reducible." Our graphical description of "reducible," when specialized to scalar flowchart schemes, does not differ much from the characterization of "reducible" given by [HU].

We expect to treat the case of all finite Γ-flows in a sequel.

The paper is organized into three parts. Part I is introductory. Part II deals with Γ*Ac* and Part III with Γ*Red*. The section headings constitute an outline:

I 1. Brief description of $\mathscr{F}_\Gamma$, $\circ$, $+$, $\dagger$,

2. Brief description of content,

3. Entry, exit, labeled digraphs,

4. On the choice of operations,

5. Discussion of proof ideas,

6. Theory-algebra terminology;

II 1. The categories **Map** and **Sur**,

2. Definition and examples of flow theories,

3. Block permutations,

4. Normal factorization of a flow in ΓAc,

5. The characteristic freeness property of ΓAc,

6. ΓAc satisfies ***P1-P4***,

7. Pointed flow theories;

III 1. Definition and examples of scalar iteration flow theories,

2. The component decomposition of reducible flows,

3. The scalar anti-iterate in Γ***Red***,

4. The nested $\dagger$ index ν,

5. The characteristic freeness property of Γ***Red***,

6. The case of vector atoms.

I Section 3 Entry, exit, labeled digraphs.

An entry, exit labeled digraph F:n → p is obtained from its "corresponding" begin, exit labeled digraph by adjoining n "new" vertices to be called *entries* (and to be sharply distinguished from the begin vertices), simply ordering them, adjoining one outedge per entry and directing the ith entry outedge to the ith begin i.e. vertex ib_F for each $i \in [n]$; that's all. Thus entries have no inedges and each entry has exactly one outedge. The requirement that this correspondence between the two graphical versions of "flow" preserve the defined operations implicitly gives the meaning of $\circ$, +, † for entry, exit, labeled digraphs. There are four illustrations below, where $\perp \in \Gamma_0$, $\gamma \in \Gamma_2$ and (2,1,1,2): [4] → [2] is the surjection which takes 1,2,3,4 respectively into 2,1,1,2.

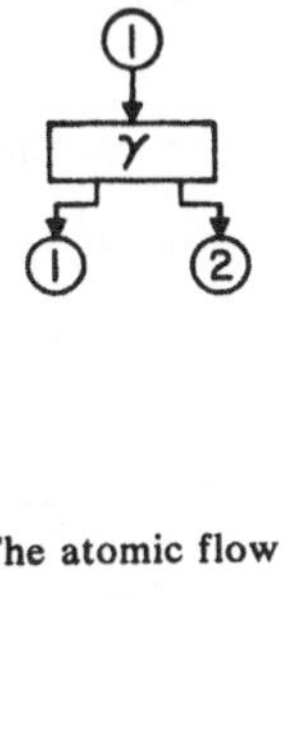

Figure 1. The atomic flow γ:1 → 2

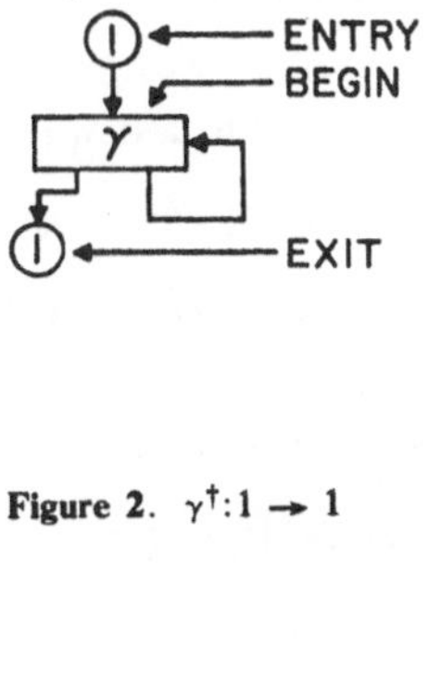

Figure 2. $\gamma^{\dagger}$:1 → 1

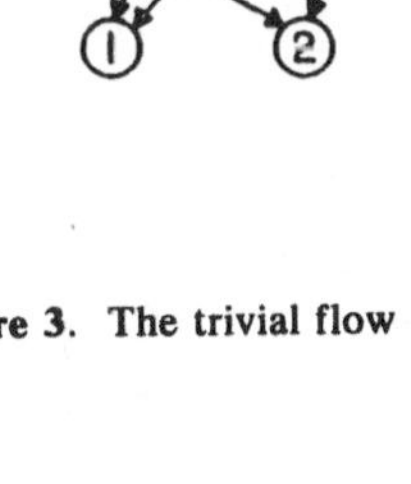

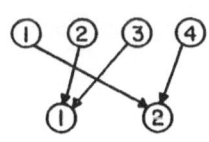

Figure 3. The trivial flow (2,1,1,2):4 → 2

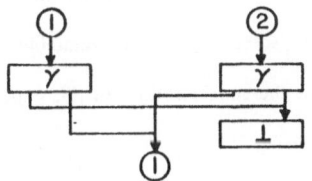

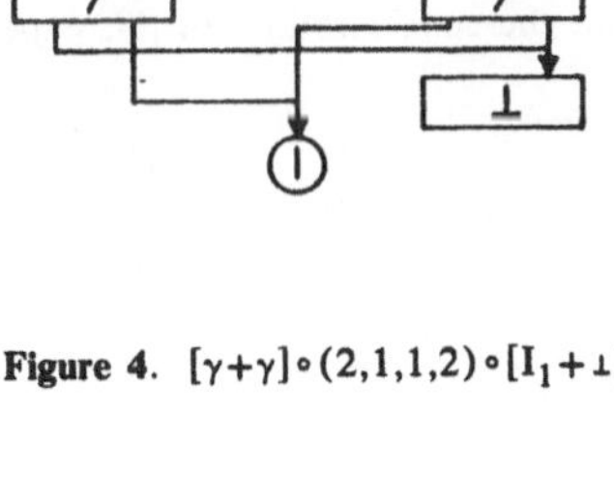

Figure 4. $[\gamma+\gamma]\circ(2,1,1,2)\circ[I_1+\perp]$:2 → 1

I **Section 4** On the choice of operations.

It has already been noted that the flows in ΓAc are exactly those which can be built from Γ and **Sur** by a finite number of applications of $+$ and $\circ$. It is not difficult to convince oneself that this is at least a reasonable generalization of flows constructed from Γ and **Sur** by means of a finite number of applications of $\circ$ and the "if ... then ..." construct of programming languages. The "if ... then ..." construct may be defined in terms of our primitives as follows: given $\pi \in \Gamma_2$, and flows $F_i:1 \rightarrow 1$, $i \in [2]$, produce $\pi \circ [F_1+F_2] \circ (1,1)$, where $(1,1)$ is the unique surjection $[2] \rightarrow [1]$. If one interprets π as a predicate π' on X and one interprets $\omega \in \Gamma_1$ as an operation ω' on X, then

(4.1) $(\pi \circ [F_1+F_2] \circ (1,1))'$

is an operation on X but $(\pi \circ [F_1+F_2])'$ is a "combined" operation and predicate on X; the operation being (4.1) and the predicate π'; therefore after $(\pi \circ [F_1+F_2])'$ is applied to an argument $x \in X$ one has available both the outcome of the "test" π' and a new $x' \in X$.

Let $F:1 \rightarrow 2$ be a flow in ΓAc and let F' be its interpretation as a combined operation and predicate on X. Then the interpretation $F^{\dagger\prime}$ of $F^{\dagger}:1 \rightarrow 1$ is: apply F' to $x \in X$; if the outcome is 1, halt; if not apply F' to the new $x' \in X$ etc.

If $G:1 \rightarrow 1$, then the familiar while-do and while not-do programming constructs may be expressed by $(\pi \circ [I_1+G])^{\dagger}$ and $(\pi \circ (2,1) \circ [I_1+G])^{\dagger}$; which is which of course depends upon which element of [2] one correlates with "true."

This is sufficient to exemplify the semantics and indicate the relationship of these operations to some familiar programming constructs without involving ourselves with details irrelevant for the considerations of this paper.

I Section 5 Discussion of proof ideas.

The idea behind the proof of the freeness of ΓAc is as follows. First express F as $f \circ F'$ where $f \in \mathbf{Sur}$ does "as much identifying as possible," i.e. if $F' = g \circ F''$, $g \in \mathbf{Sur}$ then g is injective (so that g is bijective); once f has been chosen, F' is determined by the requirement that $F = f \circ F'$. If one imposes a total order on the begin vertices of F and if jf = k (jf is the value obtained by applying f to j), then the jth entry outedge ends with the kth begin vertex of F (cf. I Section 3). It is inessential but convenient to require as well that in the imposed total order all the non-exit begins precede all the exit begins of F. We may then express F' as $[(\Sigma A)+I_q] \circ G$, where A is the sequence of labels of the non-exit begins and q is the number of exit begins of F. The flow F may then be expressed as $f \circ [\Sigma A+I] \circ G$ giving rise to the idea of "normal factorization" of F.

The proof of the freeness of ΓRed exploits the freeness of ΓAc. The programming idea of replacing a "high level" instruction by a "subroutine" is sharpened as follows. To each $F \in \Gamma Red$, there is a $F\alpha \in \Omega Ac$ (the elements of Ω may be thought of as "high level" instructions) and a map $\beta: \Omega Ac \rightarrow \Gamma Red$ such that $(F\alpha)\beta = F$ for all F.

The freeness of ΓRed means: given a map φ_0 from Γ into an appropriate "theory" T, there is a unique extension φ of φ_0 to ΓRed which preserves "the structure" of ΓRed. To prove this the following idea is used. The definition of φ on F will be determined once the definition of φ on $\omega\beta$ is known for each "high level" instruction $\omega \in \Omega$ in $F\alpha$; if the begin of $\omega\beta$ has inedges, there is a flow $(\omega\beta)^{\dagger} \in \Gamma Red$ such that $(\omega\beta)^{\dagger\dagger} = \omega\beta$; assuming φ defined on $(\omega\beta)^{\dagger}$, it is extended to $\omega\beta$ by defining $(\omega\beta)\varphi = ((\omega\beta)^{\dagger}\varphi)^{\dagger}$. To carry out this idea a "nested $\dagger$ index" ν is defined on ΓRed having the property that $F\nu > (\omega\beta)^{\dagger}\nu$, where ω is as immediately above and, in the case where ω is in $F\alpha$ but $\omega\beta$'s begin has no inedges, $F\nu \geq (\omega\beta)\nu$. The proof of the freeness of ΓRed is by induction on ν.

I Section 6 Theory-algebra terminology.

The word "theory" is frequently used to mean a set of sentences. In formal contexts the set of sentences may be required to be closed under logical consequences. The phrase "algebraic theory" as introduced by Lawvere [WL] and used in categorical algebra is compatible with this usage. The word "algebraic" in the phrase "algebraic theory" in effect means the form of the sentences is equational. In the language of universal (i.e. general) algebra, however, Lawvere's algebraic theories may be termed multi-sorted or heterogeneous algebras.

Our use of the word "theory" is also compatible with its use as a set of sentences. In fact, our flow theories, pointed flow theories, and scalar iteration flow theories are all "algebraic" in the sense used in the phrase "algebraic theory." In informal discourse, such as in the title and abstract, however, using the language of universal algebra, we say "algebra" rather than "theory."

II THE ACYCLIC CASE

II Section 1 The categories Map and Sur

The category **Map** has $\{[n] \mid n \in N\}$ as its class of objects. The set of morphisms of **Map** with source $[n]$ and target $[p]$ is the set of all functions $f:[n] \to [p]$ with domain $[n]$ and values in $[p]$. If $x \in [n]$, we write xf for the value of f applied to x and if $g:[p] \to [q]$ is a function, we write $f \circ g:[n] \to [q]$ for the composite.

Given a pair of functions $f_i:[n_i] \to [p_i]$, $i \in [2]$, we define the function $f_1+f_2:[n_1+n_2] \to [p_1+p_2]$ as follows:

$$x \xmapsto{f_1+f_2} xf_1, \text{ if } x \in [n_1],$$

$$n_1+x \xmapsto{f_1+f_2} p_1+xf_2, \text{ if } x \in [n_2].$$

In non-sensitive situations we will often omit brackets. Now, + and $\circ$ are related by, where $g_i:p_i \to q_i$ for $i \in [2]$,

(1.1) $(f_1+f_2) \circ (g_1+g_2) = (f_1 \circ g_1 + f_2 \circ g_2)$.

In particular, it may be verified using (1.1) that

(1.2) $I_n+I_p = I_{n+p}$, where $I_n:n \to n$ is the identity function and

(1.1') $[f+I_q] \circ [I_p+h] = f+h = [I_n+h] \circ [f+I_r]$, where $h:[q] \to [r]$.

In non-sensitive contexts the subscripts on occurrences of "I" are often omitted.

The operation + maps a morphism $(f_1,f_2):(n_1,n_2) \to (p_1,p_2)$ in the category **Map**×**Map** to a morphism $f_1+f_2:n_1+n_2 \to p_1+p_2$ in **Map**. By virtue of (1.1) and (1.2), this map is a functor $+$:**Map**×**Map** $\to$ **Map**.

The operation + also enjoys the following "monoidal" properties:

(1.3) $(f_1+f_2) + f_3 = f_1 + (f_2+f_3)$

(1.4) $(f+I_0) = f = I_0+f$.

By virtue of satisfying (1.1) - (1.4), (**Map**,$+$,I_0) is called a *strict monoidal category* (cf. [SM] p. 158). The operation + is not, however, commutative. In fact,

(1.5) *block permutation axiom* $\pi(n_1,n_2) \circ (f_2+f_1) = (f_1+f_2) \circ \pi(p_1,p_2)$ where

$$\pi(n_1,n_2) : [n_1+n_2] \to [n_1+n_2]$$

is the permutation which maps $[n_1]$ bijectively onto $n_2+[n_1]$ via $i \mapsto n_2+i$ and maps $n_1+[n_2]$ bijectively onto $[n_2]$ via $n_1+i \mapsto i$. It is easy to see that

(1.6) $\pi(n_2,n_1) = (\pi(n_1,n_2))^{-1}$.

Thus (1.5) is equivalent to

(1.7) $(f_2+f_1) \circ \pi(p_2,p_1) = \pi(n_2,n_1) \circ (f_1+f_2)$ as well as

$$f_2+f_1 = \pi(n_2,n_1) \circ (f_1+f_2) \circ \pi(p_1,p_2)$$

For example, in the case that $n_1=3$, $n_2=4$:

$3 \xmapsto{\pi(3,4)} 7 \xmapsto{f_2+f_1} p_2 + 3f_1$,

$3 \xmapsto{f_1+f_2} 3f_1 \xmapsto{\pi(p_1,p_2)} p_2 + 3f_1$;

$4 \xmapsto{\pi(3,4)} 1 \xmapsto{f_2+f_1} 1f_2$,

$4 \xmapsto{f_1+f_2} p_1 + 1f_2 \xmapsto{\pi(p_1,p_2)} 1f_2$.

The class of surjections in **Map** is closed under $\circ$ and +; the identities of **Map** are surjections; thus this class is a subcategory of **Map**. Let **Sur** be the subcategory of **Map** consisting of all the surjections. Then (**Sur**, +, I_0) is a sub-monoidal category of (**Map**, +, I_0).

In accordance with our motivating considerations (cf. Introduction), the focus of our attention shifts from **Map** to **Sur** in the next section.

II Section 2 Definition and examples of flow theories

By a *flow theory* T *over* **Sur**, we mean a strict monoidal category which extends* (**Sur**, +, I_0), which satisfies (1.5) and which has the same class of objects as **Sur**.

Definition The category **FloS** has as its objects flow theories over **Sur**. A morphism $\varphi:T \to T'$ in **FloS** is a functor which maps $a:n \to p$ in T onto $a\varphi:n \to p$ in T' (so that $I_n\varphi = I_n$ and $(a \circ b)\varphi = a\varphi \circ b\varphi$, where $b:p \to q$) and which satisfies in addition

(2.1) $(a_1+a_2)\varphi = a_1\varphi + a_2\varphi$, for all a_1,a_2 in T,

(2.2) $f\varphi = f$ for all $f \in$ **Sur**.

Examples As noted in [ES] (cf. p. 339, Proposition 5.1) the collection of all isomorphism classes of multi-begin, multi-exit Γ-flowchart schemes, where Γ is the disjoint union of a family $\{\Gamma_n \mid n \in N\}$ of sets, is a flow theory over **Sur** with respect to ∘ (composition) and + (separated pairing). Actually, [ES] uses ⊕ rather than +. Also, any subcollection of flowchart schemes containing the accessible trivial ones and closed under + and ∘ is also an object of **FloS**; this includes e.g. ΓAc whose morphisms are the accessible, acyclic Γ-flowchart schemes. [A trivial flowchart scheme $n \xrightarrow{f} p$ is determined by a mapping $f:[n] \to [p]$. The associated bijection is an isomorphism of the two strict monoidal categories. The justification for calling these examples objects of **FloS** relies on identifying a trivial flowchart scheme with the mapping which determines it. If this identification is rejected one has recourse to the definition of the footnote; the definition of **FloS** may then be appropriately modified.]

---

* Alternatively, "extends an isomorphic copy of" but this trivial generalization involves conceptual and notational complications which, in the considerations of this paper, have no redeeming characteristics.

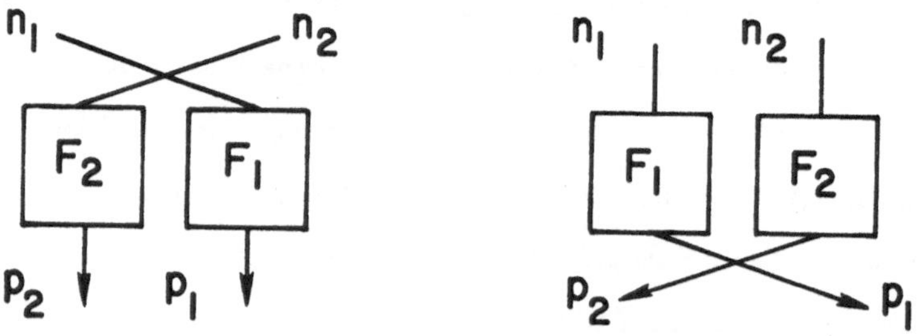

Figure 5. The Block Permutation Axiom applied to flows F_1,F_2

Categories C which are extensions of **Map** (with the same class of objects) amount to algebraic theories in the sense of Lawvere (cf. [EW] or [CE-AT]; the origin is [WL]) provided that coproducts in **Map** are coproducts in C. In such an algebraic theory one may define + by the requirement that the following diagram commutes

(2.3)

$$\begin{array}{ccc} [n_i] & \xrightarrow{\;f_i\;} & [p_i] \\ \Big\downarrow{\scriptstyle \iota_i} & & \Big\downarrow{\scriptstyle \kappa_i} \\ [n_1+n_2] & \xrightarrow{\;f_1+f_2\;} & [p_1+p_1] \end{array}$$

for $i\in[2]$, where $\iota_1:[n_1] \to [n_1+n_2]$ is inclusion and $\iota_2:[n_2] \to [n_1+n_2]$ is n_1-translated inclusion i.e.

$$j\in[n_2] \longmapsto^{\iota_2} n_1+j.$$

Similarly, κ_1 is inclusion and κ_2 is p_1-translated inclusion. With this definition of +, (C, +, I_0) is an object of **FloS**. To see that (1.7) is satisfied, we note first

(2.4) $\pi(n_2,n_1) = (\iota_2,\iota_1)$, the source-pairing of ι_2,ι_1 so that (1.7) becomes

(2.5) $(f_2+f_1) \circ (\kappa_2,\kappa_1) = (\iota_2,\iota_1) \circ (f_1+f_2)$.

Using (2.3) and the corresponding commutative rectangle for f_2+f_1:

$$\iota_1\circ(f_2+f_1)\circ(\kappa_2,\kappa_1) = f_2\circ\kappa_1\circ(\kappa_2,\kappa_1) = f_2\circ\kappa_2 = \iota_2\circ(f_1+f_2) = \iota_1\circ(\iota_2,\iota_1)\circ(f_1+f_2)$$

and

$\iota_2\circ(f_2+f_1)\circ(\kappa_2,\kappa_1) = f_1\circ\kappa_2\circ(\kappa_2,\kappa_1) = f_1\circ\kappa_1 = \iota_1\circ(f_1+f_2) = \iota_2\circ(\iota_1,\iota_2)\circ(f_1+f_2)$ from which it follows that (2.5) holds.

Other examples are a consequence of universal algebraic considerations. To be more specific, the category **FloS** has e.g., binary products and enjoys the following property:

every morphism $\varphi: T \to T'$ admits a factorization $T \xrightarrow{\psi} U \xrightarrow{\psi'} T'$, i.e. $\varphi = \psi \circ \psi'$ where ψ is surjective and ψ' is injective. The morphism ψ induces a "congruence" $\sim$ on T such that $T/\sim \approx U$.

The block permutation axiom (1.5), (1.7) admits a generalization to a relationship between the sum of a finite sequence **a** of morphisms and the sum of the sequence $g \circ \mathbf{a}$, where g is a permutation. The next section is devoted to the formulation and proof of this generalization. In a first reading of the next section, it is suggested that one merely note the form of the Block permutation theorem and its Corollary and then pass to the next section. In this first reading, the definition of **g#p** may be ignored; what is important is that **g#p** is a permutation which depends on, and only on, the permutation g and the sequence **p** of non-negative integers.

II Section 3 Block permutations

In this section we have often omitted the composition symbol.

Let $\mathbf{n} : [r] \to N$ be a function or a finite sequence $(n_1, n_2, \cdots, n_r)$ of non-negative integers, let $g : [r] \to [r]$ be a permutation. The *induced block permutation* $g\#\mathbf{n} : [\Sigma_j(j\mathbf{n} \mid j\in[r])] \to [\Sigma_j(j\mathbf{n} \mid j\in[r])] = [n_1+n_2 + \cdots + n_r]$ is defined by the requirement that $g\#\mathbf{n}$ bijectively maps $\Sigma_d(d\mathbf{n} \mid d<i) + [i\mathbf{n}] = 1\mathbf{n}+2\mathbf{n} + \cdots + (i-1)\mathbf{n} + [i\mathbf{n}] = n_1+n_2 + \cdots + n_{i-1}+[n_i]$ onto $\Sigma_d(dg\mathbf{n} \mid d<ig^{-1}) + [i\mathbf{n}]$ for every $i\in[r]$ by the rule for each $x\in[i\mathbf{n}]$, $\Sigma_d(d\mathbf{n} \mid d<i) + x \overset{g\#\mathbf{n}}{\longmapsto} \Sigma_d(dg\mathbf{n} \mid d<ig^{-1}) + x$. In particular, if (2,1) is the non-trivial permutation on [2], then

$$[n_1] \overset{(2,1)\#\mathbf{n}}{\longmapsto} n_2 + [n_1]$$

$$n_1 + [n_2] \overset{(2,1)\#\mathbf{n}}{\longmapsto} [n_2].$$

To see that the function $g\#\mathbf{n}:[n_1+n_2 + \cdots + n_r] \to [n_1+n_2 + \cdots + n_r]$ is well defined, it is sufficient to show that $i<j\in[r]$ implies that

(a) $(\Sigma_d(d\mathbf{n} \mid d<i) + [i\mathbf{n}]) \cap (\Sigma_d(d\mathbf{n} \mid d<j) + [j\mathbf{n}]) = \phi$.

From $\Sigma_d(d\mathbf{n} \mid d<j) = \Sigma_d(d\mathbf{n} \mid d<i) + \Sigma_d(d\mathbf{n} \mid i\le d<j)$ we infer that (a) holds iff (b) holds, where

(b) $[i\mathbf{n}] \cap (i\mathbf{n} + \cdots + (j-1)\mathbf{n} + [j\mathbf{n}]) = \phi$.

Let $x\in[i\mathbf{n}]$ and $y\in i\mathbf{n} + \cdots + (j-1)\mathbf{n} + [j\mathbf{n}]$. Then, $x\le i\mathbf{n}<y$. Thus (b) holds and $g\#\mathbf{n}$ is well-defined.

To show that $g\#\mathbf{n}$ is a permutation, it is sufficient to show that $(ig^{-1}<jg^{-1}) \Rightarrow (\Sigma_d(dg\mathbf{n} \mid d<ig^{-1}) + [i\mathbf{n}]) \cap (\Sigma_d(dg\mathbf{n} \mid d<jg^{-1}) + [j\mathbf{n}]) = \phi$.

This is equivalent to $ig^{-1} < jg^{-1} \Rightarrow [i\mathbf{n}] \cap (i\mathbf{n} + \Sigma(dg\mathbf{n} \mid ig^{-1} < d < jg^{-1}) + [j\mathbf{n}]) = \phi$.

As before, we see that the implication does indeed hold.

Let $r \in N$, let g and h be permutations on $[r]$ and let $n_i \in N$ for $i \in [r]$. Also let $r_i \in N$, g_i be a permutation on $[r_i]$ and $\mathbf{n}_i:[r_i] \rightarrow N$ be a function for $i \in [2]$.

Block Manipulation Theorem Block permutations enjoy the following properties:

(a) if $n_i=1$ for all $i \in [r]$, then $g\#\mathbf{n} = g^{-1}$;

(b) $I\#\mathbf{n} = I$, i.e. if g is the identity then so is $g\#\mathbf{n}$ (on $[\Sigma(n_i \mid i \in [r])]$);

(c) $(g \circ h \# \mathbf{n}) = (h\#\mathbf{n}) \circ (g\# h \circ \mathbf{n})$;

(d) $(h\#\mathbf{n})^{-1} = (h^{-1} \# h \circ \mathbf{n})$;

(e) $(g_1+g_2) \# (\mathbf{n}_1, \mathbf{n}_2) = (g_1\#\mathbf{n}_1) + (g_2\#\mathbf{n}_2)$. (Note that $(g_1+g_2)^{-1} = g_1^{-1}+g_2^{-1}$).

Proof (a) For each $j \in [r]$, $x \in [jn]$, $\Sigma(dn \mid k<j)+x \xmapsto{g\#\mathbf{n}} \Sigma(dgn \mid d<jg^{-1}) + x$. Substituting $x = 1 = dn = dgn$, we obtain $j \xmapsto{g\#\mathbf{n}} jg^{-1}$, i.e. $g\#\mathbf{n} = g^{-1}$.

(b) Immediate from the definition.

(c) For each $j \in [r]$ and for each $x \in [jh\mathbf{n}]$:

$$\Sigma_d(d\mathbf{n} \mid d<jh) + x \xmapsto{h\#\mathbf{n}} \Sigma_d(dh\mathbf{n} \mid d<j) + x \xmapsto{gh\#\mathbf{n}} \Sigma_d(dgh\mathbf{n} \mid d<jg^{-1}) + x$$

and

$$\Sigma_d(d\mathbf{n} \mid d<jh) + x \xmapsto{gh\#\mathbf{n}} \Sigma_d(dgh\mathbf{n} \mid d<jhh^{-1}g^{-1}) + x)$$

$$= \Sigma_d(dgh\mathbf{n} \mid d<jg^{-1}) + x.$$ Thus (c) follows.

(d) This follows by substituting h^{-1} for g in (c) and then using (b).

(e) $[g_1 + g_2]$ maps $r_1 + [r_2]$ bijectively onto itself. Let $\mathbf{n} = (\mathbf{n}_1, \mathbf{n}_2)$ and $g = g_1 + g_2$. For each $j \in [r_2]$ and each $x \in [(r_1+j)\mathbf{n}] = j\mathbf{n}_2$, $\Sigma_d(d\mathbf{n} \mid d<r_1+j) + x \xmapsto{g\#\mathbf{n}} \Sigma(dg\mathbf{n} \mid d<(r_1+j)g^{-1}) + x$.

But $\Sigma_d(d\mathbf{n} \mid d<(r_1+j)) = \Sigma_d(d\mathbf{n}_1 \mid d \in [r_i]) + \Sigma(d\mathbf{n}_2 \mid d<j)$ while

$$\Sigma(dg\mathbf{n} \mid d<(r_1+j)g^{-1} = \Sigma_d(dg_1\mathbf{n}_1 \mid d \in [r_1]) + \Sigma_d(dg_2\mathbf{n}_2 \mid d<jg_2^{-1})$$

$$= \Sigma_d(d\mathbf{n}_1 \mid d \in [r_1] + \Sigma_d(dg_2\mathbf{n}_2 \mid d<jg_2^{-1}).$$ Thus (e) holds in this case.

The other case is easier.

The *block permutation axiom* for flow theories T may now be restated. Let $\mathbf{n} = (n_1,n_2)$, $\mathbf{p} = (p_1,p_2)$, let $g = (2,1)$ be the nontrivial permutation of [2] and let $a_i : n_i \to p_i$, $i \in [2]$, be a morphism in T.

Block permutation axiom $a_2+a_1 = (g\#\mathbf{n})^{-1} \circ [a_1+a_2] \circ (g\#\mathbf{p})$.

Now let $\mathbf{a}:[r] \to T$, $\mathbf{n}:[r] \to N$, $\mathbf{p}:[r] \to N$ be functions and let $i\mathbf{a}:i\mathbf{n} \to i\mathbf{p}$ for each $i \in [r]$. Let $\Sigma g\mathbf{a}$ abbreviate $1g\mathbf{a}+2g\mathbf{a} + \cdots + rg\mathbf{a}$.

Block permutation theorem $\Sigma g\mathbf{a} = (g\#\mathbf{n})^{-1} \circ \Sigma \mathbf{a} \circ (g\#\mathbf{p})$.

Proof We show first the formula holds for g of the form $I_s + (2,1) + I_t$, for appropriate $s,t \in N$. It is then convenient to express $\mathbf{a}$ as $(\mathbf{a}_1, \mathbf{a}_2, \mathbf{a}_3)$ where the length $|\mathbf{a}_1| = r$, $|\mathbf{a}_2| = 2$, $|\mathbf{a}_3| = t$. Similarly, $\mathbf{n} = (\mathbf{n}_1, \mathbf{n}_2, \mathbf{n}_3)$, $\mathbf{p} = (\mathbf{p}_1, \mathbf{p}_2, \mathbf{p}_3)$, where $|\mathbf{n}_1| = s = |\mathbf{p}_1|$, $|\mathbf{n}_2| = 2 = |\mathbf{p}_2|$, $|\mathbf{n}_3| = t = |\mathbf{p}_3|$. From (e) and (b) of the block manipulation theorem we obtain

$$g\#\mathbf{n} = [I_s+(2,1)+I_t] \# (\mathbf{n}_1, \mathbf{n}_2, \mathbf{n}_3)] = I+((2,1)\#\mathbf{n}_2)+I$$

and $(g\#\mathbf{n})^{-1} = I + ((2,1) \# \mathbf{n}_2)^{-1} + I$ and

$$\begin{aligned}
(g\#\mathbf{n})^{-1} \circ \Sigma\mathbf{a} \circ (g\#\mathbf{p}) &= [I +((2,1)\#\mathbf{n}_2)^{-1}+I] \circ [\Sigma\mathbf{a}_1+\Sigma\mathbf{a}_2+\Sigma\mathbf{a}_3] \circ [I+((2,1)\#\mathbf{p}_2)+I] \\
&= \Sigma\mathbf{a}_1 + ((2,1) \# \mathbf{n}_2)^{-1} \circ \Sigma\mathbf{a}_2 \circ ((2,1)\#\mathbf{p}_2) + \Sigma\mathbf{a}_3 \\
&= \Sigma\mathbf{a}_1 + \Sigma(2,1)\mathbf{a}_2 + \Sigma\mathbf{a}_3 \quad \text{by the block permutation axiom} \\
&= \Sigma g\mathbf{a}.
\end{aligned}$$

Since every permutation on [r] is expressible as a composite of permutations of the form $I + (2,1) + I$ (i.e. transpositions of consecutive integers), the proof may now be completed by induction on the minimum number of transpositions required to express the permutation, say, f. Assume then that the formula holds for any sequence $\mathbf{a}$ of morphisms and for permutations g and h on [r] and let $f = gh$. To show the formula holds for f, we calculate

$$\Sigma g(h\mathbf{a}) = (g\#h\mathbf{n})^{-1} \circ \Sigma h\mathbf{a} \circ (g\#h\mathbf{p})$$

$$= (g\#h\mathbf{n})^{-1} \circ (h\#\mathbf{n})^{-1} \circ \Sigma\mathbf{a} \circ (h\#\mathbf{p}) \circ (g\#h\mathbf{p})$$

$$= (gh\#\mathbf{n})^{-1} \circ \Sigma\mathbf{a} \circ (gh\#\mathbf{p}),$$ by (c) of the block manipulation theorem.

Corollary Taking $\mathbf{n}$ so that $i\mathbf{n} = 1$ for all $i \in [r]$ we obtain:

$$\Sigma g\mathbf{a} = g \circ \Sigma\mathbf{a} \circ (g\#\mathbf{p})$$

II Section 4 Normal factorization of a flow in ΓAc

The following properties of the flow theory ΓAc (the proofs of ***P1-P4*** are postponed to Section 6) are sufficient to prove that ΓAc has a characteristic freeness property. The next section is devoted to formulating and proving this freeness property. Let $F:n \to p$ be a morphism in ΓAc.

Definition By an *alternating factorization* of F is meant the length three sequence

(4.1) $$n \xrightarrow{f} r+q \xrightarrow{\Sigma A+I} t+q \xrightarrow{G} p$$

such that

(4.2) $F = f \circ (\Sigma A + I_q) \circ G$ where $A:[r] \to \Gamma$ is a function, $\Sigma A = 1A + 2A + \cdots + rA$ and $f \in$ **Sur**.

A *normal factorizaton* of F is an alternating factorization of F in which $r+q$ is minimum and among all $r+q$-minimum factorizations, r is maximum.

We are now prepared to tabulate the relevant properties of ΓAc.

P1. There is a function $\delta:\Gamma Ac \to N$, called *degree*, which satisfies

P1.1 $(G \circ H)\delta = G\delta + H\delta$ for composition compatible morphisms G,H,

P1.2 $(G_1+G_2)\delta = G_1\delta + G_2\delta$,

P1.3 $\gamma\delta = 1$ for $\gamma \in \Gamma$,

P1.4 $f\delta = 0$ for $f \in$ **Sur**.

P2. If $F \notin$ **Sur** and (4.1) is a normal factorization of F, then $r>0$ (so that, by P1.1, P1.2, P1.3, $F\delta>0$) and so $G\delta < F\delta$.

P3. If

(4.3) $$n \xrightarrow{f} r+q \xrightarrow{\Sigma A+I} t+q \xrightarrow{H} p$$

and

$$(4.1) \quad n \xrightarrow{f} r+q \xrightarrow{\Sigma A+I} t+q \xrightarrow{G} p$$

are alternating factorizations of F, then $G = H$.

Before stating ***P4***, it is convenient to introduce the following notation. Let $\mathbf{p}:[r] \to N$ be the function obtained by composing the function $\mathbf{A}:[r] \to \Gamma$ with the target function $\partial:\Gamma \to N$, i.e. $\mathbf{p} = \mathbf{A} \circ \partial$.

P4. If

$$(4.1) \quad n \xrightarrow{f} r+q \xrightarrow{\Sigma A+I} t+q \xrightarrow{G} p$$

is a normal factorization of F and

$$(4.1') \quad n \xrightarrow{f'} r'+q' \xrightarrow{\Sigma A'+I} t'+q' \xrightarrow{G'} p$$

is an alternating factorization of F, then there exist a permutation $g:[r] \to [r]$, a surjection $x:[q'] \to [r-r'+q]$ and a function $\mathbf{B}:[r-r'] \to \Gamma$ such that

P4.1 $f = f' \circ [I+x] \circ [g+I]$ or equivalently, since $g^{-1}+I = [g+I]^{-1}$,

$f \circ [g^{-1}+I] = f' \circ [I+x]$

P4.2 $g \circ \Sigma\mathbf{A} \circ (g\#\mathbf{p}) = \Sigma\mathbf{A}' + \Sigma\mathbf{B} = \Sigma(g \circ \mathbf{A})$.

The following property of ΓAc is a consequence of ***P3*** and ***P4*** (as well as the fact that ΓAc is a flow theory over **Sur**).

P5 If (4.1) is a normal factorization and (4.1') an alternating factorization of F then

P5.1 $G' = [I_{t'}+x] \circ [I_{t'}+\Sigma\mathbf{B}+I_q] \circ [h^{-1}+I_q] \circ G$ where $h = g\#\mathbf{p}$ and $\mathbf{B}$ is as in ***P4***.

In the case that (4.1') is normal as well, x becomes a permutation on [q] and ***P5.1*** specializes to

P5.2 $G' = [h^{-1}+x] \circ G$.

Proof $F = f \circ [\Sigma \mathbf{A} + I] \circ G$

$= f \circ [g^{-1} + I] \circ [g \circ \Sigma \mathbf{A} \circ h + I] \circ [h^{-1} + I] \circ G$ by (1.1)

$= f' \circ [I + x] \circ [\Sigma \mathbf{A}' + \Sigma \mathbf{B} + I] \circ [h^{-1} + I] \circ G$ by **P4.1** and **P4.2**

$= f' \circ [I + x] \circ [\Sigma \mathbf{A}' + 1] \circ [I + \Sigma \mathbf{B} + I] \circ [h^{-1} + I] \circ G$ by (1.1')

$= f' \circ .[\Sigma \mathbf{A}' + 1] \circ [I + x] \circ [I + \Sigma \mathbf{B} + I] \circ [h^{-1} + I] \circ G$ by (1.1')

But $F = f' \circ [\Sigma \mathbf{A}' + I] \circ G'$ as well so that ***P5.1*** by ***P3***.

In the case that (4.1') is normal, $r=r'$ and $q=q'$ so that x is a surjection from $[q]$ to $[q]$ i.e. a permutation. The formula ***P5.2*** follows from (1.1') and the observation that when (4.1') is normal, that $\Sigma \mathbf{B} = I_0$.

The following properties are also formal consequences of the properties already observed. ***P8*** below will be cited in the proof of the next section; ***P6*** and ***P7*** are intended to provide a graceful transition to ***P8***.

P6. The first factor f in a normal factorization of F may be characterized, up to a permutation on the right, by the following two properties.

P6.1 $f \in \mathbf{Sur}$ and f is a left factor of F.

P6.2 if $f' \in \mathbf{Sur}$ and f' is a left factor of F, then there exists a unique $f'' \in \mathbf{Sur}$ such that $f = f' \circ f''$.

Proof According to ***P4.1***, the f'' whose existence is asserted in ***P6.2*** is $[I+x] \circ [g+I]$. The uniqueness is automatic.

If ***P6.1*** and ***P6.2*** hold with f and f' interchanged, it is easy to see that f'' is a permutation; thus ***P6*** is established.

Definition A morphism F in ΓAc will be said to be be *bijective* if

$$f \in \mathbf{Sur} \wedge F = f \circ F' \Rightarrow f \text{ is a bijection.}$$

(Note that if $F \in \mathbf{Sur}$ is bijective, then F is a bijection.)

P7 Let $F = f \circ F'$, $f \in \mathbf{Sur}$.

P7.1 If f is the first factor in a normal factorization of F, then F' is bijective.

P7.2 If F' is bijective, then f is the first factor in a normal factorization of F.

Proof of P7.1 Let $f' \in \mathbf{Sur}$ be a left factor of F'. Then $f \circ f'$ is a left factor of F and by ***P6.2*** $\exists! f''$ such that $f \circ f' \circ f'' = f$. By ***P3***, $f' \circ f'' = I$, from which it follows that f' is a bijection.

Proof of P7.2 Let F' be bijective and suppose $F = f' \circ F''$ where f' is the first factor of a normal factorization of F, then by ***P6.2***, $f' = f \circ f''$ for a unique $f'' \in \mathbf{Sur}$ so that $f \circ F' = F = f \circ f'' \circ F''$ and by ***P3***, $F' = f'' \circ F''$. Since F' is bijective, f'' is a bijection and so f is the first factor in a normal factorization of F.

P8 If (4.1) is a normal factorization of F and $e \in \mathbf{Sur}$, then

$$(4.4) \quad m \xrightarrow{e \circ f} r+q \xrightarrow{\Sigma A + I} t+q \xrightarrow{G} p$$

is a normal factorization of $e \circ F$.

Proof According to ***P7.1***, $F' = [\Sigma A + I] \circ G$ is bijective. Since $e \circ F = (e \circ f) \circ F'$, it follows from ***P7.2*** that $e \circ f$ is the first factor in a normal factorization of $e \circ F$. Let this normal factorization be

$$(4.4') \quad m \xrightarrow{f'} r'+q' \xrightarrow{\Sigma A' + I} t'+q' \xrightarrow{G'} p$$

so that, by ***P6.2***, for some $g \in \mathbf{Sur}$, $f' = e \circ f \circ g$ and so $e^{-1} \circ f' = f \circ g \in \mathbf{Sur}$, where $e^{-1}:[n] \to [m]$ is a left inverse (an injection) of e. From the two factorizations (4.4) and (4.4') of $e \circ F$ we obtain, since (4.4') is normal, $r+q \geq r'+q'$; we also obtain

$$f \circ [\Sigma A + I] \circ G = (e^{-1} \circ f') \circ [\Sigma A' + I] \circ G'$$

from which we infer, since (4.1) is a normal factorization of F, $r'+q' \geq r+q$. Hence $r'+q' = r+q$ and similarly using again the normality of (4.4') ana (4.1) we obtain $r = r'$ and hence by comparison with (4.4') we see that (4.4) is a normal factorization of $e \circ F$.

To explain the term "normal factorization," we remark that iteration of the process of normal factorization (4.1) by normal factoring G until a "residue G" is obtained which is in **Sur**, leads to a "normal form" for F

(4.5) $F = f_0 \circ [\Sigma A_1 + I] \circ [\Sigma A_2 + I] \circ \cdots \circ [\Sigma A_s + I] \circ f_s$

which is unique "up to interspersed permutations" (cf. **P5.2**). The number s is the length of the longest path from a begin vertex to a vertex of outdegree 0 in the flow F.

II Section 5 The characteristic freeness property of ΓAc

Let T be a flow theory and let $\varphi_0:\Gamma \to T$ be a source and target preserving function.

Extension Lemma There is a unique extension $\varphi:\Gamma Ac \to T$ of φ_0 such that

(5.1) $F\varphi = F$ if $F \in \mathbf{Sur}$

(5.2) $F\varphi = f\circ[\Sigma(A\circ\varphi_0) + I]\circ G\varphi = f'\circ[\Sigma(A'\circ\varphi_0) + I]\circ G'\varphi$ if both (4.1) and (4.1') are normal factorizations of F.

(5.3) $(e\circ F)\varphi = e\circ F\varphi$ for all permutations e on $[n]$.

Proof of existence by induction on $F\delta$. In the case that $F\delta=0$, we define $F\varphi$ by (5.1) so that (5.1) and (5.2) hold. Since $(e\circ F)\delta = 0$, (5.3) is immediate.

Now assume that $F\delta>0$ and assume that (4.1) and (4.1') are normal factorizations of F. Then, by ***P1*** and ***P2***, $G\delta = G'\delta<F\delta$ and we may assume inductively that (5.1), (5.2) and (5.3) hold for both G and G'.

A calculation analogous to the first three lines of the proof of ***P5*** in the previous section* shows that

$$f\circ[\Sigma(A\circ\varphi_0) + I]\circ G\varphi = f'\circ[\Sigma(A'\circ\varphi_0) + I]\circ[h^{-1}+x]\circ G\varphi$$

and, according to ***P5.2***, $G' = [h^{-1}+x]\circ G$. By inductive assumption, $G'\varphi = [h^{-1}+x]\circ G\varphi$ and

$$f\circ[\Sigma(A\circ\varphi_0) + I]\circ G\varphi = f'\circ[\Sigma(A'\circ\varphi_0) + I]\circ G'\varphi$$

so that we may define $F\varphi$ by the requirement that (5.2) be valid. To complete the proof of existence, we must show (5.3) holds. But (5.3) is an immediate consequence of ***P8*** or of the observation that the permutation e induces an obvious bijection between alternating factori-

---

* $F = f\circ[\Sigma A\circ\varphi_0 + I]\circ G\varphi$
$= f\circ[g^{-1}+I]\circ[g\circ\Sigma(A\circ\varphi_0)\circ h + I]\circ[h^{-1}+I]\circ G\varphi$ by (1.1)
$= f'\circ[I+x]\circ[\Sigma A'\circ\varphi_0 + I]\circ[h^{-1}+I]\circ G\varphi$ by ***P4.1, P4.2*** and $\Sigma\mathbf{B} = I_0$ since $r=r'$
$= f'\circ[\Sigma A'\circ\varphi_0 + I]\circ[I+x]\circ[h^{-1}+I]\circ G\varphi$
$= f'\circ[\Sigma A'\circ\varphi_0 + I]\circ[h^{-1}+x]\circ G\varphi$

zations of F and $e \circ F$. Specifically, this bijection maps the factorization (4.1) of F onto the factorization

$$(5.4) \quad n \xrightarrow{e \circ f} r+q \xrightarrow{\Sigma A+I} t+q \xrightarrow{G} p$$

of $e \circ F$; thus, if (4.1) is normal then so is (5.4).

The proof of uniqueness of φ is by a trivial induction on $F\delta$ using ***P2***..

Freeness Theorem The (unique) extension $\varphi:\Gamma Ac \to T$ of $\varphi_0:\Gamma \to T$ satisfying (5.1) and (5.2), is a morphism in the category **FloS**.

Proof We prove $(F \circ H)\varphi = F\varphi \circ H\varphi$ by induction on $F\delta + H\delta$.

Case 1 $F\delta=0$, $H\delta \geq 0$. In this case the result follows by ***P8*** and (5.1).

Case 2 $F\delta>0$. Let

$$(4.1') \quad n \xrightarrow{f'} r'+q' \xrightarrow{\Sigma A'+I} t'+q' \xrightarrow{G'} p$$

be a normal factorization of F so that

$F = f' \circ [\Sigma A'+I] \circ G'$, $F\varphi = f' \circ [\Sigma(A' \circ \varphi_0) + I] \circ G'\varphi$ by the Extension Lemma, $G'\delta < F\delta$ by ***P2*** and ***P1***, and

$$(5.5) \quad n \xrightarrow{f'} r'+q' \xrightarrow{\Sigma A'+I} t'+q' \xrightarrow{G' \circ H} p$$

is an alternating factorization of $F \circ H$. Let

$$(4.1) \quad n \xrightarrow{f} r+q \xrightarrow{\Sigma A+I} t+q \xrightarrow{G} p$$

be a normal factorization of $F \circ H$. A calculation exactly analogous to that in the proof of ***P5*** shows

$$(5.6) \quad (F \circ H)\varphi = f' \circ [\Sigma(A' \circ \varphi_0) + I] \circ [I+x] \circ [I+\Sigma(\mathbf{B} \circ \varphi_0) + I] \circ [h^{-1}+I] \circ G\varphi$$

Moreover, by ***P5.1***

$$(5.7) \quad G' \circ H = [I_{t'}+x] \circ [I_{t'}+\Sigma\mathbf{B}+I_q] \circ [h^{-1}+I_q] \circ G$$

and, since $G'\delta+H\delta < F\delta+H\delta$, the induction assumption is applicable. By four uses of the inductive assumption

(5.8) $(G' \circ H)\varphi = G'\varphi \circ H\varphi = [I+x] \circ [I+\Sigma\mathbf{B}+I]\varphi \circ [h^{-1}+I] \circ G\varphi.$

On the other hand

(5.9) $F\varphi \circ H\varphi = f' \circ [\Sigma(\mathbf{A}' \circ \varphi_0) + I] \circ G'\varphi \circ H\varphi$

$= f' \circ [\Sigma(\mathbf{A}' \circ \varphi_0) + I] \circ [I+x] \circ [I+\Sigma\mathbf{B}+I]\varphi \circ [h^{-1}+I] \circ G\varphi$ by (5.8).

Comparison of (5.6) with (5.9) shows that we require the

Observation $I+\Sigma(\mathbf{B} \circ \varphi_0) + I = [I+\Sigma\mathbf{B}+I]\varphi$

Now, it is obvious that for permutations e_1, e_2 provided by the Block Permutation Theorem

(5.10) $m \xrightarrow{e_1} m \xrightarrow{\Sigma\mathbf{B}+I} m' \xrightarrow{e_2} m'$

is a normal factorization of $I+\Sigma\mathbf{B}+I$, where $m = t'+(r-r')+q$, $m' = t+q$. From the Extension Lemma

(5.11) $[I+\Sigma\mathbf{B}+I]\varphi = e_1 \circ [\Sigma(\mathbf{B} \circ \varphi_0) + I] \circ e_2$

$= I+\Sigma(\mathbf{B} \circ \varphi_0) + I.$

Thus, the Observation has been verified and the proof that φ preserves $\circ$ is complete. It remains to prove that φ preserves $+$.

Because $F+H = [F+I] \circ [I+H]$, because φ preserves $\circ$ and because of the Block Permutation Theorem, it is sufficient to prove $[F+I_t]\varphi = F\varphi+I_t$. The proof is by induction on $F\delta$. If $F\delta>0$ and

(4.1) $n \xrightarrow{f} r+q \xrightarrow{\Sigma\mathbf{A}+I} t+q \xrightarrow{G} p$

is a normal factorization of F, then by ***P1*** and ***P2***, $G\delta<F\delta$ and by (1.1)

$$F+I_t = [f+I_t] \circ [\Sigma\mathbf{A}+I_q+I_t] \circ [G+I_t].$$

Thus, $[F+I_t]\varphi = [f+I_t]\varphi \circ [\Sigma\mathbf{A}+I_q+I_t]\varphi \circ [G+I_t]\varphi$

In the case that $G\delta>0$, it follows that $(\Sigma\mathbf{A})\delta < F\delta$, $[\Sigma\mathbf{A}+I_{q+t}]\varphi = (\Sigma\mathbf{A})\varphi + I_{q+t}$ and $[G+I_t]\varphi = G\varphi+I_t$ so that $[F+I_t]\varphi = [f+I_t] \circ [(\Sigma\mathbf{A})\varphi + I_q + I_t] \circ [G\varphi+I_t] = f \circ [(\Sigma\mathbf{A})\varphi+I_q] \circ G\varphi + I_t = F\varphi + I_t$. In the case $G\delta=0$, $(\Sigma\mathbf{A})\delta>1$, one may make use of a factorization $\Sigma\mathbf{A} = [1\mathbf{A}+2\mathbf{A} + \cdots + (r-1)\mathbf{A}+I_1] \circ [I_1+r\mathbf{A}]$ and the inductive assumption.

II Section 6 ΓAc satisfies *P1-P4*

Proof of P1 For any accessible flowchart scheme F, define $F\delta$ as the number of non-exit vertices in F. Then ***P1.1 - P1.4*** are an immediate consequence of the definitions involved. (Note that we are identifying $\gamma \in \Gamma_n$ with the flowchart scheme $\gamma{:}1 \to n$ which corresponds to it. More strictly, there is an injection $\Gamma \to \Gamma\mathrm{Ac}$ which defines this correspondence.)

Definition By an *essential begin* of F, we mean a begin vertex of F which is not the target of an edge of a begin vertex of F.

Proof of P2 First we construct a normal factorization of b_F. Choose an enumeration $v_1, v_2, \cdots, v_r, \cdots, v_{r+q}$ of $[n]b_F$ (the set of distinct begin vertices of F) in such a way that the first r elements enumerate, without repetition, the set V_R of non-exit, essential begins of F. Define $A{:}[r] \to \Gamma$ by $iA = v_i\lambda$, the label of v_i, so that

(6.1) $\Sigma A = v_1\lambda + v_2\lambda + \cdots + v_r\lambda$.

Define f by:

(6.2) $if = j$ if $ib_F = v_j$, for $i \in [n]$.

The underlying labeled, directed exit graph of G is obtained from F by deleting the begins of A and their outedges. It remains to define b_G.

For $k \in [t]$, let $kb_G = v \in Gv$, if the kth edge of ΣA (i.e. the edge of ΣA whose target is ΣA's k^{th} exit) has v as its target. For $j \in [q]$, define $(t+j)b_G = v_{r+j}$. Then

$$(4.1) \qquad n \xrightarrow{\ f\ } r+q \xrightarrow{\ \Sigma A + I\ } t+q \xrightarrow{\ G\ } p$$

is a factorization of F. We claim this factorization is normal.

From the definition of composition in ΓAc, if (4.1') is an alternating factorization of F then $r'+q' \geq \|[n]b_F\|$, the cardinality of $[n]b_F$. Moreover, every begin vertex of $\Sigma A'$ must be a non-exit essential begin. It follows that the factorization (4.1) is normal.

In the remaining case, i.e. $F\delta = 1$, one may make use of the Extension Lemma.

Since $F \notin Sur$, the set $V_R \neq \phi$; thus, $r = \|V_R\| > 0$. Q.E.D.

Proof of P3 For $j \in [q]$, we note that $(t+j)b_G = ((r+j)f^{-1})b_F = (t+j)b_H$ where f^{-1} is a left inverse of f. For $k \in [t]$, the vertex which is the target in F of the kth edge of ΣA is $kb_G = kb_H$. (The kth edge of ΣA is the edge whose target in ΣA is the kth exit.) Thus, $b_G = b_H$.

Moreover, the underlying directed, labeled, exit graph of G (and of H) is obtained from the underlying, directed, labeled, exit graph of F by deleting from F the begins of ΣA and their outedges.

Hence, $G = H$.

Proof of P4 Assume

$$(4.1) \quad n \xrightarrow{f} r+q \xrightarrow{\Sigma A+I} t+q \xrightarrow{G} p$$

is a normal, and

$$(4.1') \quad n \xrightarrow{f'} r'+q' \xrightarrow{\Sigma A'+I} t'+q' \xrightarrow{G'} p$$

an alternating, factorization of F. Choose a permutation $g:[r] \rightarrow [r]$ in such a way that $g \circ A$ extends the function A'. Such a g exists since the set of begin vertices of ΣA is the set V_R of non-exit, essential begins of F while the set $V_{R'}$ of begin vertices of $\Sigma A'$ is a subset of V_R. Then, from the Corollary at the end of Section 3, $\Sigma(g \circ A) = g \circ \Sigma A \circ h$ and

$$(6.3) \quad n \xrightarrow{f \circ [g^{-1}+I]} r+q \xrightarrow{g \circ \Sigma A \circ h+I} t+q \xrightarrow{[h^{-1}+I] \circ G} p,$$

where $h = g\#\mathbf{p}$ and $\mathbf{p} = A \circ \partial$, is another normal factorization of F. Having defined g, it is easy to see that there exists a surjection $x:[q'] \rightarrow [r-r'+q]$ satisfying

$$(6.4) \quad f \circ [g^{-1}+I_q] = f' \circ [I_{r'}+x].$$

Indeed, for each $j \in [q']$, $\exists k \in [n]$ $(kf'=r'+j)$. Moreover, for all ℓ

$kf' = r'+j = \ell f' \Rightarrow [kb_F = (t'+j)b_{G'} = \ell b_F \notin V_{R'} \wedge \exists z(k(f\circ[g^{-1}+I]) = r'+z = \ell(f\circ[g^{-1}+I]).]$.

Thus, we may define for all $j\in[q']$,

(6.5) $jx = z$, if $kf' = r'+j$ and $k(f\circ[g^{-1}+I]) = r'+z$.

Inasmuch as, if $kb_F\in V_{R'}$, $(kf)[g^{-1}+I] = (kf)g^{-1} = kf' = (kf')[I+x]\in[r']$,

it follows (6.4) holds if x is defined by (6.5).

Since $g\circ\mathbf{A}$ extends $\mathbf{A}'$, there exists a (unique) sequence $\mathbf{B}:[r-r'] \to \Gamma$ such that $g\circ\mathbf{A} = (\mathbf{A}',\mathbf{B})$, the concatenation of the two sequences. Thus ***P4.2*** and the proof of ***P4*** is completed.

II Section 7 Pointed flow theories

As indicated in the Introduction (cf. ...), the results of earlier studies suggest that scalar iteration should be defined for all morphisms, even the identity I_1. With the simple intent of minimizing the amount of special attention which will have to be given to $I_1{}^{\dagger}$, we say a flow theory T over **Sur** is a pointed flow theory over **Sur** if T has a morphism $1 \to 0$ and one such, $\perp : 1 \to 0$, is distinguished. There are no conditions on $\perp$.

Definition The category **PFlS** has as its objects pointed flow theories over **Sur**. A morphism $\varphi : T \to T'$ in **PFlS** is a morphism in **FloS** which, in addition, preserves $\perp$.

We say the disjoint union Γ of the family $\{\Gamma_n \mid n \in N\}$ is pointed if $\perp \in \Gamma_0$ is distinguished. Thus, when Γ is pointed, ΓAc is a pointed flow theory. It is a consequence of the freeness theorem of II6 that when Γ is pointed, that

(7.1) the pointed flow theory ΓAc is freely generated by Γ (as a pointed flow theory).

III SCALAR ITERATION

III Section 1. Definition and examples of scalar iteration flow theories

By a *scalar iteration flow theory* T *over* **Sur**, we mean a pointed flow theory over **Sur** which, in addition, is equipped with an operation, called *scalar iteration*, which takes a morphism $a:1 \to n+1$ in T into a morphism $a^{\dagger}:1 \to n$ in T and which satisfies $I_1{}^{\dagger} = \perp$,

(1.1) $c^{\dagger\dagger} = (c \circ [I_p+(1,1)])^{\dagger}$, for all $c:1 \to p+2$ in T, where $(1,1):[2] \to [1]$ is in **Sur**,

and

(1.2) $a^{\dagger} \circ b = (a \circ [b+I_1])^{\dagger}$, for all $a:1 \to n+1$ and $b:n \to p$ in T.

Definition The category **SiFS** has as its objects scalar iteration flow theories over **Sur**. A morphism $\varphi:T \to T'$ in **SiFS** is a morphism in **PFIS** which satisfies, in addition

(1.3) $a^{\dagger}\varphi = (a\varphi)^{\dagger}$, for all $a:1 \to n+1$ in T.

Examples The collection of all multi-begin, multi-exit Γ-flowchart schemes, is a scalar iteration flow theory over **Sur** with respect to ∘ (composition), + (separated pairing) and † (scalar iteration). Also, any subcollection of flowchart schemes containing the accessible trivial ones (i.e. **Sur**) and closed under +, ∘, and † is also an object of **SiFS**; this includes e.g. ΓRed, the reducible Γ-flowchart schemes. This class may be described as the smallest which contains Γ (i.e. the atomic flowchart schemes corresponding to the elements of Γ), which contains **Sur** and is closed with respect to ∘, + and †. It may also be described, graph-theoretically, as the class of finite accessible flowchart schemes which satisfy the "circuit dominator" property (cf. [ES]). The iteration theories of [BEW2] provide additional examples.

III Section 2 The component decomposition of reducible flows

Our plan for proving the characteristic freeness property of Γ*Red* involves making use of the characteristic freeness property of ΓAc (for any pointed Γ). This plan is inspired by the *component* (cf. [ES] Section 10-12) *decomposition* of flows $F \in \Gamma Red$ which is succinctly formulated immediately below and proved later in this section.

Component Decomposition Theorem In the category **FloS**, the identity morphisms $I_{\Gamma Red}$ and $I_{\Omega Ac}$ factor as follows:

$$(2.1) \quad \Gamma Red \xrightarrow{\alpha} \Omega Ac \xrightarrow{\beta} \Gamma Red\ ,\ \Omega Ac \xrightarrow{\beta} \Gamma Red \xrightarrow{\alpha} \Omega Ac$$

for suitable Ω where $\Omega\beta$ is the set of non-trivial "component flows." That is, in **FloS** (but not in **SiFS**) ΩAc and ΓRed are isomorphic! Before defining "component flow", Ω, α and β, we recall the definition of "component."

Let $F : n \to p$ be a Γ-flowchart scheme. A *component* $C \subseteq Fv$ of F is an equivalence class of the equivalence relation which holds between vertices v, v' of F when and only when there are paths in F from v to v' and v' to v. In the case that $F \in \Gamma Red$ and C contains only non-exits, C determines (up to a bijection) a non-trivial *component flow* $C':1 \to n$, where $n = \|E\|$, the cardinality of the set E of edges of F which end outside of C but which begin in C, as follows. The set of non-exit vertices of C' is C while the set E of *external edges* of C (or C') is in bijective correspondence with E', the set of exits of C'. The underlying digraph of C' has as its edges all edges of F which begin in C; the ith external edge of C ends in the ith exit of C'; the internal edges of C begin and end in C' as in F. The determination of the begin of C' exploits the reducibility of F: it is the vertex $v_C \in C$ which is encountered by all begin paths of F which end in C; it is the *dominator* of C. The definition of C' is completed by choosing arbitrarily, a bijection between E and $[n]$. Thus, if C'' is obtained by a different choice of bijection, $C'' = C' \circ e$ where e is a bijection on $[n]$; under these circumstances we say the flows C' and C'' are *component-equivalent*. If C contains exits, then clearly it consists of exactly one exit and C' is I_1; this is the only trivial component flow.

Let $\mathscr{C}$ (explicitly $\mathscr{C}_\Gamma$) be a set of non-trivial component flows consisting of exactly one element from each non-trivial component-equivalence class of (Γ-)reducible flows. Thus if C_1', C_2' are in $\mathscr{C}$ they are not component-equivalent and $\mathscr{C}$ is maximal, with respect to inclusion, among all classes of non-trivial component flows with this property. Let Ω be a set "of symbols" in bijective correspondence, with $\mathscr{C}$; Ω inherits the target indexed disjoint union structure of $\mathscr{C}$. For $C' \in \mathscr{C}$, let $\omega_{C'} \in \Omega$ be the corresponding element of Ω under the bijective correspondence; if the scalar flow C' has target n then so does the atomic flow corresponding to $\omega_{C'}$.

We turn now to defining α and β. Let $F{:}n \to p$ be in Γ*Red*, let C be a non-trivial component of (the underlying digraph of) F and let $C' \in \mathscr{C}$ be the component flow of F determined by C. Recall that the choice of C' in effect orders the set E' of exits of C'. Let $F\alpha{:}n \to p$ in Ω*Ac* be the flow which is obtained by "replacing" each C' by $\omega_{C'}$. More explicity Fα is defined as follows:

(i) $F\alpha v = \{v_C \mid C$ is a (possibly trivial) component of F and v_C its dominator$\}$;

(ii) the label of v_C in Fα is $\omega_{C'}$, if C' is non-trivial; in the contrary case v_C is an exit of F and of Fα; the ordering of the exits of Fα is inherited from their ordering in F;

(iii) if $C'{:}1 \to r$ is a non-trivial component flow of F, then the ith outedge of $v_{C'}$ in Fα, $i \in [r]$ ends in the target in F of the ith external edge of C'; for this to make sense, the observation is required that these targets are component dominators (cf [ES], 10.2);

(iv) the begin function $b_{F\alpha}:[n] \to F\alpha v$ is defined by $ib_{F\alpha} = ib_F$ for all $i \in [n]$; for this to make sense the observation is required that the begin vertices of F are component dominators.

Thus α has been defined and it remains to define β. Note first that components of distinct flows may nevertheless yield the same component flow. The characteristic freeness property of ΩAc in **PFIS** means that β is uniquely determined by its restriction to Ω and that any target-compatible map $\Omega \to \Gamma Red$ may be extended to a morphism in **PFIS**. Hence, the assignment

(2.2) $\omega_{C'} \overset{\beta}{\longmapsto} C'$

is sufficient to determine β. It is obvious from the definitions involved that $F = (F\alpha)\beta$, i.e. $I_{\Gamma Red} = \alpha \circ \beta$ and, similarly, that $I_{\Omega Ac} = \beta \circ \alpha$. To complete the proof of the Component Decomposition Theorem requires an argument that α is a morphism in **PFIS** i.e. α preserves $+$, $\circ$, and $\perp$ and keeps **Sur** "pointwise" fixed. But this is a simple consequence of the definitions involved; in particular, the simple observation may be recalled that a component of F_1+F_2 is either a component of F_1 or of F_2 but not both; a similar statement holds for $F \circ G$ (when F, G are composition-compatible).

We call the function β *replacement* and $\alpha = \beta^{-1}$, the inverse of β in **PFIS**, *inverse replacement*. It is worth noting that

(a) the function β is *ideal* in the sense that a non-trivial flow is taken by β into a non-trivial flow; of course, α, too, has this property;

(b) non-trivial component flows F which have zero indegree (the *indegree* of a scalar flow F is the indegree of its begin vertex) are of the form $\gamma \circ f$, where $\gamma \in \Gamma$ and $f \in$ **Sur** is a bijection; thus, there is a bijection between Γ and the subset $\Omega^{(1)}$ of Ω corresponding to (non-trivial) zero indegree component flows.

Before leaving this section, we wish to note that the notions "component flow" and "zero indegree scalar flow" are algebraic; more specifically, flow theoretic.

Proposition 2.1 A flow $F:1 \to p$ has exactly one non-trivial component iff it is non-trivial and $\circ$-irreducible i.e. $F = G \circ H \wedge G \notin \mathbf{Sur} \Rightarrow H \in \mathbf{Sur}$. The flow F has positive indegree iff it is expressible as $G^{\dagger}, G \neq I_1$.

Proof Note that

(2.3) $F = G \circ H$, for some (unique) H, where G is the component flow determined by b_F. Thus, F has exactly one non-trivial component $\Leftrightarrow H \in \mathbf{Sur} \Leftrightarrow F \neq I_1 \wedge F$ is $\circ$-irreducible.

The second assertion is trivial.

Proposition 2.2 A flow $F:1 \to p$ is a component flow iff F is $\circ$-irreducible and $L(F)$ holds where $L(F)$ is equivalent to the statement that for all flows $H:1 \to q$,

(2.4) $H \circ \mu_q = F \circ \mu_p \Rightarrow \exists h \in \mathbf{Sur}[H = F \circ h]$

where $\mu_q:[q] \to [1]$ and $\mu_p:[p] \to [1]$ are in **Sur**.

Proof The condition $H \circ \mu_q = F \circ \mu_p$ means that H and F differ at most in the destination of their external edges; they are "otherwise graphically isomorphic." The flow F is a component flow iff it has exactly one component and the number of exits of F is equal to the number of its external edges. The result now follows readily from the further observation that $H \circ \mu_q = F \circ \mu_p$ implies there is a *unique* bijection between the set of external edges of H and the set of external edges of F.

III Section 3 The scalar anti-iterate in ΓRed

If $G:1 \to p$ is a Γ-flowchart scheme, not necessarily reducible, the scalar anti-iterate $G^{\downarrow}:1 \to p+1$ is obtained from G by redirecting all edges which end in b_G to a "new" exit, called exit $p+1$. The proofs of the following observations follow readily from the definitions involved.

Ob.1 Scalar anti-iteration may be characterized by: $G^{\downarrow\uparrow} = G \wedge G^{\downarrow}$ has zero indegree.

Ob.2 If G is accessible (resp., reducible) and has positive indegree, then $G^{\downarrow}$ is accessible (resp., reducible) and $G^{\downarrow} \neq I_1$.

Thus, in Γ*Red*, the scalar anti-iterate of a *positive indegree morphism* G may be algebraically characterized by ***Ob.1***, but in Γ*Red* the scalar anti-iterate of a zero indegree morphism *doesn't* make sense because $G^{\downarrow}$ is not accessible. As a formal consequence of this formal characterization of scalar anti-iteration, we infer

Ob.3 if $F:1 \to p+1$ has zero indegree and $F \neq I_1$, then $F^{\uparrow}$ has positive indegree and $F^{\uparrow\downarrow} = F$; moreover,

$$F^{\uparrow} = G \circ H \wedge G \neq I_1 \Rightarrow G \text{ has positive indegree} \wedge F = G^{\downarrow} \circ [H+I_1].$$

The following observation will also be required in the proof of the characteristic freeness property of Γ*Red*.

Ob.4 Suppose $F:1 \to p+1$. Then $F^{\uparrow}$ has as a left factor the component flow of $b_{F^{\uparrow}}$ and, if $F \neq I_1$, it has positive indegree.

Proof In the case $F \neq I_1$, $F^{\uparrow}\alpha \in \Omega Ac$ has as a left factor an atomic flow. If $F^{\uparrow}\alpha = \omega \circ J$, then $F^{\uparrow} = (F^{\uparrow}\alpha)\beta = \omega\beta \circ J\beta$; that $\omega\beta$ has positive indegree follows from the fact that $F^{\uparrow}$ does. The case $F = I_1$ is trivial.

III Section 4 The nested $\dagger$ index ν

This section completes the preparation for the proof of the characteristic freeness property of ΓRed. Hereafter, as a device for abbreviation, we use the *signature of indegree* (*SIN*) function defined on all scalar flows as follows:

$F^{SIN} = 1$ if F has positive indegree, $F^{SIN} = 0$ otherwise.

Theorem There is a unique function $\nu : \Gamma Red \to N$, called the *nested* $\dagger$ *index*, which satisfies the property $Q(\nu)$ defined by

(4.1) $F\nu = 1 + \max\{G^{\dagger}\nu \mid G$ is a component flow of F and $G^{SIN} = 1\}$ where we define $\max \phi = 0$.

That the name assigned to ν is not altogether inappropriate is suggested by the following special cases. If $F \in \Gamma Ac$ or $F = I_1{}^{\dagger}$, $F\nu = 1+\max \phi = 1+0 = 1$ and if $\gamma:1 \to 2$ is atomic, $\gamma^{\dagger}\nu = 2$, $(\gamma^{\dagger} \circ \gamma^{\dagger})\nu = 2$, $(\gamma \circ [\gamma^{\dagger} + \gamma^{\dagger}])\nu = 2$, $(\gamma^{\dagger} \circ \gamma^{\dagger})^{\dagger}\nu = 3$, $(\gamma \circ [\gamma^{\dagger} + \gamma^{\dagger}])^{\dagger}\nu = 3$ etc.

Uniqueness Proof Let $\nu':\Gamma Red \to N$ satisfy $Q(\nu')$. Then assuming ν and ν' agree on $[i]\nu^{-1}$, $i \geq 1$, by virtue of $Q(\nu)$ and $Q(\nu')$, they also agree on $[i+1]\nu^{-1}$. Also according to $Q(\nu)$ and $Q(\nu')$, ν and ν' agree on $[1]\nu^{-1}$. Since $\cup\{[i]\nu^{-1} \mid i \geq 1\} = \Gamma Red$, $\nu = \nu'$.

Existence Proof We define $[i]\nu^{-1}$ for $i \geq 1$ by the requirement that $Q(\nu)$ be satisfied. Explicitly, $F\nu = 1$ if

G is a component flow of $F \Rightarrow G^{SIN} = 0$

and assuming $[i]\nu^{-1}$ is defined, we define

(4.2) $(i+1)\nu^{-1} = \{F \mid \max\{G^{\dagger}\nu \mid G$ is a component flow of $F \wedge G^{SIN} = 1\} = i\}$ so that

$F\nu = i+1 \Leftrightarrow \max\{G^{\dagger}\nu \mid G$ is a component flow of $F \wedge G^{SIN} = 1\} = i$.

It is sufficient then to prove $\mathscr{R} \overset{Df}{=} \cup\{[i]\nu^{-1} \mid i \geq 1\} = \Gamma Red$.

Closure of $\mathscr{R}$ with respect to $+, \circ$ is straightforward, since if $F\nu$ and $G\nu$ are defined then by (4.2) and the fact that a component of $F+G$ is a component of F or of G, $[F+G]\nu$ is also defined; similarly for $F \circ G$. Also $\mathscr{R}$ clearly contains Γ, Sur and $I_1{}^{\dagger}$. It remains only to show that $\mathscr{R}$ is closed under scalar iteration. Suppose $F:1 \to p+1$, $F \neq I_1$, $F \in [i]\nu^{-1}$.

We will show that $F^{\dagger} \in [i+1]\nu^{-1}$ from which it will follow that $\mathscr{R}$ is closed under scalar iteration and so $\mathscr{R} = \Gamma Red$ and the proof of the existence of a ν satisfying (4.1) will be complete. That $F^{\dagger} \in [i+1]\nu^{-1}$ is a consequence of the assertion $F\nu \leq F^{\dagger}\nu \leq 1+F\nu$ embodied in the Lemma below as (4.9).

Let $F{:}1 \to p+1$ be in ΓRed so that $F\alpha{:}1 \to p+1$ is in ΩAc and, assuming $F \neq I_1$, $(F\alpha)^{\dagger}{:}1 \to p$ is in $\Omega Red \backslash \Omega Ac$, i.e. in ΩRed but not in ΩAc.

Observation 1 Assuming $F \neq I_1$, the set of vertices of the component in $F^{\dagger}$ of $b_{F\dagger}$ is exactly $\{v \in Fv \mid \exists$ a begin path of F ending with exit $p+1$ of F which meets $v \neq$ exit $p+1\}$.

Observation 2 Let C be a component of F and $v \in C$.

Then, v is met by a begin path of F ending with exit $p+1$ of F iff v_C (the dominator of C) is met by such a path.

Lemma Assume $F \neq I_1$ so we may write $(F\alpha)^{\dagger} = G \circ H$, where $G{:}1 \to n$ is in $\mathscr{C}_{\Omega}$. Then $G \in \Omega Red \backslash \Omega Ac$ and $G^{\ddagger} \in \Omega Ac$. Furthermore

(4.3) $F = G^{\ddagger}\beta \circ [H\beta + I_1]$,

(4.4) $F^{\dagger} = (G^{\ddagger}\beta)^{\dagger} \circ H\beta$ and $(G^{\ddagger}\beta)^{\dagger} \in \mathscr{C}$,

(4.5) $G^{\ddagger} = \omega \circ J$ for a unique $\omega \in \Omega$, $J \in \Omega Ac$; for this ω and J,

(4.6) $(G^{\ddagger}\beta)^{SIN} = 1 \Rightarrow G^{\ddagger}\beta \neq (G^{\ddagger}\beta)^{\dagger\ddagger} = (\omega\beta)^{\ddagger} \circ [J\beta + I_1] \circ [I_n + (1,1)]$,

(4.7) $\omega\beta\nu \leq J\beta\nu \Rightarrow (G^{\ddagger}\beta)^{\dagger}\nu = 1 + G^{\ddagger}\beta\nu$,

(4.8) $\omega\beta\nu > J\beta\nu \Rightarrow (G^{\ddagger}\beta)^{\dagger}\nu = G^{\ddagger}\beta\nu \wedge (\omega\beta)^{SIN} = 1 = (G^{\ddagger}\beta)^{SIN}$;

(4.9) $F\nu \leq F^{\dagger}\nu \leq 1 + F\nu$.

Proof of (4.3), (4.4), (4.5). Since $F \neq I_1$, $(F\alpha)^{\dagger}$ and G have positive indegree; thus $G^{\ddagger}$ makes sense. Assume $(F\alpha)^{\dagger} = G \circ H$. Then, using $(F\alpha)^{SIN} = 0$, $F\alpha = (F\alpha)^{\dagger\ddagger} = (G \circ H)^{\ddagger} = G^{\ddagger} \circ [H+I_1]$ and, since $G^{\ddagger}$ and H are in ΩAc, $F = (F\alpha)\beta = G^{\ddagger}\beta \circ [H\beta+I_1]$ which gives (4.3).

From (4.3), using III(1.2), we infer $F^{\dagger} = (G^{\ddagger}\beta)^{\dagger} \circ H\beta$. To see that $(G^{\ddagger}\beta)^{\dagger}$ is the component flow of $b_{F^{\dagger}}$, we use Observations 1, 2 immediately above keeping in mind that v_C is a vertex of $F\alpha$; hence (4.4).

Assertion (4.5) is a consequence of $G^{\ddagger} \neq I_1$.

We shall require the following consequence of the fact that $(G^{\ddagger}\beta)^{\dagger} \in \mathscr{C}$:

(4.10) $(G^{\ddagger}\beta)^{\dagger}\nu = 1 + (G^{\ddagger}\beta)^{\dagger\ddagger}\nu$.

Proof of (4.6). Assume $(G^{\ddagger}\beta)^{SIN} = 1$. Using (4.5) and III(1.2):
$G^{\ddagger}\beta = \omega\beta \circ J\beta = ((\omega\beta)^{\ddagger} \circ [J\beta+I_1])^{\dagger}$ so that using III(1.1)

(4.11) $(G^{\ddagger}\beta)^{\dagger} = ((\omega\beta)^{\ddagger} \circ [J\beta+I_1])^{\dagger\dagger} = ((\omega\beta)^{\ddagger} \circ [J\beta+I_1] \circ [I_n+(1,1)])^{\dagger}$.

Since $((\omega\beta)^{\ddagger} \circ [J\beta+I_1] \circ [I_n+(1,1)])^{SIN} = 0$, (4.11) implies
$(G^{\ddagger}\beta)^{\dagger\ddagger} = (\omega\beta)^{\ddagger} \circ [J\beta+I_1] \circ [I_n+(1,1)]$.

Proof of (4.7). Assume $\omega\beta\nu \leq J\beta\nu$. There are two cases.

Case 1 $(G^{\ddagger}\beta)^{SIN} = 0$ so that $(G^{\ddagger}\beta)^{\dagger\ddagger} = G^{\ddagger}\beta$ and the conclusion (of the implication) follows from (4.10).

Case 2 $(G^{\ddagger}\beta)^{SIN} = 1$. Using (4.6), $(G^{\ddagger}\beta)^{\dagger\ddagger}\nu = \max((\omega\beta)^{\ddagger}\nu, J\beta\nu) = J\beta\nu$ since it follows from $\omega\beta \in \mathscr{C}$ that

(4.12) $\omega\beta\nu = 1 + (\omega\beta)^{\ddagger}\nu$.

From (4.5), $G\downarrow\beta = \omega\beta \circ J\beta$ so that $G\downarrow\beta\nu = \max(\omega\beta\nu, J\beta\nu) = J\beta\nu = (G\downarrow\beta)\uparrow\downarrow\nu$.

Thus in this case too, (4.7) follows from (4.10).

Proof of (4.8) Assume $\omega\beta\nu > J\beta\nu$. Thus, $\omega\beta\nu > 1$, $(\omega\beta)^{SIN} = 1$, $(\omega\beta)\downarrow$ makes sense and $(\omega\beta)\downarrow\nu \geq J\beta\nu$. Also $G\downarrow\beta = \omega\beta \circ J\beta$ so that $(G\downarrow\beta)^{SIN} = 1$. Thus, $G\downarrow\beta\nu = \max(\omega\beta\nu, J\beta\nu) = \omega\beta\nu$ and, again using (4.6),

$(G\downarrow\beta)\uparrow\downarrow\nu = \max((\omega\beta)\downarrow\nu, J\beta\nu) = (\omega\nu\beta)\downarrow\nu$, so that using (4.10) and (4.12) we infer $(G\downarrow\beta)\uparrow\nu = G\downarrow\beta\nu$.

Proof of (4.9) From (4.3) and (4.4), we obtain:

$F\nu = \max(G\downarrow\beta\nu, H\beta\nu)$, $F\uparrow\nu = \max((G\downarrow\beta)\uparrow\nu, H\beta\nu)$ so that the result follows from (4.7) and (4.8).

III Section 5. The characteristic freeness property of ΓRed.

The isomorphism between ΓRed and ΩAc maps the monoidal subcategory of ΓRed consisting of $[r]\nu^{-1}$ isomorphically onto $\Omega^{(r)}Ac$, where $\Omega^{(r)} = \{\omega_G \in \Omega \mid G \in \mathscr{C}_\Omega, G\nu \in [r]\}$. Let

$$[r]\nu^{-1} \xrightarrow{\alpha_r} \Omega^{(r)}Ac \xrightarrow{\beta_r} [r]\nu^{-1}$$

be the restrictions of α and β so that $\alpha_r \circ \beta_r$ and $\beta_r \circ \alpha_r$ are identities, $r \geq 1$. In the case $r=1$ we obtain

$$\Gamma Ac \xrightarrow{\alpha_1} \Omega^{(1)}Ac \xrightarrow{\beta_1} \Gamma Ac.$$

By virtue of the isomorphism $[r]\nu^{-1} \approx \Omega^{(r)}Ac$, the characteristic freeness property of $\Omega^{(r)}Ac$ translates into a corresponding property of $[r]\nu^{-1}$. Specifically this means that if T is an object of **PFIS**, every map $G \mapsto G' \in T$, where $G \in \mathscr{C}_\Omega$, $G\nu \in [r]$, has a unique extension to a morphism $[r]\nu^{-1} \to T$ in **PFIS**

Now let T be a scalar iteration flow theory over **Sur** i.e. an object in **SiFS**.

Freeness Theorem For any source and target preserving function $\varphi_0 : \Gamma \to T$, there is a unique extension $\varphi : \Gamma Red \to T$ to a morphism in the category **SiFS**.

Proof Let $\varphi_1 : \Gamma Ac \to T$ be the unique extension of φ_0 to a morphism in **PFIS** and assume $\varphi_r : [r]\nu^{-1} \to T$ is a morphism in the category **PFIS**. In order to define an extension $\varphi_{r+1} : [r+1]\nu^{-1} \to T$ of φ_r in **PFIS**, it is sufficient to define φ_{r+1} on those $G \in \mathscr{C}_\Omega$ such that $G\nu = r+1$. We define For $G \in \mathscr{C}_\Omega$ and $G\nu = r+1$,

(5.1) $\quad G\varphi_{r+1} = (G^{\dagger}\varphi_r)^{\dagger}$

and define $\varphi : \Gamma Red \to T$ to be the common extension of φ_r for all $r \geq 1$. We have immediately

(5.2) $\quad G \in \mathscr{C} \wedge G\nu > 1 \Rightarrow G\varphi = (G^{\dagger}\varphi)^{\dagger}$

(5.3) $\quad \varphi : \Gamma Red \to T$ is a morphism in **PFIS**.

It remains only to prove φ preserves $\dagger$

Using the notation and (4.3) and (4.4) of the Lemma III4 we have

$(F\varphi)^{\dagger} = (G^{\dagger}\beta\varphi \circ [H\beta\varphi + I_1])^{\dagger}$

$= (G^{\dagger}\beta\varphi)^{\dagger} \circ H\beta\varphi$ by III(1.2);

$F^{\dagger}\varphi = (G^{\dagger}\beta)^{\dagger}\varphi \circ H\beta\varphi$, so that to prove $F^{\dagger}\varphi = (F\varphi)^{\dagger}$, it suffices to prove $(G^{\dagger}\beta)^{\dagger}\varphi = (G^{\dagger}\beta\varphi)^{\dagger}$. What we have from (5.2) since $(G^{\dagger}\beta)^{\dagger} \in \mathscr{C}$ and $(G^{\dagger}\beta)^{\dagger}\nu > 1$ (by (4.10)) is:

$$(G^{\dagger}\beta)^{\dagger}\varphi = ((G^{\dagger}\beta)^{\dagger\dagger}\varphi)^{\dagger}.$$

Thus, we wish to prove

(5.4) $((G^{\dagger}\beta)^{\dagger\dagger}\varphi)^{\dagger} = (G^{\dagger}\beta\varphi)^{\dagger}$.

Case 1 $(G^{\dagger}\beta)^{SIN} = 0$. In this case $(G^{\dagger}\beta)^{\dagger\dagger} = G^{\dagger}\beta$ so (5.4) holds.

Case 2 $(G^{\dagger}\beta)^{SIN} = 1$. Then

$G^{\dagger}\beta\varphi = \omega\beta\varphi \circ J\beta\varphi$

$= ((\omega\beta)^{\dagger}\varphi)^{\dagger} \circ J\beta\varphi$ since $\omega\beta \in \mathscr{C}$ and $\omega\beta\nu > 1$

$= ((\omega\beta)^{\dagger}\varphi \circ [J\beta\varphi + I_1])^{\dagger}$ by III(1.2).

So

$(G^{\dagger}\beta\varphi)^{\dagger} = ((\omega\beta)^{\dagger}\varphi \circ [J\beta\varphi + I_1])^{\dagger\dagger}$

$= ((\omega\beta)^{\dagger}\varphi \circ [J\beta\varphi + I_1] \circ [I_n + (1,1)])^{\dagger}$ by III(1.1)

$= ((G^{\dagger}\beta)^{\dagger\dagger}\varphi)^{\dagger}$ by (4.6) of Lemma III4.

This proves (5.4) and with it the existence part of the Freeness Theorem.

The uniqueness is obvious since Γ*Red* is generated from Γ and **Sur** by means of $\circ$, $+$, $\dagger$.

III Section 6 The case of vector atoms

In II (resp., III) we showed that the described algebra of concrete flows ΓAc (resp., ΓRed) generated from scalar atoms and surjections $[n] \twoheadrightarrow [p]$ by means of $\circ$, $+$ (resp., and $\dagger$) may also be described purely equationally. In this context the word "concrete" should be understood to mean a set-theoretic or geometric entity. As a consequence then of general algebraic considerations ΓAc (resp., ΓRed) may be described as a quotient of the algebra of expressions (based on these descriptions) by the congruence generated by the defining equations for a flow theory (resp., scalar iteration flow theory) over **Sur**. [The "over **Sur**" corresponds to equations such as:

(6.1) $(2,1,2)\circ(2,1) = (1,2,1)$;

each (ordered) pair of surjections gives rise to an equation of this form. Thus "over **Sur**" adds an infinite list of axioms to the five (resp., seven) explicitly stated axioms for "flow theory" (resp. for "scalar iteration flow theory"). There is a simple rule for determining the equation corresponding to a given pair of surjections viz: apply the second (of the given pair) of surjections to the sequence of numbers describing the first; this gives a sequence of numbers which describes the composite surjection. If the given pair of surjections is:

$(2,1,2),(2,1)$,

according to the calculation rule, one applies the surjection $(2,1)$ to the sequence

$2, 1, 2$;

thus (6.1) is obtained.

Again, as a consequence of general algebraic considerations flow theories and scalar iteration flow theories freely generated by a set Γ of "vector" atoms also exist even though our description of concrete flows did not admit them. This circumstance suggests a problem roughly dual to the one with which we began. Namely: given an algebra of "abstract" flows described as the flow theory ΓAc (resp., scalar iteration flow theory ΓRed) over **Sur** freely generated by Γ, where Γ is the disjoint union of the doubly indexed sets $\Gamma_{r,s}$, where $r, s \in N$, find a concrete model of ΓAc (resp., ΓRed) which extends the concrete model in the case of scalar atoms. We shall generalize the notion of entry, exit, labeled digraphs briefly described in I3. Let $\gamma_{r,s} \in \Gamma_{r,s}$. The figure below depicts $[\gamma_{3,2}+\gamma_{1,0}+\gamma_{1,1}]\circ(2,1,3)\circ\gamma_{3,2}:5 \rightarrow 2$.

The case that $r=0$ and $s>0$ will not be permitted in order to preserve the rule that atomic flows are accessible.

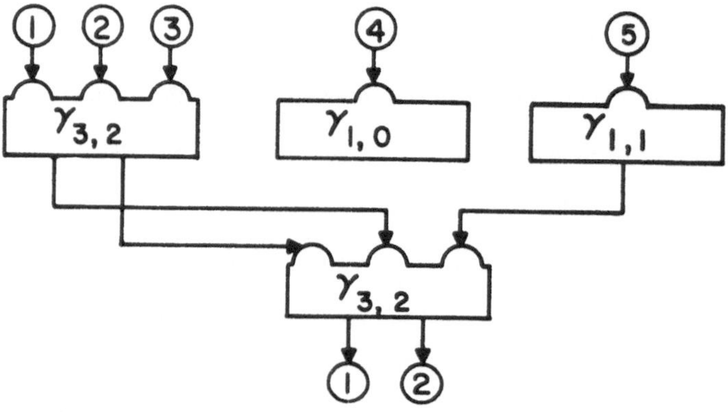

Figure 6 $[\gamma_{3,2}+\gamma_{1,0}+\gamma_{1,1}]\circ(2,1,3)\circ\gamma_{3,2}:5 \rightarrow 2$.

The algebra developed in II and III together with the nature of the result we're after and a desire for conciseness, essentially determines the generalized notions we're about to define.

Definition A *vector-vertexed directed graph* consists of a set V of vertices, a set E of edges, a function which assigns to each $v \in V$ a positive integer v©, its *component* or *hump* (cf. Figure 6) *count*, a source function $E \rightarrow V$ and a target function

$$E \rightarrow \{(v,i) \mid v \in V,\ i \in [v©]\}.$$

Thus, a directed edge of a vector-vertexed graph begins at a vertex and ends at some specific component of a vertex. If the meaning of "digraph" is extended to abbreviate "locally ordered (cf. I(i)) vector-vertexed directed graph" then the definition of Γ-flow given in I Section 1 still applies in the case that Γ is the disjoint union of a doubly indexed family of sets $\Gamma_{q,r}$ $(q>0,\ r\geq 0)$. In the case that v is an exit vertex, we require v©=1. The case of scalar atoms treated earlier should be identified with the satisfaction of the condition:

$$q \neq 1 \Rightarrow \Gamma_{q,r} = \phi.$$

The notion "path" may be taken as a sequence of edges $(e_1, e_2, \cdots, e_n)$ such that if e_k ends with (v,i) then e_{k+1} begins with v for all $1 \leq k < n$.

The notion of a graphical reducible Γ-flow is the same as before provided that "appropriate" modifications are made. Explicitly, conditions (a) and (b) of I Section 2 become:

(a) for every vertex v and for every component i∈[v©], there is a begin-path which ends in v, component i;

(b) for every closed path C, there is a scalar vertex v_C (i.e. v_C©=1) of C such that every begin-path to a vertex component of C meets v_C. The algebraic description of (algebraic) reducible flows reads exactly the same as before but the possibility of vector atoms is admitted. Thus we claim without detailed proof that the theorem which states *the class of isomorphism identified reducible graphical* Γ*-flows is exactly the class of* Γ*-flows which may be constructed from* Γ *and* **Sur** *by means of a finite number of applications of* ∘, + *and scalar iteration* still holds.

We claim further that the freeness theorems in II5, II7 and III5 hold in the case of vector atoms and that only obvious modifications in the displayed proofs are required to justify the claims.

These graphical Γ-flows, where Γ is a set of vector atoms, are essentially the same as the "deterministic" *c*-graphs of [JR] when these latter are appropriately labeled.

References

[AC] F. E. Allen and J. Cocke, Graphic-theoretic constructs for control flow analysis, IBM Research Report, RC-3923 (July 1972).

F. E. Allen, Control flow analysis, Proc. ACM SIGPLAN Symposium on Compiler Optimization, *SIGPLAN Notices* **5** (7) (July 1970) 1-19.

J. Cocke, Global common subexpression elimination, Proc. ACM SIGPLAN Symposium on Compiler Optimization, *SIGPLAN Notices* **5** (7) (July 1970) 20-24.

[BEW1] S. L. Bloom, C. C. Elgot, J. B. Wright, "Solutions of the iteration equation and extensions of the scalar iteration operation," IBM Research Report RC-7029 (March, 1978), also in *SIAM Journal of Computing* (February, 1980).

[BEW2] S. L. Bloom, C. C. Elgot, J. B. Wright, "Vector iteration in pointed iterative theories," IBM Research Report, RC-7322, (September, 1978).

[BT] S. L. Bloom and R. Tindell, Algebraic and graph threoretic characterizations of structured flowchart schemes, *Theoret. Comput. Sci.* **9** (1979), to appear.

[CE-AT] C. C. Elgot, "Algebraic theories and program schemes," IBM Research Report, RC-2925 (June, 1970). Also in *Proceedings* of Symposium on the Semantics of Algorithmic Languages (Ed. Erwin Engeler). (Springer Verlag, 1971).

[CE-SP] C. C. Elgot, "Structured programming with and without GO TO statements," IBM Research Report, RC-5626 (September, 1975). *IEEE Transactions on Software Engineering, SE-2*, No. 1 (March, 1976) 41-54. Erratum and Corrigendum (September, 1976).

[ES] C. C. Elgot and J. C. Shepherdson, "A semantically meaningful characterization of reducible flowchart schemes," IBM Research Report, RC-6656 (July, 1977). *Theoretical Computer Science* **8**, (3) (June, 1979) 325-357.

[EW] Samuel Eilenberg and Jesse B. Wright, Automata in general algebras, *Information and Control* **11** (4) (1967).

[HU] M. S. Hecht and J. D. Ullman, Flowgraph reducibility, *SIAM J. Comput.* **1** (2) (June 1972) 188-202.

M. S. Hecht and J. D. Ullman, Characterization of reducible flow graphs, *J. Assoc. Comput. Mach.* **21** (3) (July 1974) 367-375.

[JR] J. D. Rutledge, Program Schemata as Automata. 1, *J.CSS*, **7** (6) (December, 1973) 543-578.

[SM] S. Mac Lane, *Categories for the Working Mathematician*, (Springer-Verlag, New York, Heidelberg, Berlin, 1971).

[WL] F. William Lawvere, Functorial semantics of algebraic theories, *Proceedings of the National Academy of Sciences*, **50** (5) (1963) 869-872.

ON COORDINATED SEQUENTIAL PROCESSES

Calvin C. Elgot
Raymond E. Miller

Mathematical Sciences Department
IBM Thomas J. Watson Research Center
Yorktown Heights, NY 10598

ABSTRACT: The mutual exclusion problem introduced (as far as we know) by Dijkstra in 1965 is intriguing because on the one hand it appears to be typical of a broad class of important systems problems and on the other poses difficult problems of mathematization. There has been extensive attention to this (and other synchronization problems) since that date. Nevertheless our approach to the problem differs from all the others in at least two important ways. First: we take the problem to be "construct from n uncoordinated sequential processes, n coordinated ones which satisfy (1) mutual exclusion, (2) fairness and (3) simulation." Second: our mathematical model can directly represent simultaneous actions. Since we do not constrain the synchronization in some of the usual ways, however, the realm of possible solutions is much broader than considered elsewhere. The main solution we offer is simple in that each process need read only one binary variable (or register), viz. its own, and write in one, viz. its neighbor's. The emphasis of our approach is in the mathematical formulation of the problem, and the proof that the solution satisfies the desired properties, rather than in the novelty of the solution itself. The other solutions offered serve to emphasize the enormous differences in implementation associated with what we call the "conceptual solution," and the fact that instructions (or atomic actions) don't compose i.e. a sequence of two (single entry - single exit) instructions of a process is not system equivalent to any single instruction. Contrary to many other solutions to the mutual exclusion problem, the solutions given here do not employ a *test and set* primitive i.e. an indivisible test and set instruction.

Acknowledgments We are indebted to Hartmut Ehrig whose comments on an early version of the paper resulted in substantial improvements in the exposition. We owe a debt, too, to Martha Pierce who not only typed the paper but also located some misprints and compiled the Glossary.

1. Historical Background

Modern computing systems contain many different types of operational units and have complex programming systems to control the flow of information within and between these units. Much of the activity within the systems is synchronized by centralized clock mechanisms, but often the time required for various actions, or processes, cannot be predicted ahead of time. This leads to the problem of controlling the interaction of these processes, including their access to the various resources of the system, by appropriate programming.

There is an extensive literature on various aspects of the synchronization of interacting processes. Many different programming language primitives have been proposed for synchronization, such as semaphores with their P and V operations, conditional critical regions, and monitors. Programming solutions to particular synchronization problems, such as mutual exclusion, dining philosophers, readers-writers, and producer-consumers, have also been devised. Synchronization for systems of processes has been modelled in various mathematical ways, and the complexity of synchronization has been analyzed for certain situations.

This paper takes a new look at one of the most extensively studied synchronization problems, namely the "mutual exclusion problem."* Our treatment of this problem differs from previous work in a number of significant ways. First, we formulate the problem differently. We assume that a system of n uncoordinated sequential processes is given. The problem is then to modify this system in a way which coordinates it. Second, to formulate the results in a sharp mathematical way, we propose a new model for synchronization problems and develop a rather general notion of a computation which, unlike previous models, can directly represent simultaneous actions. After introducing sufficient definitions and terminology, this approach allows us to give a precise succinct statement of the problem, clearly

---

* This use of quotes is intended to point out that a key phrase is regarded as vague. Later, in many cases, such phrases are given more precise meaning.

separated from any solution, and allows us to prove, perhaps even elegantly, that our solutions satisfy the problem specifications.

Intuitively, the mutual exclusion problem consists of finding a way of providing for the appropriate interaction of n processes, each of which requires, at some points in the process, exclusive access to a resource common to all the processes. The exclusiveness of the access must be guaranteed by the synchronization mechanism. Also, each process must be guaranteed some sort of "fair" access to the common resource. In addition to guaranteeing "mutual exclusion" and "fairness" the synchronization mechanisms of the processes must be such as not to interfere with the intended computational tasks; i.e. the coordinated processes must in some sense "simulate" the uncoordinated processes. A more precise understanding of what is meant by "process," "exclusive access," "fairness," "simulation," etc., will emerge later.

The first reference to the mutual exclusion problem appears to be that of Dijkstra [Dij-1] in a very short, one-page, article. In this paper Dijkstra gives a brief definition of mutual exclusion and then proposes a programming solution for controlling process access to the common resource. He then gives an argument that this solution provides mutually exclusive access and is fair in a certain sense. This paper was quickly followed by a short letter from H. Hyman [Hym] stressing the practical importance of this problem and proposing a much simpler solution for the situation in which there were only two processes. Knuth then followed with another short note [Knu]. He pointed out that the solution proposed by Hyman would not work, and that Dijkstra's solution, although quite ingenious, would allow the possibility of a given process being prevented from getting access to the resource, when trying, simply because other processes were also trying. The success of the other processes getting access and the particular times of access of these other processes resulted in the access by the one process being frozen-out forever. Knuth's programming solution to this problem covered this new aspect of fairness not discussed previously by Dijkstra.

There are many other papers which deal with mutual exclusion, and we will not try to review them all here. A few are worth mentioning, however, to give a better idea of the types of work done.

A further extensive seventy-page work of Dijkstra [Dij-2] gives a series of five "solutions" to the two-process mutual exclusion problem. After each solution is proposed the reasons for its failure to be a solution are discussed and this leads to the next proposed solution. Finally, the fifth solution is given along with an argument that it is correct. This paper was probably most instrumental in raising interest in the mutual exclusion problem and in pointing out some of the difficulties of the problem. Another aspect of the problem is the question of how complex the solution must be. There are many aspects to this question of complexity including the length of the program parts required for process interaction, the complexity of the interaction instructions, etc.. One important type of complexity problem has been studied extensively -- that of trying to devise solutions which use a minimum, or near minimum, number of communication variable states to solve the problem, and also to obtain lower bounds on the number of such states required [R&P, P&F, BFJLP].

One of the qualities of the synchronization problem literature, including the mutual exclusion literature, that has bothered some workers is the extremely close tie between the statement of the problem and desirable features of a solution and that of the solution itself. This lack of separation makes it difficult to tell exactly what the problem is versus what the properties of the particular solution are. This has led to work on more formally defining mutual exclusion and its desired properties independent of any particular solution to the problem [C&H, M&Y-1, BFJLP].

Even though this paper presents a new look at mutual exclusion it does have ties with the previous literature. Our model of synchronization problems is similar in some respects to that of [M&Y-1] but it differs in several essential ways. Our notion of a computation differs significantly from [M&Y-1] and others in that a computation can directly represent simultaneous action of more than one process, rather than having to represent this through sets of computation sequences. Like [M&Y-1] we provide a sharp distinction between problem statement and solution, but this distinction is achieved in a different way. Also, our model assumes that a communication variable can take on only one of two values, *0* or *1*.

Numerous aspects of simultaneity have been considered by other workers, for example [M&Y-2], [L-1], [L-2] and [ERR].

Another connection can be seen with the complexity literature. In [BFJLP] a lower bound of $n+1$ states is obtained for mutual exclusion in a system of n processes which realizes "bounded waiting" and some other fairness criteria. Our solution, which uses n variables uses exactly $n+1$ states of these variables, and thus appears to attain the [BFJLP] lower bound. Our model differs from that of [BFJLP], however, and on closer inspection it turns out that their bound does not apply to our model. This can be seen in a second solution of ours to mutual exclusion, using a Gray coding of communication states, where we require only n states for a solution when n is even, and $n+1$ states when n is odd. There are numerous differences between our approach and that of [BFJLP]. For example, they assume processes can halt outside of the regions using the resource, whereas we assume that processes will pass their turn to access a resource on to another process if they do not wish to access the resource themselves in the "near future." Our round-robin type solutions require such turn passing. A crucial difference between the approaches, however, appears to be that [BFJLP] assume that the variables introduced to insure mutual exclusion can only be accessed and changed during the entering and exiting protocols for resource use. With this assumption these protocols seem to require considerable complexity. We do not make this restrictive assumption, thereby allowing solutions not allowed by [BFJLP], and find that rather simple entering and exiting protocols are possible. The tradeoff is that these "coordination variables" need also be changed by processes when they are in loops outside the entering and exiting regions. Certainly, in some situations this may be preferable to having complex protocols, and in any particular case, which approach is preferable should be considered.

2. Outline

While one of the main objectives of writing this paper was to formulate and prove correct a solution to the mutual exclusion problem in a sharply mathematical way, we have been mindful of a hoped-for broader audience than theoretical computer scientists or mathematicians. The audience we have tried to reach is problem oriented -- those who have an interest in systems synchronization problems -- and is intended to include, for example, systems engineers and programmers as well as the two groups mentioned above. For this audience the descriptive sections 3 and 11, as well as the construction of the unary solution, as depicted in Figures 5.1 through 5.6, should be of particular interest. Also, at critical points, we have attempted to paraphrase formal statements by informal statements and have tried to use terms suggestive of their meaning.

The organization of the paper is indicated by recording the section headings:

1. Historical background.
2. Outline.
3. Informal statement of the problem.
4. The system (G,K) of uncoordinated sequential programs (SP).
5. The construction of the system (P,K) of coordinated (SP) -- the unary solution.
6. The consensus operator, direct transition relation and system computation.
7. The LEMMA.
8. Preliminaries for The THEOREM.
9. The THEOREM.
10. The conceptual solution and its specialization to Gray solutions.
11. Application of the solution to master-slave and message sending contexts.
12. Concluding remarks and an open problem.

The paper has both pure and applied mathematical content. The pure mathematical content resides mainly in sections 7 and 9 where substantive concepts and arguments are brought together in the proof of the LEMMA and the THEOREM. The applied mathematical content resides primarily in section 6 where a new and rather broad, completely mathematical, model of concurrent computation is given, and secondarily in the simplicity of the unary solution to the mutual exclusion problem. The functional simplicity is a consequence of the fact that each process has its own communication variable (c.v.) and need only read its c.v. and write into its neighbor's. Many solutions to the mutual exclusion problem exploit a *test and set* primitive, i.e. an (*indivisible*) *test and set instruction*, similar to Dijkstra's P instruction. The solutions given here do not.

3. *Informal statement of the problem*

Our statement of "the mutual exclusion problem" -- even the informal statement -- differs significantly from earlier ones. We assume n (where n is a positive integer) *un*coordinated sequential processes are given; each of these "processes" is described (as completely as is required) by a flowchart program* which from time to time "communicates" with or, more specifically, "writes" in some "resource" (say, a file), and it is assumed that each of these programs is written as if it had sole access to that resource.

"The problem" then becomes: modify each of these n processes in such a way that

(3.1) (*mutual exclusion*) no two processes communicate with the resource at the same time;

(3.2) (*primary fairness*) at every instant of time each process has a forthcoming "opportunity" or "turn" to communicate with, or write in, the resource;

(3.3) (*secondary fairness*) if process i has the opportunity to communicate with, or write in, the resource then after a bounded amount of time, it either exploits or yields that opportunity.

(3.4) (*simulation*) the modified process "performs the same task" as the process which it modifies.

The mechanism for effecting this modification is often referred to as a "protocol."

In many discussions of this problem, simulation is not mentioned. Presumably this is so because of an intuitive awareness that it plays little or no role in the design of coordination.

It is assumed that when a process is executing an instruction not accessing the resource, then the contents of the resource may undergo change as a result of interaction with the

---

* By a "flowchart program" we mean a flowchart scheme, (more precisely, a *flowscheme* in the sense of [E&S]), together with an interpretation of the "atomic flows."

process's external environment. When the process is accessing the resource, however, no such external change is possible.

In order for (3.2) to be at all possible we must assume something further. One possible assumption is:

(3.5) each uncoordinated process has the property: the duration of each instance of communication with the resource is finite.

A still more basic assumption for (3.2) as well as for the coordinating strategy to hold is the following:

(3.6) each *departure*** execution of an instruction requires "only" a finite amount of time.

A stricter requirement is necessary for (3.3) to obtain. Thus, rather than (3.5) or (3.6) we assume:

(3.7) there is a single upper bound for the departure execution time of all instructions.

The solution then takes the form of a system of n coordinated sequential processes.

** The execution of an instruction of a flowchart program results in (among other things) a choice for the location (a vertex) of the next instruction. The execution is a *departure execution* if this new vertex is different from the old. Even though, intuitively, *all* executions take place in a finite amount of time, the weaker assumption (3.6) suffices. The significance of the weaker assumption is, however, that our notion of total state (introduced in section 6) suffices, even though it fails to discriminate between "no execution" and "execution and return to same vertex."

4. *The uncoordinated system of sequential programs (or asynchronous system of sequential programs)*

For our formal (i.e. mathematical) considerations the given data is assumed to be a family $(G,K) \overset{Df}{=} \{(G_i,K_i) \mid i \in [n]\}$, where $[n] = \{1,2,...,n\}$, G_i is a flowchart program, where $G_i\mathbf{v}$ is the set of all vertices of G_i, and $K_i \subseteq G_i\mathbf{v}$. Let $N_i = G_i\mathbf{v} \setminus K_i$ be the complement of K_i with respect to $G_i\mathbf{v}$. The vertices in K_i are "contact" vertices with the "world" external to the system (cf. Figure 5.5). Playing a smaller role in our considerations is a family $\{Y_i \mid i \in [n]\}$ of sets (Y_i is the set of *data states* of G_i, representing the data values computed by G_i) and a set R of *resource states*, representing the values of the common resource. We will nevertheless continue to refer to the given data, briefly, as (G,K).

The program G_i is the part of the ith process of central interest to us. Each vertex of G_i has correlated with it an instruction; in one kind of application of these ideas, the vertices in K_i (the *critical* vertices of G_i) are supposed to be the ones whose correlated instructions make reference to a "resource." The presence of the resource in the model is reflected through the set of critical vertices and the set R as both input and output. We call $\mathrm{G} = \{G_i \mid i \in [n]\}$ the *uncoordinated system of sequential programs.*

Since it is assumed that each of the n programs is written as if it has sole access to the resource, any potential for conflict among the uncoordinated programs is neglected.

We assume that for each $i \in [n]$:

(4.1) the begin vertex of G_i is in N_i.

Considerations of the kinds of applications we have in mind for K_i suggest the following assumption, for each $i \in [n]$:

(4.2) there is no closed path Π in G_i such that $\Pi\mathbf{v} \subseteq K_i$.*

---

* $\Pi\mathbf{v}$ is the set of vertices of the path Π.

That is, there are no loops in the program G_i which lie totally within the critical section. An alternative is the following weaker assumption:

(4.3) the nature of K_i's instructions and input is such that if G_i's control is in K_i at time t, then there is a $t' > t$ such that G_i's control will be in N_i at time t'.

Remark. Presumably, in most applications of these ideas, G_i would have no vertices of outdegree *0*. The vertices of outdegree *0* are correlated with "exit" or "halt" instructions.

Assumption (4.2) or (4.3) is required to insure that one of the programs does not enter its K_i vertices and stay there "forever," thus (assuming (3.1) is satisfied) preventing other programs access to the resource. Figure 5.5 is a simplified schematic of G_i, indicating that N_i may contain closed paths but that K_i does not, as specified by (4.2). To simplify slightly our considerations, we assume

(4.4) any edge of G_i which ends in an exit begins in N_i.

The program G_i has correlated with it a *next vertex function* ν_i and *a result function* ρ_i. Let $G_i\mathbf{v}^-$ be the set of non-exit vertices of G_i. If $v_i \epsilon G_i\mathbf{v}^-$ and $z_i \epsilon Z_i \overset{\text{Df}}{=} Y_i \times R$ then ν_i has the form

$$G_i\mathbf{v}^- \times Z_i \xrightarrow{\nu_i} G_i\mathbf{v}, \qquad (v_i,z_i) \longmapsto^{\nu_i} v_i'$$

and $(v_i,z_i)\nu_i = v_i'$ is the target of an edge of G_i which begins with v_i.

(4.5) If $v_i' \neq v_i$, then the execution of "the instruction at v_i" in response to z_i is a *departure execution* (cf. footnote **, section 3).

Our convention is that the function ν_i applied to the argument (v_i,z_i) is written with the argument on the left. This convention is used consistently throughout the paper.

The form of ρ_i is $G_i\mathbf{v}^- \times Z_i \xrightarrow{\rho_i} Z_i$. Let (v_i,ρ_i): $Y_i \times R \rightarrow Y_i \times R$ be the function obtained from ρ_i by fixing the first argument to be $v_i \in G_i\mathbf{v}^-$. We assume

(4.6) for each $v_i \in G_i v^-$, there exists $\rho_i^{(1)} : Y_i \rightarrow Y_i$ and $\rho_i^{(2)} : R \rightarrow R$ such that $(v_i, \rho_i) = \rho_i^{(1)} \times \rho_i^{(2)}$. That is, we assume that the result function ρ_i can be decomposed into a result function $\rho_i^{(1)}$ for the data states and $\rho_i^{(2)}$ for the resource.

The word "uncoordinated"in "*uncoordinated system*" reflects intent. The word "system" rather than "family" is intended as a reminder of the presence of the resource which is common to all the processes. This notion as well as the companion notion of a coordinated system (introduced in the next section) is directed at synchronization problems in general. The additional data K is used in our formulation of the particular synchronization problem "mutual exclusion."

5. *The construction of the system (P,K) of coordinated sequential programs for the unary solution*

Making use of the data (G,K) we shall construct, for each $i \in [n]$, a flowchart program, P_i by extending G_i and call this particular solution the "unary solution." The vertices of P_i not in G_i will be called the *interaction vertices* and will have explicitly stated instructions correlated with them. We also use "P_i" to refer to the underlying pointed directed graph of the program P_i. ("Pointed" refers to P_i having a designated begin vertex.) As will become clear, the extension of the G_i to P_i consists of adding protocols to the G_i to give a round robin discipline of access to the critical regions K_i.

The construction: P_i is obtained from G_i by inserting:

(a) a copy of the single entry–single exit *wait program* ω_i (cf. Figure 5.1 where the notation ω_i:*1* → *1*, is used to denote single-entry single-exit) in each edge from N_i to K_i, i.e. from a vertex in N_i to a vertex in K_i, (cf. Fig. 5.6).

(b) a copy of the *augment program* α_i (cf. Fig. 5.2) in each edge from K_i to N_i, (cf. Fig. 5.6).

(c) copies of β_i (the *test and augment program* cf. Fig. 5.3) in such a way that all closed paths Π in G_i with $\Pi v \subseteq N_i$ are interrupted by a copy of β_i, (cf. Fig. 5.6).

(d) Redirect all edges of G_i which end in an exit to the begin of a single copy of $(\omega_i \circ \alpha_i)^{\dagger}$, (cf. Fig. 5.4), this notation is from [E&S].

The passage from G_i to P_i is illustrated by Figures 5.5 and 5.6.

Remark 5.1. The letter "x" which occurs in Figures 5.1-5.3 is a "new" (i.e. not part of the given data (G_i,K_i); $i \in [n]$) vector variable which ranges through the set $\{0,1\}^n$.

Remark 5.2. The begin vertex of P_i is understood to be that of G_i.

Remark 5.3. In the case that the underlying pointed directed graph of G_i is reducible (cf. [E&S]), we may be more definite about interrupting closed paths by a copy of β_i. In this case (c) may be replaced by:

(c') a copy of β_i in each backedge of G_i. (A directed edge $e:v_1 \to v_2$ is a *backedge* if each begin path in G_i to the vertex v_1 must first pass through the vertex v_2.)

Remark 5.4 Although in most applications the G_i would probably not contain exit vertices, the use of the $(\omega_i \circ \alpha_i)^{\dagger}$ is a simple way of handling exits when they do occur. The problem is that when a program G_i enters its exit it halts, but other programs in the system may still be running. Thus in the coordinated system if it is P_i's "turn" there must be some mechanism to pass the turn on to P_{i+1}. Another approach to solving this problem is to design program P_i in the following way: just before exiting it modifies program P_{i-1} so that thereafter P_{i-1} passes the turn to P_{i+1} rather than P_i.

Explanation of the instructions. In Figures 5.1 - 5.3 the letter x is a variable whose values are in $\mathbf{2}^n$, where $\mathbf{2} = \{0,1\}$. More accurately, x may be regarded as a function variable which depends upon time t ($t \in R$, where R is the set of real numbers). Then, for each $k \in [n]$, the instruction "change $x\pi_k$ to 0" may be understood to mean: if at time $t \in R$, the value of x is x_t then at time $t' > t$, where t' is the instant at which execution of the instruction is terminated, the value of x is $x_{t'}$, where $x_{t'}$ is characterized by the following conditions

$$\begin{cases} x_{t'}\pi_i = x_t\pi_i & \text{for all } i \neq k,\ i \in [n] \\ x_{t'}\pi_k = 0 \end{cases}$$

Here we are assuming that only this one instruction is being executed. Since it is a "change" instruction the value of x need never be read. The change of value, however, is assumed to occur exactly at time t' which is the termination time of the instruction. Actions of other instructions between t and t' are possible, either for reading or writing variables, and no confusion of the value of $x\pi_k$ can exist.

$\pi_i : \mathbf{2}^n \rightarrow \mathbf{2}$ is projection onto the ith component; $x_t\pi_i$ is the result of applying π_i to $x_t \epsilon \mathbf{2}^n$. Hence we may correlate with the instruction "change $x\pi_k$ to 0" the function $\mathbf{2}^n \xrightarrow{\rho_k} \mathbf{2}^n$ such that $x_t\rho_k = x_{t'}$. Note that this use of ρ is consistent with the previous use as the result function of an instruction. The interaction vertices of the program P_k control coordination.

The letter π will be used later for other projections also, where the subscript will indicate the target of the projection.

The meaning of the instruction "$x\pi_k \overset{?}{=} 1$" should now be clear as well. This instruction may be correlated with a function of the form $\mathbf{2}^n \longrightarrow$ {"yes", "no"}, in analogous fashion.

The set of interaction vertices of all n programs P_i taken together is partitioned into four sets: ωA, $0A$, $1A$, βA as indicated in Figures 5.1-5.3. Thus ωA *(resp.* $0A$, βA*)* is the set of all begin vertices of copies of wait programs (resp. augment programs, test and augment programs), one begin vertex for each copy.

If $A = \omega A \cup 0A \cup 1A \cup \beta A$ is the set of interaction vertices of the system P, then $A_i \overset{Df.}{=} A \cap P_i v$ is the set of interaction vertices of P_i.

We call the system (P,K) formed by our construction the "unary solution" to the mutual exclusion problem because each test function in the solution depends only upon a single communication variable.

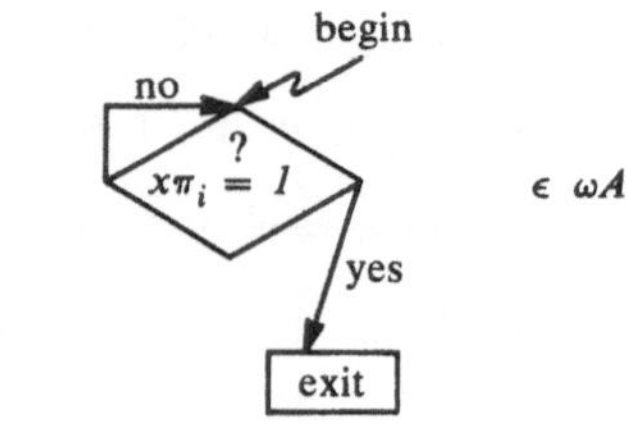

Figure 5.1 ω_i*:1→1, wait*

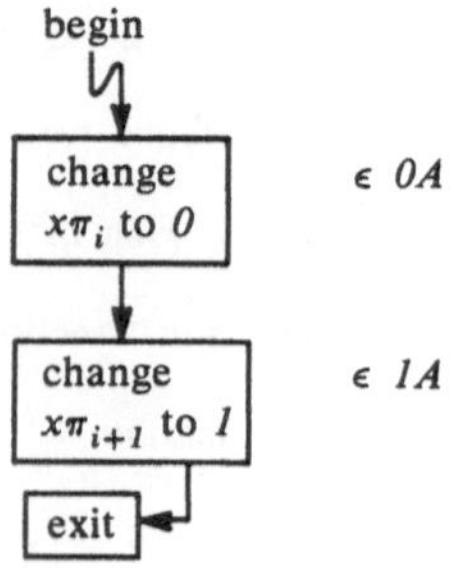

Figure 5.2 α_i ***1→1, augment***
(+ is modulo n)

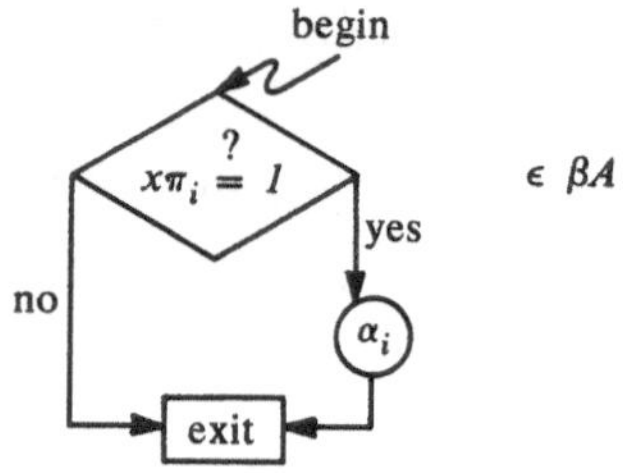

Figure 5.3 β_i*:1→1, test & augment*

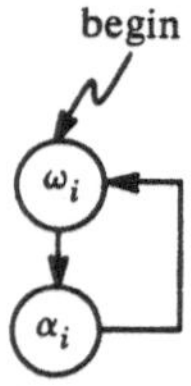

Figure 5.4 $(\omega_i \circ \alpha_i)^\dagger$*:1→0*
Repeatedly wait and augment

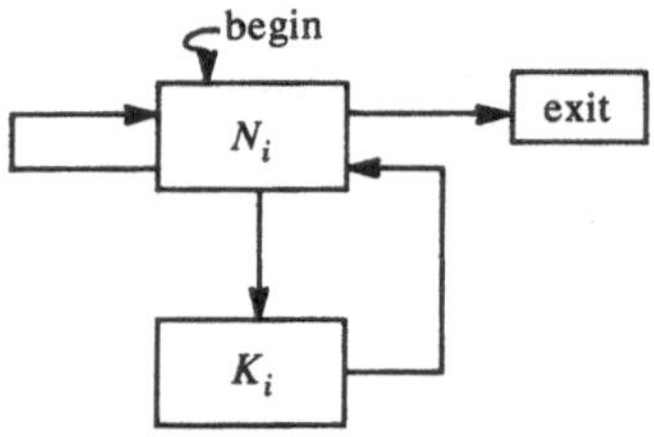

Figure 5.5 G_i*:1→*p_i
(p_i is the number of exits of G_i).

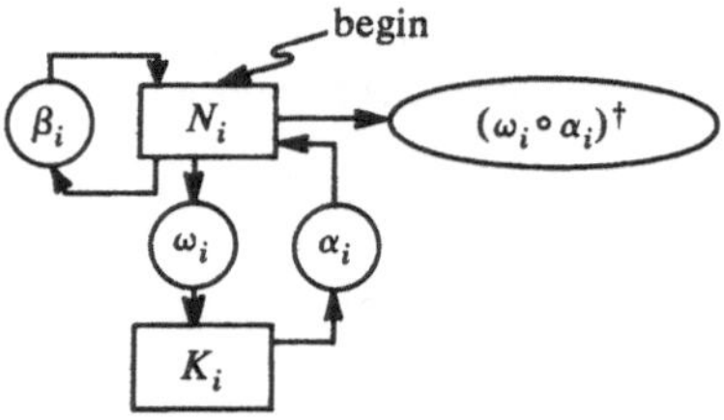

Figure 5.6 P_i*:1→*p_i

Remark $\omega_i \circ \alpha_i$ is obtained from ω_i and α_i by redirecting the edge of ω_i which ends with its exit to the begin vertex of α_i and deleting ω_i's exit. $(\omega_i \circ \alpha_i)^\dagger$ is obtained from $(\omega_i \circ \alpha_i)$ by redirecting the edge which ends with its exit to its begin. These operations are discussed in full in [E&S].

6. The consensus operator, direct transition relation and system computation

In order to separate the purely mathematical considerations from the application to the mutual exclusion problem, as well as to divorce the mathematical modeling of coordinated systems of sequential processes from the particular problem and solution discussed in previous sections, we formulate the following notion.

Definition 6.1. An *abstract system* P *(more fully,* (P, A, ρ, ν)*) of* n *coordinated sequential processes* consists of the following data:

(P1) a set $X = \mathbf{2}^r$ of *communication states of* P (where $\mathbf{2} = \{0,1\}$ and $\mathbf{2}^r$ is the cartesian product of $\mathbf{2}$ with itself r-times) and a set $Z = (\Pi Z_i \mid i \in [n])$, where $Z_i = Y_i \times \mathrm{R}$; and for each $i \in [n]$*

(P2) a pair, (P_i, A_i), where P_i is a pointed directed graph and $A_i \subseteq P_i\mathrm{v}$ (is the *subset of interaction vertices* of the set $P_i\mathrm{v}$ of all vertices of P_i);

(P3) (a) a *result function*

$$A_i \times X \xrightarrow{\rho_i} X;$$

(b) a *next vertex function*

$$A_i \times X \xrightarrow{\nu_i} P_i\mathrm{v};$$

subject to the two constraints that

(1) for each $v_i \in A_i$, $x \in X$, there exists an edge $e{:}v_i \rightarrow v_i'$ (i.e. an edge which begins with v_i and ends with v_i'), where $v_i' = (v_i,x)\nu_i$ is the image of (v_i,x) under ν_i and;

* Constructing the set Z as the cartesian product of the sets Z_i, $i \in [n]$ means that the set R is repeated n times in Z. This arises from the fact that we assume the given system to be uncoordinated rather than coordinated. The resource constraint (P3)(d) has the same effect as assuming the resource occurs but once in the coordinated system.

(2) the *static loop property*** $(v_i,x)\nu_i = v_i \Rightarrow (v_i,x)\rho_i = x$;

(c) a next vertex function and a result function

$$V_i \times Z_i \xrightarrow{\nu_i} P_i\mathrm{v}, \qquad V_i \times Z_i \xrightarrow{\rho_i} Z_i$$

where V_i is the set of all non-exit vertices in $P_i\mathrm{v}\backslash A_i$, subject to the constraint that for each $v_i \epsilon V_i$, $z \epsilon Z$, there exists an edge $e{:}v_i \rightarrow (v_i,z)\nu_i$.

No confusion will result from thus assigning double duty to "ν_i".

(d) The *resource constraint*

$$z\pi_i\pi_R = z\pi_j\pi_R \text{ for all i,j} \in [\text{n}].$$

(Thus when the ith process changes the state of the resource, from the point of view of the jth process, j ≠ i, the external environment makes the change. The research constraint property insures that all n copies of the resource have identical values.)

We say P satisfies the *static loop property* if for each i, the vertices of P_i in A_i satisfy this property. Notice that the P constructed from G in section 5 satisfies the static loop property.

The number r may be called the number of *communication registers* of P or the number of *communication variables* of P although some would regard this terminology as biased toward a particular "implementation" of X.

We should point out that we may correlate with each vertex $v_i \epsilon A_i$ an *abstract instruction* derived from (a) and (b) viz., the pair of functions described immediately below,

(a') the function $X \rightarrow X$, $x \mapsto (v_i, x)\rho_i$

** The static loop property simply says that if we do not move to a new instruction then the values of the coordination variables do not change. This assumption plays no role in our argument. The point of the assumption is that our notion of computation is realistic or, at least reasonable, in the presence of the assumption that interaction vertices v_i satisfy (2) for all $x \epsilon X$. The crux concerning reasonableness lies with the definition of consensus function, in which changes in X occur only when there is a transition to a new vertex.

(b') the function $X \rightarrow P_i\mathbf{v}, \quad x \mapsto (v_i, x)\nu_i$

this is the (abstract) instruction *at* v_i.

Let $V = P_1\mathbf{v} \times P_2\mathbf{v} \times \cdots \times P_n\mathbf{v}$ (the set of *vector vertices* of the system P) and let $S = V \times X$, $T = S \times Z$, where $Z = (\Pi Z_i \mid i \in [n])$. We call T the set of *total states* of P.

Since, however, in considerations which ignore simulation the focus may be on S, rather than T, we give S a name as well, viz. the set of *interaction states* of the system P.

In preparation for defining the transition relation of the system P and thence "a computation of the system P," we turn our attention to the crux of the transition relation viz.,

Definition 6.2. the *consensus function*, (more specifically, *the change-dominant consensus function*).

(6.1) $$V \times X \times V \xrightarrow{\mathbf{c}} X$$

of the system P. Let $x' = (v, x, v')\mathbf{c}$. Then, for each $m \in [r]$,

$x'\pi_m \neq x\pi_m$ iff there exists $i \in [n]$ such that

(6.2) $$[v\pi_i \in A_i \wedge v'\pi_i \neq v\pi_i \wedge (v,x)\rho_i\pi_m \neq x\pi_m]$$

The role of the consensus function is to abjudicate disputes. More specifically, if several interaction instructions act simultaneously and some do not tend to change the value of the mth communication variable but at least one does, then the change occurs. This is unambiguous since communication variables are 2-valued.

For convenience in defining the direct transition relation of the system P, let $t = (v,x,z)$, $t' = (v',x',z')$, where:

$t \in T$ is (or may be regarded as) the present total state of system P, $v \in V$ is then the present vector vertex of the system P, $x \in X$ is then the present communication state of the system P, etc. while $t' \in T$ is a possible next total state of the system P.

Definition 6.3. The *T-direct transition relation* τ of the system P is defined by the following requirement, where $t \in T$, $z = (z\pi_1, z\pi_2, \cdots, z\pi_n)$ the resource constraint, i.e. (P3)(d), holds and $z' = (z'\pi_1, z'\pi_2, \cdots, z'\pi_n)$:

[$t\tau t'$ (i.e. t stands in the τ-relation to t') IFF (6.3) and (6.4) hold.]

(6.3) The "new" communication state $x' = (v, x, v')$c, where c is the consensus function of the system P.

(6.4) For each $i \in [n]$,

(a) in the case $v\pi_i \in A_i$,

(i) $z'\pi_{Y_i} = z\pi_{Y_i}$

(ii) $z'\pi_i\pi_R = z\pi_i\pi_R$ if $v\pi_j \in A_j$ for all $j \in [n]$, and

(iii) $v'\pi_i \in \{(v\pi_i,x)\nu_i, v\pi_i\}$,

(b) in the case $v\pi_i \in V_i$,

(i) $z'\pi_i \in \{(v\pi_i,z\pi_i)\rho_i, z\pi_i\}$ and

(ii) $v'\pi_i \in \{(v\pi_i,z\pi_i)\nu_i, v\pi_i\}$.

Here the T-direct transition relation allows for the possibility of more than one process "acting" in the single step. Condition (6.3) specifies that any change in the communication state is done according to the change-dominant consensus. The next vertex vector v' and the next result state z' are specified by condition (6.4). If an interaction instruction of process i is active then (6.4)(a) asserts (in part) that the Y_i part of the result state does not change and the next instruction for process i is either unchanged, indicating for example that the instruction has not completed acting, or is changed in accord with the ν function.

The S-direct transition relation τ is defined by projection from the T-direct transition relation τ, i.e.

$$s\tau s' \Longleftrightarrow \exists\, t,t' \; [t\tau t' \wedge s = t\pi_s \wedge s' = t'\pi_s],$$

where $\pi_s : T \rightarrow S$ takes (s,z) into s.

Definition 6.4. An S-computation (or *interaction computation*, resp. a T-computation) of system P is a function

$$\mathsf{P} \xrightarrow{\kappa} S, \qquad \text{resp. } \mathsf{P} \xrightarrow{\kappa} T,$$

where P is the set of positive integers, which satisfies

(6.5) κ is a τ-*sequence*, i.e.

for all $j \in \mathsf{P}$, $j\kappa \ \tau \ (j+1)\kappa$

(or if one confounds τ with the set of pairs for which it holds, for all $j \in \mathsf{P}$, $(j\kappa, (j+1)\kappa) \in \tau$);

(6.6) $1\kappa\pi_V\pi_i$ is *the initial vertex* of P_i for each $i \in [n]$ so that 1κ has the form (v,x), resp. (v,x,z).

We note in the case that κ is a T-computation that $\kappa \circ \pi_s : \mathsf{P} \rightarrow S$ is an S-*computation* of system P. Insofar as the mutual exclusion and fairness properties are concerned, consideration of S-computations suffice.

An interaction computation κ of P may be viewed as an $(n+r) \times \mathsf{P}$ matrix $M = (M_{i,j} \mid i \in [n+r], j \in \mathsf{P})$ as follows: Suppose $(v,x) = j\kappa$, then

$$M_{i,j} = v\pi_i \quad \text{if} \quad i \in [n]$$

$$M_{i,j} = x\pi_{i-n} \quad \text{if} \quad i > n.$$

Then each column of M is, in transposed form, an interaction state of P. Specifically, if $j\kappa = (v,x)$ and (v,x) is viewed as a row matrix then the jth column of M is the transpose of (v,x). Similarly, a T-computation may be viewed as a $(2n+r+1) \times \mathsf{P}$ matrix.

In the case of the unary solution P (cf. the figures of section 5), we note that $r = n$ and the result function ρ_i (cf. (P3)(a)) of P satisfies:

(6.7) *Case* 1. $v\pi_i \in 0A \Rightarrow (v\pi_i, x)\rho_i = (x\pi_1, \cdots, x\pi_{i-1}, 0, x\pi_{i+1}, \cdots, x\pi_n)$

Case 2. $v\pi_i \in 1A \Rightarrow (v\pi_i, x)\rho_i = (x\pi_1, \cdots, x\pi_i, 1, x\pi_{i+2}, \cdots, x\pi_n)$

Case 3. $v\pi_i \notin 0A \cup 1A \Rightarrow (v\pi_i, x)\, \rho_i = x.$

In the light of this section, it is worth reiterating that in our notion of computation one can directly represent the intuitive notion of more than one process acting simultaneously. This differs from many other formalisms for treating synchronization and parallelism questions in which the formal notion of a computation is some sort of sequence which allows for only one "atomic action" to occur at any moment in time.

7. *The LEMMA.*

Let $S = V \times X$ be the set of interaction states of the unary solution (which we'll call (P,K)) described in section 5. For $x \in X$ let 0 abbreviate: for all i, $x\pi_i = 0$. In order to state the LEMMA we require a definition.

Definition 7.1. A state $s = (v,x) \in S$ as well as a total state $t = (v,x,z)$ is *legitimate** iff (v,x) satisfies the following five properties: for all $i,k \in [n]$

Q1 $\quad v\pi_i \in K \cup 0A \Rightarrow x\pi_i = 1$

Q2 $\quad v\pi_i \in 1A \Rightarrow x = 0$

Q3 $\quad x = 0 \Rightarrow \exists i(v\pi_i \in 1A)$

Q4 $\quad v\pi_i, v\pi_k \in 1A \Rightarrow i = k$

Q5 $\quad x\pi_i = 1 = x\pi_k \Rightarrow i = k.$

Let $L \subseteq S$ be the set of all legitimate states of (P,K) and let $L\pi_X = \{x \mid (v,x) \in L\}$ be the set of legitimate communication states. We note immediately from Q1 and Q5, if $s \in L$, *then* s *satisfies*

Q6 (mutual exclusion, cf. (3.1)) $\quad v\pi_i, v\pi_k \in K \Rightarrow i = k$

It is also easy to see that for all n

Q7 $\quad x \in L\pi_X \Leftrightarrow$ Q5(x), i.e. x satisfies Q5, so that the cardinality $\| L\pi_X \| = 1+n.$

Again, it is easy to see (from the definition of "legitimate") that, there is a unique function $\varphi : L \rightarrow [n]$ the *focus function*, satisfying

Q8 $\quad s\varphi = i \Leftrightarrow x\pi_i = 1 \vee v\pi_i \in 1A.$

Q9 $\quad x = 0 \Rightarrow v\pi_{s\varphi} \in 1A.$

LEMMA. The transition relation $\tau \subseteq S \times S$, as well as $\tau \subseteq T \times T$, of (P,K) preserves legitimacy, i.e.

---

* We borrow the term "legitimate state" from Dijkstra [Dij-3] where it is used in a similar way.

(7.1) if $s \in L \wedge (s, s') \in \tau$ then $s' \in L$.

Proof. If Q is the conjunction of the five properties Q1 - Q5, then (7.1) may be restated as

(7.2) if $Q(s) \wedge (s, s') \in \tau$ then $Q(s')$.

Let $s' = (v', x') \in S$. (To help follow the proof the reader may wish to have the figures of Section 5 within view.) Assume $Q(s)$ and $(s,s') \in \tau$. It is required to show $Q(s')$ which we do in five steps corresponding to the five properties. For each $m \in [n]$, let $I_m = \{i \in [n] \mid v\pi_i \in A_i \wedge v'\pi_i \neq v\pi_i \wedge (v,x)\rho_i\pi_m \neq x\pi_m\}$; cf. (6.2).

1. In order to verify $Q1(s')$ assume $v'\pi_i \in K \cup 0A$. It is, then, required to show that $x'\pi_i = 1$.

   (a) Suppose first that $v'\pi_i = v\pi_i$ so that $v\pi_i \in K \cup 0A$. From $v\pi_i \in K \cup 0A$ and $Q(s)$ (specifically, $Q1(s)$), we infer that $x\pi_i = 1$. Since the instruction of the ith process is "inactive" (more formally, $i \notin I_i$, cf. section 6) and it is the only process which is potentially active relative to the ith interaction "variable" or register (so that $I_i = \phi$), we conclude $x'\pi_i = x\pi_i = 1$, which was to be shown.

   (b) Suppose now that $v'\pi_i \neq v\pi_i$. It follows that if $v'\pi_i \in K$, then $v\pi_i \in \omega A \wedge x\pi_i = 1$; while if $v'\pi_i \in 0A$ then $v\pi_i \in \beta A \cup K$. If $v\pi_i \in \beta A$ then also $x\pi_i = 1$, while if $v\pi_i \in K$, we infer $x\pi_i = 1$ from $Q(s)$. As before $I_i = \phi$, so that $x'\pi_i = x\pi_i = 1$, which was to be proved.

2. In order to verify $Q2(s')$, assume $v'\pi_i \in 1A$ and prove $x' = 0$.

   (a) Suppose first that $v'\pi_i = v\pi_i$ so that $v\pi_i \in 1A$ and from $Q2(s)$, we infer $x = 0$. A change from 0 to 1 at communication register $i+1$ (i.e. $x\pi_{i+1} = 0 \wedge x'\pi_{i+1} = 1$) can be effected only by the instruction at $v\pi_i$ if it is in $1A$ and it acted in this "interval" (i.e. $v'\pi_i \neq v\pi_i$). Since this is not the case, $I_{i+1} = \phi$ so that $x'\pi_{i+1} = x\pi_{i+1} = 0$. Moreover, utilizing $Q4(s)$, in its equivalent form, $v\pi_i \in 1A \wedge k \neq i \Rightarrow v\pi_k \notin 1A$ we conclude $k \neq i \Rightarrow x'\pi_{k+1} = x\pi_{k+1}$. Thus, $I_{k+1} = \phi$ for all $k \neq i$. Hence $x' = x = 0$, which was to be proved.

(b) Suppose now that $v'\pi_i \neq v\pi_i$. Since $v\pi_i \neq v'\pi_i \in 1A$, we conclude (see the figures for α_i and β_i in section 5) that $v\pi_i \in 0A$ which, in turn, implies by Q1(s) that $x\pi_i = 1$. Since $v'\pi_i \neq v\pi_i$ and it is a property of the solution (P,K) that no instruction other than one from P_i, can effect the ith communication register, we infer $i \in I_i$ so that $x'\pi_i = 0$. Moreover, for $k \neq i$, as a consequence of Q5(s), $I_k = \phi$ and so $x'\pi_k = x\pi_k = 0$. It follows that $x' = 0$.

3. In order to verify Q3(s'), assume $x' = 0$ and prove $\exists i(v'\pi_i \in 1A)$.

(a) Suppose first that $x = 0$. We infer from $x = 0 \wedge$ Q3(s) that $\exists i(v\pi_i \in 1A)$. Let $v\pi_i \in 1A$. If $v'\pi_i \neq v\pi_i$, then $i \in I_{i+1}$ so that $x'\pi_{i+1} = 1$. But $x' = 0$. This contradiction compels the conclusion $v'\pi_i = v\pi_i$ so that $v'\pi_i \in 1A$.

(b) Suppose now that $x \neq 0$; in particular, suppose $x\pi_i = 1$ (for some i). Since $x'\pi_i = 0$ as well (and the only "change $x\pi_i$ to 0" instructions are associated with vertices in P_i), we infer $v\pi_i \in 0A$ and $v'\pi_i \neq v\pi_i$. It follows that $v'\pi_i \in 1A$.

4. In order to verify Q4(s'), assume $v'\pi_i, v'\pi_k \in 1A$, and prove $i = k$.

(a) In the case $v'\pi_i = v\pi_i \wedge v'\pi_k = v\pi_k$, we conclude from Q4($s$) that $i = k$.

(b) In the case $v'\pi_i \neq v\pi_i$ so that $v\pi_i \in 0A$, we infer from Q1(s) that $x\pi_i = 1$ and infer from Q2(s) that $v\pi_k \notin 1A$. Thus $v\pi_k \in 0A$ and from Q1(s), $x\pi_k = 1$. Finally, we infer from Q5(s) that $i = k$, which was to be proved.

5. In order to verify Q5(s'), assume $x'\pi_i = 1 = x'\pi_k$ and prove $i = k$.

(a) If $x\pi_i = 1 = x\pi_k$, then we infer from Q5(s) that $i = k$.

(b) Assume now that $x\pi_i = 0$. Since $x'\pi_i = 1$, we conclude $v_{i-1} \in 1A$ and from Q2(s) infer that $x = 0$. Hence also $v_{k-1} \in 1A$ and by Q4(s), we infer $i = k$.

8. *Preliminaries for The THEOREM*

We formalize the property which was stated informally as (3.6).

Let S be the set of interaction states of a system P and let κ:P $\rightarrow$ S be a function.

Definition 8.1. The function κ satisfies the *finite delay property** if it satisfies the NEGATION of: there exists $i \in [n]$, $j \in$ P such that for all $j' \in$ P, letting $j\kappa = (v,x)$,

(8.1) $(j'+j)\kappa\pi_V\pi_i = v\pi_i$

(8.2) $(j'+j)\kappa\pi_X = x$

(8.3) $(v\pi_i, x)\nu_i \neq v\pi_i$ (i.e. the "intended execution" of the instruction located at $v\pi_i$ is a departure execution).

The function κ is a *finite delay computation* if it is a computation which satisfies the finite delay property. Thus κ satisfies the finite delay property if each of its intended departure execution instructions actually takes place. (In this connection see also Observation 8.1 below.)

To facilitate succinct formulation of the THEOREM we also use the following, where $D = \Sigma P_i v^-$, is the disjoint union of the family $\{P_i v^- \mid i \in [n]\}$ of non-exit vertices of P_i, $D^\wedge$ is the set of all subsets of D and $\mathrm{P}_\infty = \mathrm{P} \cup \{\infty\}$.

Definition 8.2. Let $\mu_{\kappa,i} : \mathrm{P} \times D^\wedge \rightarrow \mathrm{P}_\infty$ be a function which satisfies, where $j \in$ P, $U \in D^\wedge$: if $j\kappa\pi_V\pi_i \in U$ then

$(j,U)\mu_{\kappa,i} = (\mu j')[j<j' \wedge j'\kappa\pi_V\pi_i \notin U]$ if $(\exists j')[j<j' \wedge j'\kappa\pi_V\pi_i \notin U]$;

$(j,U)\mu_{\kappa,i} = \infty$ otherwise.

(In the above "$\mu j'$" means "the least j' such that.")

Notice $\mu_{\kappa,i}$ depends upon a computation κ and $i \in [n]$. If U is the set of vertices v_i, where v_i is the vertex of the ith process which is active at the jth step of the computation κ, then

---

* Apparently, the first explicit formulation of this property appears in [M&B] in connection with asynchronous circuits!

$(j,U)\mu_{K,i}$ is the smallest integer greater than j such that the vertex of the ith process active at j' is distinct from v_i, if such a j' exists; otherwise the value is ∞.

Remark. The values of "U" used in (8.5) below are the values of the function $U : [6] \to$ D. Our discussion could have been carried out using instead a function

$$U_i : [6] \to P_i v^- \subseteq D, \qquad \text{where } mU_i = mU \cap P_i v^-.$$

The latter course permits dispensing with D altogether while the former course permits dispensing with the subscript "i" and is more general.

Now, let $(G,K) = \{(G_i,K_i) \mid i \in [n]\}$ be as in section 4, let $K \in D^\wedge$ be $K_1 \cup K_2 \cup \cdots \cup K_n$, let (P,K) be the system constructed from (G,K) by the construction of section 5, let $S = V \times X$ be the set of interaction states of (P,K), let $L \subseteq S$ be the set of legitimate states of (P,K) and let τ be the direct transition relation of (P,K).

The proof of The THEOREM (stated below) will be facilitated by articulating the idea behind the construction of (P,K) from (G,K). In particular, it will be helpful to distinguish six (cf. Figure 9.1) mutually exclusive, exhaustive properties of legitimate states. To this end we introduce

$$\text{the } \textit{distinguish function} \qquad L \xrightarrow{\delta} [6]$$

by recalling from Q8 of the previous section, where $s = (v,x) \in L$

(8.4) $\quad s\varphi = i \Leftrightarrow x\pi_i = 1 \vee v\pi_i \in 1A$

and requiring

(8.5)	$s\delta=$	if $\upsilon\pi_{s\varphi} \in$
	1	N
	2	ωA
	3	K
	4	$0A$
	5	$1A$
	6	βA

Let $U : [6] \rightarrow D$ be defined by the table of (8.5) so that $1U = N$, $2U = \omega A$, etc.

That δ is well-defined follows from the fact that the (indexed) family

$$\{mU \mid m \in [6]\}$$

is an *indexed partition* (i.e. an indexed family of pairwise disjoint subsets, possibly empty, of a set whose union is the whole set) of D.

The six properties of legitimate states mentioned above correspond to the six subsets $1\delta^{-1}, 2\delta^{-1}, \cdots$ of L induced by δ

Observation 8.1. With respect to Figure 9.1, observe that the transitions (indicated by arrows) out of rectangle m, $m \in [6]$ are exactly the possible transitions out of m according to the construction of (P,K) from (G,K), cf. Figure 5.6. *Note particularly there is no edge which begins at box 2 which ends at box 2 even though a wait vertex (Figure 5.1) has such an edge* . A more complicated consequence of that construction is:

of the à priori possible simple "cycles" of Figure 9.1, i.e. simple closed paths, only the three indicated are indeed possible in the sense that "simple δ-cycles of κ" (defined immediately below) trace the closed path.

Definition 8.3 A finite sequence

(8.6) $\quad (j_0, j_1, \cdots j_\ell)$

of elements of P is a *δ-cycle of κ* (where κ is a τ-sequence of legitimate states of (P,K), *of length* ℓ, if, where

$$s_k = (v_k, x_k) = j_k\kappa, \text{ for } 0 \leq k \leq \ell,$$

(8.7) $s_0\delta = s_\ell\delta$

(8.8) $(j_k, U_k)\mu_{\kappa,i} = j_{k+1}$ for $0 \leq k < \ell$,

where U_k is defined by the requirement $v_k\pi_i \in U_k$ and where $i = s_k\varphi$.

The δ-cycle of κ (8.6) is *simple* if the assumption that $(j_k, j_{k+1}, \ldots, j_{k'})$ is a δ-cycle of κ, where $0 \leq k < k' \leq \ell$, implies that $k=0$ and $k'=\ell$.

The δ-cycle of κ (8.6) is a *major cycle of* κ if it is a simple δ-cycle of κ and if for some k, $0 \leq k < \ell$, $j_k\kappa\delta = 2$ and $j_{k+1}\kappa\delta = 3$, cf. Figure 9.1. Clearly the length ℓ of a major cycle of κ is 5. In similar manner the *minor cycles of* κ may be defined as indicated by Figure 9.1.

In the context of the above "language," we can now give precise meaning to (3.1) and (3.2) where $\kappa : \mathsf{P} \to S$ is a function and S is the set of interaction states of (P,K).

(8.9) (*Mutual exclusion*). κ satisfies mutual exclusion iff $j\kappa$ satisfies mutual exclusion (cf. Q6) for each $j \in \mathsf{P}$.

Using the fact that in the context of (P,K) "opportunity for process i to communicate, at time $j\iota$ with the resource" may be interpreted as meaning $j\kappa \in L \wedge j\kappa\pi_X\pi_i = 1$, we formulate: κ satisfies fairness iff for each $i \in [n]$ both

(8.10) *primary fairness* (cf. (3.2) $\forall j \in \mathsf{P}\ \exists \ell \in \mathsf{P}\ [\ell > j \wedge \ell\kappa \in L \wedge \ell\kappa\pi_X\pi_i = 1]$;

and

(8.11) *secondary fairness* (cf. (3.3) and the Remark on Fairness below)

if $j\kappa \in L \wedge j\kappa\pi_X\pi_i = 1 \wedge j\kappa\delta = 1 \wedge x = (\mu j')[j' > j \wedge j'\kappa\delta \neq 1]$

then $x\iota - j\iota$ is bounded by the maximum time required for P_i's control to traverse a closed path of P_i in the subset N of the vertices of (P,K).

Remark on Fairness. (a) The formulation (8.11) of secondary fairness corresponds to the "yields" part of (3.3) not the "exploits" part of (3.3). The "exploits" part of (3.3) is, however, satisfied (by the system (P,K)) because of assumptions (3.7) and (4.2).

(b) There is no attempt in this paper to formalize or even discuss "fairness" in anything like full generality. The kind of fairness one has in mind in large part determines the design of the coordinated system. It should be clear, however, that some of the fairness disciplines (or protocols) discussed in the literature e.g. "FIFO fairness" (i.e. roughly speaking, the discipline, according to which, the first process in a waiting state is the first process to access the resource) are not satisfied by the (P,K) of section 5.

In the context of the main theme of this paper, the role of simulation is minuscule. Nevertheless we clarify simulation below in (8.12). We note first that *simulation in the sense intended is achieved by the following properties of the construction of (P,K) from (G,K)*:

(8.12) (a) The vector variable x which occurs in the instructions at interaction vertices does not occur in the instructions at non-interaction vertices (cf. Remark 5.1).

(b) The begin vertex of P_i is the same as the begin vertex of G_i (cf. Remark 5.2).

(c) If one erases from the vertex sequence of a path of P_i all occurrences of interaction vertices of P_i then a vertex sequence of a path of G_i results; all vertex sequences of paths of G_i arise in this way. Moreover, the instruction sequence associated with the vertex sequence is the same in both cases.

9. The THEOREM

The THEOREM. Let τ be the direct transition relation of the system (P,K), where (G,K) satisfies (4.2).

(1) If

$$\mathsf{P} \xrightarrow{\kappa} S$$

is a τ-sequence, i.e. $(j\kappa, (j+1)\kappa) \in \tau$ for all $j \in \mathsf{P}$, such that

$$1\kappa \in L,$$

then κ satisfies mutual exclusion (8.9).

(2) If κ is a τ-sequence satisfying $1k \in L$ and the finite delay property, then κ satisfies fairness (8.10) and (8.11).

(3) If, where κ is as in (2),

$$U = m\delta^{-1}\pi_V\pi_i,\ m \in [6],$$

then $U \subseteq P_i v^- \subseteq D$ and

$$(j,U)\mu_{\kappa,i} < \infty$$

i.e.

(9.3) $\quad (\exists j')\ [j<j' \wedge j'\kappa\pi_V\pi_i \notin U]$.

Condition (9.3) can be viewed informally as stating that in Figure 9.1, for each $m \in [6]$, at least one of the transitions out of rectangle m does indeed take place.

Remark. It is worth observing that "fairness" is a *global property* of κ in that it involves relating interaction states $j\kappa$, $j'\kappa$ at widely different instants of time (more precisely, no à priori bound can be correctly assigned to $j'-j$), while "mutual exclusion" is a *local property* of κ. This distinction is reflected both in the statement and in the proof of the theorem.

Proof. Observation 1. Since $1\kappa \in L$ by assumption, it follows from the LEMMA that $j\kappa \in L$ for all j.

Proof of (1). Immediate from Observation 1 and Q6 of section 7.

Proof of (9.3). We consider six cases.

Case m=1, i.e. $U=N\pi_V\pi_i$. According to the construction of (P,K) from (G,K), (cf. Figure 5.6), every positive length closed path Π of G_i such that $\Pi v \subseteq N\pi_i$ is "interrupted," in P_i, by an occurrence of β_i (cf. Figure 5.3). Hence there is no positive length closed path Π of P_i such that $\Pi v \subseteq N\pi_i$ and the result follows, *and with it (8.11) as well*, by using the finite delay property repeatedly.

Case m=2, i.e. $U = \omega A\pi_V\pi_i$. The outcome of the test (cf. Figure 5.1) is "yes" since $i=s\varphi$ and $s \in L$. Hence, using the finite delay property, the transition indicated by the yes-edge of ω_i to the begin of α_i (which is in $0A$) or to K (cf. Figures 5.6 and 5.2) does indeed take place.

Case m=3 (9.3) is an immediate consequence of (4.2) and repeated use of the finite delay property of κ.

Case m=4,5,6 (9.3) is an immediate consequence of a single application of the finite delay property and, in the *Case m=6*, the observation that as in *Case m=2*, the "yes" edge (cf. Figure 5.3) is selected.

It remains to prove (8.10). But (8.10) is an easy consequence of the observation that if (8.6) is a simple δ-cycle of κ then

$$1 + j_0\kappa\delta \overset{n}{\equiv} j_\ell\kappa\delta$$

(or equivalently, regarding $[n]$ as a set of representatives of Z_n, the integers modulo n,

$$1 + j_0\kappa\delta = j_\ell\kappa\delta)$$

Thus the sequence of δ-values corresponding to n consecutive simple δ-cycles of κ, begins and ends with the same element of $[6]$.

Corollary. If the system (G,K) of uncoordinated sequential programs satisfies (4.2) then the system (P,K) of coordinated sequential programs satisfies mutual exclusion (8.9) and fairness (8.10), (8.11).

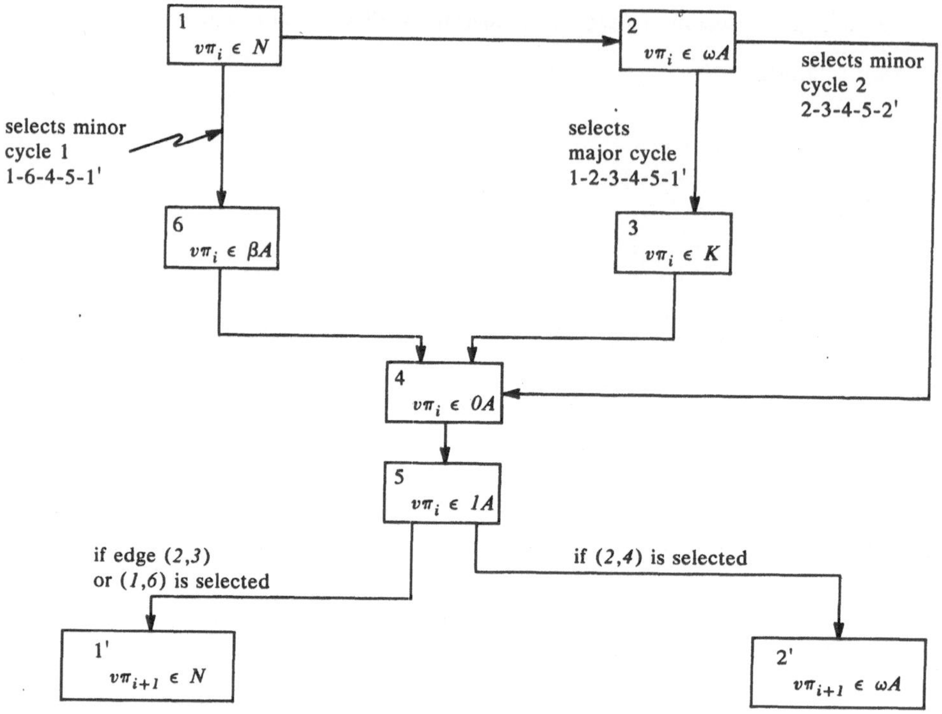

Figure 9.1. The three cycles of distinguished properties of legitimate states

Remarks. Where $s = (v,x)$, $L \xrightarrow{\varphi} [n]$, $L \xrightarrow{\delta} [6]$:
$i = s\varphi$ and $m \in [6]$ (in the upper lefthand corner of the rectangles) indicates $s\delta = m$.

10. The conceptual solution and its specialization to Gray solutions

In the unary solution the basic approach for synchronizing access is clear; namely, a "round-robin" discipline where the communication variables are used to allow one process to pass the turn on to the next process. We employ this same approach in the conceptual solution, but with an arbitrary encoding of communication variables. Here we simplify the augmenting by viewing it as a single step, and assume our solution has r communication variables, where r may be less than n.

To obtain a rough description of the conceptual solution from the rough description of the *unary solution*, (P,K), embodied by the figures of section 5, it is only necessary to make the following modifications in those figures, where $\theta : [n] \rightarrow X = 2^r$ is an injective function, r is a nonnegative integer and $i \in [n]$.

(10.1) Change the test instructions of Figures 5.1 and 5.3 to $x \stackrel{?}{=} i\theta$.

(10.2) Change the augment program, Figure 5.2, to the single instruction (single exit) program whose instruction is (cf. Figure 10.1)

change the value (content) of "x" from $i\theta$ to $(i+1)\theta$.

With this new meaning assigned to "ω_i" and α_i", new meanings are (derivatively) taken on by β_i" and "$(\omega_i \circ \alpha_i)^\dagger$".

In view of the precise description of the unary solution, this rough description of the conceptual solution (probably) suffices. We remark, however, that r is the number of communication (binary) variables (or "abstract" registers).

Remark. It is worth noting that the description given of the new augment program -- in contrast to that given for the new wait program -- is not explicit at least in one respect. The result functions (i.e. the ρ_i's) are *implicitly* determined by the function θ and so the degree of complexity of the ρ_i's (as measured, for example, by the number of variables which must be "read") is hidden from view. We no longer can claim, as in the unary solution, that "reading" requires looking at only one boolean variable or that writing to change the value of x can be done by changing only one boolean variable. In the Gray solutions described later in this section, however, the reading requires reading all the variables simultaneously but the writing requires changing only one variable at a time, so some degree of simplicity is maintained.

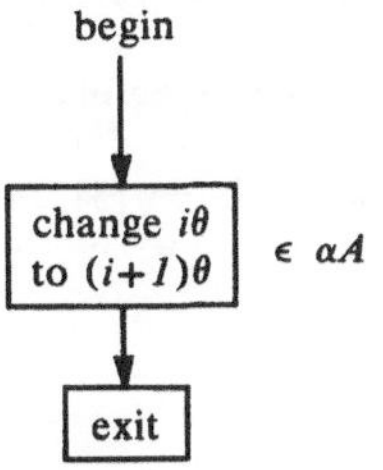

Figure 10.1. $\alpha_i : 1 \to 1$, *conceptual augment*

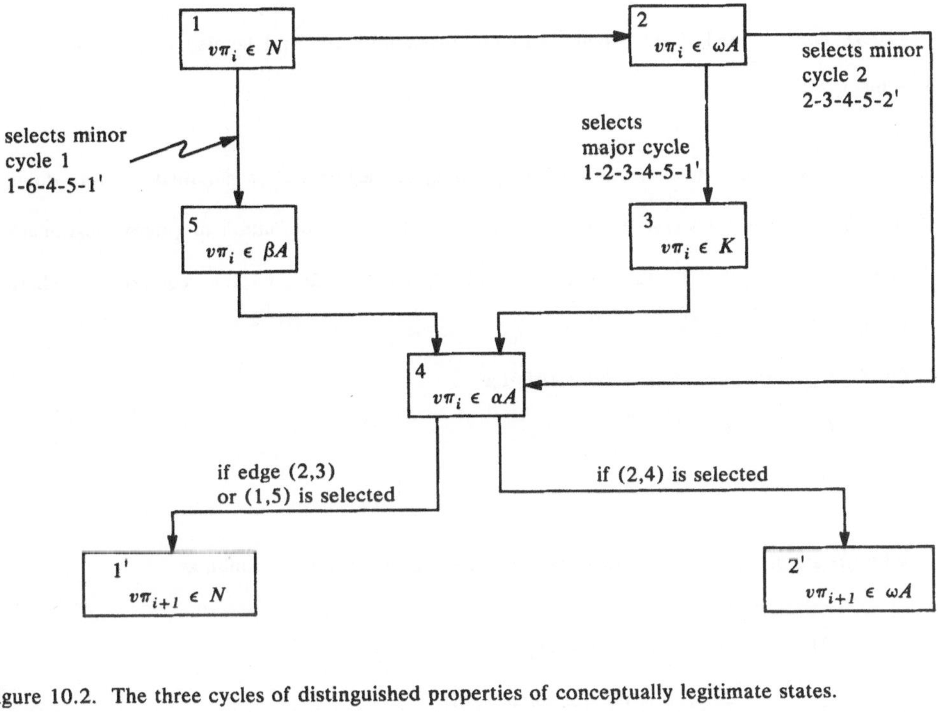

Figure 10.2. The three cycles of distinguished properties of conceptually legitimate states.

Remarks. Where $s = (v,x)$, $L \xrightarrow{\varphi} [n]$, $L \xrightarrow{\delta} [5]$:
$i = s\varphi$ and $m \in [5]$ (in the upper lefthand corner of the rectangles) indicates $s\delta = m$.

New Corollary. The conceptual solution satisfies mutual exclusion and fairness.

Proof sketch. It is only necessary to redefine "legitimate" (and hence the set L of legitimate states), $\varphi : L \to [n]$, $\delta : L \to [5]$ and Figure 9.1 in such a way that (cf. Figure 10.2)

(10.1) the transition relation τ of the conceptual solution preserves legitimacy;

(10.2) $s\varphi = i \Leftrightarrow x = i\theta$,

(10.3) the "new" δ agrees with the "old" on 1,2,3 (cf. (8.9)), $s\delta=5$ if $v\pi_{s\varphi} \in \beta A$ and $s\delta=4$ if $v\pi_{s\varphi}$ is a begin vertex of the "new" α_i;

(10.4) boxes 4 and 5 (cf. Figure 9.1) are fused into one and labeled 4 while box 6 is relabeled 5.

It only remains to indicate how the meaning of "legitimate" is modified. Before doing that, however, we will restate conditions Q1 and Q5 in the definition of (unary) legitimacy given in section 7. To this end, we define $\theta_u : [n] \to 2^n$ by the requirement, where $(v,x) \in S$.

(10.5) $i\theta_u$ is the unique $x \in 2^n$ satisfying

(a) $x\pi_i = 1$

(b) $k \neq i \Rightarrow x\pi_k = 0$

With the aid of θ_u we may now restate Q1 and Q5 in equivalent fashion as

Q1' $v\pi_i \in K \cup 0A \Rightarrow x = i\theta_u$,

Q5' $x=0$ or $x = i\theta_u$ for some $i \in [n]$.

From Q1' and Q5', we obtain the definition of "legitimate" in the sense of the conceptual solution.

Definition 10.1. *Conceptual legitimacy.* A state $(v,x) \in S$ is *conceptually legitimate* iff it satisfies both

(a) $v\pi_i \in K \cup \alpha A \Rightarrow x = i\theta$

(b) $x = i\theta$ for some $i \in [n]$.

It is a consequence of (a) and the injectiveness of θ, that

(10.6) *a conceptually legitimate state satisfies mutual exclusion* Q6.

The proof that the transition relation of the conceptual solution preserves legitimacy is rather easier than the corresponding proof in the unary case because the notion is simpler.

This leads to the *ostensibly paradoxical*

(10.7) *observation*: the proof that τ preserves legitimacy is simpler in the conceptual case than in the unary case; yet, it seems to be the case that, the conceptual solution specializes (by taking $\theta = \theta_u$) to the unary solution.

The explanation is not difficult. The specialization is not valid because (each copy of) the unary augment program α_i (for each $i \in [n]$) has two (non-exit) instructions while in the conceptual case α_i has just one (nonexit) instruction and, from the point of view of the system (rather than the individual process)

(10.8) *instructions don't compose*, i.e. the two instructions of the unary augment program α_i cannot be replaced by one instruction which is functionally equivalent.

The reader may convince himself of the validity of (10.8) by convincing himself that the two (apparently independent) instructions of the unary α_i do not commute. To explain this point in greater detail, let α_i' be obtained from α_i by interchanging the order of α_i's nonexit instructions. If now one replaces α_i by α_i' in the unary solution, then coordination fails. On the other hand, the functional equivalents of α_i and α_i' are the same. We return to

(10.8) in section 12. There are, however, perfectly valid specializations of the conceptual solution to what we shall call "the Gray solutions".

An injection $\theta : [n] \rightarrow \mathbf{2}^r$ is called a *mod n Gray coding* provided that $i\theta$ and $(i+1)\theta$ differ in at most one component position. (The + is, of course, modulo n.)

Let n be even. Suppose $n=2m,\ m\geq 1$ and $r = \lceil \log_2 n \rceil$ so that $r\geq 1$. The *reflected (Gray coding) injection* $\rho : [n] \rightarrow \mathbf{2}^r$ is well known. It is defined, when n is a positive power of 2, by the requirements; in the case $n=2$ (so that $m=1,\ r=1$) by $1\rho=0,\ 2\rho=1$ and when $r\geq 2$:

(10.9) $(m-\ell)\theta_G\pi_k = (m+\ell+1)\theta_G\pi_k$ for $0\leq \ell<m,\ k \in [r-1]$,

(10.10) $i\theta_G\pi_r = 0,\ (m+i)\theta_G\pi_r=1$ for $i\in[m]$.

The mod n reflected Gray code for the cases $n=8$ (so that $m=4,\ r=3$) and $n=6$ (so that $m=3=r$) is indicated by the table below. The table for the case $n=8$ is obtained by ignoring the two horizontal dotted lines. The table for the case $n=6$ is obtained from the table for $n=8$ by deleting $m\rho$ and $(m+1)\rho$ which appear between the horizontal dotted lines. The solid horizontal line represents a mirror in which the bits to the left of the vertical dotted line are reflected. The existence of such codings should be obvious from the example (in the presence of the above remarks) Table 10.1.

0	0	0
1	0	0
1	1	0
0	1	0
0	1	1
1	1	1
1	0	1
0	0	1

Table 10.1

In an implementation of a Gray solution to the mutual exclusion problem, the augment program α_i has the effect of changing the value of only one communication variable. This desirable property is shared by the unary solution but not by the specialization of the conceptual solution obtained by taking $\theta = \theta_u$.

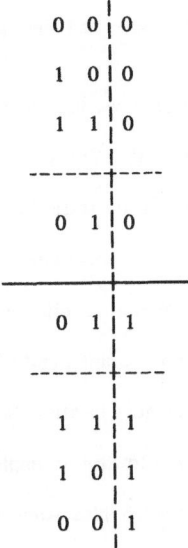

11. Application of the solution to master-slave and message-sending contexts

In both the unary and Gray solutions we use communication variables which are shared by the various processes in particular ways. The solutions prescribe the functional transition from one legitimate state to a next legitimate state, but we have not gone into the details on how the communication variables and the transitions would be realized in an actual system. One might envision, however, that there is a machine register or memory location that is used to hold the value of each communication variable, and that the interaction instructions include access to the appropriate registers or memory locations when they read or change values of the communication variables. This approach toward an implementation then has one program for each of the n processes of the solution plus some commonly accessible storage (either registers or memory) to hold the values of the communication variables.

When one faces the issue of an actual implementation, however, other possible forms of solution come to mind. One natural, and apparently simple, solution to mutual exclusion is to provide a separate process (the "master" process) to control all interaction between the other processes (the "slave" processes). Within such a "master-slave" context, whenever a slave process needs access to the common resource it would request access via the master process. The job of the master process is then to insure mutual exclusion and some form of fairness. Clearly either a unary solution or a Gray solution could be used to provide the basis for a master-slave solution. In such an approach the master process would use the values of the communication variables to control the sequencing. On the other hand, it is not at all clear that a solution based on the master-slave concept would be useful in implementing mutual exclusion on a system not using a master process. Moreover, it appears as though the master-slave approach might be quite inefficient since it centralizes all interaction into the master process, thus possibly causing a bottleneck in the system operation. Whether the master process actually were a bottleneck or not would depend upon many aspects of the particular problem. One might envision the master process as having various kinds of capabilities: to queue requests from the slave processes; to rapidly shift the round robin discipline from one process to the next; to automatically skip processes when they have terminated, or to have other efficiency features built into the implementation.

Another form of process interaction that has received considerable attention is a "message-sending" system. Our work on this paper was, in fact, originally stimulated by a conversation with Pamela Zave in which she called attention to some of this work [Z&F]. Roughly, a message-sending system is a collection of processes each of which resides in a different part of a physical system (possibly with the parts distantly separated and connected by a network), where these processes are meant to cooperate with each other to solve a problem of mutual interest. To provide cooperation among the processes one imagines that "messages" are generated within the processes and sent to other processes. Since the system may have distantly separated parts it may be inappropriate to store communication variables in a place that can be accessed by any process, at any time. The use of messages, say of process i sending a message to process j, when cooperation between i and j is required, is an approach that appears quite different from the use of common communication variables. The unary solution has, apparently, a message-sending interpretation. This is based on the particularly simple functional character of the communication variable usage. The ith communication variable could be located in the same part of the system as the ith process, so that the ith process has ready access to reading its own variable. Then, when process $i-1$ wishes to change the value of communication variable i (in the *1A* portion of process $i-1$) this would be realized, in more detail, within the system by sending a message to process i to change variable i.

Although we have been rather informal in our discussion in this section (it is beyond the scope of this paper to present detailed formulations for master-slave and message-sending systems) it should be clear that the unary solution can be interpreted both in the master-slave and message-sending contexts, and that the Gray solutions also can be interpreted in the master-slave context. It is not so clear how the Gray solutions could be interpreted in the message-sending context, however, since the reading of communication variables is somewhat more complex than in the unary solution. From this discussion, however, one should notice that for both the unary and Gray solutions whenever one is considering an actual implementation of a solution in a practical system there is considerable detail that needs further specification.

12. Concluding remarks and an open problem

It is often the case that discussions of synchronization problems are carried out in nondeterministic settings which appear excessively nondeterministic in that there is no logical reason for the presumed degree of nondeterminism. In order to avoid such excess we have formulated the given and constructed programs G_i and P_i, $i \in [n]$, as deterministic rather than nondeterministic programs. Specifically, the next vertex and result relations are single-valued and total i.e. functional. On the other hand, the direct transition relation τ of the system P is not functional. Notice, however, that it doesn't differ very much from being a function. As a specific example (cf. (6.4)(a)) the ith component $v'\pi_i$ of the next vertex vector is either $v\pi_i$, the ith component of the present vertex vector, or determined by the function ν_i. Note too that the new communication statė (cf. (6.3)) is given by a function.

The attempt made here to simultaneously cope with

(a) a general model for coordinated sequential programs, viz. that given in section 6, and

(b) the solution to a particular problem, viz. "mutual exclusion" as formulated here has resulted in the "loose end" (4.6) which deals with the resource. The assumption (4.6) is never used. It remains, nevertheless, because it would be useful in a completely detailed formulation of (P,K) which sought to accurately reflect the role of the resource.

There are a number of open problems which suggest themselves. For example, to what extent does the notion of computation used here apply to other solutions of the mutual exclusion problem as well as to solutions to other synchronization problems. A problem of a different character, in that it would involve a new notion of system computation, revolves around the unpleasant circumstance (10.8) that instructions don't compose.

Open problem. Provide a solution to the mutual exclusion problem in which instructions do compose (in the sense detailed soon after "(10.8)") or prove no such solution exists.

GLOSSARY

	page	section
abstract instruction	17	6
abstract system	16	6
augment program	12	5
backedge	12	5
change-dominant consensus function	18	6
communication registers	17	6
communication state	16	6
communication variables	17	6
conceptual legitimacy	38	10
consensus function	18	6
construction	12	5
coordinated sequential process	16	6
data states	9	4
departure execution	8,10	3,4
distinguish function	26	8
finite delay computation	25	8
finite delay property	25	8
finite sequence	27	8
flowscheme	7	3
focus function	22	7
global property	30	9
Gray coding	39	10
indexed partition	27	8
(indivisible) test and set	6	2

References

[BFJLP] Burns, J. E., M. J. Fischer, P. Jackson, N. A. Lynch, and G. L. Peterson, "Shared data requirements for implementation of mutual exclusion using a test-and-set primitive," *Proceedings* of the 1978 International Conference on Parallel Processing, August, 1978. Also, University of Washington, Department of Computer Science Technical Report No. 78-08-03.

[C&H] Cremers, A. and T. Hibbard, "An algebraic approach to concurrent programming control and related complexity problems," University of Southern California, Computer Science Department Technical Report, November, 1975.

[Dij-1] Dijkstra, E. W., "Solution of a problem in concurrent program control," *Communications of the ACM, 8*, No. 9 (September, 1965) 569.

[Dij-2] Dijkstra, E. W., "Co-operating sequential processes," in *Programming Languages*, (F. Genuys, Ed.) New York, Academic Press, 1968, 43-112.

[Dij-3] Dijkstra, E. W., "Self-stabilizing systems in spite of distributed control," *Communications of the ACM 17*, No. 11 (November 1974) 643-644.

[E&S] Elgot, C. C. and J. C. Shepherdson, "A semantically meaningful characterization of reducible flowchart schemes," IBM Research Report RC-6656, July, 1977. *Theoretical Computer Science 8*, No. 3 (June 1979) 325-357.

[ERR] Elgot, C. C., A. Robinson, and J. A. Rutledge, "Multiple control computer models," IBM Research Report RC-1622 (May, 1966) and *Systems and Computer Science*, University of Toronto Press, 1967.

[Hym] Hyman, H., "Comments on a problem in concurrent program control," *Communications of the ACM, 9*, No. 1 (January, 1966) 45.

[Knu] Knuth, D. E., "Additional comments on a problem in concurrent programming control," *Communications of the ACM, 9*, No. 5 (May, 1966), 321-322.

[L-1] Lamport, L., "Concurrent reading and writing," *Communications of the ACM, 20*, No. 11 (November, 1977) 806-811.

[L-2] Lamport, L., "Time, clocks, and the ordering of events in a distributed system," *Communications of the ACM, 21*, No. 7 (July, 1978) 558-565.

[M&B] Muller, D. E. and W. S. Bartky, "A theory of asynchronous circuits," Proceedings of an International Symposium on the Theory of Switching, Vol. 29 of the Annals of the Computation Laboratory of Harvard University, Harvard University Press, 1959, 204-243.

[M&Y-1] Miller, R. E. and C. K. Yap "On the formal specification and analysis of loosely connected processes," *Proceedings* of the International Conference on Mathematical Studies of Information Processing, Kyoto, Japan, August, 1978, 33-66. An earlier version of this work is included in IBM Research Report RC-6716, "Formal specification and analysis of loosely connected processes," September 1, 1977.

[M&Y-2] Miller, R. E. and C. K. Yap "On formulating simultaneity for studying parallelism and synchronization," *JCSS* (April, 1980).

[P&F] Peterson, G. L. and M. J. Fischer, "Economical solutions for the critical section problem in a distributed system, Extended abstract." *Proceedings* of the Ninth Annual ACM Symposium on Theory of Computing, May, 1977, 91-97. Also, University of Washington, Department of Computer Science Technical Report No. 77-02-03, February 25, 1977.

[R&P] Rivest, R. L. and V. R. Pratt, "The mutual exclusion problem for unreliable processes: Preliminary report," *Proceedings* of the 17th Annual IEEE Symposium on Foundations of Computer Science, October, 1976, 1-8.

[Z&F] Zave, P. and D. R. Fitzwater, "Specification of asynchronous interactions using primitive functions," Technical Report, Department of Computer Science, University of Maryland, 1977.